COMPOSITE MATERIALS AND PROCESSING

COMPOSITE
MATERIALS AND
PROCESSING

M. Balasubramanian

CRC Press
Taylor & Francis Group
Boca Raton   London   New York

CRC Press is an imprint of the
Taylor & Francis Group, an **informa** business

CRC Press
Taylor & Francis Group
6000 Broken Sound Parkway NW, Suite 300
Boca Raton, FL 33487-2742

First issued in paperback 2017

Version Date: 20130801

ISBN 13: 978-1-138-07687-7 (pbk)
ISBN 13: 978-1-4398-7935-1 (hbk)

Library of Congress Cataloging-in-Publication Data

Balasubramanian, M.
    Composite materials and processing / M. Balasubramanian.
        pages cm
    Includes bibliographical references and index.
    ISBN 978-1-4398-7935-1 (hardback)
    1. Composite materials. I. Title.

TA418.9.C6B3544 2014
620.1'18--dc23   2013028659

Visit the Taylor & Francis Web site at
http://www.taylorandfrancis.com

and the CRC Press Web site at
http://www.crcpress.com

Contents

Preface

The use of composite materials, because of their light weight and high performance, has increased manifold over the years, starting from aerospace to building applications. Composite materials can be made to suit any property requirements for any application. A proper selection of reinforcement, matrix material, and composition will result in the formation of composites with specific properties. Many conventional materials are currently being replaced with composite materials. Their wide use has also been facilitated by the development of new materials, improvements in manufacturing processes, and the availability of new analytical tools.

The processing of composites will also play a major role in achieving specific properties. Since the composites are made of two or more heterogeneous materials, conventional processing methods used for the respective matrix materials may not be suitable. To realize the full potential of a composite material, an appropriate processing method should be selected and optimum processing conditions followed. Moreover, it is necessary to understand the science behind the processing method so that appropriate processing parameters can be selected. Hence, the aim of this book is to provide comprehensive information about the science of processing various composites using different processing methods. The basis for the performance of different reinforcements and matrix materials is described. The technological advancements in processing methods are also highlighted. Different processing methods can be selected depending on the cost and quality of the final products. In some cases, production volume can be a deciding factor in the selection of a processing method. Hence, essential information regarding the advantages and disadvantages of each processing method is presented for suitable selection. Although more emphasis is given on the processing of composites, important properties and applications of various composites are also described in this book.

At present, carbon–carbon composites are considered very important materials. A separate chapter is devoted to these composites, providing a detailed account of various processing methods, properties, and applications.

Research for the development of new or improved nanocomposites is ongoing. The main advantage of nanocomposites is the requirement for a merely small volume of reinforcement to significantly improve the properties. The last chapter deals with the processing of different nanocomposites.

This book is an outcome of the courses I have taught to graduate and undergraduate students for the last 12 years at IIT Madras, as well as short-term courses for practicing engineers in composite industries. It will be useful for materials scientists, engineers, and graduate students, as well as for practicing engineers in composite industries. The details given in this book regarding various processing methods of composites, the important points to be considered for different processing methods, and the selection of composite material and processing methods for any particular application will be very useful in producing quality composites.

I thank my research students, who have helped me directly or indirectly. I especially thank the reviewers of the chapters, who provided critical comments that enabled improvement of the book to its present form. I thank the Center for Continuing Education, IIT Madras, for financially supporting the manuscript preparation and the administrative staff at IIT Madras for their encouragement and support.

I also thank the staff at CRC Press, especially Gagandeep Singh and Amber Donley, for their invaluable assistance during the book production process. Last, but not the least, I would like to thank my wife and son for their help and support throughout this endeavor.

Author

Dr. M. Balasubramanian is currently a professor in the Department of Metallurgical and Materials Engineering, Indian Institute of Technology, Madras, India. His area of specialization is composite materials, including nanocomposites. He is also actively involved in the research of ceramics and composites. He has more than 17 years of teaching experience and has taught a broad spectrum of courses related to materials science, particularly to composite materials, during his long stint as an academician. He has organized more than ten training programs for those working in the composite industry and those planning to start their own businesses. This book is the outcome of the vast knowledge he has accumulated over the years in his interaction with the composite industry and during his career as a teacher. Dr. Balasubramanian has published more than 75 research papers and has written 4 chapters in different books. A life member of the Indian Ceramic Society, FRP Institute, and the Indian Institute of Metals, he is also an elected member of the Indian Institute of Ceramics.

Abbreviations

AAS	Atomic absorption spectroscopy
AES	Auger electron spectroscopy
ASTM	American Society for Testing and Materials
BMC	Bulk molding compound
BMI	Bismaleimide
CBN	Cubic boron nitride
CEC	Cation exchange capacity
CFRP	Carbon fiber–reinforced plastics
CMC	Ceramic matrix composite
CNT	Carbon nanotubes
COF	Coefficient of friction
COV	Coefficient of variation
CSM	Chopped strand mat
CTBN	Carboxyl-terminated butadiene acrylonitrile
CTE	Coefficient of thermal expansion
CVD	Chemical vapor deposition
CVI	Chemical vapor infiltration
DGEBA	Diglycidyl ether of bisphenol A
DICY	Dicyandiamide
DMA	Dynamic mechanical analysis
DMMT	Dodecylamine-treated montmorillonite
DSC	Differential scanning calorimetry
DTA	Differential thermal analysis
EDS	Energy dispersive spectroscopy
FCC	Face-centered cubic
FOD	Foreign object damage
FRP	Fiber-reinforced plastics/polymer
FTIR	Fourier transform infrared spectroscopy
GC	Gas chromatography
GFRP	Glass fiber–reinforced plastics
GSM	Grams per square meter
HBN	Hexagonal boron nitride
HCP	Hexagonal close packing
HDT	Heat distortion temperature
HEXA	Hexamethylenetetramine formaldehyde
ICP	Inductively coupled plasma
ILSS	Interlaminar shear strength
IOC	Unmodified clay

IPE	Isophthalic polyester
LSS	Laminate stacking sequence
MDA	4,4′-Methylenedianiline
MEKP	Methyl ethyl ketone peroxide
MMC	Metal matrix composites
MMT	Montmorillonite
NDI	Nondestructive inspection
NMR	Nuclear magnetic resonance
OMC	Organomodified clay or organoclay
OMMT	Organomodified montmorillonite clay
PACVD	Plasma-assisted chemical vapor deposition
PAI	Polyamide-imide
PAN	Polyacrylonitrile
PBI	Polybenzimidazole
PBT	Polybenzothiazole
PEEK	Polyether ether ketone
PEI	Polyetherimide
PES	Polyethersulfone
PI	Polyimide
PNC	Polymer nanocomposite
PP	Polypropylene
PPS	Polyphenylene sulfide
PS	Polystyrene
PSU	Polysulfone
PTFE	Polytetrafluroethylene
PVA	Polyvinyl alcohol
PVC	Polyvinyl chloride
RH	Relative humidity
RIM	Reaction injection molding
RRIM	Reinforced reaction injection molding
RT	Room temperature
RTM	Resin transfer molding
SAE	Society of Automotive Engineers
SCRIMP	Seemann composites resin infusion molding process
SEM	Scanning electron microscope
SIMS	Secondary ion mass spectroscopy
SMC	Sheet molding compound
SRIM	Structural reaction injection molding
TEM	Transmission electron microscopy
TGA	Thermogravimetric analysis
TGDDM	Tetraglycidyl-4,4′-diaminodiphenylmethane
TMA	Thermomechanical analysis

TOS	Thermal oxidative stability
TPES	Thermoplastic polyester
UDC	Unidirectional fiber composite
WAXD	Wide-angle x-ray diffraction
WRM	Woven rowing mat
XRD	X-ray diffraction

1

Introduction to Composites

Materials play an important role in the development of technology and the evolution of modern civilization. Ancient civilizations have been named after the materials used in that period. During the Stone Age (10,000 BC to 3,000 BC), people used only the materials found around them, such as stone, wood, and bone. They used these materials to make weapons to kill the animals for their food. During the Bronze Age (3000 BC to 1000 BC), people were able to extract copper from its ore. During the Iron Age (1000 BC to AD 1860), the extraction of iron from its ores signaled another major development. Heat treatment processes were developed during this period. In the Steel Age (AD 1860 onward), Bessemer and open hearth processes for the production of steel were developed. The general use of steel as a construction material started during this period. From 1950 onward, the era is named the Silicon Age. The development of silicon has led to the development of electronics, computers, and automation. It is very clear from the history that the development of materials is the prime factor for the development of civilization.

The designer puts his new ideas into practice through the use of new materials, since the performance of conventional materials is limited. Hence, there is a need to develop new materials for modern technological applications. The new materials should have high performance efficiency and reliability. They should be light in weight and also show a combination of properties. It should be possible to use them at the extreme environments, such as high temperature, high pressure, low temperature, low pressure, and highly corrosive. Over the years, so many new materials have been developed to meet the technological requirements. In some cases, it may not be possible to meet the stringent property requirements by using a single type of material. Hence, a combination of materials was thought of and thus composite materials have evolved. It is possible to get the best properties of constituents from a composite material. Predictions suggest that the demand for composites will continue to increase steadily. In the last 50 years, there has been a rapid increase in the production of synthetic composites, especially the fiber-reinforced polymer (FRP) composites. In recent times, the metal- and ceramics-based composites are also making a significant contribution.

1.1 Definition

There is no universally accepted definition for composite materials. Definitions in the literature differ widely. But, no one can deny that composite materials are made up of at least two distinct phases. The following definition covers a broad range of multiphase materials and at the same time excludes conventional multiphase materials. Composite materials are the combination of two or more materials in such a way that certain improved or desired properties are achieved. In a composite, the dispersed phase(s) is distributed in a continuous medium called matrix. The only condition is that the dispersed phase should retain its original identity after processing and/or during service/reprocessing. This can be explained with a simple example. Plastics are known for their good processability and low cost. However, it is very difficult to use the plastics for any load-bearing applications because of their poor mechanical properties. The Young's modulus of plastics is only a few GPa and strength is usually less than 50 MPa. Synthetic fibers, such as glass fibers, have good mechanical properties; the modulus is about 70 GPa and the strength is of the order of 3000 MPa. A composite made with the glass fiber, say 50 vol%, and plastics can have good mechanical properties. The modulus and strength values of this composite will be more than 30 GPa and 1000 MPa, respectively. This composite material is suitable for any load-bearing applications. Hence, the mechanical properties of plastics are improved by forming a composite with glass fiber and the glass fiber retains its original identity in the processed composite. This definition of composite material includes natural composites, as well as in situ composites. Actually, all natural composites are in situ composites, in which the reinforcement phase develops during the formation of matrix phase. This definition excludes age-hardenable alloys and other multiphase alloys, since the dispersed phase is not stable during service/reheating.

The properties of the composites are a function of the properties of the constituents, their relative proportion, and the geometry, distribution, and orientation of the dispersed phase. The main factors are the properties and the relative amount of constituents. Currently, numerous materials with widely differing properties are available. Hence, it is possible to get the desired properties in the final composite by selecting the right combination of materials in their proper proportion.

1.2 Brief History of Composites

Nature has been making composites for millions of years and hence the history of composites is perhaps as old as that of life on earth. Most of the plant parts have embedded fiber structures for better mechanical properties.

However, composites have been made by human beings only for the last 2000 years or so. Israelites made bricks using clay and plant straw, which are the earliest examples for the man-made composites. The Samurais of Japan made swords using laminated metals during the fifteenth century. Concrete, filled rubber, and phenolic resins were developed during the early twentieth century. The development of a process to manufacture glass fibers has led to the development of composites during the Second World War. The combination of glass fiber and plastics resulted in an incredibly strong material called FRPs. This material is used for making radomes of aircrafts during the early stages of development. The military applications of polymer matrix composites (PMCs) during the Second World War led to large-scale commercial exploitation after the war, especially in the marine industry, during the late 1940s and early 1950s. The first commercial boat hull was introduced in 1946. The rapid growth in composite science and technology happened during the 1950s in the United States and Europe. The composites industry began to mature in 1970s. During this period DuPont introduced the aramid fiber in the name of Kevlar. Carbon fiber production also started during this time. Many high-performance composites have been produced with these fibers. Composite materials have fully established themselves as workable engineering materials and are now very common everywhere, particularly in structural applications. They have gained wide acceptance and, wherever possible, conventional materials are being replaced by the composites materials. At present, the aircraft, automobile, marine, sports, electronic, chemical, and medical industries are quite dependent on FRPs.

Even now, the technology of composites is evolving with the development of nano-reinforcements, new matrix materials, production methods, design softwares, etc. The major concern of the today's world is energy conservation. The operating efficiency of any machinery can be increased when the light materials are used. At present, wind energy is one of the main sources of energy in many countries. Huge windmill blades are being made from polymer composites. There is no doubt that the composites will continue to play a major role in the modern world for energy conservation/generation.

1.3 Classification

Composites can be broadly classified into natural composites and synthetic composites. Some of the examples of natural composites are wood, bone, etc. Wood is a composite made of strong and flexible cellulose fibers in lignin matrix. The constituents of bone are hard and brittle hydroxy apatite platelets and strong yet soft protein called collagen. Although the constituents are present in fine scale in the natural composites, it is very difficult to tailor the properties of natural composites. The synthetic composites are

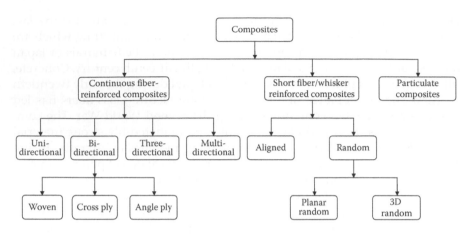

FIGURE 1.1
Classification of composites based on dispersed phase.

man-made composites. In synthetic composites, there is sufficient flexibility of selecting a suitable reinforcement and a matrix from the wide variety of reinforcements and matrices so that composites with the desired properties can be made.

Based on the dispersed phase, the composites can be classified as fiber-reinforced composites (FRCs) and particulate composites (Figure 1.1). The FRCs can be fabricated as single layer or as multilayer composites. When the layers in a multilayered composite are made with the same type of fiber reinforcement, then it is called a "laminate." If two or more types of fiber reinforcements are used in different layers, then it is a hybrid composite. The whole composite will have a single layer in the single-layer composites. This also includes composites having the same orientation and properties in each layer. Long or short fibers are used as reinforcements in single-layer composites. Better load transfer is possible with the fiber reinforcements and hence high mechanical properties are realized. However, the properties are not same in all the directions. Usually, the mechanical properties are high along the fiber direction. Isotropic composites can be easily made with short fibers. During processing, the fibers may align in a particular direction in some of the processing methods and then the composites become anisotropic. However, the composites are isotropic in most of the cases when the short fibers are used as the reinforcement.

Hybrid composites are made with two or more types of reinforcements. Commonly they are made with glass/carbon, glass/aramid, and aramid/carbon fibers. A recent development is glass/natural fiber hybrid composites. Hybrid composites take advantage of the properties or features of each reinforcement type. For example, in the glass/carbon hybrid composite, the low cost of glass fiber and high modules of carbon fiber are taken into consideration to form the low cost and high modules composite. Hybrid composites can be made with alternate layers of the two types of fibers, one layer of the second type at middle plane, or by incorporating the second fiber type at selected areas. Similarly, carbon/aramid hybrid composites have increased modulus and compressive strength over an all-aramid composite and increased toughness over an all-carbon composite. The coefficient of thermal expansion (CTE) of each fiber type in a hybrid composite needs careful consideration to ensure that high internal stress are not developed in the laminate during cure.

Whiskers are single crystal, short fibers with extremely high strength. The diameter of whiskers ranges from 0.1 to a few micrometers. The strength is almost close to the theoretical strength because of the absence of crystalline imperfections such as dislocations. Hence, it is possible to improve the strength of a matrix material significantly by using whiskers. However, there are some issues to be resolved before using whiskers. Most of the whiskers are expensive. Moreover, it is difficult and often impractical to incorporate whiskers uniformly into a matrix. The handling of whiskers is a problem because of their fine nature. Alignment in the matrix is another problem. Some of the whiskers are classified as carcinogenic materials. Unless there is a specific need, whiskers are seldom used.

Particulates are nearly equi-dimensional materials. They will not improve the strength of matrix in most of the particulate composites. Hence, it is not appropriate to call them as reinforcements. There are a few exceptions; very fine particles in a metal matrix can improve the strength by hindering dislocation motion and the tough particles can arrest the crack propagation and improve the strength of ceramics. Particulates are added mainly to modify certain physical properties of the matrix, such as functional or thermal properties. For example, particulates are added to polymers to reduce the cost of the product and to increase the modulus. However, the increase is much less than that predicted by the rule of mixtures for continuous FRCs in the fiber direction. Figure 1.2 shows Young's modulus of composites as a function of reinforcement volume fraction for different forms of reinforcement, namely, continuous fiber, whisker, and particle. It should be noted that the reinforcement efficiency goes down from continuous fiber to particle.

Based on the matrix material, the composites are classified into PMCs, metal matrix composites (MMCs), and ceramic matrix composites (CMCs). The classification of composites based on matrix material is shown in Figure 1.3. The three types of composites differ in the manufacturing method adopted, mechanical behaviors, and functional characteristics. Since the matrix materials undergo physical or chemical change, the processing method to be used for making the composites has a direct bearing on the matrix system used. The temperature at which the matrix materials are processed determines the choice of the dispersed phase because the reinforcement should neither undergo any chemical reaction or physical change nor have any change in its properties. The temperature level at which a particular composite can be used is determined by the temperature resistance of the matrix material.

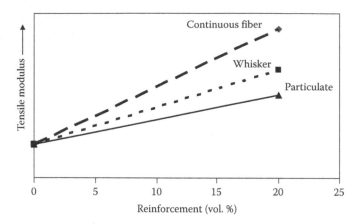

FIGURE 1.2
Tensile modulus as a function of reinforcement volume fraction for continuous fiber, whisker, and particulate-reinforced composites. (Reprinted from Chawla, K.K., *Composite Materials: Science and Engineering*, 2nd edn., Springer-Verlag, Inc., New York, 1998, p. 193. With permission.)

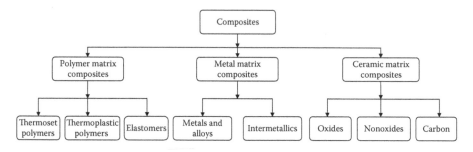

FIGURE 1.3
Classification of composites based on matrix materials.

The matrix material can be a thermoset polymer, a thermoplastic polymer, or an elastomer in the PMCs. Thermoset polymers are very commonly used because of the processing advantage. Currently thermoplastic polymers are gaining importance because of their relatively high toughness values and the possibility of post-processing. To meet the specific property requirements, a wide variety of thermoplastic polymers are available. PMCs are suitable for making products, which are used at ambient temperature. There are some special polymers, which can be used up to 250°C. In any case, PMCs are not suitable for applications where the service temperature is more than 350°C. The success of PMCs, largely as replacement for metals, results from the much improved mechanical properties of the composites compared to the plastic matrix materials. The good mechanical properties of the composites are a consequence of utilizing the high-strength and high-modulus fiber reinforcement.

Metals or metallic alloys are used as the matrix material in MMCs. Mainly lightweight metals and alloys such as, aluminum, titanium, and their alloys are used. In some special applications, heavy metals such as copper and cobalt are used. MMCs can be suitable for applications where the service temperature is up to 1200°C. At present, short fibers or particulates are mainly used as the dispersed phase because of the processing advantage. Metals and alloys are also reinforced with continuous fibers to improve modulus and strength significantly. A major problem for the MMCs is corrosion.

Most of the MMCs are still under development and only a few components are made commercially. Recent interest on MMCs has concentrated on transport applications and consequently the light metal-based MMCs, particularly aluminum and its alloy-based MMCs, have received the most attention. The relatively low Young's modulus of aluminum and its alloys can be significantly improved by the incorporation of reinforcement. Metals and alloys inherently have good ductility and toughness. The reinforcements improve Young's modulus but at the expense of ductility.

Many oxide and nonoxide ceramic materials are used as matrix materials in CMCs. CMCs are useful for high-temperature applications, where the service temperatures are above 1200°C. These materials are very expensive because most of the CMCs are processed at high temperature. In some cases, there is a need to apply high pressure at that high temperature to get a quality product.

Based on the size of dispersed phase, the composites can be classified into macrocomposites, microcomposites, and nanocomposites. The size of dispersed phase in the macrocomposites is in the millimeter level. The best example for macrocomposite is concrete. Concrete is made up of cement, sand, and gravel. Microcomposites consist of dispersed phase in the micrometer level. Most of the composites currently used are microcomposites. Glass FRPs (GFRPs) and carbon FRPs (CFRPs) are some of the common examples. At least one dimension of the dispersed phase is at the nanometer level in the nanocomposites. Carbon nanotube-reinforced plastics and nanoclay-reinforced plastics are the typical examples for the nanocomposites. The micro- and nanocomposite materials are inhomogeneous at the level of dispersed phase but homogeneous at the macro-level. That means they behave like a single component material at the macro-level.

Irrespective of the type of composite material, the interface plays a major role in controlling the properties. The constituents of the composites are separated by well-defined interfaces. Interfacial bond strength is very important for the coherent behavior of the composites. Greater bond strength is ensured by the use of coupling agent and/or mechanical interlocking of the constituents. It is not necessary to have better interfacial bond strength always; sometimes a tailored interface is more beneficial than strong interface. A weak interface would be preferred for the energy-absorbing systems. Debonding followed by fiber pullout absorbs a lot of energy during fracture processes in those systems. Crack deflection by debonding is another major contributor for the improved fracture toughness.

1.4 Advantages of Composites

The prime advantage of composites is their high specific stiffness and strength. Therefore, the component weight can be drastically reduced by using composites. Figure 1.4 shows the specific strength and modulus values of composites and conventional materials. For example, if a component is to be made for a particular load-bearing capacity, the GFRP component will weigh only one-fourth of steel component. Similarly, if stiffness is the criterion for the selection of material, the CFRP component will weigh only one-tenth of the steel component. Weight reduction is a major concern in aerospace and automobile sectors. Hence, the composite materials will have a clear edge over the other conventional materials in these sectors.

The second advantage of composites is their energy efficiency. Most of the composites currently used are polymer-based composites. The polymer composites can be produced at ambient temperature or slightly above ambient temperature and may be a few hundred degrees above ambient temperature. Hence, very little energy is required for the production of composites. As mentioned earlier, the extensive usage of polymer composites in

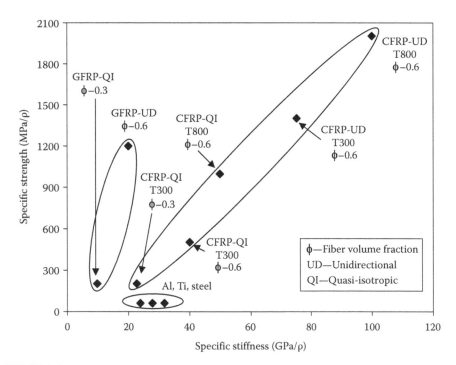

FIGURE 1.4

Specific strength and stiffness values of conventional materials and composites. (Adapted from Brandt, J., http://www.netcomposites.com/images/CompositeN_Aerospace.pdf, 2012. With permission.)

aircrafts and automobiles will reduce the total weight and thus increase the fuel efficiency. Composites are energy efficient because of these two reasons.

Generally polymers have better weather resistance than metallic materials. In PMCs, only the polymer matrix is exposed to the environment. Hence, the PMCs will also have good weather resistance.

It is possible to align the fibers in any particular direction in a FRC during manufacturing. The final composite will have directional properties and the mechanical properties will be high along the fiber direction. In certain structures, the requirement of mechanical properties is high, only in a particular direction, and it may not be critical in other directions. Composites are more suitable for these kinds of applications.

It is possible to make composites with desired properties. A suitable reinforcement, matrix, and processing method can be selected from the wide variety of reinforcements, matrices, and processing methods. A theoretical estimation of composite properties for any type of reinforcement and matrix with different fiber orientation is very easy now, because of the availability of numerous software programs.

Complex shapes can be made very easily with composites. Processing methods of composites, at least polymer composites are matured enough to produce any complex shape. Most of the polymer composites are processed at ambient temperature or slightly above ambient temperature (~150°C). Hence, the processing equipment and tools need not be very critical.

It is possible to produce composites with combination of desired properties. In some of the applications, like automobile body parts, it is necessary to have good mechanical properties with better thermal insulation and aesthetics. An automobile body part made with polymer composites will have all these properties. For the covers of electronic components, it is desirable to have some electrical conductivity to avoid static charge development. By incorporating a conducting material in the polymer, it is possible to produce a composite with some electrical conductivity.

1.5 Disadvantages of Composites

Composites are more expensive than conventional materials on a cost to cost basis. The composites are approximately 5 and 20 times costlier than aluminum and steel, respectively, on weight basis. However, the performance levels of composites are high. Composites find a place only when the high performance is a prime factor during the selection of material.

The chances of formation of defects at the interface are high, since composites are made with entirely different kind of materials. Unless great care is exercised during processing, defects are inevitable.

Most of the FRCs are anisotropic in nature. Accidental high-stress application in the transverse direction to the fibers may damage the product.

The production rate of composites is generally low. Composites may not be suitable for high volume production industries like automobile industries. The problem of slow production rate is overcome to some extent with the thermoplastic composites and molding compounds. Currently, FRC components can be made at rates comparable with conventional material components.

The selection of suitable material among the conventional metallic materials is very easy because of the availability of widely accepted database on the properties. However, similar database on the properties of different composites is not available. This is another disadvantage for the composites.

Recycling is another hurdle for the wide usage of composites. The recycling of composites is difficult compared to the conventional metallic materials. The good weather resistance of polymer matrix is a hindrance during recycling. At present various options are available for the recycling of composites but these options are little expensive.

1.6 Properties of Composites

The properties of a composite are a function of the properties of constituent phases and their relative proportions, size, shape, distribution, and orientation of the dispersed phase. These characteristics are illustrated in Figure 1.5.

The proportion of constituents can be expressed either by weight fraction or by volume fraction. The weight fraction is relevant to fabrication and the volume fraction is commonly used in property calculations. Weight and volume fractions are related to each other through density (ρ).

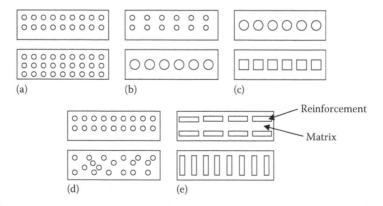

FIGURE 1.5
Schematic of geometrical and spatial characteristics of reinforcements in composites: (a) concentration, (b) size, (c) shape, (d) distribution, and (e) orientation.

$$W_r = \frac{w_r}{w_c} = \frac{\rho_r}{\rho_c} V_r \qquad (1.1)$$

where
the subscripts r and c refer to the reinforcement and composite, respectively
W and V are weight and volume fractions, respectively
w is weight

Similarly,

$$W_m = \frac{w_m}{w_c} = \frac{\rho_m}{\rho_c} V_m \qquad (1.2)$$

where the subscript m refers to matrix. ρ_c can be determined, if either volume fractions or weight fractions of the constituents are known. When the volume fractions of the constituents are known, then

$$\rho_c = \rho_r V_r + \rho_m V_m \qquad (1.3)$$

When the weight fractions of the constituents are known, then

$$\frac{1}{\rho_c} = \frac{W_r}{\rho_r} + \frac{W_m}{\rho_m} \qquad (1.4)$$

Equation 1.3 gives the density of the composite by the volume fraction adjusted sum of the densities of the constituents. This equation is not only applicable to density but, under certain conditions, may be applicable to other properties of composites. A generalized form of the equation is

$$X_c = X_r V_r + X_m V_m \qquad (1.5)$$

where
X represents any particular property
the subscripts c, r, and m refer to the composite, reinforcement, and matrix, respectively

This equation is known as the rule of mixtures or the law of mixtures.

The size of the dispersed phase also influences the properties of composites. In general, the smaller the size, the better will be the mechanical properties, because of size effect; that is, the size of defect is restricted and it should be smaller than the size of the material. The strength of any material is inversely proportional to the size of defect. The surface area of a material depends on the size and shape. For any particular volume, the smaller reinforcement has more surface area. The contact area between reinforcement and matrix increases with surface area, resulting in better properties. However, it is very difficult to produce continuous fibers with diameters less than a micrometer. Very fine short fibers/whiskers and particles have the tendency to agglomerate and it is very difficult to get a uniform distribution.

Ideally, a composite should be homogeneous, that is, it should have the uniform distribution of the constituents, but this is difficult to achieve during manufacturing. Homogeneity determines the extent to which a representative volume of the material may differ in physical and mechanical properties from the average properties of the material. Nonuniform distribution should be avoided as much as possible, otherwise different regions of a composite will have different properties. Those properties that are governed by the weakest part of the composite will be reduced by the nonuniformity of the system. For example, failure in a nonuniform composite will initiate in the area of lowest strength, thus adversely affecting the overall strength of the composite.

The orientation of the reinforcement in a particular direction within the matrix affects the isotropic properties of the composite. When the reinforcement is in the form of equiaxed particles, the composite behaves essentially as an isotropic material. When the dimensions of reinforcement are unequal, the composite can behave as an isotropic material only when the reinforcement is randomly oriented. For example, a randomly oriented, short FRC will have isotropic properties. In some cases, the manufacturing process may induce orientation of the reinforcement and hence loss of isotropy, that is, the composite is said to be anisotropic. In components manufactured from continuous FRCs, such as unidirectional or cross-ply laminate, anisotropy may be desirable as the laminate can be arranged in such a way that the highest strength is along the direction of maximum service stress. Indeed, a primary advantage of these composites is the ability to control the anisotropy by design and fabrication.

Most of the properties of a composite material are a complex function of other factors. The constituents of a composite usually interact in a synergistic way to determine the properties of the composites that are not fully accounted for by the rule of mixtures. The chemical and bonding characteristics of an interface are particularly important in determining the properties of the composite. The interfacial bond strength should be sufficient for load transfer from the matrix to the fibers, and only then will the composite have better strength than the unreinforced matrix. On the other hand, if the toughness of the composite is more important than strength, then the interface should readily fail to allow toughening mechanisms such as debonding and fiber pullout to take place.

1.6.1 Fiber-Reinforced Composites

FRCs are technologically the most important composites. Exceptionally high specific strength and modulus values are realized in FRCs. The fiber orientation, content, and the distribution all have a significant influence on the strength and other properties. Continuous fibers are normally aligned, whereas discontinuous fibers may be aligned, partially aligned, or randomly aligned.

In an FRC, the fibers are the load-bearing members. The matrix material binds the fiber reinforcements together into a solid. It transfers the load to the fibers when the composite is subjected to loading. It also serves to protect the reinforcements and the texture, color, and functional properties of composites are imparted through the modification of the matrix.

Mechanical response of this type of composite depends on the stress–strain behaviors of the fiber and matrix, the fiber content, and the direction in which load is applied. The properties of aligned fiber composites are highly anisotropic, that is, dependent on the direction in which they are measured. Consider the stress–strain behavior for the situation wherein the tensile stress is applied along the fiber direction. Usually the fibers are considered as totally brittle materials and the matrix is considered to have reasonably good ductility. The stress–strain behaviors of the fiber, matrix, and composite are shown in Figure 1.6. In the initial stage, both fibers and matrix deform elastically in the composite. When the applied stress is more than the matrix yield stress, the matrix deforms plastically while the fibers continue to deform elastically. This stage is also linear for the composite but with a diminished slope. The onset of composite failure begins when the fibers start to fracture approximately at the fiber fracture strain. In the ideal case, all the fibers fail at the fiber fracture strain and the matrix cannot take the load and the whole composite fails suddenly. The stress in the matrix during the composite

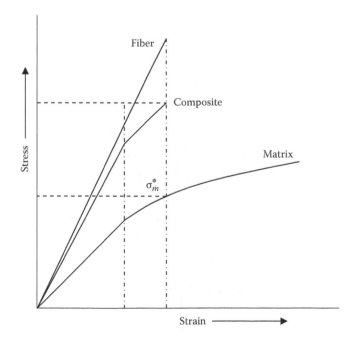

FIGURE 1.6
Stress–strain behavior of fiber, matrix, and composite. (Reprinted from Callister, W.D. Jr., *Materials Science and Engineering,* John Wiley & Sons, New York, 2007, p. 549. With permission.)

failure is σ_m^*. The composite failure is not catastrophic in the real situation, since not all fibers fracture at the same time and even after the failure of some fibers, the matrix holds the broken fibers and continues to deform plastically.

1.6.1.1 Elastic Behavior under Longitudinal Loading

Consider a composite made with continuous and aligned fibers loaded in the fiber direction (Figure 1.7). It is assumed that the fiber–matrix interfacial bond is very strong, such that the deformation of both matrix and fibers is the same, that is, isostrain condition. The load acting on the composite (F_c) is equal to the sum of the load carried by fibers (F_f) and matrix (F_m):

$$F_c = F_f + F_m \tag{1.6}$$

According to the definition of stress, $\sigma = F/A$, hence $F = \sigma A$

$$\sigma_c A_c = \sigma_f A_f + \sigma_m A_m \tag{1.7}$$

where σ_c, σ_f, and σ_m and A_c, A_f, and A_m are the respective stresses and cross-sectional areas of the composite, fibers, and matrix.

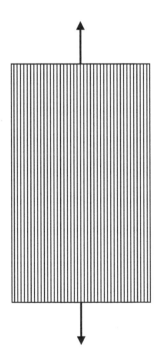

FIGURE 1.7
An aligned and continuous FRC loaded in the fiber direction.

On dividing this equation by the total cross-sectional area of the composite,

$$\sigma_c = \sigma_f \frac{A_f}{A_c} + \sigma_m \frac{A_m}{A_c} \tag{1.8}$$

Since the composite, fiber, and matrix lengths are equal, A_f/A_c is equivalent to volume fraction of fibers (V_f) and A_m/A_c is equivalent to volume fraction of matrix (V_m). Hence, the Equation 1.8 can be written as

$$\sigma_c = \sigma_f V_f + \sigma_m V_m \tag{1.9}$$

It has been assumed that the strains in the composite, fiber, and matrix are equal, that is,

$$\varepsilon_c = \varepsilon_f = \varepsilon_m \tag{1.10}$$

On dividing each term in Equation 1.9 by its respective strain,

$$\frac{\sigma_c}{\varepsilon_c} = \frac{\sigma_f}{\varepsilon_f} V_f + \frac{\sigma_m}{\varepsilon_m} V_m \tag{1.11}$$

When the deformation in the composite, fiber, and matrix is assumed to be elastic, then $\sigma_c/\varepsilon_c = E_c$, $\sigma_f/\varepsilon_f = E_f$, and $\sigma_m/\varepsilon_m = E_m$, where E is the modulus of elasticity. Hence, the Equation 1.11 becomes

$$E_c = E_f V_f + E_m V_m \tag{1.12}$$

This equation can be modified as

$$E_c = E_f V_f + E_m(1 - V_f) \tag{1.13}$$

for the composite consisting of only fibers and matrix, that is, $V_f + V_m = 1$.

Thus, the composite modulus is equal to the volume fraction weighted average of the moduli of elasticity of the fiber and matrix. By increasing the volume fraction of fibers, it is possible to increase the modulus of the composites.

On similar lines, it can be shown that the ratio of the loads carried by the fibers and matrix is

$$\frac{F_f}{F_m} = \frac{E_f V_f}{E_m V_m} \tag{1.14}$$

Provided that the response of the composite remains elastic, this proportion will be independent of the applied load. The higher the modulus and the volume fraction of fibers, the higher will be the load carried by the fibers.

The reinforcement is considered effective, when it carries large proportion of the load. For a given fiber–matrix system, the volume fraction of fibers in the composite must be maximized if the fibers are to carry a higher proportion of the composite load. However, above 80 vol% of fibers, the composite properties usually begin to decrease because of the inability of the matrix to wet and infiltrate the bundles of fibers. Incomplete wetting results in poor bonding and the formation of voids.

1.6.1.2 Elastic Behavior under Transverse Loading

Consider a composite made with continuous and aligned fibers loaded in the transverse direction, that is, 90° to the fiber direction (Figure 1.8). Under this situation, the stress in the composite, fiber, and matrix is the same, that is,

$$\sigma_c = \sigma_f = \sigma_m \tag{1.15}$$

This is termed as an isostress state. The deformation in the composites is the summation of the deformations in the fibers and the matrix:

$$\delta_c = \delta_f + \delta_m \tag{1.16}$$

FIGURE 1.8
An aligned and continuous FRC loaded in the transverse direction.

The deformation in the constituents or the composite can be written as the product of the strain and the corresponding cumulative thickness, so

$$\delta_c = \varepsilon_c t_c$$

$$\delta_f = \varepsilon_f t_f$$

$$\delta_m = \varepsilon_m t_m$$

Hence, Equation 1.16 can be written as

$$\varepsilon_c t_c = \varepsilon_f t_f + \varepsilon_m t_m \tag{1.17}$$

On dividing this Equation 1.17 by t_c,

$$\varepsilon_c = \varepsilon_f \frac{t_f}{t_c} + \varepsilon_m \frac{t_m}{t_c} \tag{1.18}$$

The thickness fraction is equal to the volume fraction for a composite with uniform length, hence

$$\varepsilon_c = \varepsilon_f V_f + \varepsilon_m V_m \tag{1.19}$$

When the constituents are assumed to deform elastically, the strain can be written in terms of the corresponding stress and the elastic modulus, that is, $\varepsilon = \sigma/E$:

$$\frac{\sigma_c}{E_c} = \frac{\sigma_f}{E_f} V_f + \frac{\sigma_m}{E_m} V_m \tag{1.20}$$

Since $\sigma_c = \sigma_f = \sigma_m$

$$\frac{1}{E_c} = \frac{V_f}{E_f} + \frac{V_m}{E_m} \quad \text{or} \quad E_c = \frac{E_f E_m}{V_f E_m + V_m E_f} = \frac{E_f E_m}{V_f E_m + (1 - V_f) E_f} \tag{1.21}$$

1.6.1.3 Longitudinal Tensile Strength

Failure of a continuous FRC is a relatively complex process, since several failure modes are possible. The mode that operates in a specific composite depends on fiber and matrix properties and the nature and strength of the fiber–matrix interfacial bond. When the failure strain of matrix is higher than that of fibers (which is the usual case), then fibers will fail before the matrix. Once the fibers are fractured, the load that was borne by the fibers is transferred to the matrix. However, the matrix cannot take the whole load, and it will fail immediately, without further deformation.

Only above a certain volume fraction of fibers will the composite strength be higher than the matrix strength. That fiber volume fraction is called critical fiber volume fraction. Let us consider a composite made up of brittle fibers and a ductile matrix. When the fiber content is sufficiently high, the composite will fail immediately after the failure of the fibers. The ultimate strength of composite, according to the rule of mixture, is given as

$$\sigma_c^u = \sigma_f^u V_f + \sigma_m^*(1 - V_f) \tag{1.22}$$

where
σ_c^u is the ultimate strength of composite
σ_f^u is the ultimate strength of fiber
V_f is the volume fraction of fiber
σ_m^* is the matrix stress at the fiber fracture strain ε_f

There is real strengthening by the fibers only when the ultimate strength of composite exceeds the ultimate strength of matrix material σ_m^u. Thus,

$$\sigma_c^u = \sigma_f^u V_f + \sigma_m^*(1 - V_f) \geq \sigma_m^u \tag{1.23}$$

This equation defines a critical fiber volume fraction that must be exceeded to get the strengthening effect by the fibers. On rearranging this equation,

$$V_f, \text{ that is, } V_{crit} = \frac{\sigma_m^u - \sigma_m^*}{\sigma_f^u - \sigma_m^*} \tag{1.24}$$

The composite will not fail at the stress indicated by Equation 1.22, if the fiber content is lower than certain minimum (V_{min}). Even after the failure of all fibers, the composite can take the load. The composite will fail at a stress $V_m\sigma_m^u$. There is a decrease in the strength of composite on increasing the fiber content till V_{min} is reached. The ultimate strength of the composite when the fiber content is less than V_{min} is given by

$$\sigma_c^u = \sigma_m^u(1 - V_f) \tag{1.25}$$

Equation 1.22 is applicable only when the fiber content is above V_{min}:

$$\sigma_c^u = \sigma_f^u V_f + \sigma_m^*(1 - V_f) \geq \sigma_m^u(1 - V_f) \tag{1.26}$$

Hence,

$$V_f, \text{ that is, } V_{min} = \frac{\sigma_m^u - \sigma_m^*}{\sigma_f^u + \sigma_m^u - \sigma_m^*} \tag{1.27}$$

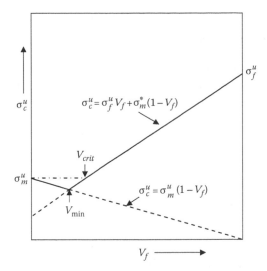

FIGURE 1.9
Tensile strength of unidirectional composite as a function of fiber volume fraction. (Reprinted from Agarwal, B.D. et al., *Analysis and Performance of Fiber Composites*, 3rd edn., John Wiley & Sons, New York, 2006, p. 76. With permission.)

The ultimate strength of a composite as a function of fiber volume fraction based on Equations 1.22 and 1.25 is shown in Figure 1.9. The solid lines indicate the range of applicability of the earlier equations. The critical fiber volume fraction and the minimum fiber volume fraction are also indicated in the figure. V_{min} and V_{crit} exist because the strength of the matrix is higher than the stress at the fiber fracture strain. For any particular fiber, the V_{min} and V_{crit} increase with the increase in matrix ultimate strength. High-strength matrix materials require more amounts of fibers to get the strengthening effect, that is, V_{crit} is high for the high-strength matrix materials.

1.6.1.4 Transverse Tensile Strength

The strength of a continuous and unidirectional fibrous composite is highly anisotropic and it is normally designed to take the load in the fiber direction. However, during service, transverse tensile loads may also be encountered. This may lead to premature failure, since the transverse strength of a unidirectional fiber composite is extremely low, sometimes lower than the tensile strength of the matrix. Actually in this case the reinforcing effect of the fibers is negative. A variety of factors influence the transverse strength and these factors include the properties of fiber and matrix, interfacial bond strength, and defects, such as voids. Usually the transverse strength of unidirectional composite is improved by modifying the properties of matrix.

1.6.1.5 Discontinuous Fiber-Reinforced Composites

Even though the reinforcement effect is lower for the discontinuous fibers than the continuous fibers, discontinuous and aligned fiber composites are becoming more important in the commercial market. These short fiber composites can have moduli of elasticity and tensile strengths that approach 90% and 50%, respectively, of their continuous fiber counterparts.

There are some advantages in using short fibers as the reinforcement in composites, such as easy fabrication and isotropic properties. In an FRC, the fibers are the load-bearing member. However, the load is not directly applied to the fibers. The load is applied to the matrix material and the matrix translates the load to the fibers through the fiber ends and cylindrical surface. The end effects can be neglected in the long FRCs, since the length of the fiber is very high compared to stress transfer length and the long FRC will be considered continuous FRCs. However, the end effects cannot be neglected in the short FRCs. The stress transfer at the fiber ends is negligible because of the yielding of the matrix adjacent to the fiber end or the separation of the fiber end from the matrix as a result of large stress concentrations. There is a minimum length at which the maximum applied stress is transferred to the fiber. That minimum length is called load transfer length (l_t). It can be given as

$$\frac{l_t}{d} = \frac{(\sigma_f)_{\text{max}}}{2\tau_y} \quad (1.28)$$

where
$(\sigma_f)_{\text{max}} = (E_f/E_c)\sigma_c$, σ_c is the applied stress
d is the diameter of the fiber
τ_y is the matrix yield stress in shear

The variations of fiber stress for different fiber lengths are shown in Figure 1.10. When the fiber length is shorter than the load transfer length, maximum fiber stress is not realized in the fibers. However, when the fiber

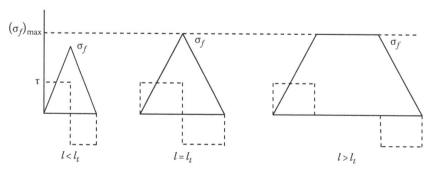

FIGURE 1.10
Variations of fiber stress and interface shear stress for different fiber lengths.

length is much greater than the load transfer length, the maximum fiber stress is realized in the major portions of the fiber and in this case the composite can be considered continuous FRC. The load transfer length varies with the applied stress. It is possible to define a critical fiber length, at which the maximum fiber stress corresponds to the ultimate strength of the fiber:

$$\frac{l_c}{d} = \frac{\sigma_f^u}{2\tau_y} \tag{1.29}$$

The small length adjoining the fiber ends is stressed less than the maximum fiber stress. This may affect the strength and modulus of short FRCs significantly, since there are numerous fiber ends within the composite. It is generally recommended that the fiber length should be at least 50 times greater than the critical fiber length to minimize the end effects.

The longitudinal strength of discontinuous and aligned fibers with uniform distribution of fibers of length $l > l_c$ is given by the equation

$$\sigma_c^u = \sigma_f^u V_f \left(1 - \frac{l_c}{2l}\right) + \sigma_m^*(1 - V_f) \tag{1.30}$$

If the fiber length is less than the critical fiber length ($l < l_c$), then the longitudinal tensile strength is given by the equation

$$\sigma_c^u = \left(\frac{l\tau}{d}\right) V_f + \sigma_m^*(1 - V_f) \tag{1.31}$$

where
   d is the fiber diameter
   τ is the smaller of either the interfacial bond strength or matrix shear strength

Consider an FRC, in which discontinuous fibers are randomly arranged. An adapted rule of mixtures can be utilized for this composite:

$$E_c = KE_f V_f + E_m V_m \tag{1.32}$$

where K is a fiber efficiency parameter that depends on the dimensions of the fibers and the E_f/E_m ratio, usually K values are in the range of 0.1–0.6. For the composites with randomly oriented fibers also, the modulus increases in some proportions of the volume fraction of fiber.

The efficiency of fiber reinforcement for several situations is given in Table 1.1. The efficiency is taken as unity for an aligned fiber composite in the fiber direction and zero in the perpendicular direction. For applications involving multidirectional applied stresses, randomly oriented

TABLE 1.1

Reinforcement Efficiency of FRCs for Several Fiber Orientations
and at Various Directions of Stress Application

Fiber Orientation	Stress Direction	Reinforcement Efficiency
All fibers parallel	Parallel to fibers	1
	Perpendicular to fibers	0
Fibers randomly and uniformly distributed within a specific plane	Any direction in the plane of the fibers	3/8
Fibers randomly and uniformly distributed within three dimensions in space	Any direction	1/5

Source: Reprinted from Krenchel, H., *Fiber Reinforcement*, Akademisk Forlag, Copenhagen, Denmark, 1964, p. 33. With permission.

discontinuous FRCs are preferred. Even though the reinforcement efficiency is one-fifth of the aligned composite, the mechanical properties are isotropic. The consideration of orientation and length for a particular application depends on the level and the nature of the applied stress as well as fabrication cost. The production rates of discontinuous FRCs are very high and the intricate shapes that are not possible with continuous fibers can be formed with discontinuous fibers. Moreover, fabrication costs are considerably lower than those of continuous and aligned FRCs.

1.6.2 Particulate Composites

Unlike the fiber reinforcements, particulates have aspect ratio close to unity; that is, particulates have almost same dimensions in all directions. Hence, load transfer to particulates is not effective as compared to fibers. Based on reinforcement mechanism, the particulate composites can be subdivided into large-particle composites and dispersion-strengthened composites. The term "large" indicates that particle–matrix interactions cannot be treated on the atomic or molecular level. In the large-particulate composites, the hard and stiff particles tend to restrain the movement of the matrix phase. As a result, the matrix transfers some of the applied stress to the particles, which bear a fraction of the load. The degree of improvement in mechanical properties depends on the strength of bonding between particles and matrix.

The volume fraction of the particulates influences the properties of composites; usually the modulus values are increased with increasing particulate content. Two mathematical expressions are available to show the dependence of elastic modulus on the volume fraction of particulates. These rules of mixture equations predict that the elastic modulus of a particulate composite should lie between an upper limit represented by Equation 1.33

$$E_c = E_p V_p + E_m V_m \tag{1.33}$$

and a lower limit represented by the Equation 1.34

$$E_c = \frac{E_p E_m}{V_p E_m + V_m E_p} \qquad (1.34)$$

where
  E and V denote the elastic modulus and volume fraction, respectively
  the subscripts c, m, and p represent composite, matrix, and particulate, respectively

These equations are based on the rule of mixtures of FRCs along the fiber direction (upper limit) and perpendicular to the fiber direction (lower limit). The upper limit of rule of mixtures predicts very high modulus values for the particulate composites. However, in the real case, the modulus values of polymers are increased only by a factor of 2–3 rather than 10 as expected from the rule of mixtures. This is because Young's modulus represents the strain per unit stress, and the strain contribution from different phases cannot simply be added as the rule of mixtures assumes. The rigid particles in fact concentrate stress at the interface with their soft embedding matrix, and the increased strain in the surrounding matrix outweighs the reduced strain in the region occupied by the particles. Hence, the experimental modulus values of particulate composites lie between the upper and lower limits (Figure 1.11).

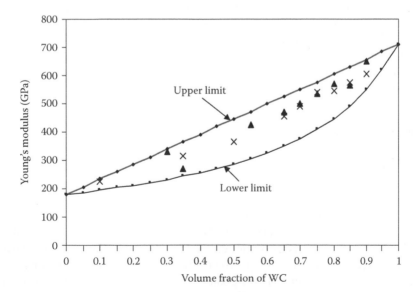

FIGURE 1.11
Theoretical and experimental modulus values of WC–Co composites. Solid lines correspond to theoretically estimated values based on the rules of mixtures and the discrete points are experimental values from Kwon and Dharan (1995).

Particulates are also added to the polymers to improve certain functional properties. Aluminum powder is widely used in polymers to improve their appearance, but it may also induce useful levels of electrical conductivity. Aluminum nitride is an effective electrical insulator and at the same time is also a thermal conductor. These aluminum nitride particles find use in the plastics molded around integrated circuits, where they efficiently dissipate heat generated from the circuits.

In the dispersion-strengthened composites, the particles are much smaller with the size lying between 10 and 100 nm. Particle–matrix interactions occur at the atomic or molecular level, which lead to strengthening the composites. The matrix bears the major portion of applied load and the small dispersed particles hinder or impede the motion of dislocations in crystalline materials. Thus, plastic deformation in metallic materials is restricted and thereby improves the yield and tensile strengths, as well as hardness. Another common example of dispersion-strengthened composite is carbon black-reinforced rubber. Automobile tires contain about 15–30 vol% of carbon black, which has particle size in the range of 20–50 nm. These carbon black particles enhance tensile strength, toughness, and tear resistance of rubber. For better improvement in properties, the particles must be evenly distributed in the rubber matrix and must form a strong bond with rubber. Carbon black particles tend to cluster into strongly bonded and irregularly shaped aggregates. At the processing temperatures, the polymer chains are able to burrow in by their typical wriggling motion. There they anchor to the carbon surfaces strongly by a combination of physical and chemical bonding. Carbon black improves the strength of rubber by a factor of up to 10, with significant increase in stiffness and wear and ultraviolet resistance.

It is evident from the preceding discussion that many factors play a role in controlling the properties of composites. This should be kept in mind when studying theoretical models, as all the factors are seldom accounted for during the development of models.

1.7 Applications

Composites are finding applications in all sectors. A rough estimate indicates that there are more than 60,000 composite products available in the market. A study made by DuPont predicts that in the year 2030, 50% of the engineering products will be made of composites. The annual growth rate of composites is 4%–7%. North America is the major consumer of composites, and Europe and Asia stand second and third, respectively (Table 1.2). The civil engineering sector is the major consumer of composites. Automobile, electrical/electronics, and mechanical industries are the other major consumers of composites. Apart from these sectors, composites also find applications in aerospace, marine, sports, and medical sectors. Figure 1.12 shows the market shares of the major composite sectors. Although the volume of composites used in the

TABLE 1.2

Composite Consumption in the World

	North America	Europe	Asia
Composite consumption (million tons)	3.4	2.1	1.5
Share of global composite consumption (%)	48	30	22

Source: Reprinted from Biron, M., *Thermosets and Composites*, Elsevier, Amsterdam, the Netherlands, 2003, p. 36. With permission.

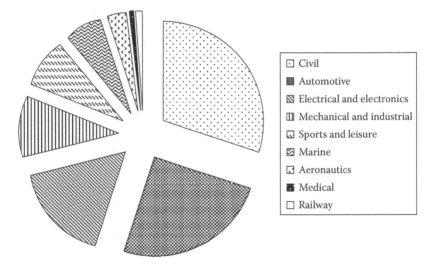

- ⊡ Civil
- ▓ Automotive
- ▨ Electrical and electronics
- ⬚ Mechanical and industrial
- ◪ Sports and leisure
- ▧ Marine
- ◩ Aeronautics
- ■ Medical
- ☐ Railway

FIGURE 1.12
Volume of composites used in different sectors. (Reprinted from Biron, M., *Thermosets and Composites*, Elsevier, Amsterdam, the Netherlands, 2003, p. 41. With permission.)

aerospace sector is small, the value of composites used in this sector is very high because of the usage of high-performance advanced composites.

The full potential of composites is realized in the aerospace sector. High specific stiffness and specific strength are the important properties that have led to the wide usage of composites in aerospace sector.

Polymer composites are more suitable for marine vessels and offshore structures because of their better weather resistance. Moreover, it is possible to make marine vessels with fewer joints using composites.

Polymer composites cannot compete with the conventional building materials. However, wood replacement is a suitable option. Doors, door frames, partition walls, etc. are being made with polymer composites. Polymer composites also find applications as aesthetic roofing in car parking areas and architectural domes in many buildings.

Polymer composites are good electrical insulators. Many types of electrical insulators and the substrates of printed circuit boards are being made with glass fiber-reinforced epoxy composites.

Polymer composites find an important place in wind energy. Very large windmill blades are being made with polymer composites. Many mechanical and automobile components are made with polymer composites. Some of the examples are leaf spring, automobile bodies, containers, etc. Polymer composites are the choice of material for many sport goods because of their high mechanical properties coupled with light weight. Tennis rackets, golf shafts, fishing rod, and polls are some of the examples. Moreover, body support systems for the physically disabled persons are made with CFRPs. The covers of many consumer durable products are currently being made with GFRPs.

MMCs are composites of relatively new origin and the manufacturing methods are much more complex than that of PMCs. There are many processing issues to be solved before inducted them into commercial applications. As a result, MMCs have registered a slow progress. The current applications of MMCs are mainly in aerospace and automobile sectors. Aerospace structures, turbine engine blades, landing gears, cutting tools, and piston, driving shaft, and connecting rods of automobiles are some of the prominent applications.

The development of CMCs is still in the infant stage. The processing of CMCs is even more complicated because of the requirement of very high processing temperatures. It is very difficult to produce good-quality composites because of adverse chemical reaction between the fibers and matrix at these high temperatures. Like PMCs, the initial impetus for the development has come from aerospace industries, where there is a great need for materials with high specific strength and modulus, and the capability to withstand under severe high temperatures or corrosive environment. At present, these materials are very expensive to make, hence their use is limited to applications that can utilize their special characteristics, such as high temperature resistance and high wear resistance. With the development of lower-cost fibers and more cost-effective manufacturing techniques, it is conceivable that CMCs will find commercial applications in the near future.

Questions

1.1 What are the advantages of composites? Even though there are many advantages with composites, they are not very common as conventional metallic materials; give reasons.

1.2 Combination of properties can be achieved in composites. Explain.

1.3 What is the difference between matrix and dispersed phases in a composite material? What should be the important mechanical characteristics of matrix and dispersed phases for FRCs?

1.4 What is a hybrid composite? List two important advantages of hybrid composites over single-type fiber composites.

1.5 Unidirectional composites may not be useful or cost effective in some applications. Explain. What are the ways to overcome those problems?

1.6 A burn-out test was performed to determine the volume fractions of glass fiber in a glass fiber-reinforced epoxy composite. The following observations were made:

Weight of empty crucible = 49.651 g

Weight of crucible and a piece of composite = 52.182 g

Weight of crucible and glass after burn-out = 51.448 g

Find out the weight and volume fractions of glass fibers and epoxy resin. The densities of fibers and resin are 2.54 and 1.2 g cm^{-3}, respectively. If the density determined experimentally was 1.82 g cm^{-3}, determine the void content in the composite.

1.7 Calculate the ratios of fiber stress to composite stress for unidirectional, continuous FRCs with V_f of 0.1, 0.3, 0.5, and 0.7. Assume that the composites are loaded in the fiber direction and the moduli of fiber and matrix are 380 and 3 GPa, respectively.

1.8 Estimate the maximum and minimum thermal conductivity values for a composite that contains 80 vol% TiC particles in a nickel matrix. Assume thermal conductivities of 27 and 67 W (m K)$^{-1}$ for TiC and Ni, respectively.

1.9 A large-particle composite consisting of tungsten particles within a copper matrix is to be prepared. If the weight fractions of tungsten and copper are 0.80 and 0.20, respectively, estimate the upper limit for the specific stiffness of this composite (ρ_W = 19.3 g cm^{-3}; ρ_{Cu} = 8.9 g cm^{-3}; E_W = 407 GPa; E_{Cu} = 110 GPa).

1.10 A continuous and aligned FRC is produced with 55 vol% aramid fibers in 45 vol% of polycarbonate matrix. The modulus of elasticity and tensile strength of aramid fibers are 125 GPa and 3300 MPa, respectively, and the elasticity and tensile strength of polycarbonate are 2.1 GPa and 66 MPa, respectively. Also, the stress on the polycarbonate matrix when the aramid fibers fail is 33 MPa. For this composite, compute the longitudinal tensile strength and the longitudinal modulus of elasticity.

1.11 Assume that the composite described in Problem 1.10 has a cross-sectional area of 400 mm^2 and is subjected to a longitudinal load of 43,400 N.

a. Calculate the fiber–matrix load ratio.

b. Calculate the actual loads carried by both fiber and matrix phases.

c. Compute the magnitude of the stress on each of the fiber and matrix phases.

d. What strain is experienced by the composite?

1.12 Is it possible to produce a continuous and aligned aramid fiber–epoxy matrix composite having longitudinal and transverse moduli of elasticity of 36 and 5.3 GPa, respectively? Why or why not? Assume that the elastic moduli of the aramid fiber and epoxy are 125 and 3.4 GPa, respectively.

1.13 For a continuous and aligned FRC, the moduli of elasticity in the longitudinal and transverse directions are 52 and 5.57 GPa, respectively. If the volume fraction of fibers is 0.4, determine the moduli of elasticity of fiber and matrix.

1.14 In an aligned and continuous carbon FRP composite, the fibers are to carry 93% of a load applied in the longitudinal direction. Determine the volume fraction of fibers that will be required, if the modulus values of carbon fiber and polymer matrix are 240 and 3 GPa, respectively. What will be the tensile strength of this composite? Assume that the matrix stress at fiber failure is 56 MPa and tensile strength of carbon fiber is 3600 MPa.

1.15 It is desired to produce an aligned carbon fiber–epoxy matrix composite having a longitudinal tensile strength of 400 MPa. Calculate the volume fraction of fibers necessary, if the average fiber diameter and length are 0.01 and 0.5 mm, respectively, the fiber fracture strength is 3.6 GPa, the fiber–matrix bond strength is 24 MPa, and the matrix stress at composite failure is 9 MPa.

1.16 A rectangular cross-sectional beam subjected to a bending moment is made of steel and is 12 cm in width and 6 mm in thickness. If the width of the beam is held constant, calculate the beam thickness when it is made from carbon–epoxy composite to provide the equivalent stiffness and the weight difference between steel and composite beams of 1 m length. (Given E_c = 240 GPa; ρ_c = 1.7 g cm^{-3}; E_s = 200 GPa; ρ_s = 7.8 g cm^{-3}.)

1.17 Discuss the significance of the critical fiber length with reference to composites with discontinuous fibers. From the following data, calculate the critical length of the fibers and hence the tensile strength of the composite. (Assume the following fiber and matrix properties: Fiber–diameter = 12 μm, tensile strength = 3000 MPa, length = 5 mm, volume fraction = 0.5; Matrix–shear strength = 100 MPa, tensile stress at failure strain = 250 MPa.)

1.18 A continuous and aligned glass FRC consists of 60 vol% of glass fibers having a modulus of elasticity of 70 GPa and 40 vol% of polyester resin, when hardened shows a modulus of elasticity of 3.2 GPa. Calculate the modulus of elasticity of this composite in the longitudinal and transverse directions. If the cross-sectional area is 300 mm^2 and a stress of 60 MPa is applied in the longitudinal direction, determine the magnitude of load carried by the constituents.

1.19 Find out the minimum and critical volume fraction of glass fibers needed to reinforce epoxy resin ($\sigma_f^u = 1.9\,\text{GPa}$; $\sigma_m^u = 80\,\text{MPa}$; $\sigma_m^* = 60\,\text{MPa}$). Calculate the composite strength with minimum and critical volume fraction of fibers. (Assume that the strength of matrix is unaltered in the composite.)

Reference

Kwon, P. and C.K.H. Dharan. 1995. *Acta Metall. Mater.* 43: 1141.

Bibliography

Agarwal, B.D., L.J. Broutman, and K. Chandrashekhara. 2006. *Analysis and Performance of Fiber Composites*, 3rd edn. New York: John Wiley & Sons.
Anderson, J.C., K.D. Lever, P. Leevers, and R.D. Rawlings. 2003. *Materials Science for Engineers*. Cheltenham, U.K.: Nelson Thornes.
ASM Handbook. 2001. *Composites*, Vol. 21. Materials Park, OH: ASM International.
Biron, M. 2003. *Thermosets and Composites*. Amsterdam, the Netherlands: Elsevier, p. 41.
Callister, W.D. Jr. 2007. *Materials Science and Engineering*. New York: John Wiley & Sons.
Campbell, F.C. 2004. *Manufacturing Processes for Advanced Composites*. Oxford, U.K.: Elsevier.
Chawla, K.K. 2012. *Composite Materials: Science and Engineering*. New York: Springer-Verlag, Inc.
Hull, D. and T.W. Clyne. 1996. *An Introduction to Composite Materials*, 2nd edn. New York: Cambridge University Press.
Krenchel, H. 1964. *Fiber Reinforcement*. Copenhagen, Denmark: Akademisk Forlag, p. 33.
Matthews, F.L. and R.D. Rawlings. 1994. *Composite Materials: Engineering and Science*. Cambridge, U.K.: Woodhead Publishing Limited.

2

Dispersed Phase

The dispersed phase can be in the form of long fibers, short fibers, whiskers, flakes, sheets, or particulates. Among these forms, fiber forms are widely used in the composites because of their superior properties and load transfer characteristics. A fiber can be defined as an elongated material, mostly with a circular cross-section having a more or less uniform diameter of less than 250 μm and an aspect ratio (length to diameter ratio) of more than 100 (Chawla and Chawla 2006). Fiber form is stronger than bulk form due to the size effect, that is, the defect size in the fiber is restricted to the size (diameter) of fibers. Figure 2.1 shows the strength of different material types in three forms, namely, bulk, fiber, and whisker forms. Whiskers are monocrystalline short fibers. The diameter of whiskers is 0.1 to a few micrometers and length can be up to a few millimeters. Silicon carbide whiskers are the common whisker materials used in composites. Unless there is a specific need, whiskers are rarely used because of the reasons given in the previous chapter.

The fibrous materials can be broadly grouped into natural and synthetic fibers. The plant kingdom is a rich source of natural fibers, and these fibers are mainly based on cellulose. Some of the natural fibers are cotton, flax, jute, hemp, sisal, coir, and ramie. Animal kingdom also supplies many natural fibers. These fibers are mainly based on proteins. Hair, wool, and silk are some of the natural fibers. The silk fibers produced by spider are considered as the toughest material.

Although natural fibers are relatively cheap and produced from renewable sources, they are not widely used because of their poor mechanical properties compared to synthetic fibers. The other problems with natural fibers are poor moisture resistance, nonuniform dimensions and properties, incompatibility with polymer matrices, and poor weather resistance. Hence, natural fiber-reinforced polymer composites are used only in applications where the load-bearing capacity is not very critical. A major advantage of natural fibers is their biodegradability. Biodegradable composites can be made by using natural fibers with biodegradable polymer matrices.

Unlike natural fibers, synthetic fibers are high-performance fibers. The modulus and strength values are very high for most of the synthetic fibers.

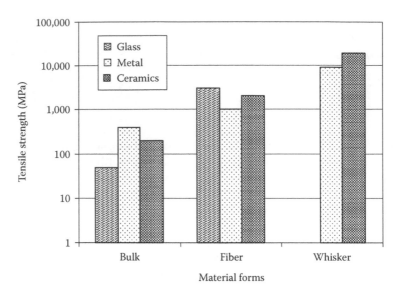

FIGURE 2.1
Tensile strength of materials in different forms.

Some of the common synthetic fibers include glass, boron, carbon, aramid, alumina, and silicon carbide. Glass fibers are the most widely used synthetic fibers. At present boron fibers are seldom used, since carbon fibers of equivalent properties with good flexibility are available at a cheaper rate. Most of the high-performance composites are made with carbon fibers. Aramid fibers are used only in polymer matrices, since they are based on organic polymers and cannot withstand the processing temperatures of metals and ceramics. The attractive property of aramid fiber is its high-damping characteristics. Because of this property, bullet proof vests are made with this fiber. Ceramic fibers are mainly used in metal matrix composites (MMCs) and ceramic matrix composites (CMCs). They are relatively expensive, but their high temperature resistance is very good.

Many particulate materials are derived from the natural minerals. Some particulate materials are also prepared using synthetic routes. Particulate materials are available from a few hundred micrometers to nanometer level. Particulates are added to polymers, mainly to reduce cost and to metal and ceramic matrices, to improve some specific properties. Calcium carbonate, titanium oxide, and clays are the common particulate materials added to polymer. Alumina, silicon carbide, and zirconia are the common particulate materials used in metal and CMCs. Since the fibers are the most important type of reinforcements in composites, more emphasis is given for this type of reinforcement.

2.1 Fiber Reinforcements

The important characteristics of fibers are small diameter, high aspect ratio, and good flexibility. Most of the synthetic fibers have diameter in range of 5–15 μm. According to Griffith's theory, the fracture strength of brittle material can be related to the defect size by the following equation:

$$\sigma = \sqrt{\frac{4E\gamma}{\pi c}} \qquad (2.1)$$

where
  E is Young's modulus
  γ is surface energy
  c is the crack length

The smaller the defect the higher will be the fracture strength. This equation is very well applicable to most of the synthetic fibers, since they are also brittle. Since the size of the defects in fibers is restricted to be less than fiber diameter, very high strength values could be realized in fibers.

The fracture strength of each fiber in a bundle is different, since the maximum size of defect in each fiber is different. Hence, there is a statistical distribution of strength values for the fibers in a bundle. The average strength of fibers in a bundle is normally represented as the strength of fiber. Consider a fiber bundle consisting of 100 individual filaments. The strength distribution of fibers is shown in Figure 2.2. The average strength value of fibers is 2500 MPa.

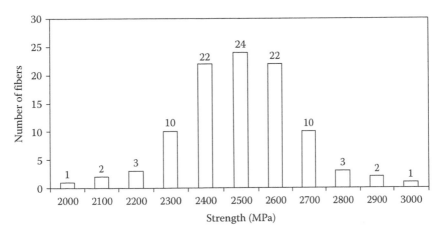

FIGURE 2.2
Strength of fibers in a bundle containing 100 filaments.

There are three models available for the failure of materials. They are the weakest link, cumulative weakening, and cumulative weakening with reduced gage length models (Figure 2.3). Most bulk materials fail by the weakest link model, that is, the failure starts at the weakest location (the place where the largest defect exists), propagates, and finally leads to the failure. Fiber bundles fail by the cumulative weakening model. After the failure of

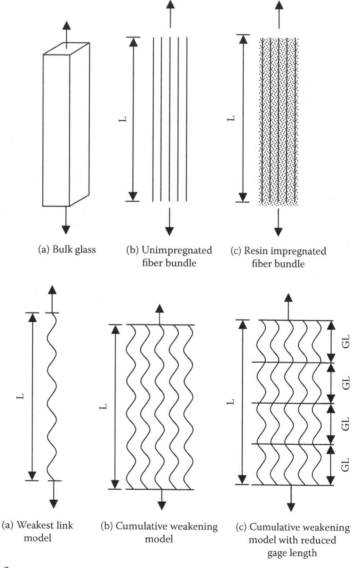

(a) Bulk glass (b) Unimpregnated (c) Resin impregnated
      fiber bundle   fiber bundle

(a) Weakest link (b) Cumulative weakening (c) Cumulative weakening
  model   model   model with reduced
                   gage length

FIGURE 2.3
Failure models of (a) bulk material, (b) fiber bundle, and (c) composite. (Courtesy of Dr. N.G. Nair, NGN Composites, Chennai, India.)

TABLE 2.1

Failure Behavior of a Fiber Bundle

Applied Load	Bundle Stress σ_B (MPa)	No. Failed n	Total No. Failed Σn	Filaments Left $100 - \Sigma n$	Actual Stress in Filaments σ (MPa)
1900 MPa × 100A	1900	0	0	100	1,900
2000 MPa × 100A	2000	1	1	99	2,020
2100 MPa × 100A	2100	2	3	97	2,165
2200 MPa × 100A	2200	3	6	94	2,340
2340 MPa × 94A	2340	10	16	84	2,785
2785 MPa × 84A	2785	78	94	6	46,415

A is the cross-sectional area of the individual fibers.

any fiber in the bundle, the remaining fibers take the load. The load carried by this fiber will now be shared by all other filaments. Correspondingly, the stress on each fiber will increase. As the load is further increased, a few more fibers will break and the load will be redistributed. The process of redistribution continues until the remaining fibers can no longer take the load and the whole bundle fails at this stage. The stress level of fiber bundle at different applied loads is given in Table 2.1. From the table, one can infer that the fiber bundle fails before the average strength value of fibers is reached due to cumulative weakening. However, the strength of fiber bundle is not governed by the weakest link. In a fiber bundle, the broken fiber ceases to carry load, and therefore, the entire length of fiber becomes ineffective. The failure process is changed when the fiber bundle is incorporated in a matrix. Although the load is released from the broken fiber ends, the portion of the fiber away from the broken ends can take the load. The load released by the broken fiber will be transferred to the adjacent fibers through the matrix. The load transfer takes place over a length called critical load transfer from the broken fiber ends. As the load is increased, more fibers will break, but any breakage of fiber at a distance of two times the load transfer length will not affect the stress distribution within the region. This restraint helps each section to carry more loads. Failure of fiber bundle eventually occurs in the segment at which the remaining fibers can no longer take the load. Thus, the bonding of fibers with matrix effectively reduces the gage length equal to two times the load transfer length. The fracture strength of composite is decided by the number of broken fibers within a gage length. For example, if there are two broken fibers in a composite and the failed regions are well separated (at two different gage lengths), it can be considered as a single-fiber breakage. Hence, the failure load of matrix-bonded fiber bundle is necessarily higher than that of unbonded fiber bundle. This model is called cumulative weakening with reduced gage length. Another common example is coir rope. It is made by twisting discontinuous coir fibers. Twisting leads to the formation of mechanical bonding between the fibers. Although the rope is made of discontinuous fibers (all fibers are broken), it can take very high loads because the fiber ends are well separated, that is, at different gage lengths.

The fibrous materials have high aspect ratio (length to diameter ratio). For effective load transfer, the aspect ratio should be of the order of 200. The diameter of common synthetic fibers is approximately 10 μm. For effective load transfer, a minimum length of 2000 μm (2 mm) is required. Even the short synthetic fibers have length greater than this. Hence, the fibers take the load applied to the composite and high mechanical properties can be realized in the composite.

The flexibility of fibrous materials is also generally good. Flexibility is a function of modulus and size. For the same material available in different forms, the modulus value is the same. Hence, only the size factor controls the flexibility. The smaller the size the better will be the flexibility. The inverse of the product of bending moment and radius of curvature is a measure of flexibility. The single fiber can be assumed as a circular rod of diameter, d. When a circular rod is subjected to bending, according to bending theory,

$$\frac{M}{I} = \frac{E}{R} \qquad (2.2)$$

where
   M is the bending moment
   I is moment of inertia (for the circular rod, it is $\pi d^4/64$)
   E is Young's modulus
   R is radius of curvature

Hence, the flexibility given as

$$\frac{1}{MR} = \frac{64}{E\pi d^4} \qquad (2.3)$$

This equation indicates that flexibility in a sensitive function of diameter of fiber. A small reduction in fiber diameter will significantly increase the flexibility. Ceramic materials are categorized as rigid materials without flexibility. However, the ceramic fibers are flexible, because of the very small diameter. Figure 2.4 shows the diameter of fibers produced from various materials having the flexibility same as that of 25 μm polyamide fiber. Flexibility is one of the important criteria for the fabrication of composites, especially the complex shapes. If the flexibility is good, then it is easy to bend the fiber to conform to the shape of the product.

2.1.1 Natural Fibers

Nowadays environmental-friendly composites are the most sought-after composites because of growing environmental concerns. The synthetic fiber-based composites are difficult to dispose after their utility. This has driven

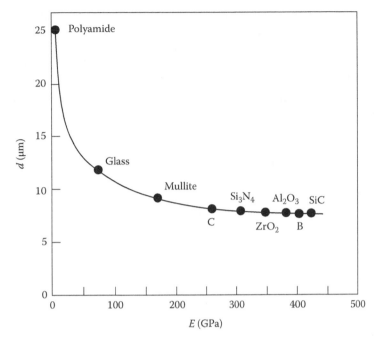

FIGURE 2.4
Diameter of fiber from different materials to have flexibility equal to that of a 25 μm diameter polyamide fiber. (Reprinted from Chawla, K.K., *Composite Materials: Science and Engineering*, 2nd edn., Springer-Verlag, Inc., New York, 1998, p. 8. With permission.)

the use of natural fiber-reinforced composites (FRCs) for less critical load-bearing applications. Different kinds of natural fibers are available in different countries. The separation of long fibers from the plants is a major issue. Nevertheless, many types of natural fibers are processed in many countries.

Based on a study conducted by U.S. consultant, Kline & Co., the North American market for natural fibers in polymer composites is over 180 million kilograms (Eckert 2000). Long natural fibers are finding increasing use in automotive components, since natural fibers cost around 30% less than the commonly used glass fiber. The weight of natural fibers is also only about half of the weight of glass fiber. Another added advantage is that the natural fiber composites are easier to recycle. Typical use of natural fiber composites include interior parts, such as door trim, package trays, and rear shelves. In addition to the automotive industry, significant markets are also emerging in other industries. The advantages of natural fibers are as follows:

- The cost of the natural fibers is low (approximately 30% lower than glass fibers).
- The density values of natural fibers are roughly one-half of the density of glass fiber. Hence, the specific strength and modulus values are comparable to glass fibers.

- It is a renewable resource and the production requires little energy. Unlike synthetic fiber production, the natural fiber plants release oxygen to the environment by absorbing CO_2. Moreover, these materials are biodegradable. Hence, the natural fibers are environmental-friendly materials.
- The polymer composites made using natural fibers can be thermally decomposed easily, whereas glass fiber causes problem in combustion furnaces.

The disadvantages of natural fibers are as follows:

- There is a wide variation in dimensions and properties. The quality of the fibers depends on the place of cultivation and weather conditions.
- The price of the fiber depends on the harvest results and agricultural policies.
- Even though the specific strength and stiffness of natural fibers are comparable to glass fibers, the actual mechanical properties are lower.
- Moisture absorption is another major problem for the natural fibers and the fibers swell on moisture absorption. This affects the durability of composites. Hence, the natural FRCs are only suitable for indoor applications.
- The maximum processing temperature of composites is limited due to the degradation of natural fibers. Hence, the natural fibers are suitable for only certain organic polymer matrices.

Cellulose is the basic compound present in the fiber cells of plants. It is a natural polymer with high specific strength and stiffness. The fiber cells can be found in the stem, leaves, or seeds of plants.

2.1.1.1 Bast Fibers

The bast of a plant consists of wood core surrounded by stem. The stem is made up of fiber bundles and each fiber bundle has many individual filaments. The fiber bundles are bonded together by lignin or pectin. The pectin is removed during the retting process and that enables the separation of individual fiber bundles. Even after the removal of pectin by the retting process, lignin will be holding the individual filaments together. Lignin affects the bonding of fiber with polymer matrix, and hence, the performance of the composite will be poor. This lignin can be removed by treating in boiling alkaline solution.

The bast of plants like flax, hemp, jute, kenaf, and ramie contains the fiber bundles. The largest producers of jute are India, China, and Bangladesh.

Jute is a cheap fiber and has reasonable strength and resistance to rot. Flax is mainly cultivated in Canada, China, India, and the United States. Flax fiber is strong and stiff. It is also a raw material for the high-quality paper industry. The world's leading producer of hemp is China, with smaller production in Europe, Chile, and Korea. The length of hemp fiber varies between 0.91 and 4.6 m. China is the leader in producing ramie. Other producers include Japan, Taiwan, Philippines, and Brazil. The fiber extraction and cleaning from ramie stem are expensive.

2.1.1.2 Leaf Fibers

In general, the leaf fibers are coarser than the bast fibers. Some of the leaf fibers are sisal and palm. Among these fibers, sisal fiber is the most important fiber, since it is relatively stiff. It is obtained from the agave plant.

2.1.1.3 Seed Fibers

Cotton, coir, and kapok are the common seed fibers. Cotton fiber is used in textile industry all over the world. Coir fiber is derived from coconut husk. It is a coarse fiber with good flexibility. Kapok fiber is a hollow fiber with a sealed tail. It is a short fiber with a smooth, silky surface. The strength of this fiber is not high.

2.1.1.4 Properties of Natural Fibers

The important properties of commonly used natural fibers are given in Table 2.2. Sisal, banana, jute, and flax fibers are cellulose rich (>65%) and show high tensile strength and modulus. Coir fiber is lignin rich (>40%) and shows poor strength and modulus but high failure strain. Despite having low strength and modulus, coir fiber significantly improves the impact resistance of thermoset polymers due to its high strain-to-failure value. The specific modulus of jute fiber is superior to glass fiber. Hence, jute FRCs can be used in place of glass FRCs, when the strength of the composites is not very important.

Due to the presence of hydroxyl and other polar groups in the natural fibers, the moisture absorption is high. It can be up to 15% in certain fibers. This moisture absorption affects the wettability with resin and interfacial bond strength. Hence, it is advisable to completely dry the fiber before it is incorporated into the polymer matrix.

Being cost effective and renewable, the natural fibers are gaining wide acceptance as reinforcement for making composites used in many applications. When the mechanical properties needed for any application are high, then it is possible to use natural fiber in combination with synthetic fibers to meet the requirements. Although the usage of natural fiber is increasing, the present scenario is that the synthetic fibers are the most widely used fibers in composites.

TABLE 2.2

Typical Properties of Commonly Used Natural Fibers and Glass Fiber

Properties	Fiber								
	E-Glass	Flax	Hemp	Jute	Ramie	Coir	Sisal	Abaca	Cotton
Density (g cm⁻³)	2.5–2.59	1.4–1.5	1.4–1.5	1.3–1.5	1.0–1.5	1.15–1.46	1.3–1.5	1.5	1.5–1.6
Diameter (μm)	<17	12–600	25–500	20–200	20–80	10–460	8–200	—	10–45
Length (mm)	—	5–900	5–55	1.5–120	900–1200	20–150	900	—	10–60
Tensile strength (MPa)	2000–3500	345–2000	270–900	320–800	400–1000	95–230	360–700	400–980	290–800
Specific strength (MPa/ρ)	770–1400	230–1430	180–640	210–615	265–1000	65–200	240–540	265–650	180–530
Young's modulus (GPa)	70–76	27.6–103.0	23.5–90.0	8–78	25–128	2.8–6.0	9–38	6–20	5.5–12.6
Specific modulus (GPa/ρ)	~29	~45	~40	~30	~60	~4	~17	~9	~6
Elongation (%)	1.8–4.8	1.2–3.3	1.0–1.8	1.5–1.8	1.2–4.0	15–51	2–7	1–10	3–10
Moisture absorption (wt.%)	—	8–12	6–12	12–14	8–17	8	10–22	5–10	7.9–8.5

Source: Data from Dittenber, D.B. and GangaRao, H.V.S., *Composites*, A43, 1419, 2012.

2.1.2 Synthetic Fibers

The majority of the composites produced today are based on synthetic fibers, since there are many drawbacks with the natural fibers. The properties of a particular type of natural fiber vary from place to place. Hence, it may not be possible to produce a natural FRC with the desired properties. Moreover, the natural fibers degrade rapidly under normal environmental conditions. However, the synthetic fibers have narrow range of properties and also they are more resistant to normal environmental conditions. Some of the commonly used synthetic fibers are glass, carbon, aramid, polyethylene, aluminum oxide, aluminum silicates, and silicon carbide.

2.1.2.1 Glass Fibers

Glass fibers are the most widely used fiber reinforcements in composites, especially in polymer matrix composites. Bulk glass has high hardness, moderate stiffness, transparency, and chemical resistance. In addition to that, glass in the fiber form has high strength and good flexibility. Many structural composites, printed circuit boards, and wide range special products are manufactured using glass fibers as reinforcement.

Glass is an amorphous material. Originally the amorphous compounds formed from silica-based compounds are called glass. At present, any amorphous compound is called glass. The major constituent of any inorganic glass forming material is silica. Apart from silica, oxides of aluminum, boron, calcium, etc. are also present in varying quantities depending on the type of glass. The glass fibers used in composites are made from the silica-based inorganic compounds. There are a variety of glass fibers available with varying chemical composition.

There are two categories of glass fibers and they are low-cost, general-purpose fibers and premium special-purpose fibers. The chemical compositions of commonly used glass fibers are given in Table 2.3. The glass fiber varieties are designated with English alphabets implying special properties. The general-purpose fibers constitute more than 90% of the glass fibers used, mainly the E-glass variety. E-glass is the most preferred fiber type for the composite industries because of the cost and processing advantages. Originally this type was developed for electrical applications because of its better electrical insulating properties. The general-purpose E-glass fiber is also produced as two variants: one with boron and another without boron. Stringent environmental regulations in many countries forced the glass fiber manufacturers to reduce boron emission, which has led to the development of boron-free E-glass. It has about 5% higher elastic modulus than boron-containing glass fiber, but the filament strengths of both types of glass fibers are almost the same. The corrosion resistance of boron-free fiber is found to be seven times higher than that of boron-containing fiber when tested in 10% sulfuric acid (Rossi and Williams 1997).

TABLE 2.3

Chemical Composition of Different Types of Glass Fibers

Fiber Type	Composition (wt.%)												
	SiO_2	Al_2O_3	B_2O_3	MgO	CaO	Na_2O/K_2O	Fe_2O_3	ZnO	BeO	TiO_2	ZrO_2	Li_2O	Other Oxides
E	52.4–53.2	14.4–14.8	8.0–10.0	4.5	17.5	0.5	0.4	—	—	—	—	—	—
E Boron free	59.0–60.0	12.1–13.2	—	3.1–3.4	22.1–22.6	0.8–0.9	0.2	—	—	0.5–1.5	—	—	—
C	64.4	4.1	4.7	3.3	13.2	9.4	—	—	—	—	—	—	BaO: 0.9
S	64.2	24.8	0.01	10.27	0.01	0.24	0.21	—	—	—	—	—	BaO: 0.2
Z	68	0.7	—	—	—	12.3	—	—	—	1.5	16.5	1.0	—
M	53.7	—	—	9.0	12.9	—	0.5	—	8.0	7.9	2.0	3.0	CeO_2: 3.0
ECR	58.0	11.6	—	2.0	21.7	1.2	0.1	2.9	—	2.5	—	—	—
D	55.7–74.5	0.3–13.7	22.0–26.5	0–1.0	0.5–2.8	0.2–2.3	—	—	—	—	—	—	—
Silica	99.99	—	—	—	—	—	—	—	—	—	—	—	—

Source: Data mainly from Wallenberger, F.T. et al., *ASM Handbook*, Vol. 21, *Composites*, ASM International, Materials Park, OH, 2001, p. 28.

The special-purpose glass fibers include S-glass, C-glass, Z-glass, D-glass, M-glass, ECR-glass, and silica fibers. Among the special-purpose glass fibers, ECR fibers offer better long-term acid resistance and short-term alkali resistance. C-glass provides better corrosion resistance to acids compared to E-glass. It is mainly used for chemical applications. High-strength glass fibers such as S-glass fibers have 10%–15% higher strength than E-glass, but the real importance of these fibers is their stability at high temperatures. The tensile strength of glass fibers is generally determined by the silicate network connectivity. The alkali ions break the network. The other important constituent of glass fiber is boron oxide. Although it can become part of the network, it is weaker than silica. Hence, the high-strength glass fibers are made with less alkali and boron contents. M-glass is a high-modulus glass fiber with modulus value higher than E-glass. However, the production is stopped in the recent times due to the toxicity of BeO. Z-glass has a high resistance to alkaline environment and is mainly used in chemical applications.

Printed circuit boards are generally made using E-glass fibers. However, the miniaturization drives the industry to use specialty fiber with low dielectric constant. The D-glass fibers have been developed for this purpose. They contain high levels of B_2O_3 (20–26 wt.%) and therefore have much lower dielectric constants than that of E-glass fibers. However, the cost of these fibers is high because of the necessity to use special processing methods to overcome the problem of high boron content.

Glass fiber is produced by melt-spinning process. The required amount of natural minerals conforming to the final glass composition is melted in a furnace. This melt is allowed to flow through a platinum alloy bushing containing several hundred individual orifices of diameter 0.793–3.175 mm. The molten glass is coming out as a continuous stream. Fibers are rapidly drawn from this highly viscous melt and then cooled. The final diameter of the fiber depends on the platinum orifice diameter and viscosity of molten glass. The viscosity is in turn controlled by the chemical composition. The diameter of glass fibers typically range from 3 to 20 μm. The fibers are drawn at velocities up to 61 m s^{-1}. Individual filaments are gathered into multifilament strands after applying a size coating. The size coating performs several functions, such as a lubricant, protective agent, binder, and coupling agent to the polymer being reinforced. A collection of strands is either wound on a tube as continuous roving or used to produce various glass fiber forms. A schematic of glass fiber manufacturing and various forms of glass fiber are shown in Figure 2.5. Some special glass fibers are produced by marble melt process. In this process, the raw materials are melted and solid glass marbles of 20–30 mm in diameter are formed. These marbles are used as raw materials to produce glass fibers by the process mentioned above.

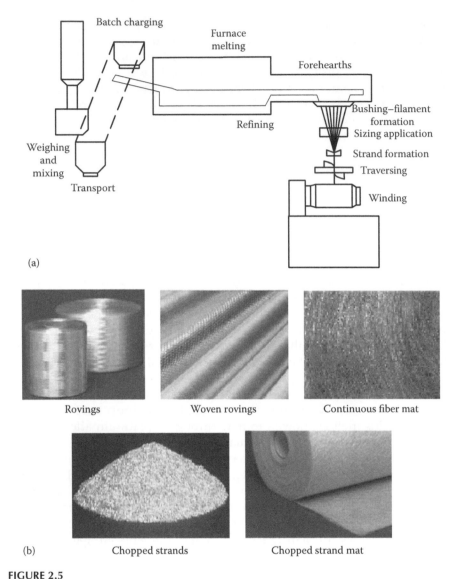

(a)

(b)

FIGURE 2.5
Schematic of glass fiber manufacturing process (a) (Reprinted from Katz, H.S. and Mileswski, J.W., *Handbook of Fillers and Reinforcements for Plastics*, Van Nostrand Reinhold Company, New York, 1978, p. 471. With permission) and various forms of glass fiber (b).

On similar lines, high-silica fibers with more than 95% silica are produced for high-temperature applications up to 1040°C. They can be made by acid leaching of E-glass fibers. Pure silica fibers containing more than 99% silica are made by dry spinning from water-glass solutions. They can be used up to 1090°C. Ultrapure silica fibers with 99.999% silica are prepared by dry spinning of high-purity tetraethyl orthosilicate gel.

2.1.2.1.1 *Important Forms of Glass Fibers*

Once the continuous glass fibers are produced, they must be converted into a suitable form for use in composites. The important forms of E-glass fibers are continuous roving, woven roving, chopped strand mat (CSM), chopped strand, continuous strand mat, yarn, and fabrics. During the melt-spinning process, fibers are collected as strands consisting of around 200 individual filaments. The rovings are a collection of continuous strands. The glass fiber rovings are generally designated by "tex." Tex is a number, which indicates the weight of rovings in grams for a kilometer length. Glass fiber rovings are available at tex values of 150, 300, 600, 1200, 2400, and 4000. Rovings are directly used in many processes, such as filament winding and pultrusion. In these processes, the roving is passed through a liquid resin bath and then wound on a mandrel or pulled through a heated die. In the spray-up process, the roving is chopped and an air-powered gun propels the chopped strands with resin to a mold. In the production of sheet molding compound (SMC), the roving is chopped onto a bed of resin paste and compacted into a sheet.

Woven roving is produced by weaving the rovings into a cloth (mat). Woven roving mats are available at different surface densities, such as 200, 360, 600, and 800 g m^{-2} (GSM). This form is widely used in hand lay-up process to improve the strength of the product and also used in panel molding process. Many weave configurations are possible to meet the requirements of the laminated composites. Plain or twill weave provides strength in two perpendicular directions, while a unidirectionally stitched or knitted fabric provides strength primarily in one direction.

Glass fiber mats can be produced using chopped strands, as well as continuous strands. A CSM is produced by randomly depositing chopped strands of 5–50 mm length onto a conveyor belt and spraying a chemical binder over that. The binder is usually a thermoplastic material. The appropriate binder can be selected depending on the required level of solubility in styrene. In some applications such as moderate corrosion-resistant liners or boat hulls made by hand lay-up process, a high styrene solubility is required, whereas a low styrene solubility is required when the composite is fabricated by cold pressing or compression molding to prevent washing in the mold during curing. CSMs are available at different surface densities, such as 300, 450, 600, and 900 g m^{-2} (GSM).

Continuous strand mat is produced in a similar manner but with continuous strand. Usually less binder is used in this mat since the continuous fibers can provide some entanglement. Continuous strand mats are used in closed mold processes to prevent fiber washout by the flowing resin. They are also used in pultrusion process to improve transverse strength.

Chopped strands are short fibers of length between 3.2 and 12.7 mm, which are cut from the dried continuous rovings or chopped in a wet state during the fiber production. They are widely used in the injection molding process. The fiber and polymer are dry blended and extruded to form composite pellets. These pellets are used for the fabrication of thermoplastic composites by injection molding. Chopped strands are also used in the preparation of bulk molding compounds with thermoset polymer resin. This molding compound is used for the fabrication of composites by compression molding process.

Milled fibers are prepared by hammer milling of chopped or cut continuous strands. Typical dimensions of the milled fibers range from 0.79 to 6.4 mm. They may not provide improved strength because of relatively low aspect ratio but provide some increased stiffness and dimensional stability to plastics. Milled fibers are mainly used in phenolic resins, reaction injection molded urethanes, and fluorocarbons.

Yarns are twisted fibers, which have better integrity during weaving process. Glass fiber fabrics are made by conventional weaving of yarns. The major characteristics of fabrics depend on the weave pattern and fabric count. Basically there are four weaving patterns and they are plain, basket, twill, and satin. More information on the weave pattern can be found in any textile book. In the context of composite fabrication, the flexibility of fabric is an important parameter for better conformity to complex contours. The satin weave gives the best conformity, followed by twill, basket, and the least by plain weave. The compactness of woven yarn cloth is indicated by the thickness of cloth in mil. For example, a 70 mil cloth has a thickness of 0.07 in. (1.78 mm).

2.1.2.2 Carbon Fibers

Carbon exists in various allotropic forms. The three important forms are graphite, diamond, and fullerenes. Carbon atoms are arranged in a hexagonal fashion in the graphite structure. It has a layer structure with closely packed and strongly bonded carbon atoms in the layer and weak van der Waals forces between the layers. Graphite is a highly anisotropic material, because of this structure. The modulus value along the plane is close to 1000 GPa and across the plane is only about 35 GPa. This is the form that exists in carbon fibers. Diamond has a cubic structure with strong covalent bonding in all the three directions. Fullerenes are made of 60 or 70 carbon atoms. They are also known as "buckyballs."

Carbon is one of the light elements with a density of 2.27 g cm^{-3}. Carbon fibers have density values ranging from 1.6 to 2.20 g cm^{-3} depending on the processing conditions. They are made from organic precursor fibers. The organic precursor fibers are special-grade polymeric fibers, which undergo large amount of plastic deformation before fracture. These fibers are converted to carbon fibers on pyrolysis. The widely used precursor fiber is polyacrylonitrile (PAN) fiber. Other precursor fibers are derived from rayon, pitch, polyvinyl alcohol, polyimides, and phenolics. There are four essential steps in the carbon fiber preparation. They are fiberization, stabilization, carbonization, and graphitization. The schematic of carbon fiber preparation is shown in Figure 2.6.

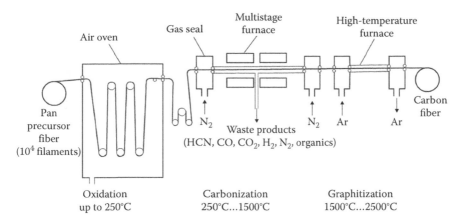

FIGURE 2.6
Schematic of carbon fiber production. (Reprinted from Baker, A.A., *Metals Forum*, 6, 81, 1983. With permission.)

Fiberization is the preparation of precursor fiber from the organic material by melt, wet, or dry-spinning method. The selection of appropriate polymer is very important. It is based on the carbon yield, molecular weight, fiber forming ability, etc. Carbon fiber yield from PAN is about 50%, pitch gives a yield of 65%, whereas the yield from cellulose based fibers are only about 20%.

Stabilization treatment is to make the precursor fiber non-melting during the subsequent high-temperature treatments. This is carried out at 250°C under normal atmospheric conditions. Wherever possible, stretching is applied during this treatment so that there is some molecular alignment along the fiber axis. Even the thermosetting rayon fiber is subjected to this stabilization without stretching to prevent the formation of tar.

Carbonization treatment is to drive out the non-carbon elements from the precursor fiber. It is normally carried out under inert atmospheric conditions between 1000°C and 1500°C with a very slow heating rate. After this treatment, the fiber has a network of hexagonal carbon ribbons. However, all the ribbons are not aligned to the fiber axis. Hence, the mechanical properties of the fibers after carbonization are poor. The ribbons are aligned during the subsequent graphitization treatment.

Graphitization treatment is carried out between 1500°C and 2500°C under inert atmospheric condition with some stretching. The ribbons get aligned due to plastic deformation at these high temperatures. The extent of carbon ribbons aligned to the fiber axis decides the final properties of the fiber. The better the orientation, the better will be the modulus of carbon fibers (Figure 2.7). The amount of stretching and graphitization temperature control the lamellar ribbon orientation. From the selection of the appropriate polymer to final graphitization temperature, each and every parameter plays a major role in determining the final properties of carbon fibers. These details are maintained as closely guarded secrets by the

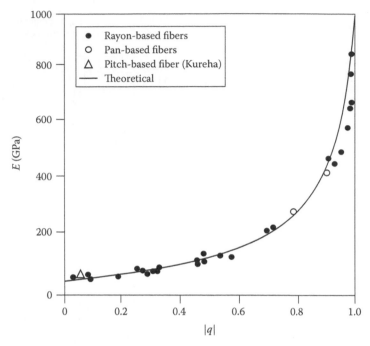

FIGURE 2.7

Variation of longitudinal elastic modulus for various carbon fibers with the degree of orientation of graphite planes. (Reprinted from Fourdeux, A. et al., *Carbon Fibers: Their Composites and Applications*, The Plastics Institute, London, U.K., 1971, p. 57. With permission.)

carbon fiber manufacturing industries. Hence, only a few industries are producing carbon fibers.

Carbon fibers are also available in many forms similar to glass fibers, which include continuous filament tow (similar to glass fiber roving), chopped fiber, milled fiber, woven fabrics, felts, veils, and chopped fiber mats. Most of the fibers produced today are spooled and then processed into other forms by secondary operations. The number of filaments in a tow can range from 1000 (1K) to more than 200 K, while most commercial-grade fibers are available in 48 K or larger filament counts.

2.1.2.2.1 Properties of Carbon Fibers

The internal structure of carbon fiber consists of tiny undulating ribbonlike crystallites, which are intertwined and oriented more or less parallel to the fiber axis. Each ribbonlike layer is made up of carbon atoms arranged in a hexagonal structure like chicken wire mesh and the layer is called graphene plane. Strong covalent bonds within a layer give high strength and stiffness. In between the layers, there is a weak van der Waals force, which gives rise to poor transverse mechanical properties or shear resistance. The width of the ribbons, the number of graphene layers in a set (thickness of a group), and the length of the ribbons determine the mechanical, electrical, and thermal properties of the carbon fiber. Larger and more oriented graphene planes generally result in higher modulus and thermal and electrical conductivity. The orientation of graphene planes toward fiber axis can be improved by increasing the final heat treatment temperature. Better orientation of ribbons can also be obtained by stretching the fiber during stabilization and carbonization treatments.

There are a variety of carbon fibers available from high modulus to high strength (Table 2.4). The PAN-based carbon fibers have a good balance in properties and that is the reason for their dominance in structural

TABLE 2.4

Properties of Commercially Available Carbon Fibers

Manufacturer	Product Name	Precursor	Filament Count	Density (g cm⁻³)	Tensile Strength (MPa)	Tensile Modulus (GPa)	Strain-to-Failure (%)
Amoco (United States)	Thornel 75	Rayon	10K	1.9	2520	517	1.5
	T300	PAN	1, 3, 6, 15K	1.75	3310	228	1.4
	P55	Pitch	1, 2, 4K	2.0	1730	379	0.5
	P75	Pitch	0.5, 1, 2K	2.0	2070	517	0.4
	P100	Pitch	0.5, 1, 2K	2.15	2240	724	0.31
HEXEL (United States)	AS-4	PAN	6, 12K	1.78	4000	235	1.6
	IM-6	PAN	6, 12K	1.74	4880	296	1.73
	IM-7	PAN	12K	1.77	5300	276	1.81
	UHMS	PAN	3, 6, 12K	1.87	3447	441	0.81
Toray (Japan)	T300	PAN	1, 3, 6, 12K	1.76	3530	230	1.5
	T800H	PAN	6, 12K	1.81	5490	294	1.9
	T1000G	PAN	12K	1.80	6370	294	2.1
	T1000	PAN	12K	1.82	7060	294	2.4
	M46J	PAN	6, 12K	1.84	4210	436	1.0
	M40	PAN	1, 3, 6, 12K	1.81	2740	392	0.6
	M55J	PAN	6K	1.93	3920	540	0.7
	M60J	PAN	3, 6K	1.94	3920	588	0.7
	T700	PAN	6, 12K	1.82	4800	230	2.1
Toho Rayon (Japan)	Besfight HTA	PAN	3, 6, 12, 24K	1.77	3800	235	1.6
	Besfight IM 60	PAN	12, 24K	1.8	5790	285	2.0
Nippon Graphite Fiber Corp. (Japan)	CN60	Pitch	3, 6K	2.12	3430	620	0.6
	CN90	Pitch	6K	2.19	3430	860	0.4
	XN15	Pitch	3K	1.85	2400	155	1.5
BASF	GY-80	PAN	—	1.96	1860	572	0.32
	GY-70	PAN	—	1.90	1860	517	0.36
	G40-700	PAN	—	1.77	4960	300	1.65

Source: Data from Krenkel, W., *Ceramic Matrix Composites*, Wiley-VCH, Weinheim, Germany, 2008, p. 71. With permission.

applications. They generally exhibit higher tensile and compressive strength, higher strain to failure, and lower modulus compared to mesophase pitch-based fibers. The microstructure of PAN-based carbon fiber has relatively good layer alignment and small stack heights, which minimize interlayer shear failure and hence improve compressive strength. Larger stack height and width and greater orientation of planes in mesophase pitch-based carbon fibers give them superior modulus and thermal conductivity and lower thermal expansion as compared to PAN-based fibers. High-performance fibers with high modulus and strength are used as reinforcements in composites, whereas low-performance fibers are used as fillers. Thermal and electrical conductivities of carbon fibers are also fairly good.

One of the important advantages of using carbon fibers in composites is superior fatigue resistance. Unlike glass or aramid fibers, carbon fibers do not undergo stress rupture but undergo complete elastic recovery upon unloading. Another important characteristic of carbon fiber is its negative coefficient of thermal expansion (CTE) in the axial direction at room temperature. It is slightly negative for low-modulus carbon fibers and more negative for high-modulus carbon fibers. However, the CTE becomes positive above 700°C. It may be possible to produce a carbon FRC with a CTE of zero over limited temperature ranges by selecting the appropriate matrix.

The purity of low-modulus carbon fibers is less than 99%, mainly because of retained nitrogen. This nitrogen can be removed by subjecting the fiber to higher heat treatment temperature, which also improves density and crystalline perfection. The increase in purity and crystalline perfection increases the electrical and thermal conductivities of carbon fibers.

As an inert inorganic material, carbon fibers are not affected by moisture, air, solvents, bases, and weak acids at room temperature. However, they undergo oxidation at elevated temperatures. The threshold temperatures for oxidation on exposure to long duration in air are 350°C and 450°C for low-modulus PAN-based fibers and high-modulus (pitch and PAN) fibers, respectively. Impurities present in the fibers tend to catalyze oxidation at these low temperatures and the oxidation resistance can be increased by increasing the purity of fibers (McKee and Memeault 1981).

Polymer resins and molten metals do not easily wet carbon fibers, because of the relatively inert and nonpolar nature of the fiber. Surface treatments given by fiber manufacturers introduce active chemical groups such as hydroxyls, carbonyls, and carboxyls on the fiber surface. These groups can facilitate the formation of chemical bonding with the polymer matrix. However, the bonding is not strong and the interface bond strength mainly depends on the number of links. In addition to this, some mechanical bonding also helps to improve the bond strength.

2.1.2.2.2 *Developments in Carbon Fibers*

All carbon fiber manufacturers are putting much effort to reduce the cost of carbon fibers. The prospects of cost reduction can stimulate interest in many new applications. On the other hand, the development of new products with carbon fiber can lead to an increase in the production and hence cost reduction. Composite materials are also becoming popular, and it is expected that in the near future more engineers learn to design with carbon FRCs. Standardization of the properties of carbon fiber is on the immediate need. Unlike glass fibers, each carbon fiber supplier has many grades, but there is no commonality between the grades produced by different manufacturers. Once the applications for carbon fibers are developed significantly, then standardization among carbon fiber producers can be expected.

2.1.2.3 Organic Fibers

Synthetic organic fibers are made from organic polymers having linear molecular chains. In general, the molecular chains in polymeric materials are neither arranged in a regular order nor fully stretched. The molecular chains are held together by very weak secondary bonds. This type of random coil structure is responsible for the poor mechanical properties of polymers. The polymer molecules undergo stretching and orientation during load application. A force sufficient to overcome the weak van der Waals force is only required to initiate deformation. This is the reason for the low modulus values for most of the polymers, which are generally less than 10 GPa. It is possible to improve the modulus values of polymers by stretching and aligning the polymer molecules in a particular direction. Very high modulus values have been realized by complete stretching and orientation of linear polymer molecules.

Drawing is a process used to align and stretch the polymer molecules. The higher the draw ratio (ratio of original diameter to final diameter), the better will be the stretching and orientation. However, the maximum draw ratio depends on the nature of polymer. For a particular polymer, the maximum draw ratio is controlled by molecular weight, molecular weight distribution, and processing conditions, such as temperature and strain rate. At very high temperatures, there will be bulk flow of material; hence, there will not be any stretching or orientation of molecules. At very low temperatures, void formation will lead to the premature failure of material. Hence, an optimum temperature should be used to get better alignment and stretching. Similarly, when the drawing is carried out at high strain rate, there may not be sufficient time available for the molecules to realign and that will lead to failure. Hence, the stain rate should be as low as possible. Stretching and orientation of molecules is the basis for the preparation of high-modulus organic fibers, such as polyethylene fiber.

There are certain polymer molecules, which have rigid rodlike structure. At certain conditions, a group of these molecules align themselves and forms liquid crystal polymer. When the liquid crystal polymers are used for the preparation of fiber, further alignment of molecules can be achieved.

Thus, it is possible to realize very high module values from these polymers. This approach is used for the preparation of aramid fiber.

2.1.2.3.1 Polyethylene Fiber

Polyethylene is a simple polymer with $-CH_2-CH_2-$ repeating units. Generally the molecules in bulk polyethylene material assume the random coil configuration; hence, the mechanical properties are poor. Once the polymer chains are aligned with full extension, then it is possible to get better properties in this polymer. The van der Waals forces existing between molecular chains in this polymer are very weak. Hence, significant overlap of chains is required to get better properties. That means the molecular chain length should be very high. In other words, the molecular weight of the polymer should be high. By careful control of the processing conditions, now it is possible to prepare polyethylene with molecular weight in the order of millions. Since there are large number of entanglements at this high molecular weight, melt spinning may not be useful to produce high-modulus fibers. The entanglements are less when the material is dissolved in a suitable solvent at low concentrations. Currently the commercial polyethylene fibers are produced from the solutions and the process is called gel-spinning process. There are only a few companies producing high-performance polyethylene fibers, which include DSM high-performance fibers in the Netherlands (Dyneema), Mitsui in Japan (Tekmilon), and Honeywell in the United States (Spectra).

The schematic diagram of gel-spinning process is shown in Figure 2.8. The solvent is selected in such a way that the high solubility of polyethylene

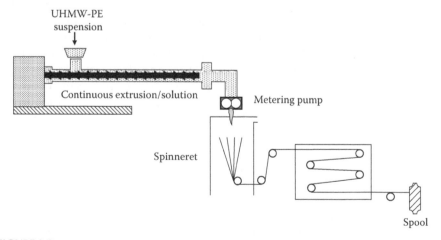

FIGURE 2.8

Schematic of gel-spinning process to produce polyethylene fiber. (Reprinted from Hearle, J.W.S., Ed., *High-Performance Fibres*, Woodhead Publishing Limited, Cambridge, U.K., 2001, p. 65. With permission.)

at high temperature is drastically reduced when the temperature decreased by a few degrees. Thus, the polymer crystallizes from the solution when the temperature is lowed. This can happen when the solution at high temperature is allowed to pass through a spinneret. The gelled fiber coming out of the spinneret is passed through a low-temperature liquid bath, where most of the solvent is removed. At this stage, it has a swollen network structure with the presence of crystalline regions at the network junctions. With this structure, it is possible to draw the fiber to very high draw ratios of the order of 200.

The spatial structure of polyethylene molecules in the gel is very important to achieve high draw ratio. It is determined by the number of entanglements in the solution, the geometry of the spinneret, and the conditions of quenching. The gelled fiber consists of very fine crystals embedded in noncrystalline material. After the extensive drawing, the randomly oriented crystals and most of the noncrystalline material are transformed into highly oriented crystalline material.

The highly crystalline polyethylene material has orthorhombic crystal structure with a density of 0.97 g cm^{-3}. It is lighter than water. The important properties of some commercially available polyethylene fibers are given in Table 2.5.

It is very difficult to bond highly crystalline polyethylene fiber with any polymeric matrix. Some surface treatments should be given to the fiber for better bonding. One of the common surface treatments given to polyethylene fiber is cold gas plasma treatment. The gas can be air, ammonia, or argon. This surface treatment removes the surface contaminants and introduces some functional groups and surface roughness.

2.1.2.3.2 Aramid Fibers

Aramid is the short form of aromatic polyamide. Aromatic polyamides contain rigid rodlike polymer molecules. Under certain conditions of concentration in a suitable solvent, these rodlike molecules form liquid crystal

TABLE 2.5

Properties of Commercial Polyethylene Fibers

Property	Spectra 900	Spectra 1000	Spectra 3000
Density (g cm^{-3})	0.97	0.97	0.97
Diameter (μm)	38	27	27
Tensile strength (GPa)	2.34–2.60	2.91–3.68	3.25–3.42
Tensile modulus (GPa)	73–79	97–133	115–122
Specific modulus (GPa/ρ)	75–81	100–137	118–125
Strain to failure (%)	3.6–3.9	2.8–3.5	3.3

Source: Data from *Product Information Sheet of Honeywell Specialty Materials,* Colonial Heights, VA.

structure. This is the structure, in which a group of rodlike molecules aligns in a particular direction, but the orientation of each group is different. During fiber spinning, the groups are aligned in a particular direction. The fiber consists of highly crystalline and oriented polymer molecules. This type of structure is responsible for the high modulus and strength values of the aramid fibers.

Only a few industries are commercially producing aramid fibers, which include DuPont (Kevlar and Nomex) and Teijin (Twaron and Technora). The basic difference between Kevlar and Nomex is that the Kevlar contains molecules with para-substituents in the benzene ring, whereas Nomex has molecules with meta-substituents in the benzene ring. The molecules are more linear in Kevlar because of the para-substitution. Kevlar fibers are made from poly(p-phenylene terephthalamide). This compound is usually synthesized by a low-temperature polycondensation reaction between p-phenylenediamine and terephthaloyl chloride.

Fiber spinning from the aramid polymer is very difficult because of the rigid rodlike molecular structure. This kind of structure in aramid is responsible for its high glass transition temperature and poor solubility in many solvents. For para-aramid, 100% sulfuric acid is used as a solvent. When the polymer content is more than the solubility limit, usually a saturated solution is formed with the undissolved polymer remaining in the liquid. In some systems, when there is right kind of polymer/solvent interaction, it is possible to dissolve more amount of polymer than the saturation limit. Additional polymer molecules form regions, in which the solvated polymer chains approach a parallel arrangement. This is the liquid crystalline state (Figure 2.9). Para-aramids form liquid crystal solutions under certain conditions of concentration, temperature, solvent, and molecular weight. The formation of liquid crystalline state is indicated by a drastic decrease in the viscosity of solution. Liquid crystal solution is optically anisotropic because of the presence of ordered molecular groups. The crystallites in the liquid crystal polymer can be observed using polarized light microscope.

The incorporation of diamines with wider distances between the two amino groups would be favorable to a lower spatial density of hydrogen bonding and rate of crystallization. Moreover, it improves the fiber drawability. This is the basis for the preparation of Technora fibers. Paraphenylene diamine reacts with 3,4'-diaminodiphenyl ether and terephthaloyl chloride in N-methyl-2-pyrrolidone/$CaCl_2$ solvent and forms aramid. The ether linkage gives more flexibility to the polymer and the compressive properties are better than Kevlar fiber. Unlike the Kevlar fiber, which is spun from the liquid crystalline polymer, Technora fiber is spun from the isotropic liquid.

Para-aramid fibers are produced from the liquid crystal solutions by dry-jet wet-spinning method (Figure 2.10). The solution is made with

FIGURE 2.9
Arrangement of molecular rods in the liquid crystalline state.

100% sulfuric acid and maintained at 80°C. A gap of 1 cm is maintained between the spinneret and coagulating bath while extruded through the spinneret. The liquid bath is maintained at 0°C–4°C. The crystal domains become elongated and oriented along the fiber axis because of shear forces acting in the spinneret and air gap. Another advantage of air gap is the possibility to maintain the solution at a higher temperature, since the spinneret is separated from the low-temperature bath. The concentration can be increased at high temperature, and hence, the fiber yield will increase. Usually, the fibers are spun at the rate of several hundred meters per minute.

The presence of aromatic rings in the para-aramid is responsible for the rigid rodlike structure. The rodlike molecules are highly oriented along the fiber axis; hence, the fiber has very high modulus values. Since the molecules are linear because of para-substitution, the packing efficiency is high. The fiber is highly crystalline due to better orientation and packing of polymer chains. There is strong covalent bonding in the fiber direction and weak hydrogen bonding in the transverse direction (Figure 2.11). The fiber is highly anisotropic because of this structure. The compressive strength of fiber is only 1/8 of its tensile strength.

The structure of Kevlar fiber has been investigated by electron microscopy and diffraction. Based on these studies, a supramolecular structure has been proposed (Dobb et al. 1980). It consists of radially arranged and axially pleated crystalline sheets. The bonding between the sheets is very weak. Each pleat is about 500 nm long. The alternating components of

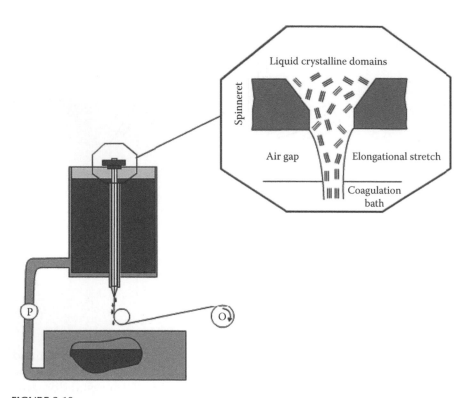

FIGURE 2.10
Schematic of dry-jet wet-spinning process to produce aramid fibers. (Reprinted from Hearle, J.W.S., Ed., *High-Performance Fibres,* Woodhead Publishing Limited, Cambridge, U.K., 2001, p. 33. With permission.)

each sheet are arranged at approximately equal but opposite angles to the plane of the section. The angle between adjacent components of the pleat is about 170°. Such a structure is responsible for the poor shear modulus of the fiber.

There are a variety of aramid fibers available with modulus values ranging from 62 to 179 GPa. The tensile strength of aramid fiber is in the range of 3.6–4.1 GPa, which is about 50% greater than that of E-glass roving. It is believed that tensile failure starts at fibril ends and propagates by the shear of fibrils. The important properties of some of the aramid fibers are given in Table 2.6. Depending on the application, a suitable fiber can be selected. For example, strain-to-failure value is important for bulletproof vests. Since the strain-to-failure value of K29 Kevlar fiber is good, this fiber is more preferable than any other Kevlar fiber. Similarly, for aerospace applications, stiffness is the main criterion. Hence, K49 or K149 Kevlar fibers are preferable for this application.

Another important characteristic of aramid fiber is its good vibration damping property. Damping is quantified by the term logarithmic

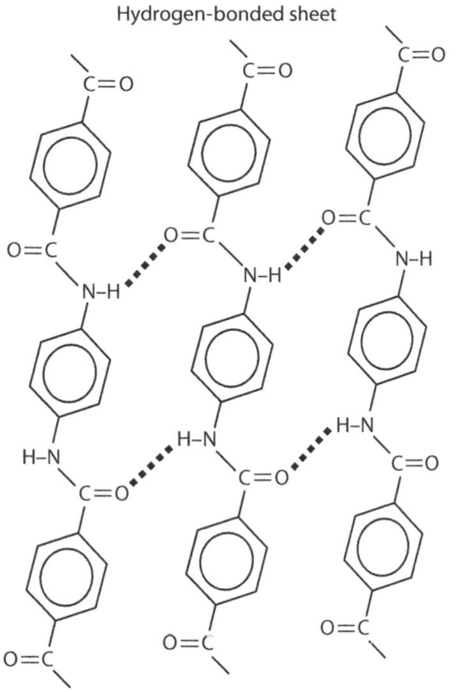

Hydrogen-bonded sheet

FIGURE 2.11
Molecular arrangement in aramid fiber. (Reprinted from DuPont, Wilmington, DE.)

decrement. Logarithmic decrement is defined as the natural logarithm of the ratio of amplitude of successive vibrations when a sudden vibration is induced in the material. It is proportional to the ratio of maximum energy dissipated and maximum energy stored in a cycle. Composites made using Kevlar fiber also have good damping property. The logarithmic decrement value of Kevlar/epoxy composite is five times higher than glass/epoxy composite.

Aramid fiber is sensitive to ultraviolet radiation. The original yellow color of the fiber turns into brown color on exposure to ultraviolet radiation. The color change is also an indicator for the degradation of the fiber. After the color change, there will be significant loss in mechanical properties. It is advisable to store the aramid fiber in a place where there is no ultraviolet radiation.

High-modulus aramid yarns show a linear decrease of both tensile strength and modulus with increasing temperature. However, more than 80% of the original properties are retained even at 180°C. Moisture absorption at room temperature reduced the tensile properties by <5%, whereas the yarn conditioned for 21 days at 180°C and 95% relative humidity had the same properties as dry yarn.

TABLE 2.6

Properties of Kevlar Aramid Fiber Yarns

Property	K 29	K 49	K 68	K 119	K 129	K 149
Density (g cm^{-3})	1.44	1.45	1.44	1.44	1.45	1.47
Diameter (μm)	12	12	12	12	12	12
Tensile strength (GPa)	2.8	2.8	2.8	3.0	3.4	2.4
Tensile strain to fracture (%)	3.5–4.0	2.8	3.0	4.4	3.3	1.5–1.9
Tensile modulus (GPa)	65	125	101	55	100	147
Moisture regain (%) at 25°C, 65% RH	6	4.3	4.3	—	—	1.5
CTE (10^{-6} °C^{-1})	−4.0	−4.9	—	—	—	—

Source: Data from *Du Pont Technical Guide*, Du Pont Advanced Fibers Systems, Richmond, VA.

Para-aramid has better resistance to fatigue. Creep rate is low and similar to that of glass fiber but less susceptible to creep rupture (Riewald et al. 1977). Although para-aramid fiber exhibits brittle behavior in tension, it exhibits nonlinear, ductile behavior under compression. At a compressive strain of 0.3%–0.5%, yield is observed due to the formation of kink bands. As a result of this behavior, the use of para-aramid fibers in applications that are subjected to high compressive or flexural loads is restricted.

2.1.2.4 Ceramic Fibers

The operating conditions of components are becoming more severe as the technology develops. One of the severe operating conditions is high temperature. Apart from the matrix materials, the fiber should also withstand this condition. Ceramic fibers are the suitable candidates for this type of environment. They usually have high modulus and high strength values and retain them at high temperatures.

Two classes of ceramic fibers are commercially available: (1) oxide fibers based on alumina–silica system, including α-alumina, and (2) non-oxide fibers, primarily SiC. These are polycrystalline fibers with small grain size (<1 μm) and generally produced with a diameter of less than 20 μm. The production costs of these fibers are significantly lower than those of single-crystal fibers or whiskers from the same material. However, the creep resistance and high-temperature stability of these polycrystalline fibers are lower, mainly due to the presence of amorphous or creep-enhancing second phases at the grain boundaries. Nevertheless, a variety of oxide and non-oxide polycrystalline fibers with sufficient performance levels are available today to reinforce metals and ceramics. The most commonly used high-temperature ceramic fibers are silicon carbide and alumina-based fibers. Some other fibers like silicon nitride, boron carbide, boron nitride, and zirconium oxide fibers are also used in certain applications.

There are three commonly used fabrication methods available for the manufacture of high-temperature ceramic fibers. They are sol–gel, polymer pyrolysis, and chemical vapor deposition (CVD) methods. The most common approach for producing polycrystalline ceramic fibers is spinning from the chemically derived precursors and heat treating to get the final ceramic fibers. Sol–gel processing is generally used for oxide fibers, while non-oxide fibers are prepared using organometallic polymer precursors. An important characteristic of these ceramic fibers is their ultrafine microstructure with grain sizes sometimes in the nanometer scale. Fine-grain sizes are preferred for better tensile strength but can be detrimental to creep resistance. These precursor-derived ceramic fibers are generally coated with a thin polymer-based sizing. These size-coated fiber tows are flexible and easy to handle, and hence, they can be woven into fabrics and tapes or braided into sleeves and other complex shapes.

2.1.2.4.1 Silicon Carbide Fiber

Among the non-oxide fibers, SiC fibers are the widely used fibers. The SiC fibers are available in various types, which range from fibers with very high oxygen content and excess carbon to the near-stoichiometric fibers. These large compositional differences affect the chemical and thermomechanical properties of SiC fibers and their composites. The SiC fibers are not commercially used as the oxide fibers, because of high production costs and the need to protect from oxidizing environments.

Polymer pyrolysis and CVD methods are used to produce SiC fibers. SiC fibers with small diameters (<20 µm) are produced by polymer pyrolysis, whereas large-diameter (>50 µm) fibers are produced by CVD. The selection of appropriate polymer is very important for the polymer pyrolysis process. Organosilane polymers are generally used, which contain C and Si atoms on their backbone chain. Molecular weight, yield, and purity are important parameters to be considered during the selection of polymer. Polycarbosilane is an ideal polymer for the preparation of SiC. The production of SiC fiber from polycarbosilane is almost similar to the production of carbon fiber. The precursor fiber is produced from the organic polymer by melt spinning. It is then stabilized by heating in air at 190°C. After the stabilization, it is pyrolyzed at 1000°C under N_2 atmosphere to form SiC fiber. This fiber is further heated to 1200°C in N_2 under stretch for better crystallization. These are the basic steps by which Nicalon fibers are produced commercially. The schematic flow diagram for the preparation of SiC fibers by polymer pyrolysis is shown in Figure 2.12. The maximum pyrolysis temperature is dictated by the fact that a small amount of oxygen is introduced into the precursor fiber during stabilization treatment and that forms oxide impurities. These impurities can react with carbon and carbides above 1200°C, tend to decompose into gases, and thereby create porosity in the fiber. These SiC fibers start degrading above 600°C because of the presence of residual carbon and oxygen. Since many CMCs are fabricated and used above 1200°C, it is desirable

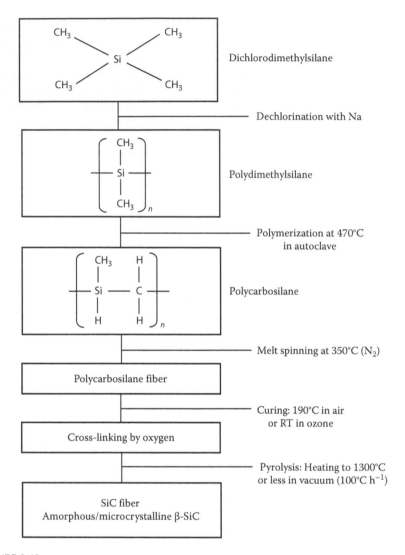

FIGURE 2.12
Flow diagram for the preparation of SiC fibers. (Reprinted from Andersson, C.-H. and Warren, R., *Composites*, 15, 16, 1984. With permission.)

to limit the oxide phases. Keeping this in mind, some advanced processing methods have been developed. In one such method, the stabilization is carried out by electron beam curing. Since the electron beam curing is carried out under high vacuum conditions, the chance of oxygen pickup is very much reduced. Another alternative to avoid oxygen pickup during curing is to use high molecular weight polymers. These high molecular weight polymers do not melt before decomposition.

In another method, the oxide-based impurities are allowed to decompose at high temperatures, and the resulting porous fibers are sintered at very high temperatures to form dense, oxygen-free, and nearly stoichiometric SiC fibers. To accelerate densification, sintering aids such as aluminum and boron are introduced in the polymer. Although the grain size of these sintered fibers is larger than the pyrolyzed fibers, it is beneficial for improved creep resistance and thermal conductivity.

Stoichiometric SiC fibers with better creep resistance can also be produced by CVD route (Figure 2.13). A substrate fiber is passed through a chamber, where the gases required for the formation of SiC are present. Trichloromethyl silane is an ideal gas for the deposition of SiC. This gas undergoes reduction reaction in the presence of H_2 at high temperature. The substrate fiber is heated by resistance heating. Mercury seals are used at both the ends of reaction chamber, which act as contact electrodes for the fiber heating. By controlling the current and frequency, it is possible to heat the filament to a particular temperature. The speed at which the fiber is moving into the chamber decides the final fiber diameter. Typically carbon fiber of 30 μm diameter is used as a substrate fiber. SiC fibers made using tungsten substrate fibers of 13 μm diameter are also available, but these fibers are not suitable for CMCs because of high-temperature reactions between SiC and W. However, these fibers can be used as reinforcements in MMCs for low- and intermediate-temperature applications. Although the CVD SiC fibers have very high strengths (6000 MPa),

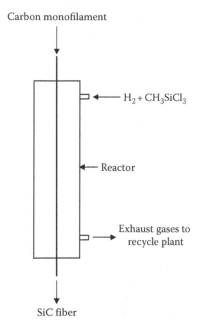

FIGURE 2.13
Schematic for the preparation of SiC fibers by CVD process. (Reprinted from Specialty Materials, Inc., Lowell, MA.)

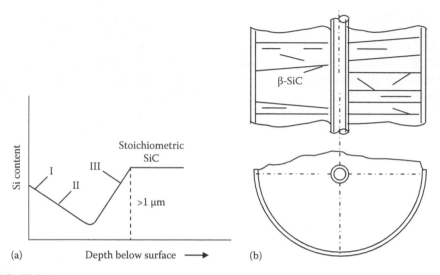

FIGURE 2.14

Surface compositional gradient of SCS SiC fiber and schematic longitudinal (a) and cross-sectional (b) view of SiC fiber prepared by CVD process. Zone I: Bondable and wettable surface, Zone II: Broad forgivability zone, and Zone III: Inner gradient necessary for maintaning filament strength. (Reprinted from Specialty Materials, Inc., Lowell, MA.)

the fabrication of composites with these fibers is difficult due to the large filament size (>50 µm). Attempts have been made to reduce the diameter and form continuous tows, but there are problems in finding suitable substrate filaments with small diameter and proper composition and a method for spreading the filaments without fiber-to-fiber welding.

Textron Specialty Materials Company is producing a series of SiC fibers by CVD process. The fibers are surface modified to have better compatibility with the matrix materials (Figure 2.14). A pyrolytic carbon coating is given to the carbon fiber substrate before the deposition of SiC. This fiber is used for the deposition of SiC. At the surface of the SiC fiber, Si and C ratio is modified in such a way that the surface is carbon rich. The carbon content is further increasing up to about 0.7 µm from the surface, then it is decreasing, and at 1 µm it matches to that of stoichiometric SiC.

2.1.2.4.2 Alumina-Based Fibers

There are some applications where the component is exposed to oxidizing atmosphere at a temperature of above 1400°C. Oxide fibers with high melting points are more suitable for these kinds of applications. Among the oxide fibers, α-alumina fiber is widely used because of its good refractory properties. Other fibers are made with alumina and varying amounts of silica. These alumina-based fibers are mainly produced by the sol–gel process. In this process, precursor materials are reacted to form a sol, which is then polymerized into an inorganic polymeric gel. The main advantages of this process are compositional and microstructure control, wide variety of shapes, and low processing temperatures. Sol–gel processing involves the formation of sol and gel. Sol is a colloid, in which fine solid particles are dispersed in a liquid. The sol is reasonably stable for long time because of the very fine nature of solid particles in the sol. The sol can be converted into a gel by the evaporation of liquid or by introducing a coagulating liquid. The gel is an interconnected network of solid particles in a liquid medium. Fibers can be prepared either from the sol or gel by wet-/dry-spinning process.

Metallo-organic and inorganic compounds are used to prepare the sol. For example, aluminum isopropoxide or aluminum secondary butoxide is used as a precursor for the alumina fiber. The metal alkoxide precursors used in the sol–gel process give better homogeneity and high-purity metal oxide products. However, the alkoxides are highly moisture sensitive and expensive. Chloride and nitrate salts of aluminum can also be used. Aluminum alkoxides form sol on hydrolysis reaction in the presence of acid/base catalyst. To form sol from inorganic salts, the salts are first dissolved in water and then small quantity of base solution is introduced so that very fine precipitates of the respective hydroxides are formed. Short fibers are produced from the sol by dry spinning. Continuous fibers are usually produced from the gel by dry spinning. A large volume of volatile materials are going out during the conversion from gel to oxide fiber. Hence, a careful control of temperature is very important. Otherwise the fiber will either break or have large amount of porosity.

Alumina exists in various polymorphic forms. Some of the forms are α, η, θ, γ, etc. Alpha (α) is the stable form of alumina. For high-temperature applications, all the transition forms of alumina should be converted into stable α-form. The transformation normally takes place at 1100°C. Since the nucleation rate of α-alumina is very slow, rapid grain growth occurs at this temperature. The mechanical properties of these fibers are poor due to the very large grain size. Very fine α-Al_2O_3 or α-Fe_2O_3 particles can be introduced in the sol to increase the nucleation rate. By adding about 4% Fe_2O_3, very fine-grain α-Al_2O_3 fiber is prepared and at the same time the transformation temperature is reduced to 975°C. Apart from reducing the grain size, the additional advantage of adding α-alumina is that it can reduce the weight loss and shrinkage of the fiber during calcination.

Another important alumina-based fiber is mullite fiber. Mullite has a chemical composition of 72 wt.% Al_2O_3 and 28 wt.% SiO_2. With this composition, mullite crystallizes before the formation of α-alumina. Sometimes a small amount of boria is also added to this composition. This boria addition lowers the mullite formation temperature and helps in sintering; hence, the strength of the fiber is increased. Although boria addition improves the room temperature mechanical properties, the high-temperature properties are affected due to the evaporation of boron compounds above 1100°C. The fiber also shows poor creep resistance due to the presence of glassy borosilicate intergranular phase.

Other alumina-based fibers contain varying amounts of silica. The silica addition lowers the stiffness of alumina and improves strength by controlling the grain growth. These high-strength and flexible fibers are more suitable for room temperature applications. These fibers lose their strength at high temperature of the order of 1000°C. Silica also facilitates creep at 900°C; hence, it is not advisable to use this fiber above 900°C. A variety of alumina-based ceramic fibers are available commercially. The properties of commercial alumina-based fibers are given in Table 2.7.

In the alumina–silica fibers, alumina exists as transition alumina form (η-Al_2O_3 or γ-Al_2O_3) and silica remains as an amorphous phase. These fibers have good strength because of the fine-grain size (10–100 nm) of transition alumina. The presence of silica preserves submicrometer transition aluminas, and therefore, fiber strength and flexibility are retained for extended periods of time above 1100°C. However, the amorphous silica affects chemical stability and creep properties at high temperatures above 1200°C. At this temperature, silica will crystallize over a period and form mullite on reaction with alumina. The transition alumina present in the fiber will also be converted to α-Al_2O_3. These transformations lead to the formation of large grains, which lead to low fiber strength. B_2O_3 addition may be helpful in preventing grain growth, but it increases fiber creep rate and reactivity.

The α-Al_2O_3 fibers have higher creep resistance than the fibers containing transition alumina and amorphous silica. Unlike the alumina–silica fibers, these fibers do not undergo shrinkage due to crystallization at high temperature. Moreover, the reactivity with oxide matrices is reduced because of better chemical stability of α-Al_2O_3, which can result in much higher strength in porous matrix composites. However, these fibers may not be suitable for applications in which significant thermal gradients exist, because of their high thermal expansion and low thermal conductivity.

TABLE 2.7

Properties of Commercial Oxide Ceramic Fibers

Fiber Type	Composition (wt.%)	Diameter (μm)	Density (g cm^{-3})	Tensile Strength (MPa)	Young's Modulus (GPa)	CTE (10^{-6} K^{-1})
Nextel 312	Al_2O_3-62.5, SiO_2-24.5, B_2O_3-13	10–12	2.7	1700	150	3.0
Nextel 440	Al_2O_3-70, SiO_2-28, B_2O_3-2	10–12	3.05	2000	190	5.3
Nextel 550	Al_2O_3-73, SiO_2-27	10–12	3.03	2000	193	5.3
Nextel 610	Al_2O_3-99+, SiO_2-0.2–0.3, Fe_2O_3-0.4–0.7	10–12	3.9	3100	380	8.0
Nextel 720	Al_2O_3-85, SiO_2-15	10–12	3.4	2100	260	6.0

Source: Data from *3M Nextel™ Technical Specifications*, 3M Center St. Paul, MN.

2.1.2.4.3 Other Non-Oxide Fibers

Apart from SiC fiber, other non-oxide fibers such as silicon nitride, boron carbide, and boron nitride are also promising high-temperature fibers. Silicon nitride fibers can be produced by CVD or polymer pyrolysis process, similar to SiC fiber production. The only difference is that the starting chemical compound is different. For the CVD fiber, the reactants are $SiCl_4$ and NH_3. In the polymer pyrolysis process, organosilazane polymers containing methyl groups on Si and N are used. It is very difficult to produce 100% Si_3N_4 fiber by polymer pyrolysis, since carbon in the polymer reacts with Si and form SiC. Hence, the fiber is not strictly a Si_3N_4 fiber, but it is a carbonitride fiber. Similar processing methods can also be used for the preparation of boron carbide and boron nitride fibers.

2.1.2.4.4 Applications of Ceramic Fibers

Oxide-based ceramic fibers are preferred for applications up to 1100°C in which high-temperature chemical resistance is an important criterion. Although they are very sensitive to surface flaws, oxide ceramic fibers are used in reactive environments such as waste incineration, hot gas filtration, and metal–matrix composites (especially aluminum-based composites). These fibers are also used in the fiber form in high-temperature thermal insulation applications, such as sleeves for pipes and electrical cables, high-temperature shielding blankets, and gasket seals. The cost of most small-diameter oxide ceramic fibers is higher than that of glass and carbon fibers but lower than that of SiC fibers.

Non-oxide fibers such as SiC_f are useful for high-temperature structural applications (>1100°C). They have better creep resistance and lower grain growth rate compared to oxide fibers, which allow better dimensional stability and strength retention. The SiC fiber can also provide greater thermal and electrical conductivity. However, the non-oxide fibers will oxidize slowly at high temperature under oxygen-containing environment. Nevertheless, silica is one among the most protective scales and it protects SiC from further oxidation. Hence, SiC fiber in general has very good oxidation resistance, and SiC-based fibers are generally preferred for applications that require long-term service under mild oxidizing conditions at temperatures higher than possible with oxide fibers. Extensive research is going on to develop SiC/SiC CMCs for use in gas turbine engines as hot-section components requiring service for thousands of hours under combustion gas environments.

2.1.2.4.5 SiC Whiskers

Whiskers are monocrystalline short fibers. They have very high strength, approaching theoretical strength because of the absence of crystalline imperfections, such as dislocations. However, the spread in the properties is high due to the nonuniform dimensions of the whiskers. The diameter of the whiskers varies from less than a micrometer to a few micrometers and the length varies from a few micrometers to a few millimeters. Hence, the aspect ratio varies from 50 to 10,000. The handling and alignment of whiskers in a matrix are difficult because of their fine nature. They are also considered as a carcinogenic material. Certain precautions should be taken while handling the whiskers. Even though whiskers can be produced from different materials, SiC whiskers are the widely used reinforcements in composites due to their attractive properties and ease of processing.

SiC whiskers were initially produced from rice husks. Rice husks are the waste by-products obtained during polishing of rice grains. In the rice husk, very fine silica particles are embedded in the cellulose network. During pyrolysis, the cellulose is converted into carbon and that reacts with silica and forms SiC.

Raw rice husks are heated at 700°C under inert atmospheric conditions to decompose the organic materials. After this coking treatment, the residue contains only silica and carbon. This residue is taken in a graphite crucible and then heated to 1500°C–1600°C. At this high temperature, the carbon reacts with silica and forms SiC whiskers as well as powders. Some residual carbon also remain with SiC and this can be removed by heating at 700°C under oxidizing atmosphere. SiC whiskers are separated from the particles by dispersing in a liquid with some surface active agents. The surface active agents selectively adsorb on the whiskers and thus whiskers can be separated from the particles. Whisker yield from this process is poor and also the whiskers have varying dimensions and properties.

Los Alamos National Laboratory in the United States has developed a process called VLS (vapor–liquid–solid) process to produce strong and stiff SiC whiskers. The schematic of VLS process is shown in Figure 2.15. In this process, a thin layer of steel particles (~30 μm diameter) is spread over a graphite substrate and heated in a furnace at 1400°C. During heating, a gas mixture containing SiO, H_2, and CH_4 is passed into the furnace chamber. At 1400°C, steel particles melt and extract Si and C from the feed gases. Once it is supersaturated, SiC is grown as whiskers from each steel particle. Any other

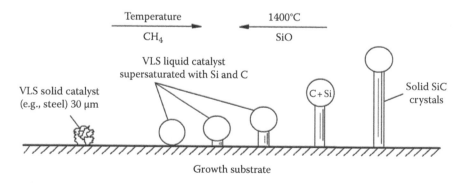

FIGURE 2.15
Schematic of VLS process to produce SiC whiskers. (Reprinted from Milewski, J.V. et al., *J. Mater. Sci.*, 20, 1160, 1985. With permission.)

transition metal or alloy particles can also be used as the catalyst, instead of steel particles. Whiskers of diameter 6 μm and length about 10 mm are usually produced by this process. The average tensile strength and modulus are of the order of 5 and 580 GPa, respectively.

2.1.3 Surface Modification of Fibers

An optimized interface between reinforcement and matrix is necessary for the composite to achieve maximum static and dynamic mechanical properties and environmental resistance. It is the interface that controls the effective load transfer from the matrix to the fiber reinforcement. A strong interface is necessary for better load transfer, which is achieved in compatible fiber and matrix materials. However, in many composite systems, the fiber reinforcement is not compatible to the matrix. Hence, it is essential to give a surface treatment to the fiber for improving compatibility. Surface treatments can introduce compatible chemical groups or modify the original surface.

The interfacial interaction between the reinforcement and the matrix is determined by the surface free energy. For better interaction, the surface energy of the reinforcement must be greater than that of the matrix. Once the fiber surface comes in contact with matrix, both physical and chemical bonds can form. Surface chemical groups present on the fiber surface can react with chemical groups in the matrix and form chemical bonds. Depending on the nature of surface groups present on the fiber surface and matrix, van der Waals attraction, hydrogen bonds, and electrostatic bonds can also form. The type and number of bonding strongly influence the adhesion between fiber and matrix. The fiber–matrix interface can be engineered through the use of fiber surface treatment and/or coupling agents.

Most of the composite processing is carried out with liquid matrix materials. A thorough infiltration of the fiber tow by the matrix should be ensured to obtain better properties in the composite. This infiltration process is controlled by the interfacial thermodynamics. Wettability of the fiber by the liquid matrix is a necessary prerequisite for thorough infiltration. The wetting of the fiber by the matrix as well as the adhesion of the matrix to the fiber is directly linked to fiber surface chemistry and matrix surface energy.

A force is acting on an atom or molecule on the surface of a liquid or solid to pull toward the interior. This force is commonly called the surface tension, and the energy corresponding to this force is called surface energy. Depending on the surface in contact, the surface molecules in a liquid can rearrange in microseconds and create a skin on the liquid. Liquids composed of nonpolar molecules (e.g., hexane) have low values of surface tension, while highly polar liquids (e.g., water) have large values of surface tension. Solids also have surface tension, but the atoms in the surface cannot rearrange spontaneously as in a liquid. Hence, the surface of a solid appears to be unaffected by any disturbance.

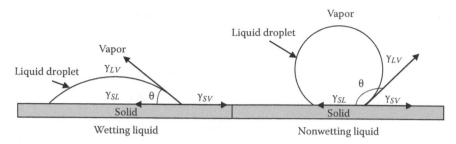

FIGURE 2.16
Contact angles of a wetting and a non-wetting liquids on a surface.

When a few drops of liquid are placed on a nonabsorbing solid surface, the liquid will spontaneously form a droplet or spread out into a film. Young's equation gives a relationship between the solid surface tension and liquid surface tension, if a droplet is formed (Figure 2.16):

$$\gamma_{SV} = \gamma_{SL} + \gamma_{LV} \cos \theta \qquad (2.4)$$

where
  γ_{SV} and γ_{LV} are surface energy of solid and liquid, respectively
  γ_{SL} is surface energy of the solid–liquid interface
  θ is the contact angle

When the contact angle is greater than 90°, the liquid is considered "non-wetting" and if it is less than 90°, then the liquid is considered "wetting." If the liquid spreads on the surface, then the contact angle is 0 and the relationship is not valid for this situation. For better wetting or spreading, the surface tension of the wetting liquid must be less than the solid surface tension.

Based on Young's equation, it is possible to find out whether a fiber surface can be wet by a liquid matrix or not. However, the real composite systems have very large assembly of small-diameter fibers. The small interstices between the fibers can create large capillary forces that aid the wetting. This capillary force is generally expressed as a pressure drop in the capillaries due to the surface tension. The liquid rise due to the capillary force can be related to the surface tension of the liquid, given as follows:

$$\Delta P = \Delta \rho g h = \frac{\gamma_{LV} - \cos \theta}{2r} \qquad (2.5)$$

where
  h is the rise of the liquid of density, ρ
  r is the radius of the capillary
  g is the gravitational constant

The contact angle controls the capillary forces; the capillary force vanishes when θ is 90° and infiltration is prevented when θ is greater than 90°. To facilitate good wetting and infiltration, the surface of fiber should be modified so that the contact angle between matrix and reinforcement is minimized.

Most polymers have low surface free energy values. Solids, on the other hand, can have very high surface free energy values in the pristine state. However, they tend to minimize their surface energy by absorbing material or growing oxides when exposed to the ambient environment. In some cases, the surface energy of solid becomes lower than that of the polymer matrix. In order to increase the solid surface free energy, surface treatment and coatings have been developed.

The surface treatment alters the surface chemistry of a material, primarily in the outermost layer, and/or creates beneficial micro-topographical features without a deliberate coating on the surface. This treatment can be carried out by using gas or liquid (base or acid) or bombardment of the surface with radiation of various types. In most cases, the surface treatment also removes some material from the surface, that is, etch the surface to some degree. The surface treatment is carried out on fibers, mainly to promote wettability and intimate contact with matrix.

Surface coating is applied to reinforcement surface after or in conjunction with surface treatments. Two types of surface coatings generally used on fiber reinforcements are size and finish coatings. Finishes for the fibers used in thermoset polymer matrices are based on the solution or water emulsion of the base resin component of the matrix without any curing agent. The thickness of finish is about 100 nm. The surface finish layer prevents actual contact between fibers and thus reduces the chance of flaw generation due to abrasion. It also protects the surface from environmental attack or contamination. Finishes are very helpful in assisting wetting and infiltration during composite processing. However, the retention of solvent in the finish and volatilization during composite processing is a potential problem.

In the composite industry, sizing means any surface coating applied to a reinforcement to protect it from damage during processing, help processing, and improve the mechanical properties of the composite. The size coating is almost always applied to glass fibers and sometimes applied to the other types of reinforcements. Typical sizing composition contains a silane coupling agent or a combination of coupling agents with film formers, binders, antistatic agents, and lubricants. The sizing is applied to glass fibers from solution at the time of drawing the fiber. The binder maintains the strand integrity and the lubricant prevents abrasive damage during handling. The antistatic agent prevents the buildup of static electricity due to rubbing during handling. Size formulations on commercial glass fibers are the proprietary of the manufacturer and are formulated to have good handling properties with good adhesion to a particular matrix material. The silanes are hydrolyzed and react with the hydroxyl groups of glass fiber surface to form siloxane bonds.

Ideally, the applied sizing should be chemically compatible with the matrix polymer and should not affect the mechanical properties of the composite. If these conditions are not met, the sizing should be removed by washing or heating the fiber, before the fabrication of composite. However, these treatments may damage the fiber or leave residues that may prevent the bonding with the matrix.

A variety of coupling agents compatible with the more widely used polyester and epoxy resins are available. However, only a few sizing materials are available for high-temperature polymer matrix materials, such as bismaleimides and polyimides, and tough thermoplastic matrices, such as polyphenylene sulfide or polyether ether ketone.

2.1.3.1 Surface Modification of Glass Fibers

Polymer matrix materials rarely adhere well to untreated glass fibers, since the glass fiber surface is hydrophilic, whereas the polymers are hydrophobic. However, some adhesion is possible owing to mechanical interlocking that arises because of the shrinkage of the matrix during polymerization and thermal contraction of the matrix from the curing temperature (generally the CTE of polymer is nearly 10 times higher that of fibers). At elevated temperatures or at high applied loads, this mechanical interlocking may be relieved due to the difference in expansion of fibers and matrix.

Coupling agents are used to make the glass fiber surface compatible with the polymer matrix. The most widely used coupling agents are alkoxy silanes, though chromium complexes and titanates can also be used. The general chemical formula of alkoxy silanes is R–Si $(-OR')_3$, where OR' is a hydrolysable group, such as $-OC_2H_5$, and R is a polymer compatible group. The alkoxy groups are hydrolyzed in aqueous medium and form trihydroxy silanols. These silanols form hydrogen bonding with the hydroxyl groups present on the glass surface. During fiber drying, condensation reaction takes place between neighboring hydroxyl groups, leading to the formation of chemically bonded polysiloxane layer with glass surface. The sequence of reactions taking place on the fiber surface is illustrated in Figure 2.17. The other end of the polysiloxane has an organo-compatible group, through which better bonding is established with the polymer matrix. There are some shortcomings with this model. If there is a strong bond, then it will fail during the curing of resin due to differential thermal contraction. Under the conditions of composite fabrication, the chances of condensation reaction between silanol and hydroxyl groups of glass are remote. Hence, a reversible bond model has been proposed by Plueddemann (1974) to account for the bonding in the cured composite. According to his model, the silanol groups are bonded to the glass through hydrogen bonding. Whenever a shear parallel to the interface occurs, the polymer with silanol groups can

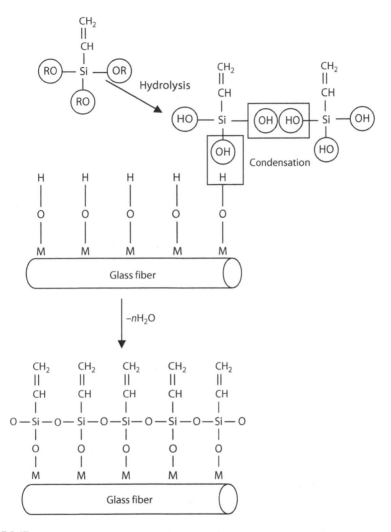

FIGURE 2.17
Bonding mechanism of a coupling agent on a glass fiber surface.

slide over glass fiber surface without a permanent bond rupture as shown in Figure 2.18. By infrared spectroscopic analysis, Ishida and Koenig (1978) have found evidence for this reversible bond mechanism. Hence, the interface is not a rigid sandwich of polymer–silanol–glass; instead, a dynamic equilibrium prevails that involves making and breaking bonds. Although this model explains the interface behavior in a better way, it is also not proved beyond doubt.

Whatever may be the bonding mechanism, it is certain that composite properties are improved when silane-treated fibers are used. For the epoxy resin matrix, γ-aminopropyltriethoxysilane, γ-glycidyloxypropyltrimethoxysilane,

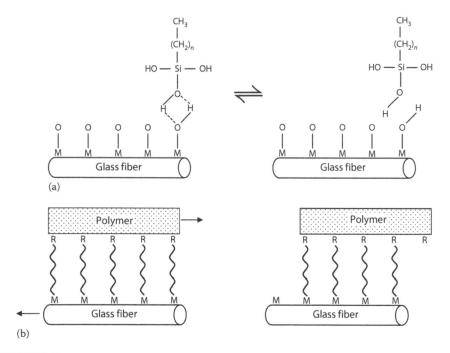

FIGURE 2.18
(a) Plueddemann's reversible bond formation and (b) shear displacement at a glass/polymer interface without permanent bond rupture. (Reprinted from Hull, D. and Clyne, T.W., *An Introduction to Composite Materials*, 2nd edn., Cambridge University Press, Cambridge, U.K., 1996, p. 151. With permission.)

and N-β-aminoethyl-γ-aminopropyltrimethoxysilane coupling agents are generally used. Silane coupling agents such as γ-methacryloxypropyltrimethoxysilane, vinyltriethoxysilane, and vinyltrist (β-methoxyethoxy)silane are used when the glass fibers are used in polyester or vinyl ester resin. It is necessary to find the type of silane present on the fiber, before using the fiber for a particular matrix.

Evidence for the fiber–matrix adhesion can be observed from the fracture surface of composites. If the adhesion is poor, the interfacial failure is characterized by clean fiber surfaces, thin cracks between fibers and matrix, and clear impressions in the matrix from which the fibers have pulled out or separated. The strong interfaces are characterized by fiber surfaces with thin layer of matrix and the absence of fiber–matrix debonding and pulled-out fiber impressions in the matrix. The tensile strength as well as the interlaminar shear strength of composites is improved by strong interface, since it allows a better load transfer between matrix and fibers. Although a strong interface leads to high strength, a relatively weaker interface may be desirable for higher energy dissipation through the debonding of fibers from the matrix, which is useful to improve fracture toughness.

2.1.3.2 Surface Modification of Carbon Fibers

Carbon fibers do not have a reactive surface as the glass fibers. The basal plane of carbon is very stable and unreactive. However, the edges and corners of these planes can be reactive. Since the basal planes are parallel to the fiber axis, the majority of surface area is covered by basal planes, which increases with increasing modulus value of the fibers. For the widely used intermediate modulus fibers, only 20% of the fiber surface is chemically active (Hammer and Drzal 1980). Attempts to increase surface activity usually result in a loss of fiber strength, because the oxidation of carbon fibers invariably produces surface flaws. Surface chemical treatments etch away the native fiber surface formed during fiber preparation and improve the adhesion with the matrix. However, excessive chemical treatment may also produce large flaws. Finishes are commonly used after the surface chemical treatment. Finishes are usually the matrix materials, which are applied from solutions to form a coating of about 100 nm thickness. Since the composition of the finish is the same as the matrix, wetting and impregnation of the fiber is enhanced. The finish coating also protects the fiber surface from damage during handling and chemical reactions until processing. The finish coated tow can also be handled with a minimum of difficulty in textile processing equipment.

2.1.3.3 Surface Modification of Polymeric Fibers

High-modulus fibers made from organic polymers have low surface energy and require some surface treatment to enhance their wettability. Since the major part of the polymeric fiber surface is unreactive, the use of coupling agents is generally not effective. Finishes can be used, but that enhance only the impregnation of the tow. Surface abrasion and corrosion problems are not severe with polymeric fibers. Chemical treatments and corona discharge, radio frequency, and microwave plasma treatments are used to alter the native surface of fibers. These treatments invariably remove low-energy contaminants from the surface and introduce some roughness to the surface. In addition to that, a few treatments can add surface chemical species to improve the wettability.

The highly oriented molecular chains in the fiber are responsible for poor interfacial strength in polymeric FRCs. This may be an advantage in composites used for impact-resistant items such as helmets or body armor, where delamination facilitates energy dissipation. However, poor adhesion will be a problem in high-strength and high-stiffness composites. Various surface treatments have been tried for aramid fiber to improve the bonding. Some of the treatments are the following:

- Bromine water treatment, which also reduces fiber strength.
- Coating of silane coupling agents used for glass fibers.

- Chemical treatments with acetic anhydride, methacryloyl chloride, sulfuric acid, etc. However, there is a reduction in fiber strength.
- Acid or base hydrolysis of aramid, which yields reactive amino groups for reaction with epoxy resin.
- Plasma treatment in vacuum, ammonia, or argon. (Plasma treatment in ammonia introduces amine groups on the fiber surface for reaction with polymer matrix materials.)

For the polyethylene fibers, a cold gas plasma treatment is a successful surface treatment. The gas used for this treatment can be air, ammonia, or argon. Surface modification during this treatment occurs by the removal of surface contaminants, addition of polar and functional groups, and introduction of surface roughness. Chemical treatment with chromic acid is another commonly used method to modify the surface characteristics of polyethylene fiber.

2.1.3.4 Surface Treatment of Natural Fibers

Even after complete drying, the bonding between the fiber and polymer matrix is poor due to the hydrophilic characteristics of the natural fiber. Hence, it is essential to modify the surface of the natural fiber for obtaining a quality composite with good performance. Alkali treatment (NaOH treatment) or mercerization is one of the widely used chemical methods for natural fiber surface modification. It removes a certain amount of lignin, wax, and oils covering the external surface of the fiber cell wall, thereby increasing the surface roughness. The important modification achieved with alkali treatment is the disruption of the hydrogen bonding in the network structure. It reduces the hydrogen bonding capacity of the cellulose by eliminating open hydroxyl groups that tend to bond with water molecules. Alkali can also dissolve hemicellulose. The removal of hemicellulose significantly reduces the ability of the fibers to absorb moisture, since it is the most hydrophilic part of natural fibers.

The surface of the natural fibers can also be modified by chemical reaction with suitable coupling agents or by giving a coating of appropriate resin. The surface treatment not only improves the bonding but also decreases the moisture absorption. Some of the common coupling agents are isopropyl triisostearoyl titanate, g-aminopropyltrimethoxysilane, sebacoyl chloride, and toluene diisocyanate.

Coatings made with phenol-formaldehyde or resorcinol formaldehyde also improves the bonding between natural fiber and polymer. It has been reported that there is 20%–40% improvement in flexural strength and 40%–60% improvement in flexural modulus of the composites made using natural fibers coated with these materials.

2.1.4 Fiber Selection Criteria

The correct selection of reinforcement is very important to get the desired properties in a composite. An improper selection may lead to composite properties less than the desired values, difficulty in fabrication, and high product cost.

The selection of a fiber for any particular composite should be made after considering certain criteria. The most important consideration is the cost. It is not advisable to use an expensive fiber for a low-cost composite product. Based on the cost considerations, the natural fibers are the cheapest. The next cheapest fibers are glass fibers and then carbon fiber and so on.

The next criterion is mechanical properties. Natural fibers are the suitable choice, if stiffness and strength are not very important. In situations where strength is very important rather the modulus, glass fibers can be used. If strength as well as stiffness is important, then carbon fibers are the appropriate choice of fibers. In some cases, damping and impact properties are very important. In those places aramid or polyethylene fibers can be used. In some of the applications, high-temperature strength is an important criterion. For these applications, SiC or alumina fibers are more suitable.

Fiber selection should be made depending on the type of matrix materials. The composite processing temperature is decided by the processing temperature of the corresponding matrix material. Generally polymeric materials are processed at a lower temperature than metallic on ceramic matrix materials. Natural fibers and synthetic organic fibers are not suitable for reinforcing metallic and ceramic materials. Even glass fibers cannot be used in metallic and ceramic matrix materials. Some of the polymeric materials are processed above 300°C and natural fibers may not be suitable for these polymers.

Processing methods also sometimes decide the selection of fibers. For processes such as filament winding and pultrusion, continuous fibers are needed. Natural fibers may not be useful for these processes, unless they are made as continuous fibers.

The directional property of the composite is also a deciding factor on the selection of fiber. For the isotropic composites, short fibers are preferred. Isotropic composites can be prepared with continuous fibers using 3D preform. However, that will add up to the cost. Unless there is a requirement of very high mechanical properties, short fibers would be a preferable choice for making the isotropic composites. It is also relatively difficult to produce continuous fiber-reinforced metal and ceramic composites. Short fibers are the preferable choice for these matrices.

Figure 2.19 shows the general use of a variety of reinforcements in different matrix materials. This is an indicator for the possibility to use a fiber in a particular matrix. This is a general guideline and there are some exceptions. Some pretreatments to the fibers or precautions are needed to address the problems of incompatibility between fiber and matrix and interfacial reactions.

Reinforcements   Matrices

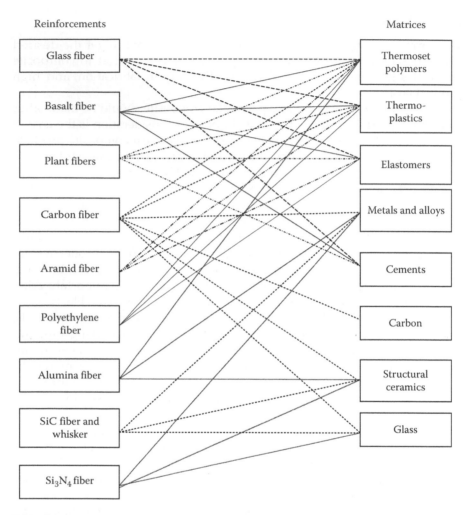

FIGURE 2.19
Possible combinations of commonly used fiber reinforcements and matrices.

2.2 Particulate Materials

Particulate materials are mainly used to modify the structure-insensitive properties, such as modulus, density, thermal conductivity, and hardness. These materials generally do not improve the structure-sensitive properties, such as strength and toughness. In polymeric materials, particulates are mainly added to reduce the cost of products and sometimes to increase the hardness and thermal stability. They are added in metal matrix, mainly to improve hardness and wear resistance. Same is the case for ceramic matrix also, but in some ceramic matrices, the particulates are added to improve the fracture toughness. The best example is zirconia-toughened alumina.

Particulates can also improve the strength of metal matrices. They improve the strength in two ways: one way is by load transfer, which may happen when the aspect ratio of particles is greater than 1, and the second way is by producing a large number of dislocations in the matrix. For this to happen, the bonding between particles and matrix should be good. Usually the thermal expansion coefficient of metals is larger than the ceramic particulates. During cooling from the high processing temperature, the matrix material tries to shrink more, but the shrinkage is restricted by the constraints imposed by the particles. This produces local tensile stress near the particles and that is responsible for the formation of dislocations. It is a well-known fact that higher dislocation density improves the strength.

There are many particulate materials used in composites. Calcium carbonate, titanium oxide, fused silica and graphite are the common particulate materials used in plastics. In metals, SiC, B_4C, TiC, WC, and Al_2O_3 are the widely used particulate materials. SiC, TiC, and ZrO_2 particles are more common in CMCs.

Purity of particulate materials is not a major concern for the plastics. Hence, most of the particulates used in plastics are the milled products of the natural minerals. Fused silica is a synthetic material, which is produced by the vapor phase hydrolysis of $SiCl_4$ in a hydrogen–oxygen flame. It is used mainly to reduce the density of plastics.

In metal matrix composites, the purity of particulates is also important. The impurities present in the particulates modify the alloy composition of the matrix and affect the bonding between particles and matrix. Hence, synthetic particulate materials are mainly used in metal matrix composites. Non-oxide particles are generally used when the composites are made by non-melting routes, whereas oxide particles are used in the composites made by casting. This is due to the less reactive nature of oxides with molten metals. Sometimes carbides such as SiC are also used in cast composites but with some surface treatment. Particle size and shape are important factors in determining the composite properties. Usually SiC particles are angular in shape. This shape is responsible for the highly abrasive nature of SiC composites. Fatigue strength is significantly improved by the use of fine particles. The distribution of SiC is uniform where the particles sizes of SiC and matrix powder are almost the same. Figure 2.20 shows the effects of particle size on various properties of composites. The characteristics of fine, medium, and coarse particles are given in Table 2.8. The shape of particle is characterized by its aspect ratio. Most of the ceramic particles have low aspect ratio; hence, the composites made with these particles are isotropic.

SiC is produced by the Acheson process using silica and coke in an electric arc furnace at 1700°C. The formed bulky material is milled to produce smaller particles. Particles of different size ranges are separated by using a classifier.

		Yield strength	Ultimate strength	Modulus	Ductility	Fatigue strength	Crack initiation resistance	Formability
Reinforcement particle size and effects	Fine	↑	↑	↑	Increase	↑	↑	Increase
	Coarse	Increase	Increase	Increase	↓	Increase	Increase	↓

FIGURE 2.20
Material properties and formability as a function of reinforcement particle size. (Reprinted from Smith, C.A., *ASM Handbook*, Vol. 21, *Composites*, ASM International, Materials Park, OH, 2001, p. 53. With permission.)

TABLE 2.8

Characteristics of Particulate Reinforcements

Relative Size	FEPA Grit Size	Particle Diameter, d_{50} (μm)	Advantages	Limitations
Fine	F1500	1.7	Greatest strength and stiffness contribution	Tendency to agglomerate
	F800	6.5	Highest fatigue resistance	Blending difficulty (powder/casting)
			Lowest resultant CTE	Lowered ductility
				High cost
				Segregation during casting common
Medium	F600	9.3	Excellent balance between properties (elevated strength and good ductility) and ease of manufacturing	Necessary for high-volume reinforcement systems
	F360	22.8		Good balance between properties and raw materials costs
				Good balance between manufacturing ease and resultant ductility
Coarse	F12	1700	Good wear resistance High ductility and ease of manufacturing Great for armor application	Lowest benefit to resultant properties

Source: Data from Smith, C.A., *ASM Handbook*, Vol. 21, *Composites*, ASM International, Materials Park, OH, 2001, p. 52.

Boron carbide is less reactive with metals than SiC. The density of boron carbide is also low. Another attractive property of born carbide is its high neutron absorption characteristic. The aluminum MMCs produced with these particles are more suitable for welding because of the less reactive nature of boron carbide.

B$_4$C is commercially produced by the carbothermic reaction of boric acid. The starting materials are boric acid and carbon (petroleum coke or graphite). The reaction is carried out at 1500°C–2500°C in an electric furnace. The large lumps produced by this process are milled to produce appropriate particle size. High-purity carbide is produced by magnesiothermic reaction of magnesium and boric acid in the presence of carbon.

Aluminum oxide is another important particulate material used in MMCs because of its relatively low cost. The reactivity of alumina is also low with the molten metals. It is more suitable for cast composites. The improvement of strength by alumina is not very high, but significant improvement in wear resistance can be obtained by using alumina particles. Alumina is the intermediate product of aluminum production from bauxite ore. As SiC and B$_4$C, large blocks of alumina are also milled to produce alumina particles of appropriate size.

Titanium carbide is preferred only when thermal stability in the presence of molten metal is very important. TiC particles are used mainly in titanium- and nickel-based alloys, where thermal stability up to 1100°C is required. When SiC or B$_4$C particles are used, rapid reaction may occur between the particulates and metallic elements, thus leading to the formation of brittle intermetallics. Once the intermetallics are formed, then it is very difficult to obtain good mechanical properties in those composites. Iron-based composites with TiC particulates are the best candidates for extrusion dies, where extremely high strength and wear resistance at high temperatures are required. TiC is commercially produced by the carbothermal reduction of TiO$_2$ using carbon black.

Fly ash is a fine powder that is collected from the combustion gases of coal-fired power plants with electrostatic precipitators and/or bag houses. It is predominantly an inorganic residue consisting of oxides, such as silica, alumina, iron oxide, and calcium oxide. The silicates are usually present as spherical particles and they are believed to be the melted products of clays, feldspars, quartz, calcite, and other common minerals in the coal. The iron oxides are usually derived from pyrite, hematite, siderite, and limonite minerals present in the coal. Low-density silicates are frequently high-alkaline silicates. The lower melting point of these silicates may facilitate gas entrapment, which leads to the formation of hollow spherical particles. Variations in fly ash bulk chemistry are due to the change in the ratio of these mineral components present in the coal.

Fly ash particles are very fine and mostly spherical. The properties of fly ash vary with the type of coal used, grinding equipment, the furnace, and the combustion process itself. It is relatively light and possesses high melting temperature and elastic modulus. Fly ash is generally a heterogeneous mixture of precipitator ash particles (solid spheres) and cenospheres (hollow spheres). The particle size of fly ash ranges from 1 to 150 μm for precipitator ash (density is 1.6–2.6 g cm^{-3}) and from 10 to 250 μm for cenosphere (density is 0.4–0.6 g cm^{-3}). This very unique and inexpensive resource with a wide range of sizes and density may be suitable for different lightweight structural and wear-resistant polymer and metal matrix components. It should be noted that there is considerable variability in the composition, size, impurity content, porosity, and density of fly ash obtained from different sources and the composite manufacturing technique must be sufficiently agile and adaptable to this variability.

In CMCs, SiC, TiC, and ZrO_2 are the widely used particulate materials. Carbide particles are mainly used in non-oxide matrix materials as well as oxide matrices, whereas ZrO_2 particles are mainly used in oxide matrices. Particulates are added in ceramic matrices generally to improve hardness and wear resistance. In addition, they improve fracture toughness in some ceramics. For example, zirconia particles are added in many oxide ceramics, mainly to improve the fracture toughness by phase transformation toughening.

2.3 Nano-Reinforcements

Nano-reinforcements in the form of fibers, tubes, whiskers, platelets, and particles significantly improve the properties of a variety of matrix materials. The main advantage of nano-reinforcements is the requirement of very low quantity of reinforcements to achieve significant improvement. This is due to the high surface to volume ratio of nano-size reinforcements. Very high mechanical properties can be realized in these reinforcements because of their fine size and that can be translated into the nanocomposites.

2.3.1 Nanofibers

A variety of nanofibers are available as reinforcements for nanocomposites. Generally the aspect ratio of nanofiber is close 1000 and they are polycrystalline. Some of them are carbon, silicon carbide, alumina, and zirconia nanofibers. The carbon nanofibers consist of graphite sheets arranged almost parallel to the fiber axis.

Carbon nanofibers are being produced by a large number of methods. The most common is the CVD method. In this process, a ceramic boat containing catalyst is placed in a quartz tubular furnace. A reaction mixture consisting of acetylene and argon is passed through the tube for several hours at temperature ranging from 500°C to 1000°C. Another method uses plasma to synthesize carbon nanofibers. A combination of slow discharge and electric arc discharge plasma would be a better option to get high yield. In this approach, a high-frequency glow–arc discharge with low flow rates of hydrocarbon gas is maintained. The advantages of this method are short reaction time and low-cost equipment.

A relatively new technique is based on electrospinning process. It seems to be a simple and most versatile technique for producing nanofibers. In this process, the precursor polymer solution (or melt) is delivered from the syringe. The solution is drawn in the form of fiber by the high potential maintained between the electrodes. The driving force for the formation of fiber is the electrostatic interaction. The applied electrostatic force exceeds the surface tension forces acting on the droplet and so it draws the jet toward the oppositely charged electrode. On evaporation of the solvent or cooling, the jet forms a thin fiber. The schematic of the process is shown in Figure 2.21.

Zirconia is a material with extreme refractoriness, high strength and fracture toughness, poor thermal conductivity, and good wear resistances. Zirconia nanofibers are being synthesized by the electrospinning process using zirconium salt and a polymer. Zirconia nanofibers with diameters of less than 40 nm have been prepared by electrospinning process using a viscous solution of zirconium oxychloride, polyvinylpyrrolidone, ethyl alcohol, and distilled water in the weight ratio of 1:1:3:3. The electro-spun fibers are calcined at 500°C to form zirconia fiber.

Alumina nanofiber is another important fiber with attractive properties. It retains its strength at high temperature. It has better corrosion resistance, chemical stability, and low thermal and electrical conductivities. Various methods such as hydrothermal, sol–gel, electrospinning, and CVD have been developed for producing Al_2O_3 nanofibers. In the hydrothermal and sol–gel

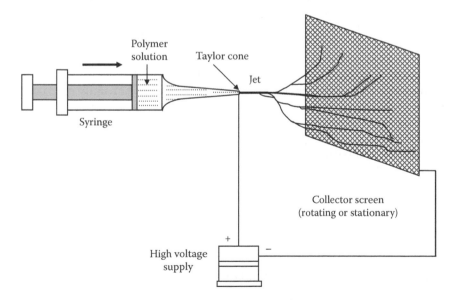

FIGURE 2.21
Schematic of electrospinning process to produce nanofibers. (Adapted from Wilkes G., Electrospinning, VirginiaTech, http://www.che.vt.edu/Faculty/Wilkes/GLW/electrospinning/electrspinning.html. With permission.)

processes, initially a sol is formed from the precursor materials. These sol particles aggregate on aging and form two-dimensional networks. With further aging, the nanosheets curl up and form nanofibers. This morphology is retained after calcination at 1200°C. Alumina nanofibers can also be prepared by the electrospinning process from the boehmite gel. The boehmite gel with suitable viscosity is used for electrospinning.

SiC is one of the ceramic materials widely used in many applications, because of its high-temperature strength, chemical resistance, and relatively high thermal conductivity. SiC nanofibers can be a suitable choice in high-temperature nanocomposites. SiC nanofiber can also be prepared by electrospinning process using polymers such as polycarbosilane. Controlled pyrolysis of the polymer fiber under inert conditions leads to the formation of SiC nanofiber.

2.3.2 Nanotubes

Among the nanotube materials, carbon nanotubes (CNTs) are the most popular material. Prior to 1980, it was thought that only two allotropes of carbon exist, namely, diamond and graphite. However, by 1996 it was clear that at least a third allotrope exists and that is the so-called buckyball or C-60 fullerene. The nickname is derived from the similarity between C-60 and the geodesic dome structure designed by the architect Buckminster Fuller. Later, other carbon molecules were discovered, which include C-72, C-76, C-84, and C-100. It was suggested that single-walled CNTs (SWCNTs) could be considered as elongated fullerenes. An SWCNT is nothing but a single graphene layer wrapped into a tubular shape, whereas multiwalled CNTs (MWCNTs) consist of several layers of graphene wrapped into tubular shapes. A single plane of densely packed carbon atoms arranged in a hexagonal configuration is referred as graphene. Some variants of such nanotubes have semiconductor characteristics, whereas others are electrically and thermally conductive, similar to metals. These materials have potential applications in many fields, including electronics, pharmaceuticals, and catalysis.

Lot of research is going on to improve the production of CNTs and to use this material in diverse applications. Several methods have been developed to synthesize CNTs. They can be classified into two main categories depending on the temperature of preparation: high-temperature processes and medium-temperature processes. High-temperature routes are based on the vaporization of graphite and medium-temperature routes are based on the CVD process.

2.3.2.1 Electric Arc Discharge Technique

An arc discharge is generated between two graphite electrodes under a low pressure of helium or argon gas (~600 mbar). The discharge temperature is in the range of 6000°C. At this temperature, graphite sublimates to form gas. During sublimation, pressure increases to high levels and ejects carbon atoms from the solid graphite. These atoms move toward the cathode and form nano-tubes. The type of nanotubes formed on the cathode depends on the presence of metal catalysts. If small amounts of transition metals such as Fe, Co, Ni, or Y are introduced in the graphite target, then SWCNTs are the major product. In the absence of such catalysts, MWCNTs are the predominant product.

Although CNTs have a wide range of potential applications, the commercial development has been constrained by the lack of a reliable large-scale manufacturing process. The availability of CNTs in commercial quantities with consistent quality at affordable cost will open up the potential for a wide range of applications. The carbon arc discharge method is a promising method for the production of CNTs in commercial quantities. Generally, a mixture of components is produced by this process, which requires a careful purification treatment. The CNTs are separated from the other carbon products and metal catalyst in a systematic manner during the purification process. The main limitation of this process is electrode erosion, which affects the flow rate of carbon. Now there is a mechanism available to move the electrodes closer depending on the erosion rate.

2.3.2.2 Chemical Vapor Deposition Process

It is a relatively low-temperature process, where the temperature is maintained between 600°C and 1200°C. A hydrocarbon undergoes decomposition reaction at high temperature and forms carbon. This carbon is deposited on the substrate in various forms. This is the basis for the preparation CNTs by the CVD process. The type of catalyst and the process parameters decide the nature of product. Nano-size metallic particles have been widely used as catalyst. Co, Fe, and Ni nanoparticles, either alone or associated with Mo, V, or W, are the most active catalysts. Depending on the maximum temperature at which there is no coalescence of catalytic particles, the carbonaceous gas can be either CO or hydrocarbons (CH_4, C_2H_2, C_2H_4, C_6H_6, etc.). Generally the carbonaceous gas is mixed with an inert gas (Ar, He, or N_2) or with H_2. The reaction temperature must be adjusted to ensure a favorable condition for the decomposition of the carbonaceous gas, without forming pyrolytic carbon on the CNTs. The gas flow must also be adjusted depending on the quantity of catalyst. Generally MWCNTs are synthesized at relatively lower temperatures (600°C–800°C) and higher temperatures (1000°C–1100°C) are preferred to synthesize SWCNTs. Most often, the processing is carried out in a tubular furnace by keeping a plate containing catalyst particles at the center. For more details on the CNTs, the reader is advised to read the book by Loiseau et al. (2006) on understanding CNTs.

2.3.3 Nanoclays

Nano-size platelets can be obtained from crystalline materials having layer structure. Some of the materials are graphite, metal chalcogenides, metal phosphates ($Zn(HPO_4)_2$), clays, and layered double hydroxides ($M_6Al_2(OH)_{16}CO_3 \cdot nH_2O$; M = Mg, Zn). Among these materials, clays are widely used to obtain nano-size platelets, since the separation of layers is relatively easy in clays. In particular, the specific groups of clays such as montmorillonite, saponite, and hectorite have mainly been used because of their excellent intercalation capabilities, which allow them to be chemically modified and made compatible with organic polymers. Another reason is that they occur ubiquitously in nature and can be obtained in pure form at low cost.

Montmorillonite clay is a hydrated aluminum silicate mineral, whose lamellae are constructed from an octahedral alumina layer sandwiched between two tetrahedral silicate layers. The silicate surfaces of montmorillonite exhibit a negative charge, and these surfaces adsorb cations such as Na^+ or Ca^{2+}. This clay adsorbs large amounts of water and polar liquids, which separate the silicate layers. In fact, montmorillonite clay can adsorb 20–30 times of water than its volume.

The layer structure of montmorillonite clay is shown in Figure 2.22. Each nanolayer consists of two tetrahedral sheets. The hydrated inorganic

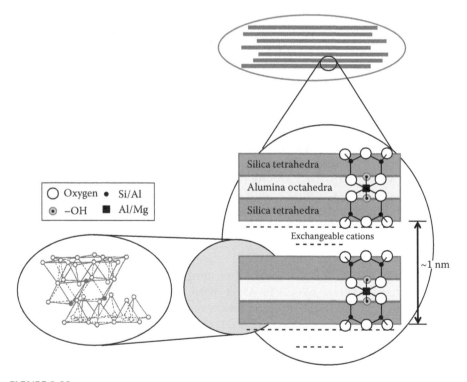

FIGURE 2.22
Structure of clays. (Adapted from Pinnavaia, T.J. and Beall, G.W., *Polymer-Clay Nanocomposites*, John Wiley & Sons, Chichester, England, 2000, p. 98. With permission.)

exchange cation (sodium or calcium) occupies the space between nanolayers. The length and width of the clay layers are of the order of 200–2000 nm and the thickness is about 1 nm.

The natural clays are hydrophilic and hence incompatible with hydrophobic organic polymers. These clays can be made compatible by the modification of layer surface. This is carried out by exchanging the surface cations with organic cations. In addition to surface modification, the organic groups increase the spacing between the clay layers and thus facilitate the easy incorporation of polymer molecules between the layers. Now there are a few companies producing these modified clays in large scale.

2.3.4 Nanoparticles

There are many metallic and ceramic nanoparticles available to produce nanocomposites. Metallic nanoparticles are mainly used in polymer and metal matrices, whereas ceramic nanoparticles can be used in any kind of matrices.

2.3.4.1 Preparation of Nanoparticles

There are two approaches to produce nanoparticles, namely, bottom-up and top-down approaches. In the bottom-up approach, atoms, molecules, or nanoparticles can be used as building blocks. By altering the size of building blocks and controlling their surface and internal chemistry and their organization and assembly, it is possible to control the size and properties of nanoparticles. These processes involve essentially highly controlled, complex chemical reactions. On the other hand, top-down approaches are inherently simpler and rely either on the removal or division of bulk material.

Bottom-up processes are based on chemical synthesis and/or highly controlled deposition and growth of materials. Chemical synthesis may be carried out in the solid, liquid, or gaseous state. Vapor phase deposition methods can also be used to produce nanoparticles. These methods are broadly classified into physical vapor deposition (PVD) and CVD methods.

PVD involves the conversion of bulk solid material into gaseous phase and the deposition of solid on cooling. By carefully controlling the parameters, the size of the deposited solid particles can be restricted to nanometer size. The conversion of solid into vapor can be effected by various means, such as thermal (resistive heating, electron beam, laser, or flame), spark erosion, and sputtering. CVD involves the reaction of gaseous species at elevated temperature and the deposition of nanoparticles on a substrate. Often nanometer size particles are used as catalyst to enhance the reaction rate.

Liquid phase chemical synthesis involves the reaction of precursor chemicals in a liquid medium (either aqueous or nonaqueous). The reaction leads to the formation of very fine solid particles as a precipitate. The nature, size, and shape of the particles can often be controlled by the reaction temperature, pH, time, and concentration. For a multicomponent product, even more careful control of coprecipitation is needed to get a chemically homogeneous product.

Sol–gel processing is one of the liquid phase synthesis methods. The details of this processing method have been discussed earlier under oxide fiber preparation. Extra care is needed while using sol–gel process to prepare nanoparticles. Generally a finely dispersed system (sol) is in a high free energy state. The colloidal material will therefore tend to aggregate due to attractive van der Waals forces, unless a substantial energy barrier exists. The magnitude of energy barrier depends on the balance of attractive and repulsive forces between the particles, as shown in Figure 2.23. Agglomeration can be prevented by increasing the repulsive force. This is possible by electrostatic or steric stabilization. A suitable surfactant should be added to the sol to control the agglomeration. The electrostatic repulsion arises as a result of the development of an electrical double layer around each particle. This generally happens in a polar liquid. Steric stabilization is more common in organic liquids. It has been generally accepted that all the adsorbed polymers provide steric stabilization, when they are anchored strongly to the particle surface and they have sufficient extension of long chain into the solution to prevent the particles to approach closer than 10–20 nm from each other. The success of the steric stabilization depends on the surface coverage, the configuration of the adsorbed polymers, and the thickness of the adsorbed layer.

The gel network structure formed by acid- or base-catalyzed hydrolysis of metal alkoxides can be tailored to favor the formation of polymeric chains with extensive branching and cross-linking or the formation of discrete spherical particles with minimum cross-linking. For the preparation of nanoparticles, base-catalyzed hydrolysis is preferred. The precursor structure is also important in controlling the morphology of sol–gel-derived powders. Nanoparticles are prepared by the sol–gel process from the xerogel or aerogel. Xerogel is obtained by normal drying of gel and aerogel is obtained by supercritical drying of gel. Generally amorphous particles are produced from the gel. Only after calcination, the amorphous particles are crystallized with the formation of agglomerates. The formation of agglomerates can be controlled by keeping moderate calcination conditions.

Although sol–gel methods are most commonly used to synthesize oxides, they can also be used to prepare carbides and nitrides. To prepare non-oxide ceramics, the hydrolysis reactions should be avoided by performing the reactions in aprotic solvents under inert atmospheres.

2.3.4.1.1 Microemulsion Synthetic Methods

Certain combinations of water, oil, surfactant, and an alcohol- or amine-based co-surfactant produce microemulsions. The surfactants are long-chain

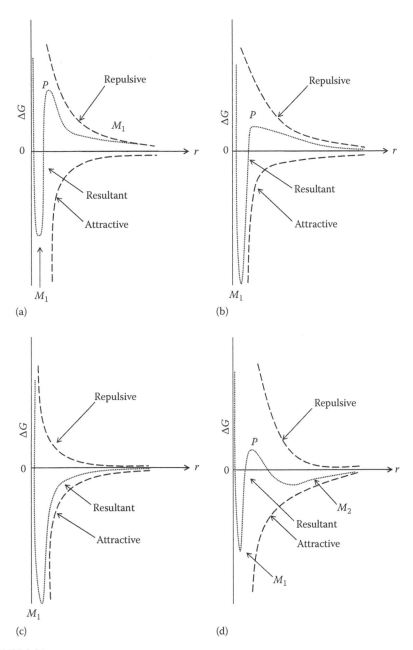

FIGURE 2.23
Schematic diagram of the possible energy balances between attractive and repulsive inter-particulate forces as a function of interparticle separation. In (a) there exists a large energy barrier (*P*) to strong aggregation of particles (minimum *M*1); this is smaller in (b) and absent in (c) owing to reduced long-range repulsive forces. Long-range attractive forces in (d) can lead to weak flocculation of particles (minimum *M*2). (Reprinted from Kelsall, R.W. et al., *Nanoscale Science and Technology*, John Wiley & Sons, Chichester, England, 2005, p. 46)

organic molecules with a hydrophilic head and lipophilic tail. Cetyltri-methylammonium bromide is a commonly used surfactant in microemul-sion techniques. The amphiphilic nature of the surfactants makes them available in both hydrocarbons and water. The surfactant forms spherical aggregates in which the polar ends of the surfactant molecules orient toward the center. The co-surfactant acts as an electronegative species to minimize repulsions between the positively charged surfactant heads. These types of spherical aggregates are commonly referred as reverse micelles. These reverse micelles can act as nano-reactors. The addition of water to the system expands the aggregates from the center since the water molecules reside at the center of the sphere. The size of the water pools in reverse micelles can be controlled by adjusting the water-to-surfactant molar ratio. Ionic salt added to the system dissociates in the aqueous cores. The system becomes complex by the salt addition and the solvated ions tend to affect both the stability of reverse micelles and the phase equilibrium.

When a reverse micelle solution containing a dissolved metal salt and a second reverse micelle solution containing a suitable reducing agent are mixed, the metal cations can be reduced to the metallic state. The reaction takes place at the core of the reverse micelle. Nanoparticles of Co, Ni, Cu, Pd, Ag, and Pt can be prepared by this process. The nanoparticles formed by this process have nearly uniform size and shape.

The synthesis of mixed oxides using reverse micelles relies on the copre-cipitation of one or more metal ions. It is almost similar to coprecipitation from aqueous solutions; the only difference is that the precipitation reac-tion proceeds at the core of micelle, which restricts the size of precipitates. Typically, precipitation of hydroxides is induced by the addition of a reverse micelle solution containing dilute NH_4OH to a reverse micelle solution con-taining aqueous metal ions. Alternatively, dilute NH_4OH can simply be added directly to a reverse micelle solution of metal salts. After the precipi-tation reaction, the solution is centrifuged to segregate the precipitates and calcined to remove water and organics. Nanoparticles of Al_2O_3, TiO_2, Fe_3O_4, SnO_2, CeO_2, etc. can be prepared by this method.

2.3.4.1.2 Hydrothermal Processing

Liquids can be brought to temperature well above their boiling points by the increase in autogenous pressures resulting from heating in a sealed ves-sel. Performing a chemical reaction under such conditions is referred as solvo-thermal processing. When the solvent is water, the process is known as hydrothermal processing.

Water is at the super critical state above its critical point (374°C and 218 atm). Super critical fluids exhibit the characteristics of both a liquid and a gas, lack surface tension as a gas, yet exhibit high viscosities, and easily dissolve chemical compounds like a liquid.

Many hydrothermal processes are not generally carried out above the critical point. In fact, high solubility and reactivity can be achieved at high temperature and pressure. Hydrothermal processing allows the preparation of many inorganic solids at a lower temperature compared to solid-state reactions. Unlike coprecipitation and sol–gel processes, there is no need of calcination treatment since the products of hydrothermal synthesis are usually crystalline. Many oxide nanoparticles like Al_2O_3, TiO_2, Fe_3O_4, SnO_2, and CeO_2 can be prepared by this method.

Top-down processes are normally carried out in the solid state. One of the important top-down processes is high-energy ball milling. Sometimes this process is referred as mechanical attrition or mechanical alloying. In this process, coarse-grained materials are crushed mechanically in rotating drums by hard steel or tungsten carbide balls. The repeated deformation in the mill can cause large reduction in grains size. Different metals can be mechanically alloyed together by cold welding to produce nanostructured alloys. It is also possible to produce composite powders containing the dispersed phase at the nanometer level.

There are two approaches for producing nanoparticles using mechanical milling: (1) milling a single phase powder and controlling the balance between fracturing and cold welding and (2) producing nanopowders using mechanochemical processes. This method is more suitable for producing metallic as well as ceramic nanoparticles. Among other problems, the biggest problem with top-down approach is the presence of imperfections at the surface structure.

2.3.4.2 Carbon Black

The oldest method of producing carbon black is by burning vegetable oil in a small lamp and collecting the carbon black accumulated on the lid. Later, natural gas was used as a source. Currently, carbon black is manufactured by furnace procedure, in which the incomplete combustion of refinery oils gives carbon black.

The structure of carbon black is similar to graphite and consists of large sheets formed by hexagonal arrangement of carbon atoms. Thus, it has strong covalent bonds within the sheets. The small primary particles are strongly connected by covalent bonds and form aggregate. The aggregate structure of carbon black is determined indirectly by measuring the amount of liquid required to fill the voids in a specified mass. This is based on the idea that irregular aggregates will pack more loosely than spherical particles and leave more voids for a liquid to fill. Generally, dibutyl phthalate is used for this purpose. The graphite layers in carbon black are arranged parallel to each other, forming concentric layers, and this arrangement is known as turbostatic structure. The distance between parallel layers of carbon black is in the range of 3.50–3.65 Å. The inner layers are less ordered than the outer layers. Carbon black undergoes graphitization on exposure to high temperature, which increases the crystallinity. In that structure, graphene layers stacked over each other regularly so that each carbon atom has another carbon atom above and below it, and the distance between carbon atoms in the adjacent layers is about 3.4 Å.

ASTM D1765 standard describes a standard classification of carbon black that involves a letter (N or S) and three numbers. The letter N indicates normal curing, and S indicates slow curing. The first number following the letter indicates the particle size range and the last two digits are arbitrarily assigned by the ASTM. For example, the first number "1" represents average size of carbon black between 11 and 19 nm, "2" represents size between 20 and 25 nm, and "9" represents size between 201 and 500 nm.

2.3.4.3 Silica

Silica exists in crystalline and amorphous forms. The amorphous form of silica is generally used in plastics and elastomers. It is manufactured as various types, such as fumed, fused, gel, and precipitated. The primary particle size of synthetic silica is typically 10–100 nm. The refractive index of silica is 1.45, which is close to the refractive indices of many polymers, making it possible to produce translucent and transparent filled polymers. The surface of silica contains hydroxyl groups, which easily tend to form hydrogen bonds with water molecules and demonstrate hydrophilic character.

The fumed silica is the most expensive among the synthetic silicas and it has some special industrial applications. Fumed silica was developed in the early 1940s and is mainly used as reinforcing filler in silicone rubber and as a thixotropic additive in polyester and epoxy resins. Small amount of fumed silica is also used in thermoplastics as a rheology control agent and a release agent. The primary particles of fumed silica are connected to several other primary particles and form a chain-like aggregate structure. This chain-like structure is responsible for the thixotropic properties observed with fumed silica. Fused silica is produced by reacting quartz with coke in an electric arc furnace at about 2000°C.

Precipitated silica is produced by the reaction of sodium silicate solution and sulfuric acid under alkaline conditions. The primary particles of precipitated silica are connected to form an open, grapelike aggregate of varying size. Precipitated silica is predominantly used as reinforcing filler in elastomers and a small amount is used as thixotropic additive in thermoset resins. It is the cheapest synthetic silica. A new generation of highly structured, precipitated silica is finding applications in silicone rubber and in many specialty plastics.

Silica gels are produced by the reaction of sodium silicate solution with sulfuric acid under acidic conditions. The primary particles of aerogel silica form a large spongelike mass that is responsible for the very high internal surface area and microporosity. One interesting application of this high internal surface area is the selective absorption of a particular compound from a solution. Silica gels are mainly used as covering agents in plastics and a small amount is used as process aids and adsorbents in plastic industries.

The properties of synthetic silica depend on particle size and shape and surface area, as well as on silanol group density. In rubber processing applications, the degree of reinforcement with silica is related to the external surface area, since it is believed that the internal surface between primary particles is inaccessible to larger polymer molecules.

Silica also exists in aggregate form, similar to carbon black. The aggregate consists of a few to a hundred primary particles. Even the aggregates join together and form agglomerate during the manufacturing process. Thus, silica exists in three particle size levels: (1) primary particles, (2) aggregates (hard to break), and (3) agglomerates (loosely attached aggregates).

Naturally occurring microcrystalline silica is used in molding compounds, adhesives, polyurethane elastomers, and silicone rubber. Fused silica exhibits low shrinkage, low thermal expansion, excellent thermal shock resistance, and good electrical insulating properties. They are used in silicone rubbers, high molecular weight fluorocarbons, epoxies, and other resins.

2.3.4.4 Zinc Oxide

Zinc oxide is a crystalline, white, or yellowish-white powder. It is prepared by either oxidation or burning of zinc. It can also be produced by pyrometallurgical techniques, in which zinc vapor reacts with oxygen and form zinc oxide. In a newer vapor synthesis method, zinc metal is vaporized and cooled in the presence of oxygen, causing nucleation and condensation of small particles of zinc oxide. The crystal structure of zinc oxide is hexagonal. Typical particle sizes of zinc oxides range from 0.1 to 0.4 μm. Particles with an average size of 36 nm have also been produced.

Zinc oxide absorbs carbon dioxide from air and also has high UV absorption characteristics. It is insoluble in water and alcohol but soluble in some acids and alkali solutions. Zinc oxide is used as a filler as well as accelerator/activator in rubber and plastics. It improves resistance to weathering and increases hardness, flame resistance, and electrical conductivity in polymers. It has been used as filler in silicone rubbers, polyesters, and polyolefins. Zinc oxide is a photochemically active material and has many applications due to this property.

2.3.4.5 Titanium Dioxide

Titanium dioxide imparts white color and has a high refractive index (~2.6) that leads to significant light scattering. Sometimes it is added to synthetic fibers to make them appear white rather than translucent or transparent. Titanium dioxide exists in different crystalline forms, and the common forms are anatase and rutile. Compounds with anatase show an outstanding bluish-white color, while rutile imparts a creamy white color. Titanium dioxide is commercially produced by sulfate and chloride processes. In the sulfate process, titanium ores are treated with sulfuric acid, and in the chloride process, they are reacting with chlorine gas to form TiO_2.

2.3.4.6 Talc

Talc is called soapstone and is a very soft material (Mohs hardness value is 1). It is a hydrated magnesium silicate. It is one of a series of lamella silicate minerals having two-dimensional silicate layers sandwiched with other mineral layers. The silicate and MgO layers of talc are electrically neutral and loosely superimposed on one another to form a crystalline material. Hence, the layers can slide readily over one another, resulting in easy cleavage and a soapy feeling. The average particle size of most industrial talc is in the range of 2–20 µm. Talc particles have flake-like shape with an aspect ratio between 10 and 30.

2.3.4.7 Calcium Carbonate

Among the methods available for $CaCO_3$ nanoparticle preparation, reactive precipitation is of high industrial importance because of its convenience in operation, low cost, and suitability for mass production. The conventional precipitation is often carried out in a stirred tank or column reactor, and the quality of the product is difficult to control and the morphology and size distribution of the nanoparticles usually change from one batch to another during production. Recently, a novel technology for the synthesis of nanoparticles has been developed, which is called high-gravity reactive precipitation (HGRP) (Chen et al. 2000). In this method, reactive precipitation takes place under high-gravity conditions. The schematic of HGRP experimental setup is shown in Figure 2.24. A rotating packed bed (RPB) reactor is used to generate acceleration higher than the acceleration due to earth gravity. The key part of the RPB is a packed rotator, in which vigorous mixing occurs under a high stress field. This generates uniform distribution of reactants almost to molecular level in the reactive precipitation process and hence yields nanoparticles with narrow size distribution. Calcium carbonate, aluminum hydroxide, and strontium carbonate nanoparticles with narrow size distribution have been successfully synthesized by means of this approach. Without the addition of particle growth inhibitors, the mean size of $CaCO_3$ particles can be adjusted in the range of 15–40 nm by controlling the operating parameters, such as gravity level, fluid flow rate, and reactant concentration.

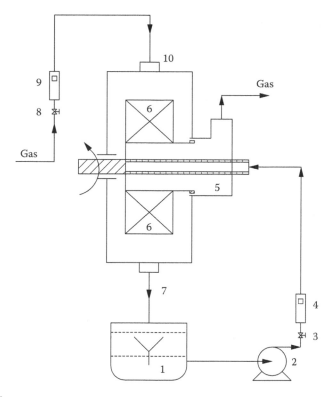

FIGURE 2.24
Schematic of experimental setup for HGRP. 1. Stirred tank, 2. pump, 3. valve, 4. rotor flow-meter, 5. distributor, 6. packed rotator, 7. outlet, 8. valve, 9. rotor flowmeter, 10. inlet. (Reprinted from Chen, J.F. et al., *Ind. Eng. Chem. Res.*, 39, 948, 2000. With permission.)

Questions

State whether the following statements are true or false and give reasons:

2.1 Aramid fibers are commonly used as reinforcements for metal and ceramic matrices.

2.2 Twenty-five micrometer diameter polyamide fiber is more flexible than 10 μm diameter SiC fiber ($E_{polyamide}$ = 1 GPa and E_{SiC} = 380 GPa).

2.3 The higher the graphitization treatment temperature, the higher will be the modulus of carbon fiber.

2.4 Carbon fiber has poor compressive properties.

2.5 Bulk material is stronger than fibrous material.

2.6 Fibers with oriented molecular chains show better modulus values.

2.7   Glass fiber is an anisotropic material.

2.8   Carbon fiber is an anisotropic material.

2.9   Graphitization treatment during carbon fiber production is to remove volatile materials.

2.10   E-glass fiber is a better electrical conductor.

2.11   Glass fiber is an amorphous material.

2.12   Fiber drawing from polymeric materials having rigid rodlike molecules is difficult.

2.13   The addition of α-Fe$_2$O$_3$ fine particles during the preparation of alumina fibers leads to better grain growth.

2.14   Whiskers are not widely used for making composites.

2.15   Fracture strength of yarn is more than the strand having same number of filaments.

Answer the following questions:

2.16   Find out the diameter of SiC fiber, having the same flexibility of 12 μm diameter glass fiber ($E_{SiC} = 380$ GPa and $E_{glass} = 70$ GPa).

2.17   The average tensile strength of as-drawn E-glass fiber is 460,000 psi and that of commercially available E-glass fiber is 280,000 psi. Compare the critical flaw size in these two types of fibers (1 psi = 6.89 kN m^{-2}).

2.18   Estimate the critical flaw size of a carbon fiber of strength 4.3 GPa (surface energy = 4.2 J m^{-2} and modulus = 230 GPa).

2.19   Compare the breaking loads and weights of Kevlar 49 and steel cables with a diameter of 6.4 mm ($\sigma_{Kevlar} = 3.63$ GPa; $\sigma_{Steel} = 1.2$ GPa; $\rho_{Kevlar} = 1.45$; $\rho_{Steel} = 7.8$).

2.20   Explain the structure and properties of carbon fiber.

2.21   Name any four methods for the improvement of interfacial adhesion in aramid fiber/polymer composite.

2.22   Distinguish a fiber and a whisker and list out the corresponding limitations.

2.23   What is the main advantage of nano-reinforcements?

2.24   What are the advantages and disadvantages of natural fibers?

References

Andersson, C.-H. and R. Warren. 1984. *Composites* 15: 16.

Baker, A.A. 1983. *Metals Forum* 6: 81.

Chawla, N. and K.K. Chawla. 2006. *Metal Matrix Composites*. New York: Springer, p. 5.

Chen, J.F., Y.-H. Wang, F. Guo, X.-M. Wang, and C. Zheng. 2000. *Ind. Eng. Chem. Res.* 39: 948.

Dittenber, D.B. and H.V.S. GangaRao. 2012. *Composites* A43: 1419.

Dobb, M.G., D.J. Johnson, and B.P. Saville. 1980. *Philos. Trans. R. Soc. Lond.* A294: 483.

Eckert, C. 2000. *Reinf. Plast.* July/August: 15.

Fourdeux, A. et al. 1971. *Carbon Fibers: Their Composites and Applications.* London, U.K.: The Plastics Institute.

Hammer, G.E. and L.T. Drzal. 1980. *Appl. Surf. Sci.* 4: 340.

Hearle, J.W.S. (Ed.). 2001. *High-Performance Fibres.* Cambridge, U.K.: Woodhead Publishing Limited.

Ishida, H. and J.L. Koenig. 1978. *J. Colloid Interface Sci.* 64: 555.

Katz, H.S. and J.W. Mileswski. 1978. *Handbook of Fillers and Reinforcements for Plastics.* New York: Van Nostrand Reinhold Company.

Kelsall, R.W. et al. 2005. *Nanoscale Science and Technology.* Chichester, England: John Wiley & Sons.

Krenkel, W. 2008. *Ceramic Matrix Composites.* Weinheim, Germany: Wiley-VCH.

McKee, D.W. and V.J. Memeault. 1981. *Chem. Phys. Carbon* 17: 1.

Milewski, J.V. et al. 1985. *J. Mater. Sci.* 20: 1160.

Pinnavaia, T.J. and G.W. Beall. 2000. *Polymer-Clay Nanocomposites.* Chichester, England: John Wiley & Sons.

Plueddemann, E.P. 1974. In *Interfaces in Polymer Matrix Composites.* New York: Academic Press, p. 174.

Riewald, P.G., M.H. Horn, and C.H. Zweben. 1977. In *OCEANS' 77 Conf. Rec.*, Vol. 1, *Third Annual Combined Conf. Mar. Technol. Soc. Inst. Electr. Electron. Eng.*, October 17–19, Los Angeles, CA, Paper 24E.

Rossi, F. and G. Williams. 1997. In *Proc. 28th AVK Conf.*, Baden-Baden, Germany, October 1–2, p. 1.

Smith, C.A. 2001. *ASM Handbook*, Vol. 21, *Composites.* Materials Park, OH: ASM International.

Wallenberger, F.T. et al. 2001. *ASM Handbook*, Vol. 21, *Composites.* Materials Park, OH: ASM International.

Bibliography

Ashok Kumar (Ed.). 2010. *Nanofibers.* Vukovar, Croatia: Intech.

ASM Handbook. 2001. *Composites*, Vol. 21. Materials Park, OH: ASM International.

Chawla, K.K. 1998. *Composite Materials: Science and Engineering.* New York: Springer-Verlag, Inc.

Hull, D. and T.W. Clyne. 1996. *An Introduction to Composite Materials*, 2nd edn. New York: Cambridge University Press.

Kelley, A. 1966. *Strong Solids.* Oxford, U.K.: Clarendon Press.

Mai, Y.W. and Z.Z. Yu (Eds.). 2006. *Polymer Nanocomposites.* Cambridge, U.K.: Woodhead Publishing Limited.

Matthews, F.L. and R.D. Rawlings. 1994. *Composite Materials: Engineering and Science.* Cambridge, U.K.: Woodhead Publishing Limited.

Omar, F., A.K. Bledzki, H.-P. Fink, and S. Mohini. 2012. Biocomposites reinforced with natural fibers: 2000–2010. *Prog. Polym. Sci.* 37: 1552.

Further Reading

Cao, G. 2004. *Nanostructures and Nanomaterials: Synthesis, Properties, and Applications.* London, U.K.: Imperial College Press.

Loiseau, A., P. Launois, P. Petit, S. Roche, and J.-P. Salvetat (Eds.). 2006. *Understanding Carbon Nanotubes.* Heidelberg, Germany: Springer-Verlag.

3

Matrix Materials

Matrix is the continuous phase in a composite and it binds the reinforcement together to form different shapes. The matrix plays a minor role in the load-bearing capacity of the composite, since it simply transfers the load to the reinforcements. However, the matrix has a major influence on the interlaminar and in-plane shear properties. The interlaminar shear strength is an important design consideration for the composite structures under bending loads, whereas the in-plane shear strength is important under torsional loads. Poor interlaminar shear strength may lead to the separation of layers and poor in-plane shear strength may lead to the debonding of fibers from the matrix. The matrix provides lateral support against fiber buckling under compressive loading and thus influences the compressive properties of the composites also. To some extent the processability and defects in a composite material depend strongly on the characteristic properties of the matrix, such as viscosity, melting point, and curing temperature (in the case of thermoset polymer matrix).

The three broad categories of matrix materials are polymer, metal, and ceramic matrices. Various matrix materials under these three categories are listed in Table 3.1. Among these materials, thermoset polymers such as epoxies and polyesters are widely used mainly because of their processing advantage. These materials are available as low-viscosity liquids at room temperature (RT) and hence it is easy to fabricate the composites by a variety of processing techniques. Metal and ceramic matrices are generally preferred for the composites used at high temperatures. However, aluminum alloy composites are gaining wide acceptance as that of polymer composites for many RT applications.

3.1 Polymer Matrix

A polymer is defined as a long-chain molecule containing one or more repeating units of atoms joined together by strong covalent bonds. A polymeric material is a collection of a large number of polymer molecules of similar chemical structure but not necessarily of equal length.

Organic polymeric materials can be broadly classified into plastics and elastomers. Plastics are the materials those generally undergo very large amount

TABLE 3.1

Types of Matrix Materials and Their Applications

Matrix	Subcategories	Examples	Applications
Polymeric	Thermoset	Epoxies	Principally used in aerospace applications
		Unsaturated polyesters, vinyl esters	Commonly used in automotive, marine, chemical, and electrical applications
		Phenolics	Used in bulk molding compounds
		Polyimides, polybenzimidazoles, polyphenylquinoxaline	For high-temperature aerospace applications (250°C–400°C)
		Cyanate ester	—
	Thermoplastic	Aliphatic polyamides, polyesters, polycarbonate, polyacetals	Used with discontinuous fibers in injection-molded articles
		Aromatic polyamide, PEEK, polysulfone, PPS, PEI	Suitable for moderately high-temperature applications with continuous fibers
Metallic	Alloys	Aluminum and its alloys, titanium alloys, magnesium alloys, copper alloys, nickel-based superalloys, stainless steel	Suitable for relatively high-temperature applications (300°C–500°C)
	Intermetallics	$MoSi_2$, TiAl	High-temperature applications (~1200°C)
Ceramic	Glass and glass ceramics	LAS	High-temperature applications, up to 1000°C
	Oxides	Alumina, zirconia, mullite	High-temperature applications, up to 1600°C
	Non-oxides	SiC, Si_3N_4	High-temperature applications, up to 1400°C

Source: Data from Mallick, P.K., *Fiber-Reinforced Composites*, 3rd edn., CRC Press, Boca Raton, FL, 2008, p. 61.

of plastic deformation before failure, whereas the elastomers undergo very large amount of elastic deformation. There are thermoset plastics and thermoplastics. Similarly elastomers can also be classified into cross-linked and uncross-linked elastomers. The detailed classification of engineering polymers is shown in Figure 3.1.

In a thermoset polymer, the polymer molecules are cross-linked through strong covalent bonds. Hence, the molecular motions are restricted in thermoset polymers and it is very difficult to plastically deform the thermoset polymers. In some cases, it may be possible to plastically deform these materials when the cross-link density (number of cross-links per unit length) is less. Low cross-link

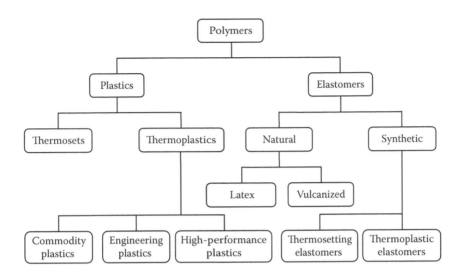

FIGURE 3.1
Classification of engineering polymers.

density allows some segments to move at high temperature on loading. Orderly arrangement of chains is not possible in thermoset polymers and so these materials are amorphous in nature. In a thermoplastic material, the polymer molecules are held together by weak van der Waals forces or hydrogen bonds. Each loosely attached segment in a polymer molecule may be in a state of random excitation. The frequency, amplitude, and the number of segmental motions increase with increasing temperature. That is why the properties of polymers are very sensitive to ambient temperature. When the temperature is increased, the amplitude of the segmental motions increases and the forces holding the polymer molecules become further weak. At a temperature called glass transition temperature (T_g), a glassy material transforms into a flexible, leathery material. At this condition, only a small force is required to deform the material. When the temperature is reduced, the molecules are frozen in the new place. Thus, theoretically it is possible to deform the thermoplastic material as many times as desired. Some depolymerization or degradation can happen during each reheating and so the quality of the polymer may degrade after every cycle.

In a thermoplastic polymeric material, the polymer molecules are frozen in space, either in a random fashion or in a mixture of random and orderly fashions. When the molecules are arranged in a random fashion, the polymer is an amorphous polymer. When the molecules are arranged in a mixture of random and orderly fashions, the polymer is called semicrystalline polymer. An amorphous thermoplastic material contains a random array of entangled molecular chains. The chemical groups within the chain are held together by strong covalent bonds, while the bonds between the molecular chains are much weaker secondary bonds. At the processing temperature, these secondary bonds break down and allow the chains to move and slide past one another.

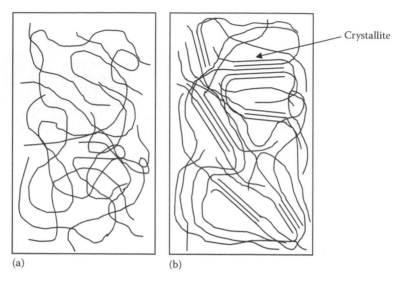

(a)   (b)

FIGURE 3.2
Molecular chain arrangement in (a) amorphous and (b) semicrystalline polymers. (Reprinted from Mallick, P.K., *Fiber-Reinforced Composites*, 3rd edn., CRC Press, Boca Raton, FL, 2008, p. 62. With permission.)

Amorphous thermoplastics exhibit good elongation, toughness, and impact resistance. As the molecular weight increases, there is an increase in melting temperature, melt viscosity, and mechanical properties. Semicrystalline thermoplastics contain tightly folded chains (crystallites) and amorphous regions. Schematic arrangements of molecular chains in an amorphous and a semicrystalline polymer are shown in Figure 3.2. Amorphous thermoplastics exhibit a gradual softening on heating, while semicrystalline thermoplastics exhibit a sharp melting. In general, the T_g and melting point of a polymer increase with increasing chain length, attractive forces between the chains, molecular rigidity and crystallinity. For a thermosetting polymer, the T_g increases with increasing cross-link density.

It is not possible to produce 100% crystalline polymer and the maximum crystallinity obtainable is about 98%. Crystallinity increases density and solvent resistance. Crystallinity also increases strength, stiffness, creep resistance, and high-temperature resistance but usually decreases toughness. Semicrystalline polymers are either opaque or translucent, while transparent polymers are always amorphous. In general, thermoplastics used as composite matrices have 20%–35% crystallinity. All thermoset resins are amorphous and the cross-links impart strength, stiffness, and temperature stability to this kind of polymers.

In a thermoplastic, crystallites form from the melt by a nucleation and growth process. The degree of crystallinity depends on the cooling rate. An amorphous structure is primarily formed at high cooling rates. Slow cooling rates are required to provide the time necessary for the nucleation and growth of crystallites. If the amorphous structure is formed because of high cooling rates, then the required crystallinity can be achieved by proper annealing treatment. The rate of crystallization depends on the annealing temperature and peak rate lies near the midpoint between the T_g and T_m.

3.1.1 Comparison of Thermoset Polymers and Thermoplastics

Prior to cure, the thermoset resin is a relatively low molecular weight liquid. Cross-links are formed during cure and rendering it into a rigid intractable solid. As the molecular weight increases during cure by the formation of strong covalent bond cross-links, the viscosity of the resin increases. The resin transforms into a gel and then finally solidifies. The high-performance thermoset resins are inherently brittle due to the high cross-link densities. However, as a result of improved toughening approaches for thermoset resins, the thermosets available today exhibit toughness comparable to thermoplastic systems. On the other hand, thermoplastics are high molecular weight polymers and they do not undergo any chemical reaction during processing. They melt and flow during processing but do not form cross-links. The molecular chains are held together by relatively weak secondary bonds. Being high molecular weight polymers, the viscosity of thermoplastics during processing is many orders of magnitude higher than that of thermosets. Generally the viscosities of thermoplastics are in the range of 10^3–10^6 Pa s and thermoset resins have viscosities of the order of 1 Pa s. Since there is no formation of cross-links during the processing of thermoplastics, they can be reprocessed; for example, they can be thermoformed into structural shapes, like metals at the processing temperature. However, multiple reprocessing will eventually degrade the material, since the processing temperatures are close to the polymer degradation temperatures. On the other hand, thermosets cannot be reprocessed due to their highly cross-linked structures. They will thermally degrade and eventually decompose at high temperatures.

The most important advantage of thermoplastics over thermoset polymers is their high impact strength. This is responsible for the better damage tolerance characteristics of thermoplastic composites. There is less chance for the formation of microcracks in thermoplastic composites because of the high strain-to-failure values of thermoplastics. The development of thermoplastic composites is relatively slow mainly because of their high melt or solution viscosities. It is very difficult to get good wetting of fibers without the application of pressure. The requirement of pressure restricts the usage of continuous fibers. Hence, continuous or long fiber-reinforced thermoplastics are not very common. In the recent times, better damage tolerance characteristics of thermoplastics drive the development of many long fiber-reinforced thermoplastics.

Since thermoplastics do not undergo chemical reactions during processing, the processing is simpler and faster. Thermoplastic composites can be fabricated in minutes, while thermosets require long durations (hours) for the completion of chemical reactions to form cross-links. However, there are some problems with the thermoplastics. They lack tackiness and they are stiff and boardy. They are processed close to their melting points. High-performance thermoplastics generally melt at high temperatures. Hence, the processing should be carried out only at high temperatures, whereas the thermoset resins can be processed at RT or at relatively lower temperatures.

Thermoplastic composite prepregs (short form of pre-impregnated fibers and more information on prepreg is available in Chapter 4) do not require refrigeration. They have infinite shelf life at ambient temperature but may require drying to remove surface moisture. Cured thermoset composite parts absorb moisture from the atmosphere that affects their elevated temperature performance. Many thermoplastic composites absorb only very little moisture and their elevated temperature performance is not much affected by moisture absorption.

Thermosets are more resistant to most of the solvents and fluids encountered in service. Many amorphous thermoplastics are very susceptible to solvents, whereas most of the semicrystalline thermoplastics are quite resistant to solvents and other fluids. Thermoplastics are more widely used than thermosets as plastics without reinforcements, whereas thermosets are the most widely used composite matrices.

The fusible nature of thermoplastics also offers a number of attractive joining options such as melt fusion, resistance welding, ultrasonic welding, and induction welding. Some of the thermoplastic composites can also be joined by solvent bonding. Another advantage of thermoplastic composites is related to health and safety issues. Since the processing of thermoplastic composites does not involve any chemical reaction of low molecular weight resin systems or solvents, there is no danger of health hazard to the workers. A detailed comparison of thermoset and thermoplastic polymers is given in Table 3.2.

3.1.2 Properties of Polymers

The mechanical properties of polymeric materials strongly depend on the ambient temperature. The general trend in the variation of modulus with temperature is schematically shown in Figure 3.3. On increasing the temperature, the modulus value is decreased by as much as five orders of magnitude. There is a sudden decrease in modulus value at the glass transition temperature. Hence, the glass transition temperature (T_g) is an important parameter for the polymeric materials. Below this temperature, they do not have either segmental or molecular mobility, and so they exist as glassy solid, that is, hard and brittle. Above this temperature, segmental mobility starts and they become leathery, that is, soft and flexible. The T_g is below RT for most polymeric materials, and hence they are soft and flexible at RT.

Above T_g, the thermoplastic materials are viscoelastic. When an external load is applied, the material undergoes instantaneous elastic deformation followed by slow viscous deformation. With increasing temperature, the viscous flow becomes dominant and the polymer is capable of undergoing large amounts of

TABLE 3.2

Comparison of Thermoset and Thermoplastic Polymer Matrix Materials

Thermoset	Thermoplastic
Widely used because composite processing is much easier with the low-viscosity liquid resin	Composite processing is difficult because the viscosity is high even above the melting temperature
Tackiness is a problem/advantage	No tackiness; hence handling is easy
Temperature and pressure requirements are less for the processing of thermoset composites	Relatively high temperatures and pressures are needed for the processing of thermoplastic composites
Limited storage life	Unlimited storage life
Long curing time because it involves chemical reaction	Most suited for automated production, since processing does not involve chemical reaction
Stepwise cure possible to permit control of viscosity and handling	Viscosity varied only by increase in temperature and/or shear rate
Phenolic resins and most polyimides emit volatiles during curing	No volatiles emitted during molding
No flow under heat and pressure after cure; scrap discarded	Soften and melt at high temperatures; remolding of scrap is usually possible
Post-curing often necessary for optimum properties	Post-molding treatment is not recommended, since shrinkage may be severe due to crystallization
Higher strength and modulus	Tougher and less brittle
Low tensile elongations	Relatively high tensile elongations
Better thermal stability and chemical resistance	Lower thermal stability and chemical resistance
Undergo less creep	Undergo more creep
Amorphous	May be semicrystalline
Post-formability is not possible	Post-formability is possible
Difficult to repair	Ease of repair by welding/solution bonding
Recycling is difficult	Can be recycled easily

Source: Courtesy of Professor Abhijit Deshpande, IIT Madras, Chennai, India.

plastic deformation. When the temperature is increased further, a highly viscous liquid is formed. However, there is no melting for the highly cross-linked thermoset polymers, instead they start charring above certain temperature. The segmental motions in these kinds of polymers depend on the number of cross-links. If the cross-link density is high, then segmental motions are not allowed and the polymer may not have glass transition temperature.

Similarly the loading rate during the testing of polymers also affects the properties of polymers. However, the effect is opposite to that due to temperature. A polymer shows brittle behavior at high loading rates and the same polymer shows ductile behavior at low loading rates. The effect of temperature and loading rate on the stress–strain behavior of polymers is schematically shown in Figure 3.4.

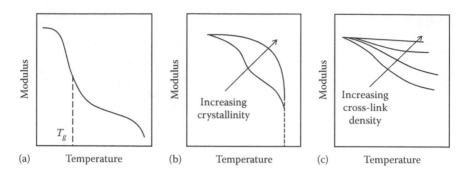

FIGURE 3.3
Variation of tensile modulus with temperature for (a) amorphous thermoplastics, (b) semi-crystalline thermoplastics, and (c) thermosetting polymer. (Reprinted from Mallick, P.K., *Fiber-Reinforced Composites*, 3rd edn., CRC Press, Boca Raton, FL, 2008, p. 64. With permission.)

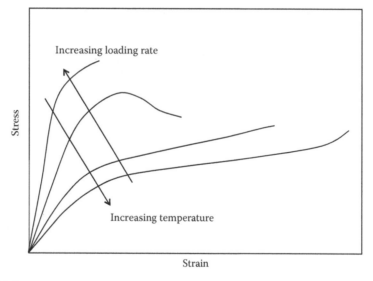

FIGURE 3.4
Effects of loading rate and temperature on the stress–strain behavior of polymeric materials. (Reprinted from Mallick, P.K., *Fiber-Reinforced Composites*, 3rd edn., CRC Press, Boca Raton, FL, 2008, p. 65. With permission.)

   The low-viscosity liquid thermoset resins are converted to solids during the fabrication of composites by a curing reaction. Catalyst or hardener is used to initiate the curing reaction. Unlike the catalyst, the hardener becomes the part of cured resin. The viscosity of resin increases with curing. The rate of viscosity increase is slow at the early stage of curing and then there is a rapid increase in viscosity once a threshold degree of cure is achieved. The time at which the rapid increase in viscosity occurs is called gel time. The gel time is an important molding parameter, since the flow of resin in the mold becomes increasingly difficult at the end of this time period.

The softening characteristics of polymers are often compared using heat deflection temperature (HDT). It is the temperature at which a rectangular bar specimen deflects to a specific value under a constant load. The ASTM D648 standard describes the determination of HDT for polymers. The HDT and T_g values are comparable to some extent, but the HDT values should not be taken as T_g values. To determine glass transition temperature, methods such as dynamic mechanical analysis (DMA) and differential scanning calorimetry should be used. Although HDT indicates the softening point of a polymer, it should not be used in predicting the elevated temperature performance of a polymer.

3.1.3 Thermoset Polymers

The majority of the polymer composites made today is based on thermoset polymers. These polymers are available as low molecular weight, low-viscosity liquids (resins). Since the viscosity is low, it is possible to get good wetting of fibers by the matrix at normal atmospheric conditions. Fiber wetting is an extremely important parameter to achieve very good mechanical properties in the composite, since fiber wetting ensures uniform spread of the resin over the fiber surface. Other advantages of thermoset polymers are better thermal stability and chemical resistance. Recently, thermoplastic composites are gaining importance because of the disadvantages of the thermoset polymers, like brittleness, long fabrication time, and limited storage life.

3.1.3.1 Polyester Resins

The most important and widely used thermoset resin is unsaturated polyester (UP) resin. This is generally known as polyester resin. The resin contains a number of C=C double bonds, through which cross-links are established. An ester is the product of reaction between an acid and alcohol. Similarly unsaturated polyester resin is produced by reacting unsaturated acid/anhydride with a dihydric alcohol. Saturated acids are added to modify the chemical structure between the cross-linking sites. The resulting polymer is dissolved in a reactive diluent, such as styrene. This reduces the viscosity of the resin and so handling becomes easy. The diluent also contains C=C double bonds through which cross-links are established between the neighboring polyester molecular chains. There are a variety of polyester resins available depending on the type of unsaturated acid, saturated acid, and alcohol. The different reactive compounds under these three groups to form polyester resin are listed in Table 3.3. Depending on the final properties needed for the polyester resin, the suitable compound from each group can be selected.

The general-purpose polyester resin is made from maleic anhydride, orthophthalic acid, and propylene glycol. The reactive diluent is generally styrene. Better performance than general-purpose polyester resin can be obtained by using isophthalic acid instead of orthophthalic acid. Terephthalic acid generally provides a higher HDT than either isophthalic or orthophthalic

TABLE 3.3

Constituents of Unsaturated Polyester Resin and Their Characteristics

Main Constituents	Typical Compounds	Characteristics
Glycols	Propylene	Lower cost, good styrene compatibility, good mechanical properties
	Ethylene	Heat resistance, lower cost, lower styrene compatibility
	Diethylene	Good toughness and flexibility
	Neopentyl	Chemical resistance, weather resistance, thermal stability, light color
Dibasic acid/anhydride	Orthophthalic anhydride	Lower cost, higher strength, low HDT
	Isophthalic acid	High strength, chemical resistance, and water resistance
	Adipic acid	Flexibility, fire retardance, high HDT
	Chlorendic anhydride	Fire retardance
Unsaturated acid/anhydride	Maleic anhydride	Lower cost, high HDT
	Fumaric acid	Higher cost; high reactivity, HDT, and rigidity
	Methacrylic acid	—
	Acrylic acid	—
Unsaturated monomer	Styrene	Lower cost, good HDT, high reactivity
	Vinyltoluene	Low reactivity, unpleasant odor
	Methyl methacrylate	Light stability
	Diallylphthalate	Good heat and electrical resistance, low volatility, high viscosity

Source: Adapted from Mark, H.F. and Kroschwitz, J.I., *Encyclopedia of Polymer Science & Technology*, Vol. 12, John Wiley & Sons, New York, 1988, p. 259.

acid, but it reduces the reactivity of the resin. Adipic acid lowers the stiffness of polyester resin, since it does not contain an aromatic ring. The flexibility can also be improved by using ethylene glycol instead of propylene glycol. The pendant methyl groups of propylene glycol restrict the rotation of polyester molecules. Maleic polymers have low reaction rates with styrene because of steric factors arising from the *cis*-arrangement. Consequently, maleate polymers cross-linked with styrene generally show less peak isotherm and remain in a rubbery state for longer periods. The final plastic is less rigid and has lower modulus and HDT than the fumarate polymer.

The properties of polyester resin strongly depend on the cross-link density. Modulus, glass transition temperature, and thermal stability of cured polyester resins are improved by increasing the cross-link density, but strain-to-failure

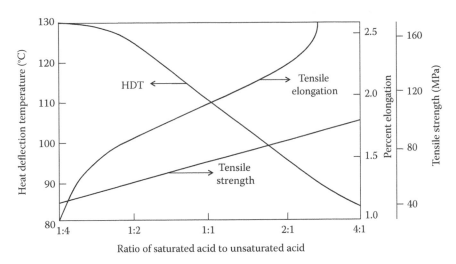

FIGURE 3.5
Effect of saturated acid-to-unsaturated acid ratio on the properties of thermoset polyester resin. (Reprinted from *Amoco Chemical Corporation Bulletin*, IP-70, Amoco Chemical Corporation, Chicago, IL. With permission.)

value and impact strength are reduced. The cross-link density is controlled by the space between unsaturated groups in the uncured polyester molecule. The space can be modified by changing the saturated acid-to-unsaturated acid ratio. For example, the space between unsaturated groups in an isophthalic resin is increased by increasing the weight ratio of isophthalic acid to maleic anhydride. The effect of saturated acid-to-unsaturated acid ratio on various properties of isophthalic polyester resin is shown in Figure 3.5.

The properties and processing characteristics of polyester resin are also controlled by the type and amount of diluent. Styrene is the most widely used diluent for polyester resin because of its low viscosity, high solvency, and low cost. Other diluents are methyl methacrylate (used for colorless products) and diallylphthalate (used for low volatile emission). When methyl methacrylate is used with styrene, the cured polyesters have superior durability, color retention, and resistance to fiber erosion. Unless there is a specific requirement of these properties, these special diluents are not preferred because of their high cost. The amount of styrene also influences the properties of polyester resin. Higher styrene content leads to a decrease in modulus value, since the styrene groups increase the space between polyester molecules. The curing time of resin is also increased with higher styrene content, because of the presence of double bonds in styrene molecule. Excessive amount of styrene leads to self-polymerization of styrene during curing and domination of polystyrene properties. Hence, it is advisable to restrict the styrene content equivalent to the number of cross-links to be formed. The property of polyester resin with varying amount of styrene is shown in Figure 3.6. Even though styrene is a widely used diluent for polyester resin, there are some drawbacks in using it.

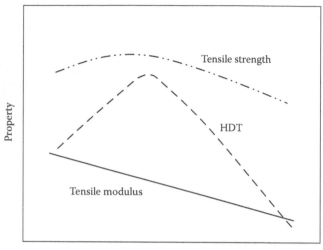

FIGURE 3.6
Effect of increasing styrene content on the properties of thermoset polyester resin. (Reprinted from Mallick, P.K., *Fiber-Reinforced Composites*, 3rd edn., CRC Press, Boca Raton, FL, 2008, p. 80. With permission.)

Styrene is a flammable solvent and excessive emission causes health hazards to the persons working with the polyester resin.

Generally trace amounts of an inhibitor, such as hydroquinone and benzoquinone, is added to the liquid resin to prevent premature polymerization during storage. The gelation time also depends on the inhibitor concentration.

The liquid resin is converted into a solid after the curing reaction. It is initiated by adding a small quantity of a catalyst. Organic peroxides and azo compounds are the common catalysts for polyester resin. The curing reaction of polyester resin is illustrated in Figure 3.7. A suitable catalyst should be selected depending on the fabrication temperature. A list of catalysts and their curing temperature is given in Table 3.4. Sometimes catalyst alone may not be sufficient to initiate the curing reaction. The reaction can be accelerated by adding an accelerator/activator or by applying heat to the fabricated component. There are many catalyst and activator combinations available for use with polyester resins, which will cause complete cure with or without heating. Cobalt naphthenate activator is available as a solution. It is used to cure the resin at RT with hydroperoxide-type catalyst. Both the gel time and curing time of a resin are affected by the activator content. This activator is not recommended for a clear or light color product since it imparts a color to the resin. Dimethyl aniline can be used in that case. However, it requires the application heat for complete cure.

The important factors controlling the curing reaction are the type and amount of catalyst and activator, and temperature. The effect of catalyst and activator contents on the curing behavior of polymer resin is shown in Figure 3.8. The reaction rate can be increased by increasing the amount of

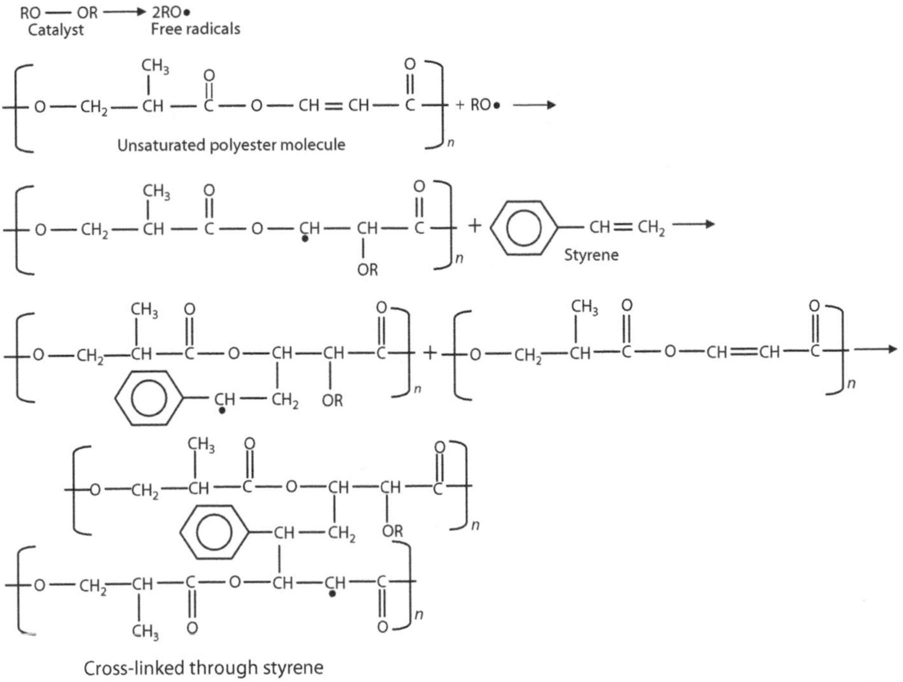

RO — OR → 2RO•
Catalyst   Free radicals

Unsaturated polyester molecule + RO• →

Styrene

Cross-linked through styrene

FIGURE 3.7
Curing reaction of thermoset polyester resin.

TABLE 3.4

Processing Temperature Range of Catalyst Systems for Polyester Resin

Catalyst	Activator	Curing Temperature (°C)
Benzoyl peroxide	Dimethylaniline	0–25
Methyl ethyl ketone hydroperoxide (MEKP)	Cobalt octoate	25–35
Cumene hydroperoxide	Manganese naphthenate	25–50
Lauroyl peroxide	Heat	50–80
t-Butyl peroctoate	Heat	80–120
Benzoyl peroxide	Heat	80–140
2,5-Dimethyl–2,5-di(2-ethylhexanoylperoxy) hexane	Heat	93–150
t-Butyl perbenzoate	Heat	105–150
Di-t-butyl peroxide	Heat	110–160
Dicumyl peroxide	Heat	130–175

Source: Data from Kolczynski, J.R. et al., *Soc. Plast. Ind. Conf.*, 22, 1A, 1967.

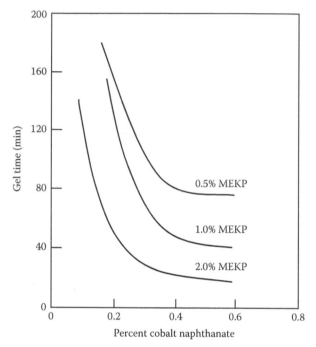

FIGURE 3.8
Effect of catalyst and activator contents on the gel time. (Reprinted from Lubin, G., *Handbook of Composites*, Springer, Berlin, Germany, 1982, p. 28. With permission.)

catalyst. Activator content and temperature also influence the reaction rate similar to the catalyst content. These parameters should be optimized to get better properties in the cured resin.

The decomposition rate of catalyst controls the curing time of polyester resin. The rate can be increased by increasing the curing temperature. However, there is an optimum temperature for a given resin–catalyst system, at which all the free radicals generated from the catalyst are utilized for curing the resin. Above this optimum temperature, free radicals are generated very rapidly and the curing reaction is very fast. Excessive amount of heat is generated during this fast reaction that deteriorates the properties of cured resin. Below this optimum temperature, the reaction rate is very slow. Hence, activators are generally added to increase the reaction rate.

The main disadvantage of polyester resin is its relatively high volumetric shrinkage during curing. This shrinkage may be beneficial for the easy release of parts from the mold, but the difference in shrinkage between the fiber and resin results in uneven depressions on the molded surface. These are called sink marks. These sink marks are undesirable for good-quality surfaces of automotive body components, which require high glossy, good appearance. One way of reducing shrinkage is by adding a thermoplastic component, which produces microvoids during curing that compensate the normal shrinkage of the polyester resin.

Heat is generated during cross-linking reaction. In small products, the heat is generated in a controlled manner but in larger thicker products the heat generation is uncontrollable and that can cause discoloration and/or cracking. Modifiers such as α-methyl styrene or copper naphthenate can be added to reduce the cross-linking rate. However, excessive amount of modifier can terminate the cross-linking and produce a rubbery product. Generally, the products made using exotherm suppressants have lower HDT and modulus because of lower cross-linkage. After the gel time, the masses begin to heat up and the hardness increases rapidly. The cross-linking reaction is ~95% complete after the heat dissipates from the mass, and post-curing at high temperatures drives the reaction to completion. Without post-cure, styrene odor may come out from the product for a few weeks.

The highly cross-linked nature of the polyester resin enhances the resistance to softening at high temperature. However, thermo-oxidative stability of these resins is poor because of the presence of vinyl groups. Aromatic constituents enhance the HDT of the polymer. Propoxylated bisphenol fumarate resin dissolved in styrene has a cast HDT as high as 130°C. Deformation at high temperatures is significantly reduced by adding inorganic fillers and fiber reinforcements, but the strength of the composite declines rapidly as the temperature approaches the HDT of the resin. Irrespective of the composition, the cross-linked polyester resin undergoes spontaneous decomposition near 300°C, even in the absence of oxygen. This is a characteristic property of all the vinyl polymers, which are decomposed into monomeric units at high temperatures, whereas non-vinyl cross-linked thermosets such as phenolic and epoxy resins tend to carbonize. Orthophthalic resins decompose at a much lower temperature than corresponding isophthalic or terephthalic resins because of the weaker ester bonding. Fillers or reinforcements improve the short-term thermal properties of polyester resin but do not reduce the loss of physical properties at exposure to high temperature for prolonged periods.

The low dipolar nature of the cross-linked polymer affects the dielectric strength and electrical resistivity. At high voltages, arcing leads to charring and the formation of conductive carbonaceous residue and finally to the failure of plastics. Hydrated fillers such as alumina trihydrate suppress the char formation and increase the resistance to failure.

The presence of styrene is responsible for the flammability of polyester resin. Flammability can be reduced by substituting the dibasic acid with chlorendic anhydride or tetrabromophthalic anhydride or the glycol component with dibromoneopentyl glycol. Brominated derivatives are much more effective than chlorinated derivatives. Further reduction is possible by introducing antimony trihalides, which are less volatile and more persistent in inhibiting the combustion process. Halogenated resins are sensitive to ultraviolet radiation and there is discoloration within a few weeks. The effect of UV radiation can be reduced by UV absorbers, such as substituted benzophenones and benzotriazoles. Acceptable resistance to UV degradation is also achieved with dibromoneopentyl glycol or dibromotetrahydrophthalic derivatives.

Orthophthalic polyester resins are generally adequate for use with non-oxidizing mineral acids and corrodents that are relatively mild. Its resistance to water of all types including sea water is more than adequate. Hence, this resin is preferred for boat building. These resins are normally not recommended for use in chemical process equipment. The isophthalic polyesters offer better chemical resistance than the general-purpose resins, at a slightly higher cost. They are used extensively in the manufacturing of underground gasoline tanks because of their resistance to gasoline and the varied conditions of ground soil corrosion.

Polyester resins lack alkali resistance due to the hydrolysis of ester linkages. The incorporation of bisglycol (the reaction product of bisphenol A and propylene oxide) reduces the concentration of ester linkages. The bisphenol polyesters have an exceptionally high degree of chemical resistance, superior to both the general-purpose and isophthalic polyester resins. Fabrication with these resins is much easier compared to epoxy or furan resins. They show superior acid resistance than epoxies. The bisphenol resins show good performance with moderate alkaline solutions and various categories of bleaching agents.

The 3-D network structure gives rise to thermoset characteristics to the polyester resin, which resist attack by corrosive chemicals, including weak alkalis, strong acids, and nonpolar solvents. Although water can effectively permeate the cross-linked network, saponification breakdown predominates only at the surface. Polyester resins are not recommended for service in strong alkali because glass fiber promotes ester disintegration by saponification.

Styrene, a major component of the polyester resin, can readily volatilize from open molds and become a hazard in the workplace. It causes irritation to the eyes and mucus membrane and acts as a depressant to the central nervous system. Efficient ventilation must be provided in the workplace. It is not recommended to carry out the fabrication of polyester resin-based components in a closed room.

There are many polyester resin suppliers available throughout the world. They can supply the specific variety of resin depending on the end use. A list of the common types of polyester resins is given in Table 3.5.

3.1.3.2 Epoxy Resins

Resins that contain epoxy group in their chemical structure are known as epoxy resins. Epoxy groups are generally three-membered rings of one oxygen atom and two carbon atoms. The epoxy resins available in the market have two or more epoxy groups in a molecule. Thus, these resins can form cross-links during curing reaction through the epoxy groups.

Epoxy resins react with many compounds and adhere well to different types of materials. These properties made epoxy a versatile material for surface coatings and adhesives. Like polyester resins, epoxy resins can also be tailor-made to suit any particular application, since there is a variety of resin systems and curing agents available.

TABLE 3.5

Types of Unsaturated Polyester Resins

Type	Description
General purpose	Standard rigid resin normally based on orthophthalic anhydride, maleic anhydride, and propylene glycol. Relatively good physical properties. It is used in a great variety of applications such as trays, boats, bath tubs, and water tanks.
Low profile	A mixture of a polyester resin and a thermoplastic additive, such as polystyrene or polymethyl methacrylate. Minimum sink marks formed due to differential shrinkage between fiber and resin. Can be supplied as a single component resin, if the thermoplastic material is compatible with the polyester/monomer solution.
Light stabilized	Contains ultraviolet absorbers. Excellent outdoor weather-resistant properties characterized by minimum discoloration to sunlight. Used in gel coats, outdoor structural panels, etc.
Surfacing	Contains wax or other barriers to permit curing in the presence of air. Can be used on vertical surfaces with the addition of thixotropic additives. Coatings resist marring and blushing in water.
Chemical resistant	Exhibits outstanding chemical and water resistance and used for the manufacture of fume hoods, reaction vessels, and tanks.
Heat resistant	High heat distortion temperatures, higher hot strength and rigidity. Triallylcyanurate is used in place of styrene.
Resilient	Greater toughness and impact strength. Some formulations have good hot strength also. Used in safety helmets, guards, gel coats, and aircraft and automotive parts. Isophthalic acid is commonly used. Isophthalic acid and glycol are combined first, and then maleic anhydride is added to give polyester chains having the reactive unsaturation at the ends.
Flame resistant	Self-extinguishing, flame resistance, moderate heat resistance, and average strength properties. Resin is prepared using halogenated dibasic acids. Further improvement in fire resistance can be achieved by adding phosphate esters and antimony oxide. Used in the manufacture of fume hoods, electrical equipment, and building panels.
Flexible	Low hardness and high flexibility and elongation. Straight-chain dibasic acid such as adipic or sebacic acid is used instead of phthalic anhydride. The use of diethylene glycol in place of propylene glycol also provides flexibility.

Source: Lubin, G., *Handbook of Composites*, Springer, Berlin, Germany, 1982, p. 18.

Depending on the type of molecule in which epoxy groups are attached, the epoxy resins can be classified as aliphatic and aromatic epoxy resins. Aliphatic diols or polyols react with epichlorohydrin and form aliphatic epoxy resins (Figure 3.9). These resins are more flexible due to the presence of long aliphatic chain between two epoxy groups. Hence, these resins are not suitable for making composites. However, these reins can be used as diluents and modifiers for highly viscous aromatic epoxy resins. Epoxidized soybean oil and castor oil also produce flexible products on curing.

Aromatic epoxy resins are produced by the reaction of aromatic compounds with epichlorohydrin. The first commercial and most widely used

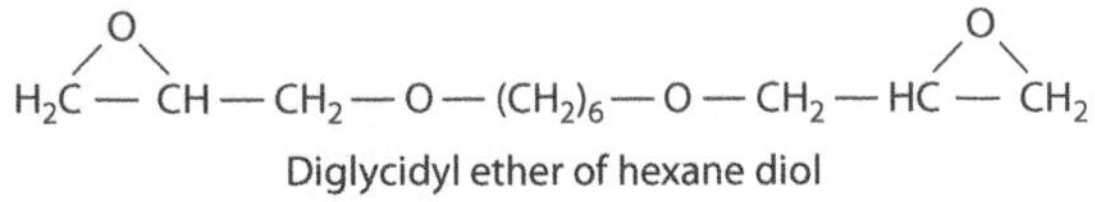

$$H_2C - CH - CH_2 - O - (CH_2)_6 - O - CH_2 - HC - CH_2$$

Diglycidyl ether of hexane diol

FIGURE 3.9
An aliphatic epoxy resin.

epoxy resins are the reaction products epichlorohydrin and bisphenol A. The reaction produces diglycidyl ether of bisphenol A (DGEBA) epoxy resin (Figure 3.10). The resins with higher molecular weight are used as adhesives and surface coatings. Lower molecular weight and low-viscosity resins are generally used for the fabrication of composites.

Novolac resins are made by the reaction of epichlorohydrin with phenol formaldehyde. These resins after reacting with a suitable curing agent can produce a product with high HDT. Another important epoxy resin is tetraglycidyl diaminodiphenylmethane (TGDDM). This resin also has the property of retaining strength at high temperature. This resin cures at high temperature. Hence, this resin can be used for making prepregs that are required for aerospace applications. The limitations of this resin are poor hot–wet performance, low strain-to-failure values, and high moisture absorption. High moisture absorption reduces its glass transition temperature and mechanical properties.

Halogen-substituted bisphenol is used for the preparation of fire retardant grade epoxy resins. Some resins are formulated by mixing two or three types of epoxy resins to suit the end application.

Epoxy resin has the following advantages over other thermoset resins:

- Absence of volatile matters during curing
- Low curing shrinkage
- Excellent resistance to chemicals and solvents
- Excellent adhesion to a wide variety of fillers, fibers, and other substances

Brittleness or low strain-to-failure value is an inherent problem of any highly cross-linked resin, including epoxy resin. Improvement in strain-to-failure value and fracture toughness is essential for making damage-tolerant composites. This can be accomplished by incorporating elastomers into the epoxy resin. One of the widely used elastomers for epoxy resin is carboxyl-terminated butadiene acrylonitrile (CTBN). This elastomer forms a second phase in the cured matrix and that restricts the formation of microcracks. Although the elastomer addition improves the toughness of resin, it reduces the glass transition temperature, modulus, and solvent resistance. Toughness improvement without the reduction in other properties is possible by incorporating a thermoplastic material. However, the uniform distribution of very fine thermoplastic particles/phase is a difficult task.

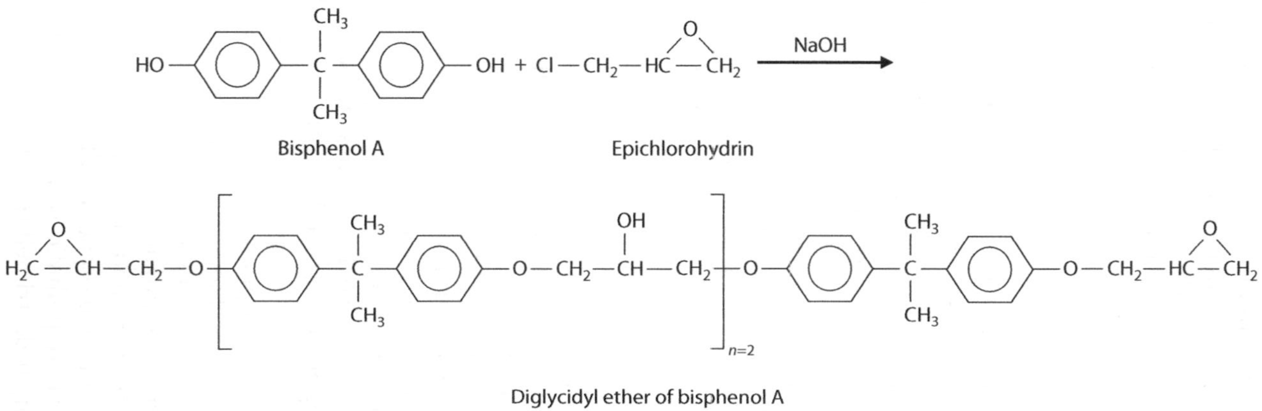

FIGURE 3.10
Formation of DGEBA epoxy resin. (Reprinted from Mallick, P.K., *Fiber-Reinforced Composites*, 3rd edn., CRC Press, Boca Raton, FL, 2008, p. 72. With permission.)

Viscosity of resin is an important parameter for the fabrication of composites. To some extent the viscosity is decided by the chemical structure of the resin. Aliphatic epoxy resins with linear chains are low-viscous liquids. The effect of chain length is insignificant for the diepoxies. However, there is an increase in the viscosity with chain length for the polyepoxies.

Aromatic epoxy resins with three or more epoxy groups per molecule are semisolids or solids at RT. Viscous liquids are made by mixing these epoxies with aromatic diepoxy resins. Generally any substitution of aromatic rings decreases the viscosity and there are some exceptions.

3.1.3.2.1 Curing Agents

Unlike polyester resin, the curing agents are the cross-linking groups for the epoxy resin. The type of curing agent also controls the final properties of cured epoxy resin. There are many curing agents commercially available. These curing agents can be broadly classified into three categories based on the reactive functional group. They are amine, acid anhydride, and acid curing agents.

Aliphatic amines are the first curing agents used for epoxy resin. These amines cure the resin at RT. The end-product properties depend on the chain length of the curing agent. The flexibility of the end product is increased when a long-chain curing agent is used. Some of the commercially used aliphatic amines are diethylene triamine (DETA), triethylene tetramine (TETA), tetraethylene pentamine (TEPA), and dicyandiamide (DICY). DICY is a hot curing agent and cures the epoxy resin at 120°C. This curing agent is suitable for components requiring long fabrication time and for making prepregs.

Aromatic amines are the widely used curing agents for epoxy resins. Some of them are m-phenylene diamine (MPDA), 4,4′-methylene dianiline (MDA), and diaminodiphenyl sulfone. These amines react very slowly at RT and require heating for fast and complete curing. The resulting cured products have good resistance to alkalis and solvents and superior mechanical properties than the products cured with aliphatic amines. These curing agents are generally solids at RT and also their solubility is poor. Usually the resin is heated to about 80°C to dissolve the required amounts of curing agents.

Acid anhydride curing agents can produce products with high HDT. These compounds are hygroscopic in nature and should be stored in airtight containers to avoid moisture absorption. These curing agents also need high temperature to cure epoxy and the resulting products have HDT as high as 200°C. Generally accelerators are used to increase the reaction rate. On curing, the anhydrides form ester linkages with epoxy groups and these linkages are stable in organic acids and some inorganic acids. However, these linkages are hydrolyzed in the presence of alkalis. The commonly used acid anhydride curing agents are phthalic anhydride, hexahydrophthalic anhydride, nadir methyl anhydride, trimellitic anhydride, maleic anhydride, and succinic anhydride.

It is necessary to add stoichiometric quality of curing agent to completely cure the epoxy resin. This quantity depends on the equivalent weights of epoxy resin and curing agents. Epoxy equivalent weight is defined as the weight of resin in grams containing one epoxy group per molecule. In the case of diglycidyl ether epoxy resins (containing two epoxy groups in a molecule), the epoxy equivalent weight is one-half of the average molecular weight.

Amine equivalent weight depends on the number of amine hydrogens. Hence, amine equivalent weight is the ratio of amine molecular weight and the number of amine hydrogens in a molecule. Similarly anhydride equivalent weight is the ratio of anhydride molecular weight and the number anhydride groups available in a molecule. To get complete cure, one equivalent of curing agent should be mixed with one equivalent of epoxy resin.

3.1.3.2.2 Curing of Epoxy Resins

Epoxy resins can cure themselves and form a homopolymer or by adding a curing agent or hardener having active hydrogens. The curing reaction is illustrated schematically in Figure 3.11. Physical and chemical changes are taking place during curing. The physical changes are viscosity and temperature and the chemical changes are linear polymerization and cross-linking.

Chemical reactions taking place during the curing of epoxy resins are exothermic, that is, heat is released. The liberated heat can cause rise in the temperature and the temperature rise is proportional to the amount of resin reacting and the rate of curing reaction. The maximum temperature attained (peak exotherm) can be controlled by controlling the rate of curing. In the case of DGEBA epoxy resin curing with primary aliphatic amines, initially all the primary amine hydrogens react with epoxy groups to form long chains. Nearly 50% of the epoxy groups are consumed and the primary amines form secondary amines. The secondary amines

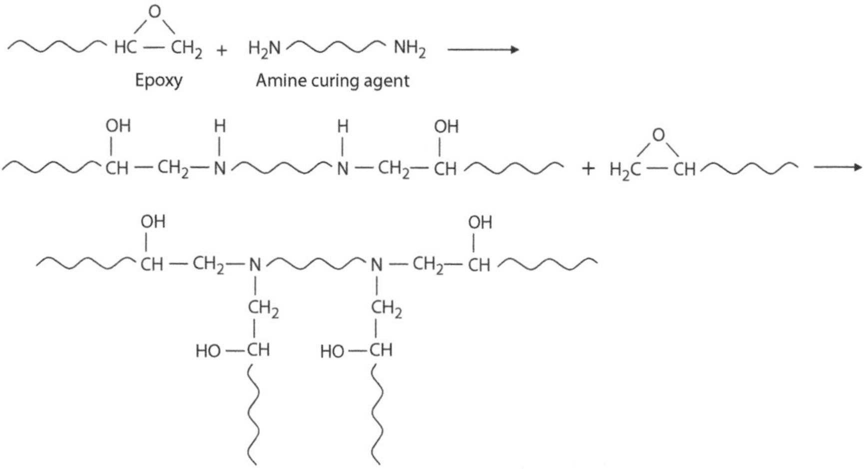

FIGURE 3.11
Curing reaction of epoxy resin. (Reprinted from Mallick, P.K., *Fiber-Reinforced Composites*, 3rd edn., CRC Press, Boca Raton, FL, 2008, p. 72. With permission.)

can react with epoxy groups only at high temperature. The heat released during the reaction of primary amines with epoxy groups is sufficient to initiate the reaction between secondary amines and epoxy groups. Cross-links are formed during this reaction. Thus, the peak exotherm is followed by gelation in epoxy resin.

The cross-linking reaction proceeds at a slower rate and it may not lead to further increase in temperature. At the initial stage of curing, the viscosity decreases due to temperature rise. The viscosity reaches to the minimum at the peak exotherm. Generally the resin is kept undisturbed at this stage for some time to allow the bubbles present in the resin to come to the surface. It is recommended to apply vacuum for the casting applications and the fabrication of high-voltage insulators. Once the cross-linking reaction starts, the viscosity increases rapidly leading to the formation of gel and then finally to the glassy polymer. The cross-linking reaction stops at the glassy state. If there are some unreacted groups present, then it is necessary to post-cure the products at high temperature. At this temperature, the chains are flexible and some more cross-linking can form and that increases the rigidity of the product as well as its HDT.

The physical properties and chemical resistances of epoxy resins depend on the extent of curing. It is very difficult to achieve 100% cure, since some unreacted groups are always entrapped within the solidified material. Moreover, it is difficult to quantitatively determine the extent of cure. However, the extent of curing can be related to the properties. The flexibility and HDT of the cured product are the indicators of the extent of curing.

3.1.3.3 Vinyl Ester Resins

Vinyl ester resin is a combination of polyester and epoxy resins. It is produced by the reaction of an unsaturated carboxylic acid with an epoxy resin. The reaction is shown in Figure 3.12. Similar to polyester resin, the cross-linking sites are the C=C double bonds and styrene is the cross-linking agent and solvent. In the vinyl ester resin, the C=C double bonds are located only at the ends of a molecule and therefore cross-linking can take place only at the ends. The cured vinyl ester is more flexible and has higher fracture toughness than polyester resin because of fewer cross-links. At the same time, the epoxy backbone imparts higher strength and chemical resistance to the resin. Another important characteristic of vinyl ester is that it contains a number of –OH groups along the length of the molecule. These –OH groups can form physical bonds (hydrogen bonds) with similar groups on a glass fiber surface. Because of this, there is excellent wet out and good adhesion with glass fibers.

The curing reaction of vinyl ester resin is similar to that of polyester resin. Same catalysts and activators can be used for the vinyl ester resin also and the curing rate is comparable to that of polyester resin. However, the volumetric shrinkage of vinyl ester resin is less and it is in the range of 5%–10%.

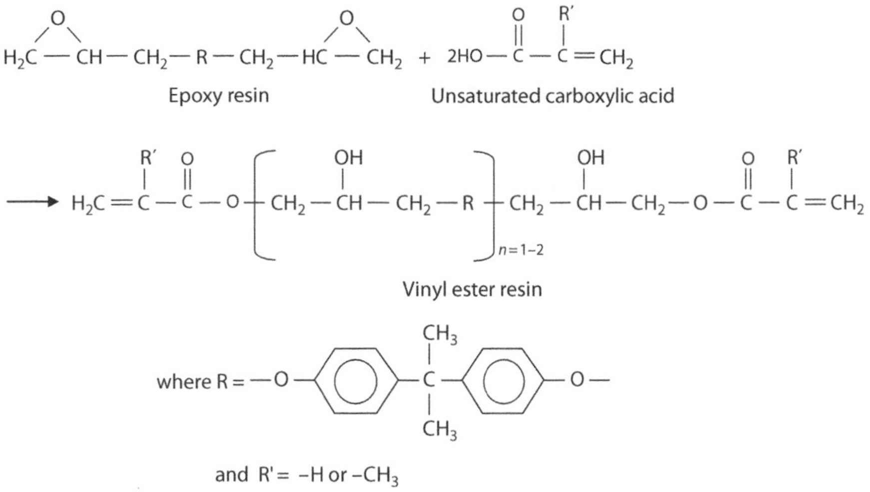

FIGURE 3.12
Formation of vinyl ester resin. (Reprinted from Mallick, P.K., *Fiber-Reinforced Composites*, 3rd edn., CRC Press, Boca Raton, FL, 2008, p. 81. With permission.)

Vinyl ester resin exhibits only modest adhesive strength compared to epoxy resin. The shrinkage volume is also higher than that of epoxy resin. The HDT and thermal stability of vinyl ester resins can be improved by using phenolic novolac epoxy resin as the starting material to prepare the resin.

3.1.3.4 Phenolic Resins

Phenolic resins (phenol formaldehyde) are produced by the reaction of phenol with formaldehyde (Figure 3.13). Phenol is reactive toward formaldehyde at the ortho- and para-positions (sites 2, 4, and 6) allowing up to 3 units of formaldehyde to attach to the ring. The initial reaction in all cases involves the formation of a hydroxymethyl phenol. The hydroxymethyl group is capable of reacting with either another free ortho- or para-position or with another hydroxymethyl group. The first reaction gives a methylene bridge, and the second forms an ether bridge. Phenolic resins are commonly used for producing plywood, printed circuit boards, paper laminates, abrasives, cloth laminates, brake linings, fiber-reinforced composites, ablative composites, etc. Different derivatives of phenol such as cresol, xylenol, or resorcinol can also be used to produce phenolic resins with some specific properties.

There are two major categories of phenolic resins: resoles or one-step resins, and novolacs or two-step resins. The phenol-to-formaldehyde ratio is less than one for the resoles. Strong bases such as NaOH, Ca(OH)$_2$, KOH, or quaternary ammonium compounds are used as catalysts. A highly water-soluble liquid resin is produced by this reaction. It can be cured by heat or change of pH or both. Resoles do not require an external catalyst and undergo self-polymerization under heat with the evolution of water. This resin is used for

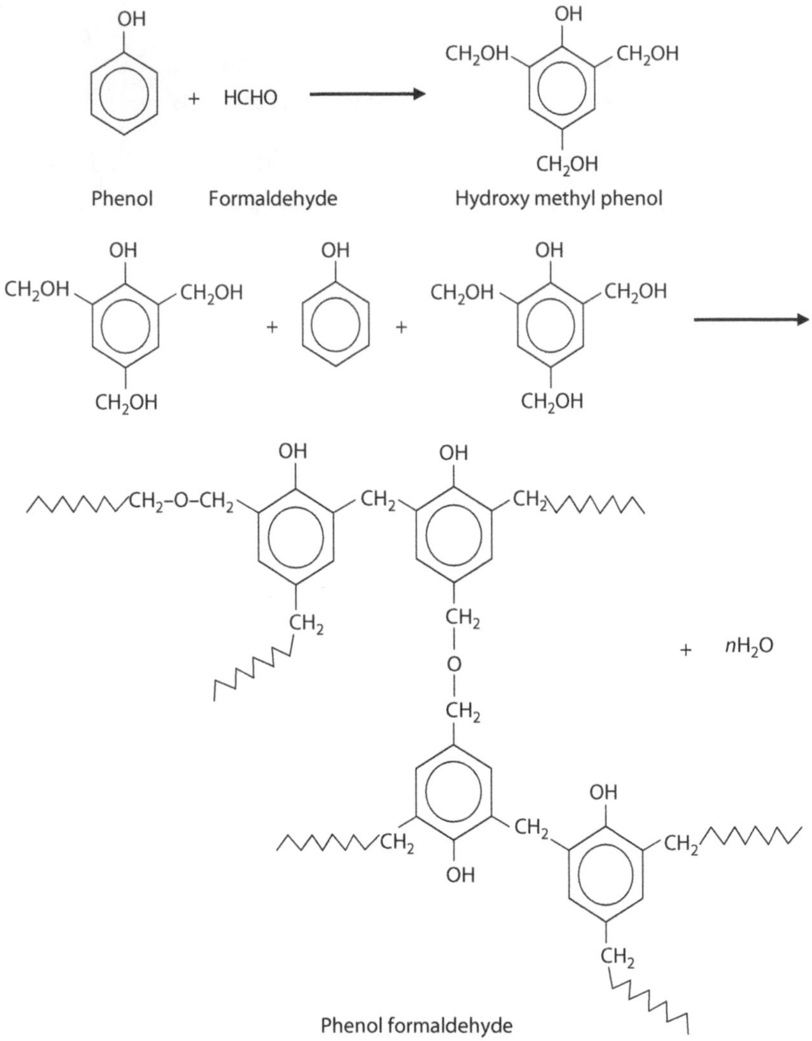

FIGURE 3.13
Formation of phenol formaldehyde resin.

casting and coating applications. Another type of resole resin is produced by using weak bases like ammonia or primary, secondary, or tertiary amines. Water produced during the reaction is removed by vacuum distillation and then the resin is dissolved in alcohol. Partial curing (B-stage curing) is possible with this resin. It is used for making industrial and decorative laminates.

The phenol-to-formaldehyde ratio is more than one for novolacs. More amount of phenol increases the rigidity of the molecule and hence leads to better thermal stability. Acid catalysts are used for producing this type of resin. Commonly used acid catalysts are formic, sulfuric, phosphoric, oxalic, or trichloroacetic acids. A solid product is obtained after the reaction, which is then ground to get fine powder. Novolac resin at this stage consists of linear polymer chains with melting temperature range of 80°C–150°C. This resin can be cured by adding 10%–15% hexamethylenetetramine formaldehyde (HEXA). High pressure is generally used to overcome the problem of the voids generation by the volatile emission.

The main advantages of phenolic resins are superior fire resistance, excellent high-temperature performance, long-term durability, and better resistance to hydrocarbons and chlorinated solvents. The main limitation is high volatile emission during curing. Unless high pressure is applied during curing, the end product will have lot of pores. Depending on the final application and processing conditions, phenolic resins are produced in a variety of physical states, such as powder, hot melt, and solution.

Resole-type phenolic resins are commonly used for the manufacture of FRCs. The use of phenolic resins in composites is growing, primarily due to low flame spread, smoke generation, and smoke toxicity properties.

The appliances, closures, and house wares industries prefer the use of resole-based phenolic molding compounds because resoles release water upon curing and are odor free. The novolac-based molding compounds are more dimensionally stable than resoles and are used for high-strength and non-household applications. Novolac powders with HEXA are commonly used in molding compounds with chopped glass and carbon fibers. They are also used as binders in brake linings, grinding wheels, foundry molds, and general molding compounds. Most of the underground railways in the United Kingdom, France, and Scandinavian countries switched to phenolic composites. Large panels (1.8 × 5.4 m) for constructing homes are manufactured using phenolic composites. They eliminate the possibility of termite damage and provide better fire safety and easier construction.

3.1.3.5 Polyimide Resins

Polyimide resins are high-temperature polymers, which can be used up to 230°C for long periods and up to 315°C for short periods. The polymers containing –OC–N–CO– groups are known as polyimides. Polyimides are classified as thermoplastic polyimides and thermosetting polyimides. Thermoplastic polyimides are derived by condensation reaction between anhydrides or anhydride derivatives and diamines. Thermosetting polyimides are derived by addition reaction between unsaturated groups of an imide monomer or oligomer.

Thermoplastic polyimides are linear long-chain polymers with high melt viscosity. Hence, composite processing with these plastics is generally carried out at high pressure and temperature. The highly aromatic nature of these polymers coupled with flexible groups, yields materials with good toughness, excellent thermal and thermo-oxidative stability, and moderate to high glass transition temperature (T_g).

TABLE 3.6

Comparison of Thermoplastic and Thermosetting Polyimides

Polyimide Type	Advantages	Disadvantages
Condensation (thermoplastic)	Reprocessability	Poor processability
	Moderate to high T_g	Volatiles released during processing
	Toughness	
	Excellent thermal and thermo-oxidative stability	High pressure required for processing
Addition type (cross-linked)	Processability	Limited reprocessability
	High T_g	Brittle
	No volatiles during processing	Poor thermal and thermo-oxidative stability
	Low pressure required in processing	

Source: Data from Scola, D.A., Polyimide resins, in *ASM Handbook*, Vol. 21, *Composites*, ASM International, Materials Park, OH, 2001, p. 105.

In the case of thermosetting polyimides, the viscosity is relatively low at the curing temperature and no volatiles are released during composite processing. The oligomer molecular weight controls the processability and the degree of cross-linking and hence the T_g. The specific advantage and disadvantage of two types of polyimides are summarized in Table 3.6.

Thermosetting polyimides on curing offer not only high temperature resistance but also high chemical and solvent resistance. However, these materials are inherently very brittle due to the high cross-link density. As a result, their composites are prone to excessive microcracking. Combining one or more tough thermoplastic polyimides with them is a possible solution to overcome the problem of brittleness without affecting the high temperature resistance. The combination produces a semi-interpenetrating network (semi-IPN) that retains the easy processability of thermoset and exhibit the good toughness of a thermoplastic polyimide. The addition of thermoplastic polyimide increases the curing time, but it is beneficial for the manufacturing of large or complex composite parts, since the addition helps in broadening the processing window (i.e., the time available for fabrication before the gelation starts).

Bismaleimides (BMIs), PMR-15 (polymerization of monomer reactants), and acetylene-terminated polyimide (ACTP) are examples of thermosetting polyimides. Among these polyimides, BMIs are suitable for applications requiring a service temperature up to 230°C. Service temperatures of PMR-15 and ACTP are 288°C and 316°C, respectively. The last two resins have exceptional thermo-oxidative stability at temperatures higher than 300°C in the presence of air.

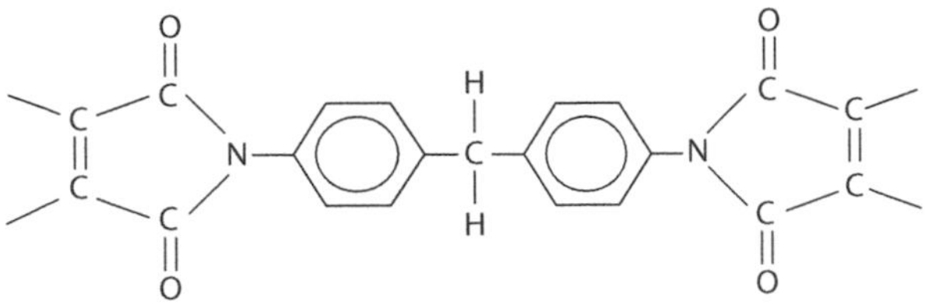

Bismaleimide (BMI)

FIGURE 3.14
BMI molecular structure. (Reprinted from Mallick, P.K., *Fiber-Reinforced Composites*, 3rd edn., CRC Press, Boca Raton, FL, 2008, p. 84. With permission.)

3.1.3.5.1 Bismaleimide Resin

BMIs are low molecular weight difunctional monomers or pre-polymers that carry maleimide terminations (Figure 3.14). The maleimide groups can undergo homopolymerization and a wide range of copolymerization reactions, to form a highly cross-linked network. These curing reactions can be initiated by the application of heat and if needed, a suitable catalyst can also be added.

These resins are gaining wide acceptance by industry because of their unique characteristics, such as retention of physical properties under hot and wet environments, almost constant electrical properties over a wide range of temperature, and nonflammability. These characteristics with their excellent processability made them popular in advanced composites and electronics. The main limitation is brittleness due to high cross-linking density. However, tougher resins are produced by incorporating a suitable comonomer.

BMI resins are used for making high-performance composites for military aircrafts and sports cars. The BMI matrix composites can retain their mechanical properties up to 290°C. The high-performance composites with BMI resins are generally made by autoclave processing. BMI resins are also suitable for resin transfer molding.

3.1.3.5.2 Polymerization of Monomer Reactants-15

Dimethyl ester of 5-norbonene 1,2-dicarboxylic acid is added to a methanol solution of benzophenone tetracarboxylic acid dimethyl ester under constant stirring at RT. 4,4′-methylene diamine is then added to the ester solution. After the complete dissolution of diamine, the resin solution can be used to impregnate fibrous materials and then cured in the temperature range of 175°C–260°C. PMR-15 is extremely resistant to most of the organic solvents. It is hydrolyzed in strong acids and bases at high temperatures.

The properties of commonly used thermosetting resin are given in Table 3.7. Polyester and vinyl ester resins are more preferred for ambient temperature applications. Epoxy and phenolic resins are useful for moderate-temperature applications up to about 150°C. In applications requiring service temperatures above this range, BMI, PMR-15, or ACTP resins can be used.

TABLE 3.7

Properties of Commonly Used Thermoset Polymer Resins

Property	UP	Vinyl Ester	Epoxy	Phenolic	BMI	PMR-15	ACTP
Density (g cm^{-3})	1.2	1.2	1.2–1.3	1.3	1.4	1.32	1.35
Tensile modulus (GPa)	4.0	3.3	4.5	3.0	4–19	3.9	4.1
Tensile strength (MPa)	80	75	130	70	70	38.6	82.7
Strain to failure (%)	2.5	4	2–6	2.5	1	1.5	1.5
Cure shrinkage (%)	5–12	5.4–10.3	1–5	—	—	—	—
CTE (10^{-6} °C^{-1})	80	50	110	10	80	—	—
T_g (°C)	80	80	180	—	230–290	340	320
Service temperature (°C)	60–200	100	90–200	120–200	250–300	315	280

Source: Adapted from Mallick, P.K., *Fiber-Reinforced Composites*, 3rd edn., CRC Press, Boca Raton, FL, 2008, p. 61.

The mechanical properties and adhesion with reinforcement are better for epoxy resins compared to polyester and vinyl ester resins. However, they are very expensive.

3.1.4 Thermoplastics

Thermoplastics are inherently tough and resistant to damage from low-velocity impacts. In addition to impact resistance, thermoplastics have excellent abrasion resistance. They exhibit attractive dielectric properties and these properties are not significantly affected by moisture absorption. In general, the high-temperature thermoplastic composites also have excellent environmental and solvent resistance.

Most of the thermoplastics are used in commodity applications, where the service temperature requirements are modest. Only a few thermoplastics are suitable for high-temperature engineering applications. Many thermoplastics have higher T_g and T_m. However, the mechanical performance of these plastics diminishes as the temperature approaches these points. Hence, the T_g or T_m must be well above the intended use temperature.

Mechanical forming or shaping of thermoplastics is carried out above T_g or near melting point. Hence, the requirement of high T_g for high-temperature applications has an adverse effect on processing. The processing temperature increases with the increase in T_g.

Most of the thermoplastic polymers are semicrystalline materials having domains of highly ordered molecular structure (crystallites). Crystalline structure development is thermodynamic and kinetic phenomena, controlled by the mobility and free energy of molecules. Cooling rate can control the crystallinity and crystallite distribution. The effects of crystallinity are similar to cross-linking; for example, the stiffness and chemical resistance of thermoplastics increase with increasing crystallinity. Unlike thermoset composites, a great degree of useful strength and stiffness may remain well above the T_g of semicrystalline thermoplastic composites. Amorphous resins lose their strength significantly above T_g, even in the presence of continuous fiber reinforcements. The formation and chemical structures of some of the most common engineering thermoplastic materials are illustrated in Figure 3.15.

High-performance thermoplastics contain aromatic rings that impart a relatively high glass transition temperature and excellent dimensional stability at high temperatures. The actual value of T_g depends on the size and flexibility of other chemical groups present in the chain.

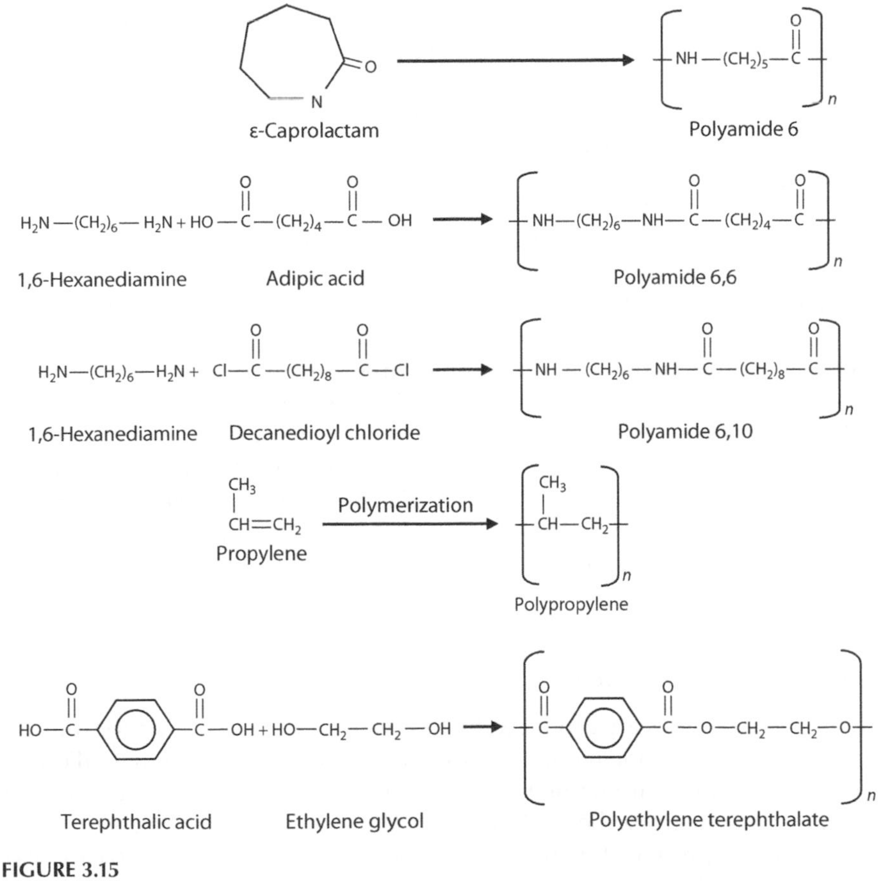

FIGURE 3.15
Formation of commonly used engineering thermoplastic matrix materials.

Polyether ether ketone (PEEK), polyphenylene sulfide (PPS), and polyether imide (PEI) are normally used for producing continuous FRCs, while polypropylene (PP) is used quite extensively with discontinuous glass fibers in automotive industry. High-performance thermoplastics, such as PEEK, PEI, and PPS, have high glass transition temperatures with good mechanical properties but are more expensive. They are usually aromatic and the presence of benzene ring is responsible for their high T_g and thermal stability. Highly aromatic thermoplastics exhibit good flame retardance because of their tendency to char and form a protective surface layer.

Since there is no steep change in density during cooling from the melt, amorphous thermoplastics have less tendency to warp or distort during rapid cooling. Many thermoplastic polyimides undergo light cross-linking during processing and are often classified as pseudo-thermoplastics. These materials are available at low and high molecular weight forms. Low molecular weight thermoplastic polyimide prepregs are normally produced using conventional thermoset prepreg equipment. However, due to the chemical inertness of some of the resin components, they must be dissolved in a high boiling point solvent to facilitate prepreg formation. Some amount of the solvent is removed during the drying operation, and an appreciable amount of solvent remains in the prepreg. This residual solvent evaporates during elevated temperature processing resulting in the formation of voids. In addition, the condensation reactions release water and ethanol, which contribute further void formation. Extensive use of breather and vacuum porting arrangements can help in volatile removal.

The high molecular weight thermoplastic polyimide prepregs are somewhat easier to process; however, voids may also form in some of these materials through additional high-temperature reactions. The low molecular weight thermoplastic polyimide prepregs are processed similar to thermoset prepregs but at much higher pressures and temperatures. The processing cycles are necessarily long, because sufficient time is needed for the buildup of required molecular weight and for the removal of volatile materials. Since many of these thermoplastic polyimides have a tendency to lightly cross-link, this small amount of cross-linking makes them much more difficult to reprocess than true melt-fusible thermoplastics. Some of the commonly used thermoplastic matrix materials are described in the following sections.

3.1.4.1 Aliphatic Polyamides

Aliphatic polyamides are produced by the reaction between aliphatic diacids and diamines. The trade name for these polyamides produced by DuPont is Nylons. Polyamides are named according to the number of carbons in the diamines and dicarboxylic acid or acid chlorides used for their synthesis. The polyamide produced by the reaction of 1,6-hexane diamines (6 carbons) and sebacic acid (10 carbons) is known as polyamide 6,10. Polyamide 6,6 is a product formed from 1,6-hexanediamine and adipic acid (6 carbons). The polymer derived from caprolactam (6 carbons) is called polyamide 6.

Polyamide is a highly crystalline material. Sulfonamides are used as plasticizers. In addition to aiding flow, the sulfonamides can retard degradation and speed up processing of polyamides. Polyamide 6,6 has a relatively high service temperature because of its high crystallinity. It undergoes a sharp transition from solid to melt. It has a combination of good properties, such as toughness, rigidity, and tribological properties. Hence, it is a suitable material for bearing and gear applications. The performance of these components is better than metals in some cases. This polymer is coming under the category of engineering thermoplastic materials due to its high performance.

The polar groups present in the molecule permit dipole association and that leads to high crystallinity. Because of this, the properties of this polymer with low molecular weight are comparable to high molecular weight amorphous polymers. It transforms into a rather low-viscosity fluid at its melting point (269°C). This low viscosity leads to some complications during molding and hence mold with close tolerances is required. In addition, very precise temperature and pressure monitoring is required to prevent flash formation due to the leakage of melt.

3.1.4.2 Polypropylene

Polyolefins are by far the largest class of synthetic polymers made and used today. The attractive properties are low cost of production, light weight, and high chemical resistance. A wide range of mechanical properties is possible through the use of copolymerization, blending, and additives.

Polyolefins are the polymers made from olefins. The density of polyolefins is less than 1.0 g cm^{-3}. Hence, the components made from the polyolefins are always lighter in weight. Since they are fully saturated, they have a high degree of chemical resistance to many of the solvents and liquids. Moreover, they are highly stable to oxidation. The weak van der Waals forces between the molecules result in lower melting temperature. Hence, these polymers are not suitable for high-temperature applications but their processability is easier because of the lower melting point.

PP is one of the most widely used engineering polyolefins. Many PP products can be found as part of daily life. The product ranges from fibers and fabrics for upholstery and carpeting to automotive parts.

The crystallinity of PP is determined by the tacticity of the PP. The important PP isomer is isotactic PP (IPP), in which the methyl groups are present on one side of the chain. With the development of catalysts, it is now possible to prepare PP having more than 99% isotacticity. The other form of PP is syndiotactic PP (SPP), where the methyl groups are on the alternate sides of the chain. The melting point of SPP is similar to IPP but has a different balance of stiffness and toughness.

3.1.4.3 Polyethylene Terephthalate

Polyethylene terephthalate (PET) is generally made from dimethyl terephthalate and ethylene glycol. It has a high melting point (~265°C) because of the presence of aromatic ring. It is highly crystalline and rigid. This polymer is difficult to process due to the presence of a large number of para-linked aromatic rings in the molecular chain. A small amount of dimethyl isophthalate is introduced during the polymerization reaction, which reduces the crystallinity and rigidity by forming meta-linkages. PET is one of the most popular commercial polymers. It has very good resistance to many chemicals and has good mechanical strength up to 175°C. PET is widely used to make textile fibers, films, and containers.

3.1.4.4 Polyether Ether Ketone

One of the most widely used high-performance thermoplastics is PEEK. PEEK is a linear aromatic thermoplastic material, consists of the repeating unit shown in Figure 3.16. It is a semicrystalline polymer with a maximum achievable crystallinity of 48%. The presence of crystalline fibers increases the crystallinity of PEEK. As in the other thermoplastic polymers, the increase in crystallinity increases both modulus and yield strength but decreases its strain-to-failure value. The glass transition temperature and melting point of PEEK are 143°C and 335°C, respectively. It can be used up to 250°C for longer durations. The melt processing of PEEK is generally carried out between 370°C and 400°C. PEEK has outstanding fracture toughness values, which is 50–100 times higher than epoxy resins. In addition, the water absorption of PEEK is only 0.5% at 23°C. PEEK has a potential to replace epoxies in many components made for aerospace applications because of these attractive properties.

3.1.4.5 Polyphenylene Sulfide

PPS molecules consist of the repeating unit shown in Figure 3.17. The crystallinity of this polymer is around 85% and the melting point is 285°C. The flexible sulfide linkages are responsible for its low T_g. This flexibility and simple structures of molecules lead to better crystallinity. It has excellent chemical

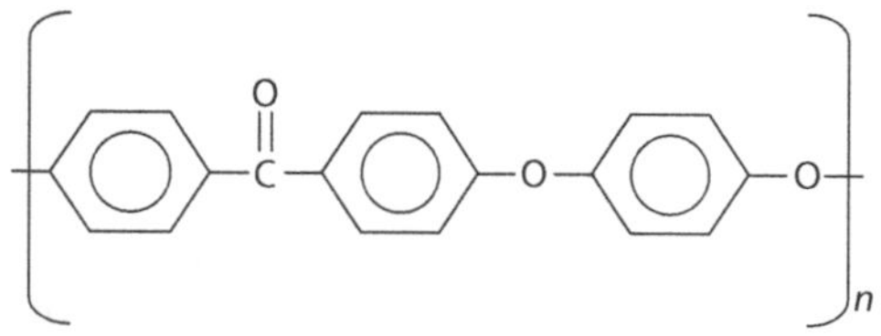

Polyether ether ketone

FIGURE 3.16

Molecular structure of PEEK. (Reprinted from Mallick, P.K., *Fiber-Reinforced Composites*, 3rd edn., CRC Press, Boca Raton, FL, 2008, p. 86. With permission.)

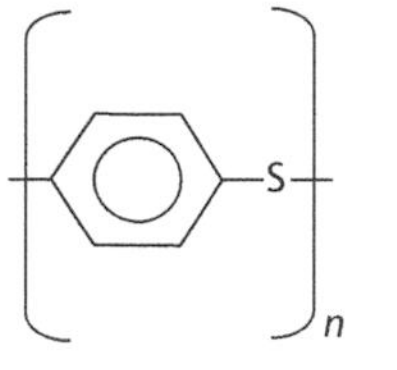

Polyphenylene sulfide

FIGURE 3.17
Molecular structure of PPS. (Reprinted from Mallick, P.K., *Fiber-Reinforced Composites*, 3rd edn., CRC Press, Boca Raton, FL, 2008, p. 87. With permission.)

resistance. This polymer is processed at temperatures between 300°C and 345°C. Its continuous use temperature is 240°C.

3.1.4.6 Polysulfone

Polysulfone (PSUL) is an amorphous thermoplastic material with the repeating unit shown in Figure 3.18. Its glass transition temperature is 185°C and continuous use temperature is 160°C. This polymer is melt processed between 310°C and 410°C. Polysulfone has high tensile strain-to-failure value (50%–100%) and excellent stability under hot and wet conditions. The chemical resistance of this polymer against mineral acids, alkalis, and salt solutions is good. However, it will swell, stress crack, or dissolve in polar organic solvents, such as ketones, chlorinated hydrocarbons, and aromatic hydrocarbons.

3.1.4.7 Thermoplastic Polyimides

Unlike thermosetting polyimides, the thermoplastic polyimides are linear polymers (not cross-linked). These polymers are derived by the condensation reaction of polyamic acid with an alcohol. Various thermoplastic polyimides can be produced by selecting different polyamic acid and alcohol. These polymers have high melt viscosity and need to be processed at relatively high temperatures. However, the main advantage is that these polymers can be reprocessed by the application of heat and pressure.

PEI and polyamide imide (PAI) are thermoplastic polyimides having the chemical structures shown in Figure 3.19. Both are amorphous polymers. The T_g values of PEI and PAI are 217°C and 280°C, respectively. These polymers

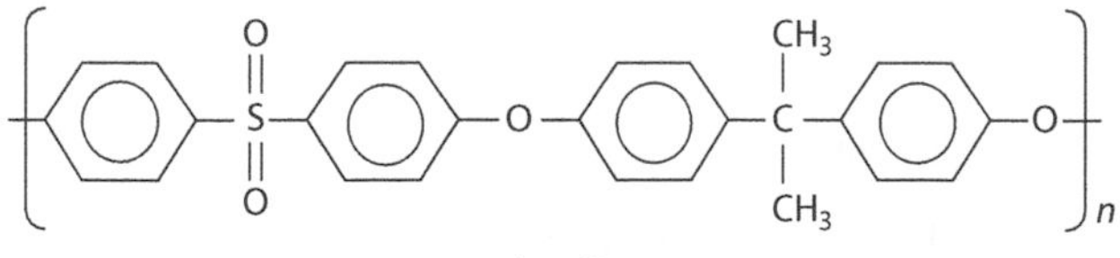

Polysulfone

FIGURE 3.18
Molecular structure of polysulfone. (Reprinted from Mallick, P.K., *Fiber-Reinforced Composites*, 3rd edn., CRC Press, Boca Raton, FL, 2008, p. 87. With permission.)

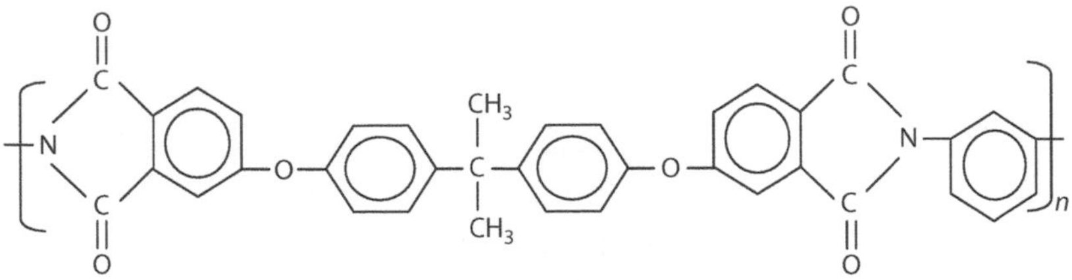

Polyetherimide (PEI)

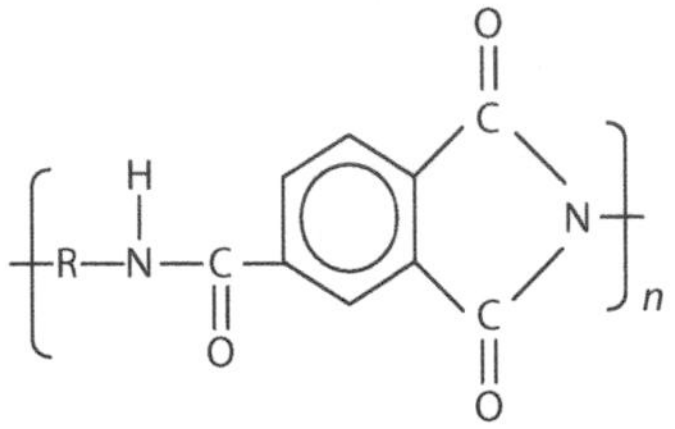

Polyamide imide (PAI)

FIGURE 3.19
Molecular structure of PEI and PAI. (Reprinted from Mallick, P.K., *Fiber-Reinforced Composites*, 3rd edn., CRC Press, Boca Raton, FL, 2008, p. 88. With permission.)

are melt processed above 350°C. Two other thermoplastic polyimides are K polymers and LARC-TPI. These polymers are generally available as pre-polymer solutions. Fiber prepregs can be made easily with these low-viscosity solutions. Polymerization reaction requires heating to 300°C or above. The glass transition temperature of K polymer is 250°C and LARC-TPI is 265°C. Both are amorphous polymers. They have excellent thermal and solvent resistance. The composites made with these polymers can be shape formed above their glass transition temperature. This is possible due to the presence of flexible chemical groups between the rigid amide rings. In LARC-TPI the presence of carbonyl groups and meta-substitution of phenyl rings having imide groups are responsible for the flexibility. The meta-substitution allows the polymer molecules to bend and flow.

The properties of some engineering and high-performance thermoplastics are given in Table 3.8. Engineering thermoplastics like PP, PET, and aliphatic polyamide are mainly used with short fibers to produce composites for ambient temperature applications. High-performance thermoplastics are mainly used in aerospace applications.

3.1.5 Elastomers (Rubbers)

The elastomers are the materials that can recover their original dimensions after substantial extension or compression. Such rubbery behavior is due to the presence of tangled long chains and flexible polymers molecules. When such a material is stretched, the individual long chains are partially uncoiled but will retract or coil up again when the force is removed. The flexibility of such

TABLE 3.8

Properties of Selected Thermoplastic Matrix Materials

Property	Polyamide 6,6	PET	PP	PEEK	PPS	PSUL	PEI	PAI	K-III	LARC-TPI
Density (g cm^{-3})	1.14	1.35	0.9	1.32	1.36	1.24	1.27	1.40	1.31	1.37
Tensile modulus (GPa)	1.6–3.8	2.8–4.1	1.1–1.6	3.24	3.3	2.48	3	3.03	3.76	3.45
Tensile strength (MPa)	95	48–72	31–41		83	—	—	186	102	138
Strain to failure (%)	15–80	30–300	100–600	50	4	75	60	12	14	5
Fracture energy (kJ m^{-2})	—	—	—	6.6	—	3.4	3.7	3.9	1.9	—
CTE (10^{-6} °C^{-1})	144	117	146–180	47	49	56	56	36	—	35
T_g (°C)	57	69	−10	143	85	185	217	280	250	265
Maximum service temperature (°C)	110	120	150	250	240	160	267	230	225	300

Source: Adapted from Mallick, P.K., *Fiber-Reinforced Composites*, 3rd edn., CRC Press, Boca Raton, FL, 2008, p. 85.

polymer molecules arises due to the ability of chemical groups present in the chain to rotate around the single bonds. Elastomers with low cross-link density undergo elastic deformation very easily. The material becomes harder when the cross-link density is high. In order to have elastomer properties for a polymer, the attraction between chains should be low and the material should be above the T_g, where local segmental mobility occurs. These properties are found in many hydrocarbon-intense polymers. In the normal, unstretched state, these polymers are amorphous with a relatively high entropy or level of disorder. When stretched, they become crystalline with a greater degree of order.

Like plastics, elastomers can be classified into thermosetting and thermoplastic elastomers. Most of the commonly used elastomers are having long chains, which are chemically cross-linked during the curing process. Once formed, this type of thermosetting elastomers cannot be softened, melted, reshaped, or reprocessed by subsequent reheating. On the other hand, thermoplastic elastomers act similar to thermosetting material at RT but can be softened and melted at high temperature.

The elastomers are generally hydrocarbon-based polymers consisting of carbon and hydrogen atoms, although some may contain other elements. Some of the common elastomers are styrene–butadiene rubber (SBR), butyl rubber, polybutadiene rubber (BR), ethylene–propylene rubber, and polyisoprene rubber (IR) (both natural and synthetic). Such rubbers are called as general-purpose rubbers because they are used in high-volume rubber products, such as tires, belts, seals, and hoses.

The common characteristics of elastomers are good elasticity, flexibility, and toughness. Beyond these common characteristics, each elastomer has its own unique properties, and additives are often added to modify/improve those properties. The additives for rubber compounding can be categorized as vulcanizing or cross-linking agents, processing aids, fillers, antidegradants, plasticizers, and other specialty additives.

The important elastomers such as natural rubber (NR), SBR, and BR are unsaturated hydrocarbons. These elastomers are subjected to sulfur vulcanization to improve the physical properties. Besides sulfur, zinc oxide, stearic acid, and accelerator are added during vulcanization. Accelerators are mainly organic compounds. There are a large number of accelerators available, but the most popular belong to thiol or disulfide type. Zinc oxide activates the accelerators by forming zinc salts. Stearic acid helps to solubilize the zinc compounds. Additional ingredients are also added to the rubber compound, to impart certain specific properties to the rubber. Softeners and extruders help in the mastication and mixing of the compound. Antioxidants are added to prevent degradation, since the unsaturated rubbers can degrade rapidly in the presence of atmospheric oxygen. Amine- or phenol-type organic compounds are generally used as antioxidants. Reinforcing fillers such as carbon black or silica are added to improve strength and abrasion resistance. Non-reinforcing fillers (clay or chalk) are mainly used to increase stiffness and reduce cost.

3.1.5.1 Natural Rubber

NR is a linear polymer with repeat units being isoprene (C_5H_8). Its glass transition temperature is about −70°C and its specific gravity is 0.93 at 20°C. Due to its regular structure, it crystallizes when stretched or stored at temperatures below 20°C. The rate of crystallization depends on the temperature and type of NR. NR contains small amounts of fatty acids and proteinaceous residues, which promote sulfur curing. Peroxide-cured NRs are found to have lower strength than that obtained by sulfur curing but have other good properties such as lower compression set, lower creep, and excellent resistance to aging with the incorporation of suitable antioxidant. NR is commercially used for the production of tires, bumpers, and thin-walled, high strength products such as balloons and surgical gloves.

3.1.5.2 Styrene–Butadiene Rubber

SBR is the most widely used synthetic rubber with the largest volume of production. It constitutes more than 50% of the synthetic rubbers produced. It is a copolymer of styrene ($C_6H_5CH=CH_2$) and butadiene ($CH_2=CH–CH=CH_2$), and a typical SBR rubber contains about 23% styrene and 77% butadiene. SBR can be produced by emulsion and solution polymerization techniques, but almost 85%–90% of the world production is by emulsion polymerization. SBR is an amorphous polymer. It does not exhibit crystallization either on stretching or cooling and hence the strength values are poor. Most SBR compound formulations use a reasonable quantity of reinforcing filler, carbon black being the commonest for improving both tensile strength and tear resistance properties. Conventional sulfur systems are the usual cross-linking agents for SBR. Being an intrinsically slower curing material compared to NR, usually an accelerator is incorporated in its compounding formulations. It is not tacky as NR, but the abrasion resistance is better than NR. The T_g value is −45°C, which is ~30°C higher than NR. Hence, its low-temperature properties are poor. The percentage elongation at the elastic limit is only about 50% and there is more heat buildup with SBR. Although it is possible to produce an all-synthetic automobile tire with this rubber, this is not suitable for truck tires because the greater mass of trucks leads to an unacceptable degree of heat buildup. The tire industry consumes almost 70% of the total SBR produced, since SBR has suitable properties for this application, such as very good abrasion resistance, aging resistance, and low-temperature properties.

3.1.5.3 Polybutadiene Rubber

This is the next important rubber after SBR. Polybutadiene is the homopolymer of butadiene ($CH_2=CH–CH=CH_2$). It is produced by the polymerization of butadiene in solution using organometallic initiators. Three types of molecular structures are possible for this polymer; they are *cis*-1,4, *trans*-1,4, and atactic. Ziegler–Natta-type catalysts can produce a polymer with very high proportions of *cis*-1,4 structure. This is more suitable for tire industries. Lithium catalysts yield variable chain structures, depending on the solvent used. The molecular weight of this polymer is lower and is not used for tires. However, they are important in the production of block copolymers for thermoplastic elastomers.

Polybutadiene is widely used as blends with other rubbers, since it is difficult to process this rubber. It is generally used for automotive tires in combination with SBR and NR. Its low-temperature properties are very good because of its low T_g (−95°C). Percentage elongation of this rubber is better than NR, which results in lower heat buildup. The *cis*-isomer also has a higher abrasion resistance, which is an added advantage for better tire tread wear resistance. Some of the other advantages of BR are high-temperature aging resistance, ozone resistance, and its ability to accommodate large quantities of fillers and oil while retaining rubber properties. Poor tack and road grip and poor tear and tensile strength as compared to those of acrylonitrile butadiene rubber (NBR) and SBRs are the limitations of BR.

3.1.5.4 Synthetic Polyisoprene Rubber (IR)

Synthetic polyisoprene is produced by polymerization of isoprene. It has four possible geometric isomers: *cis*- and *trans*-1,4-polyisoprene, 1,2-vinyl, and 3,4-vinyl. NR is 100% *cis*-1,4-polyisoprene. To produce a polymer with high *cis*-1,4 structures, Ziegler–Natta type of initiators are used. Synthetic polyisoprene may have as much as 98% stereoregular isomer. Even though the difference in stereoregularity is small, IR is more easily crystallizable. IR compounds have a lower modulus and higher elongation to failure than similar NR compounds due to less strain-induced crystallization at high rates of deformation. The most important commercial application of IR is in the preparation of chlorinated and isomerized rubbers for the surface coating industry. IR is often used in the applications where NR is being used, because of the similar mixing and processing stages.

Polyisoprene crystallizes on stretching and that increases its gum strength. The strength is lower than NR because of slightly lower *cis*-1,4 isomer content. The production of polyisoprene is relatively less because of availability of NR in sufficient quantities.

3.1.5.5 Ethylene–Propylene (Diene) Rubber

Ethylene–propylene copolymers are made by Ziegler–Natta and metallocene polymerization. They are the commercial rubbers with the lowest density. These polymers cannot be vulcanized and are also not reactive to peroxide curing. To introduce unsaturated sites for cross-linking, a non-conjugated diene termonomer such as ethylidene norbornene, 1,4 hexadiene, or dicyclopentadiene is used, which produce the rubber known as ethylene–propylene (diene) rubber (EPDM). The main advantages of EPDM are as follows:

- Excellent aging and ozone resistance
- Better low-temperature flexibility compared to that of NR compounds
- Excellent resistance to chemicals
- Excellent electrical resistivity
- Much better thermal stability than that of SBR and nitrile rubber (NBR) compounds

EPDM has very low unsaturation but sufficient for sulfur vulcanization. It has better oxidation resistance because of low unsaturation. One of the double bonds of diene group is used for copolymerization and another double bond is used for sulfur vulcanization.

When the ethylene content is <60 wt.%, EPDM is an amorphous polymer and shows no crystallization on stretching. Hence, it has poor strength and carbon black is generally used to improve the strength. The polymer is semicrystalline at higher ethylene contents. This rubber has excellent aging resistance because of the absence of in-chain unsaturation. It also has good low-temperature properties. Its excellent aging and low-temperature properties make it an ideal material of use as a sheet rubber for roofing applications. Other applications of EPDM include sealing, gaskets, hoses, cable insulations, and other products that require heat and weather resistance.

3.1.5.6 Butyl Rubber (IIR)

Butyl rubber is a copolymer of isobutylene with a small amount of isoprene (2–3 wt.%), which provides sites for cross-linking. As in the case of EPDM, higher amount of accelerator is needed for sulfur vulcanization. Since butyl rubber is largely a saturated elastomer (no double bonds), it has excellent resistance to ozone, oxygen, chemicals, and heat. Another outstanding characteristic of this rubber is its low gas permeability and hence widely used as inner tubes in tires and barrier in tubeless tires. This elastomer has sufficient chain regularity to permit crystallization on stretching. However, the modulus of this elastomer is low and carbon black is generally added to improve modulus. Although it has a T_g of $-72°C$, it is not useful for low-temperature applications because of the broad loss peak. The hysteresis loss is high and therefore it is a useful rubber for damping applications. Other industrial applications of butyl rubber are in cable insulations and jacketing, roof membranes, and pharmaceutical stoppers.

3.1.5.7 Nitrile Rubber

NBR is a copolymer of butadiene and acrylonitrile, which is prepared by emulsion polymerization technique. This is one of the most solvent-resistant synthetic elastomers. It is a polar rubber and has excellent resistance to nonpolar or weakly polar materials, such as hydrocarbon oils, fuels, and greases. However, the solvent resistance is not as high in the case of polar solvents.

NBR requires carbon black to improve strength. The low-temperature properties of this elastomer are poor because of its high T_g ($-20°C$ to $-40°C$). Hence, there should be a compromise on the low-temperature properties, when the solvent resistance is very important. Hydrogenated NBR is a more recent development in which the unsaturation of the polymer is significantly reduced, leading to better aging properties.

Acrylonitrile content in various grades of NBR varies from 20 to 50 wt.%. The properties of NBR are significantly affected by the acrylonitrile content. With increasing acrylonitrile content, the oil resistance, heat resistance, and cure rate increase, and processability becomes easier. Since NBR lacks mechanical properties, it is very important to incorporate reinforcing fillers to obtain compounds with a reasonable level of tensile strength, tear strength, and abrasion resistance. Curing of NBR is usually achieved by using an accelerated sulfur curing system; however, peroxide curing may also be used in particular instances. Sulfur has a lower solubility level in NBR than NR, which inhibits a uniform distribution in NBR matrix, leading to the formation of over- and under-cured regions. NBR has many advantages over many types of elastomers, such as excellent resistance to hydrocarbons, very good heat resistance (in the absence of air), very low level of gas permeability, moderate tensile strength, tear strength and ozone resistance, and good low-temperature flexibility.

3.1.5.8 Chloroprene Rubber

It has predominantly *trans*-1,4 configuration. Even though its solvent resistance is not comparable to NBR, it has many other advantages. It exhibits high strength without carbon black, because of the possibility of crystallization of *trans*-1,4 chloroprene on stretching. Hence, like NR latex, chloroprene latex is used in many applications, such as rubber gloves, coating, and impregnating. The trade name of this rubber produced by DuPont is Neoprene.

Some of the chlorine atoms present in the polymer chain are labile and reactive and provide sites for cross-linking. However, it needs a different vulcanizing agent. Metal oxides such as magnesium and zinc oxides are generally used as vulcanizing agents. Some amount of sulfur is also included to control the rate of vulcanization.

In addition to its excellent solvent resistance, chloroprene rubber is more resistant to oxidation or ozone attack than NR. It also has better flame resistance. The RT properties of this rubber are good, but the low-temperature properties are poor.

3.1.5.9 Silicone Rubber

Unlike the common elastomers that have carbon chain backbones, silicone rubbers contain a very flexible siloxane backbone. Silicon rubbers (polysiloxanes) consist of –Si–O– repeat units, with the organic side groups attached to the silicon atoms. The silicone chain is more flexible because the Si–O–Si bonds can rotate more freely than the C–C bond. Moreover, the Si–O–Si bond is stable at high temperatures.

Vulcanization of these elastomers is carried out by using peroxides. The peroxides can cross-link the chain by abstracting hydrogen atoms from the methyl groups. The resulting free radicals can join and form the cross-link. Some varieties of polysiloxanes contain vinyl units, which permit sulfur vulcanization.

Silicones are soft and weak rubbers. Silica fillers are generally added to improve strength, since carbon black does not work well. Even after the addition of silica fillers, the strength values are lower than the other vulcanized rubbers. Their resistance to stiffening at very low temperatures and softening at elevated temperatures is good. However, these materials may degrade at high temperature when they are exposed continuously for longer durations.

Despite the high price, silicone rubbers are very important elastomers with many industrial applications as a result of their good thermal stability, excellent resistance to ozone, oxygen and UV radiation, and good electrical insulation properties. They also show reliable and consistent properties over a wide temperature range. However, they have certain limitations, such as the low tensile strength, high gas permeability, and poor resistance to hydrocarbon oil and solvent. On account of their excellent properties, silicone rubbers are used in shaft sealing, spark plug caps, O-rings, embossing rollers and gaskets, cables, corona-resistant insulating tubing, and window and door profile seals.

3.1.5.10 Fluorocarbon Elastomers

They are the most resistant elastomers to heat, chemicals, and solvents, but are very expensive. The most common types are the copolymers of vinylidene fluoride and hexafluoropropene. The fluorine atoms impart chemical inertness to these polymers. Some hydrogen atoms must be present in the chain to maintain elastomer properties. Diamines are used for vulcanization to form cross-links by reacting with the fluorine atoms. Strength improvement can be achieved by using carbon black.

These elastomers are very useful for high-temperature applications, where the mechanical properties are not very important. These elastomers have long life at 200°C and can be heated to 315°C up to 48 h. However, their low-temperature properties are poor and they become brittle at −30°C.

3.1.5.11 Polyurethane

They are formed by the reaction of hydroxyl groups present in the low molecular weight polyesters or polyethers with organic diisocyanates. Excess diisocyanate leads to cross-linking because the diisocyanate groups can also react with the hydrogen atoms of the –NH– groups in the chains.

Polyurethane foam rubber can be made by adding water to the mixture. The isocyanate groups react vigorously with water and release carbon dioxide gas. The amine groups thus formed can also react vigorously with the isocyanate groups to continue the chain extension and cross-linking. Hence, there are simultaneous foaming, polymerization, and cross-linking reactions during the formation of foam elastomers.

Polyurethane elastomers generally exhibit good resilience and low-temperature properties. It has excellent abrasion resistance, but the solvent resistance is moderate. Its hydrolytic stability and high-temperature resistance are poor. As a castable rubber, polyurethane is used in a variety of applications, for example, footwear, toys, solid tires, and foam rubber.

3.1.5.12 Sulfur Vulcanization

Vulcanization using sulfur is a complex reaction and involves activators for the breakage of the sulfur ring (S_8) and accelerators for the formation of sulfur intermediates, which facilitate sulfur-to-double bond linkage. Vulcanization by sulfur without any accelerators takes several hours and it is not a commercially viable process. By using accelerators, the vulcanization time can be decreased to as little as 2.5 min. The most widely used vulcanization technique in various industrial applications is the accelerated sulfur curing method, since it provides better physical properties and a considerably fast cross-linking rate and it has the capability to provide delayed actions needed for processing, shaping, and forming before the formation of the vulcanized network. According to the sulfur content and the ratio of accelerator-to-sulfur content, sulfur vulcanization systems are classified as conventional, semi-efficient, and efficient.

3.1.5.13 Peroxide Vulcanization

The production of free radicals is the driving force for peroxide-based vulcanization. Radicals are atoms or molecular fragments with unpaired electrons. These radicals cause an unstable situation and these radicals readily react with reactive groups in the medium. Rubber cross-linking reaction using peroxide consists of three basic steps as follows:

1. *Homolytic cleavage*: When peroxide is heated above its decomposition temperature, the oxygen–oxygen bond ruptures. The resultant molecular fragments are called radicals, which are highly energetic, reactive species.

$$ROOR' \rightarrow RO\bullet + R'O\bullet$$

2. *Hydrogen abstraction*: Hydrogen abstraction is a process where the radical removes a hydrogen atom from a nearby polymer chain, and thus the radicals are transferred from peroxide molecular fragments to the elastomer backbone. Radicals formed from the peroxide decomposition are reactive toward hydrogen atoms in polymer chain.

$$RO\bullet + P–H \rightarrow ROH + P\bullet$$

where, P is the polymer chain.

3. *Radical coupling*: When two elastomer radials come in contact, the unpaired electrons will couple and form a covalent bond or cross-link between the elastomer chains.

$$P\bullet + P\bullet \rightarrow P–P \text{ (cross-link)}$$

The complexity of peroxide curing arises from a range of possible reactions such as β-cleavage of oxy radical, addition reaction, polymer scission, radical transfer, dehydrohalogenation, oxygenation, and acid-catalyzed decomposition of the peroxide.

Rubber compounds are usually a complex mixture of many additives (curatives, oils, fillers, anti-degradants, etc.) and each of these additives is added to impart a specific property to the final rubber compound. These additives can also affect the peroxide cross-linking reaction because of the ability of the radicals to react with these additives.

3.1.6 Polymer Matrix Materials: Selection Criteria

The first and most important consideration in selecting a polymeric matrix is the service temperature required for the component. The glass transition temperature is a good indicator of the temperature capability of a polymer. A good rule of thumb is to select a polymer, which has T_g of at least 25°C higher than the maximum service temperature of the part. Most of the polymeric materials absorb moisture, which lowers the T_g of the polymer. Hence, it is preferable to select a polymer with T_g of 50°C higher than the maximum service temperature. Generally, the high-T_g thermoset polymers are highly brittle and less damage tolerant. Toughening agents can be added, but they will increase the cost and reduce the T_g. High-temperature polymers are more expensive and more difficult to process.

The mechanical properties required for the component is the next important selection criterion. For the composites used in RT, the selection of matrix is mainly based on its mechanical properties. A high-modulus matrix is desirable for better compressive properties, whereas a high tensile strength matrix is suitable to control intraply cracking in the composites. A matrix material with better fracture toughness can have better control on crack propagation and delamination.

Some polymers wet the fibers easily and adhere to the fibers better than others and form better chemical and/or mechanical bond that increases the matrix-to-fiber load transfer capability. Resin-rich regions and brittle resin systems are prone to microcracking especially when the processing temperature is high, since cooling from high temperatures leads to a very large difference in contraction between the fibers and the matrix. Again, toughening the resin can help in preventing microcracking but often at the expense of elevated temperature performance.

Based on the processing conditions, also the polymer matrix is selected. A long pot life is desirable for processes that use neat resin, such as filament winding, resin transfer molding, and pultrusion. Frequent change of resin bath is necessary when a short-pot life resin is used and that may lead to a significant wastage of resin.

It is necessary to purchase the resin in large quantities to reduce the cost and maintain uniform quality. However, the shelf life is the deciding factor. Two-part thermoset resin systems can be stored up to 2 years at RT in airtight containers. Viscosity and chemical changes occur over a period time, and refrigeration slows down chemical reactions and extends the shelf life. Thermoset prepregs are generally stored in freezer and have shelf life of 6–12 months.

Viscosity requirements depend on the process but, generally, the lower the viscosity the easier it is to process and the better the wettability. For wet processing using thermoset resins, viscosities less than 0.1 Pa s (1000 cP) are preferred. The viscosity of resin decreases with increasing temperature, but it increases once the curing reaction starts. A thermoset resin is considered as gelled when the viscosity reaches to 100 Pa s (100,000 cP). The viscosity of resin can be modified by using nonreactive solvents, but they may affect the final properties of the component.

The production rate is decided by the cure time of the thermoset resin. Generally, high-T_g resins require longer cure times. In some cases, post-cure is required to achieve the required properties. A post-cure may not be required in most of the cases. The elimination of post-cure is a way to reduce the processing cost. Very short cure times are required for some processes, such as compression molding and pultrusion.

Other considerations of polymer matrix for composite are moisture and solvent resistance. That is, the polymer should not dissolve, swell, crack, or degrade in wet or hot–wet environment or when exposed to solvents. Some common solvents used in aircrafts are jet fuels, de-icing fluids, and paint-strippers. Similarly, petroleum fuels, motor oil, and antifreeze are common solvents used in automobiles.

3.2 Metallic Matrix Materials

Metallic matrix materials are preferred over polymer matrix materials in applications where severe environmental conditions such as high temperature are encountered. In addition to that, the metallic materials have higher strength and modulus, which are very important under transverse loading as well as compressive loading conditions. The ductility and toughness of metals are also better than the polymeric materials, particularly than the thermoset polymers.

The attractive properties of metals are mainly due to the bonding and crystal structure of metals. They generally have relatively large atomic packing factors to maximize the shielding provided by the free electron cloud. The atomic bonding in metals is metallic bonding, which is a nondirectional bonding. Consequently, there are minimal restrictions to the number and position of nearest-neighbor atoms. Hence, each metal atom has relatively large numbers of neighbors and the atomic packing is very dense; most of the metals have close-packed crystal structures, such as face-centered cubic (FCC) or hexagonal close-packed (HCP) structures.

In the FCC structure, the atoms are located at each of the corners and the centers of all the cube faces. Some of the common metals having this crystal structure are copper, aluminum, silver, and gold. The atomic packing factor for this structure is 0.74, which is the maximum packing possible for spheres having the same diameter.

Another common crystal structure found in metallic materials is body-centered cubic (BCC) structure with atoms located at eight corners of cube and a single atom at the cube center. The atomic packing factor for this structure is 0.68. Iron at RT has this structure. Other examples of metals having this structure are chromium, molybdenum, tantalum, and tungsten.

Not all metals have unit cell with cubic symmetry, and some metals have hexagonal unit cell. The top and bottom faces of the hexagonal unit cell consist of six atoms that form regular hexagon around a single atom at the center. Another plane that consists of three atoms is situated between the top and bottom planes. The atomic packing factor for this HCP structure is also 0.74. Metals having HCP structure include cadmium, magnesium, titanium, and zinc.

In general metals are ductile, and the ductility comes from the extensive plastic deformation they can undergo by slip and twinning as illustrated in Figure 3.20. Slip is a shear deformation that moves atoms in a particular set of planes by many interatomic distances. Steps are created at the surface of the crystal due to slip, but the crystal structure of the material remains the same before and after slip. The steps produced on the surface due to slip can be observed through a microscope as a group of parallel lines called slip lines. On the other hand, twinning changes the orientation of planes in the twinned part of crystal with respect to the untwined parts. The atoms move only a fraction of an interatomic distance during twinning. The deformation is predominantly by slip in many crystalline materials at ambient and

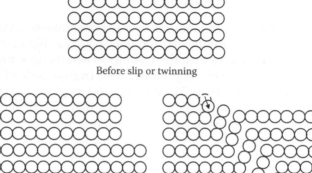

FIGURE 3.20
The slip and twinning modes of plastic deformation. (Reprinted from Raghavan, V., *Materials Science and Engineering*, 4th edn., Prentice Hall of India, New Delhi, India, 1998, p. 250. With permission.)

elevated temperatures, whereas the mode of deformation changes over to twinning at low temperatures.

The slip planes are the closest packed planes in the crystal and the directions along which slip occurs are the closest packed directions. In any crystal, the planes with the greatest atomic density are the most widely spaced planes. In the closest packed direction, the translation distance for the atoms from one minimum energy position to the next is the smallest. These are the favorable conditions for slip to take place; hence, slip predominantly occurs in the close-packed plane and in the close-packed direction. The slip plane and the slip direction that lie on it together constitute a slip system.

In metallic crystals, the slip system depends on the crystal structure. For metals having FCC structure, the {111} octahedral planes and the <110> directions constitute the slip system. There are eight {111} planes in the FCC unit cell, but the planes at opposite faces of the octahedron are parallel to each other. Hence, the number of effective slip planes is considered as four. Each (111) plane contains three <110> close-packed directions. In the family of directions, the direction that lies on a particular plane satisfies the following relation:

$$hu + kv + lw = 0$$

where
    h, k, and l are the Miller indices of the plane
    u, v, and w are the Miller indices of the direction

From this relation, it can be easily found that the combination (111) and [$\bar{1}$10] forms a slip system, but (111) and [110] will not. Therefore, the FCC lattice has 12 possible slip systems.

In the metals having HCP structure, the plane with high atomic density is only the basal plane, that is, (0001) plane. The <1120> directions are the close-packed directions. Since there is only one basal plane per unit cell and three <1120> directions lie in the plane, the HCP structure has only three slip systems. The limited number of slip systems in HCP crystals is responsible for their low ductility and extreme orientation-dependent properties. As mentioned earlier, the BCC structure is not a close-packed structure like the FCC or HCP structure. Accordingly, there is no plane in BCC structure that has similar atomic density as the (111) plane in the FCC structure and (0001) plane in the HCP structure. The {110} planes have the highest atomic density, but that is not greatly superior to atomic densities in other planes. However, the <111> direction in the BCC structure is as close packed as the <110> and <1120> directions in the FCC and HCP structures, respectively. Slip in BCC metals is found to occur on the {110}, {112}, and {123} planes and in the <111> direction. Although there are 48 possible slip systems in BCC structure, higher shearing stresses are usually required to cause slip in this structure, since the planes are not as close packed as in the FCC structure. Slip lines in BCC metals have a wavy appearance, since slip occurs simultaneously on several planes in the <111> direction.

The estimated shear stress necessary to cause slip in a perfect crystal is approximately one-sixth of the shear modulus. However, the real crystals deform at a much lower stress than that predicted by the model for a perfect crystal. This discrepancy is mainly due to the presence of dislocations in real crystals.

The dislocations are the line imperfections in crystals, that is, they are 1-D imperfection in the geometrical sense. There are two types of dislocations and they are edge dislocation and screw dislocation. Consider a perfect crystal, which is made up of vertical planes parallel to one another, and each plane extends from the top to the bottom of the crystal (Figure 3.21). If one of these vertical planes ends partway within the crystal, the end of this plane is a line defect called dislocation. In the perfect crystal, all the atoms are in equilibrium positions and the bond lengths between atoms are equal. On the other hand, in a crystal containing a dislocation, the atoms are squeezed together and are in a state of compression just above the edge of the incomplete plane. The bond lengths between the atoms in this region will be smaller than the equilibrium value. Just below the edge, the atoms are pulled apart and they are in a state of tension. The bond lengths between the atoms in this region are increased to above normal value. This distorted configuration is present all along the edge. Since the region of maximum distortion is centered around the edge of the incomplete plane, this defect is considered as line imperfection.

Another type of dislocation called screw dislocation is formed by a shear stress that is applied to shift the upper front region of the crystal by one

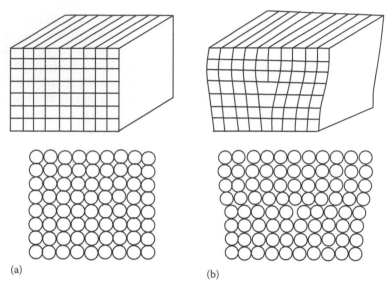

FIGURE 3.21
Planar and atomic arrangement in (a) a perfect crystal and (b) a crystal with a dislocation. (Reprinted from Raghavan, V., *Materials Science and Engineering*, 4th edn., Prentice Hall of India, New Delhi, India, 1998, p. 122. With permission.)

atomic distance to the right relative to the bottom portion (Figure 3.22). The atomic distortion associated with a screw dislocation is also linear along a line *AB* in the figure. The screw dislocation looks like a screw and the central axis of the screw is the dislocation line. Unlike edge dislocation, there is no extra plane at the screw dislocation. The atomic bonds in the region surrounding the dislocation line have undergone a shear distortion. The vertical atomic planes parallel to the side faces become discontinuous in the distorted region. The screw dislocation can also be compared to a spiral staircase, whose central pillar coincides with the screw dislocation line.

Dislocations are usually present in the real crystals and they are formed during the growth of the crystals from the melt or during prior mechanical deformation of the crystals. The density of dislocations is generally measured by counting the number of points at which they intersect a cross section of the crystal. These points can be observed as pits under a microscope after chemical etching the polished surface. The dislocation density is in the range of 10^8–10^{10} m^{-2} in an annealed crystal.

When a shear stress is applied to a crystal parallel to the slip plane, the dislocations on it will move. As they move through the crystal, each one of them causes a displacement of the part containing the extra plane with respect to the other part. In the presence of dislocations, the shear need not occur by the simultaneous displacement of atoms in the slip plane. The dislocation moves in a sequential fashion by occupying successive positions during movement. During each step of the motion, the atoms at the dislocation line are displaced

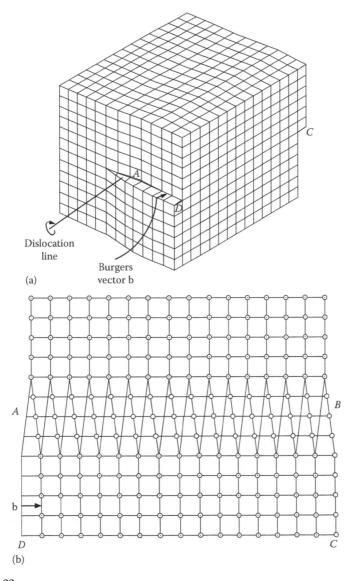

(a)

Dislocation
line

Burgers
vector b

(b)

FIGURE 3.22
(a) Planar and (b) atomic arrangement in a crystal with a screw dislocation. (Reprinted from Callister, W.D. Jr., *Materials Science and Engineering*, John Wiley & Sons, New York, 2007, p. 124. With permission.)

only a fraction of an interatomic distance, as illustrated in Figure 3.23. Hence, the incomplete plane of atoms is not bodily shifted, but the dislocation moves by small adjustments in bonding in the dislocation region. The movement of dislocation is analogous to the motion of a caterpillar. Thus, less stress is needed to cause plastic deformation in real crystals than that estimated for a

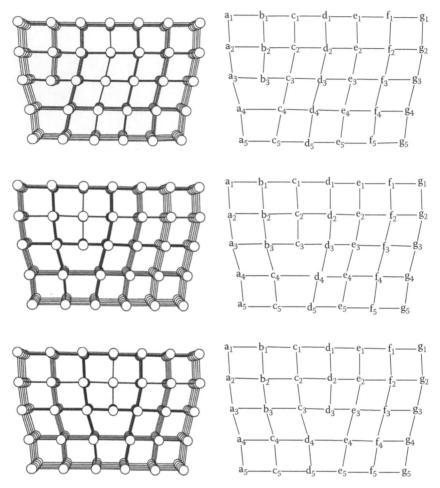

FIGURE 3.23
Successive movement of a dislocation in response to an externally applied shear stress. (Reprinted from Raghavan, V., *Materials Science and Engineering*, 4th edn., Prentice Hall of India, New Delhi, India, 1998, p. 130. With permission.)

perfect crystal. Since the dislocations are always present in bulk crystals, the yield strength values of metallic materials are found to be low. Only in special type of crystals, such as whiskers, the dislocations are absent. They show very high values of yield stress. For example, a copper whisker can withstand a stress of 700 MPa, but on the introduction of dislocations, this stress falls drastically to much lower values. Thus, the presence of relatively large number of slip systems with dislocations accounts for the better ductility of metallic materials.

A variety of metallic materials has been used as matrices in metal matrix composites (MMCs). A brief description of the commonly used metallic matrix materials is given in the following sections.

3.2.1 Aluminum Alloys

Pure aluminum is usually not considered as a matrix material because the properties are not attractive for MMC applications. Aluminum alloys find important applications in the aerospace sector because of their low density and excellent strength, toughness, and corrosion resistance. Three classes of aluminum alloys have been used for the fabrication of MMCs:

- Commercial aluminum alloys
- Aluminum–lithium alloys
- High-temperature aluminum alloys

Among these three classes of alloys, the maximum number of studies is carried out on developing MMCs with commercial aluminum alloys, because these alloys offer good properties and they are well understood. Aluminum–lithium alloys have considerable potential because of their low density and higher modulus. However, only a few studies are available on these alloy-based MMCs and it might be due to the early stage of development of these alloys. Since the high-temperature properties of commercially available aluminum alloys are limited, alloys based on Al–Fe–V–Si and Al–Sc have been used to improve the high-temperature capability of MMCs.

Commercial aluminum alloys can be further classified into wrought heat-treatable alloys, wrought non-heat-treatable alloys, and cast alloys. Wrought heat-treatable alloys include Al–Mg–Si (6000 series), Al–Cu–Mg (2000 series), and Al–Zn–Mg (7000 series) alloys. These alloys are subjected to heat treatment for strength improvement through precipitation hardening. A wide range of alloys with different strength and ductility is available in this category. These alloys have been used extensively in aerospace and other structures. Composites with improved properties have been made extensively with these alloys. Wrought non-heat-treatable alloys include Al–Mg and Al–Mn alloys. Strengthening in these alloys is achieved through solid solution strengthening and cold working. Even after the strength improvement by these mechanisms, the final strength values do not meet the requirement for aerospace applications. However, these alloys are widely used in automobile applications. These alloys are not generally preferred for the fabrication of MMCs, because of their low strength values. Casting alloys are mainly based on Al–Si, Al–Cu, and Al–Mg alloy systems. The strength and ductility of Al–Cu alloys are better than the other cast aluminum alloys, but the castability of Al–Cu alloys is not as good as Al–Si alloys. Generally, these alloys have moderate strength and ductility, and they are preferred for the fabrications of components with complex geometries. These alloys are also extensively used in MMCs.

Low-density, high-modulus aluminum alloys are based on Al–Li system. Unlike the other alloying elements, the Li addition improves modulus and at the same time decreases the density. These alloys are another important category of precipitation-hardenable alloys. Al–Li alloys generally contain some amount of Cu, Zr, and Mg besides Li. Li has high solid solubility in aluminum (4 wt.% at 610°C) and this alloy undergoes precipitation hardening by the formation of ordered, coherent δ'-Al$_3$Li precipitate. Since the binary Al–Li alloys have low ductility and fracture toughness, significant work has been carried out to develop ternary and quaternary alloys. The quaternary alloys based on Al–Li–Cu–Mg (e.g., 8090) have a good combination of strength, ductility, and toughness. A small amount of zirconium is added to these alloys, which further improves strength and toughness. However, the use of Al–Li alloys or matrices for MMCs is limited because of severe oxidation of Al–Li alloy powders during the fabrication of components by P/M technique and difficulty in casting with SiC.

The commercial aluminum alloys and Al–Li alloys are not useful at temperature above 150°C due to rapid coarsening of precipitates, which leads to loss of mechanical properties. A considerable amount of work has been carried out to develop high-temperature aluminum alloys for use up to 315°C. One such high-temperature aluminum alloy developed was Al–Sc alloy. MMCs based on this alloy have been prepared by P/M technique (Unal and Kainer 1998, Pandey et al. 2001). This alloy also has the potential for making MMCs by casting techniques.

3.2.2 Titanium Alloys

Titanium alloys are very attractive as matrices for MMCs due to their higher strength and high-temperature mechanical properties compared to aluminum alloys. Titanium is another important aerospace material. At supersonic speeds, the skin of aircraft heats up so much and aluminum alloys are not suitable for this kind of temperatures. Titanium has a relatively high melting point (1672°C) and it retains its strength at high temperatures with good oxidation resistance and hence titanium alloys can be used for this application. Titanium has a density of 4.5 g cm^{-3} and a Young's modulus of 115 GPa. For titanium alloys, the density varies between 4.3 and 5.1 g cm^{-3}, while the modulus value lies between 80 and 130 GPa. The corrosion resistance of titanium and its alloys is also very good at ambient temperature as well as at high temperatures. At present, the turbine and compressor blades of jet engines and fuselage parts are made with titanium alloys. They are, however, expensive materials compared to aluminum alloys.

Titanium exists in two allotropic forms, namely, α and β; the α-Ti has a HCP structure, which is stable below 885°C, and β-Ti has a BCC structure, which is stable above 885°C. Different alloying elements alter the transformation temperature. Aluminum is an α stabilizer, that is, it raises the $\alpha \rightarrow \beta$ transformation temperature. Many alloying elements lower the $\alpha \rightarrow \beta$ transformation temperature, that is, they stabilize the β phase. Some of the β stabilizers are Fe, Mn, Cr, Mo, V, Nb, and Ta. By varying the type and amount of alloying elements, three types of titanium alloys with α, β, or $\alpha + \beta$ structure can be produced. Titanium alloys such as Ti–6Al–4V and Ti–15V–3Cr–3Sn–3Al have been used commonly in MMCs. These alloys have good ductility at RT. Ti–6Al–4V alloy belongs to the $\alpha + \beta$ group. Generally, hot working in the $\alpha + \beta$ region is carried out for this alloy to break the structure and distribute the α phase in an extremely fine scale.

Titanium has great affinity for oxygen, nitrogen, and hydrogen. Such interstitial elements even at ppm levels can adversely affect the mechanical properties, especially brittleness is increased. Hence, the welding of titanium and its alloys requires protection from the atmosphere. Electron beam welding, which is carried out under high vacuum, is a best option for the welding of titanium alloys.

There has been considerable interest in niobium-rich Ti–Al alloys because of their good creep resistance, tensile strength, ductility, and high-temperature fatigue resistance. A new β-alloy, Ti–4.3Fe–7Mo–1.4Al–1.4V, has been developed for MMCs suitable for hot working (Saito 1995).

3.2.3 Magnesium Alloys

Magnesium is one of the lightest metals and its density is 1.74 g cm^{-3}. Magnesium and its alloys form another group of very light materials. Magnesium alloys are used in aircraft gearbox housings, chain saw housings, electronic equipment, etc. Magnesium, being a HCP metal, is difficult to cold work.

3.2.4 Copper

Copper has FCC structure and it is very easy to cold work. It is one of the best electrical conductors and also has good thermal conductivity. It can be cast and worked easily. One of the major applications of copper-based composites is niobium–copper superconductors.

3.2.5 Intermetallic Compounds

Intermetallic compounds are the compounds formed between two or more metals. The most promising intermetallics for high-temperature applications include the aluminides of nickel, titanium, and iron. Titanium aluminides have been considered for some specific applications. Many studies have been conducted to exploit the high-temperature properties of ordered Ti_3Al and TiAl (γ) intermetallics. However, their service temperature is limited to well below 1600°C. Besides high strength, creep resistance, and fracture toughness, the materials used at 1600°C should have good oxidation resistance and microstructural stability. Many silicide intermetallics have the potential to meet the requirements above 1600°C. Ti_5Si_3 and $MoSi_2$ are the most promising silicides, in which Ti_5Si_3 has lower density and $MoSi_2$ has superior oxidation resistance.

Intermetallic compounds can be ordered or disordered. Ordered inter-metallic compounds possess structures characterized by long-range order-ing. Dislocations in ordered intermetallics are much more restricted than in disordered compounds. This results in retention of strength at elevated temperatures. One of the important disordered intermetallic compounds is molybdenum disilicide. It has a high melting point and shows good stability at temperatures greater than 1200°C in oxidizing atmosphere. This intermetallic material is primarily used as heating elements in high-temperature furnaces. Its oxidation resistance at high temperature is excel-lent because a continuous film of silica readily forms over the surface that prevents further oxidation. It is a rather brittle material with moderate strength. The best high-temperature properties are achieved when the material is very pure and free of oxygen. The impurities can form a grain boundary glassy phase, which affects the high-temperature properties. However, the best RT strength is obtained by using glass-phase forming additives such as B_2O_3 or Na_2O, which enable low-temperature sintering with restricted grain growth.

Most of the intermetallics show brittle behavior at lower temperatures (below 800°C) and ductile behavior at high temperatures. Hence, the tough-ness is low for these materials at RT, while the creep rate is high at high temperatures. There have been continuous attempts to improve the ductility of intermetallics. Rapid solidification is one of the methods. Another method of improving ductility is by adding suitable elements. For example, the duc-tility of Ni_3Al is significantly improved by adding boron. With extremely small amounts of boron (0.06 wt.%), the ductility increases from about 2% to about 50%. The long-range order also has significant influence on diffu-sion-controlled phenomena, such as recovery, recrystallization, and grain growth. The activation energy for these processes is increased and hence these processes are slowed down. Thus, ordered intermetallic compounds tend to exhibit high creep resistance.

These materials can be a better choice as a matrix for high-temperature com-posites because of their good oxidation resistance and high-temperature ductility. Making composites with intermetallic matrix materials is a potential possibility to enhance the toughness. The creep resistance and toughness of these materials can be improved by adding particulate, whisker, or fiber reinforcement.

Although a variety of metallic materials has been used as matrices in MMCs, the major emphasis is being given for the development of lighter MMCs using aluminum and titanium alloys. Once the composites are made with these light materials, the resulting composites will have high specific modulus and strength. The choice of matrix material in general depends on the strength, thermal stability, density, and cost. For example, titanium alloys appear to be a suitable material for very high-strength and moder-ate-temperature applications. However, aluminum alloys can compete with titanium alloys, because of their weight and cost advantages even through the strength and temperature capability of aluminum alloys are lower.

TABLE 3.9

Typical Properties of Some Common Metallic Matrix Materials

Matrix	Density (g cm^{-3})	Tensile Modulus (GPa)	Tensile Strength (MPa)	Strain to Failure (%)	CTE (10^{-6} °C^{-1})	Melting Point (°C)
Al alloys						
2024-T6	2.78	70	580	11	23.2	
6061-T6	2.70	70	310	17	23.6	
7075-T6	2.80	70	572	11	23.6	
8009	2.92	88	448	17	23.5	
380 (as cast)	2.71	70	330	4	—	540
Ti alloy						
Ti-6Al-4V	4.43	110	1170	8	9.5	1650
Mg alloy						
AZ91A	1.81	45	235	3	26	650
Zinc–aluminum alloy						
ZA-27	5	78	425	3	26	375
Ni alloys	8.8	200	460	47	13.3	1450
Cu alloys	8.9	100	200–500	5–60	18	1080
Steel	7.85	207	400–500	15–30	11.7	1500
Intermetallics						
FeAl	5.6	263	550[a]	0.7[a]	17	1400
NiAl	5.9	190	250	0.3	13.2	1680
TiAl	3.9	94	538	0.3	—	1460
MoSi$_2$	6.3	430	325[b]	—	8.5	2030

Source: Data from Mallick, P.K., *Fiber-Reinforced Composites*, 3rd edn., CRC Press, Boca Raton, FL, 2008, p. 525.

[a] Furnace cool.
[b] Bend strength of HIPed sample.

Other factors such as ductility, fracture toughness, and fatigue resistance become more important depending on the final application. In some special applications, heavy metals and alloys are also used. The properties of some metallic matrix materials are given in Table 3.9.

3.3 Ceramic Matrix Materials

Ceramic materials form another class of materials with high modulus, high hardness, good chemical resistance, high melting points, and a capability to retain mechanical properties at high temperature. They exist as both crystalline and noncrystalline materials. Glasses are noncrystalline, while ceramics other than glasses are crystalline. Elemental materials such as carbon and boron are also sometimes considered as ceramics, because they have similar properties as ceramics, especially the high-temperature properties. The applications of carbon matrix-based composites are vast expanding and they are considered as strategic materials. Because of their growing importance, a separate chapter in this book is devoted for the carbon–carbon composites.

Ceramic materials are generally compounds formed between one or more metals and a nonmetal such as oxygen, carbon, or nitrogen. The attractive properties of ceramics are mainly due to the bonding nature. Most of the ceramics have predominantly ionic bonding and some have predominantly covalent bonding. However, none of them will have 100% ionic bonding or covalent bonding. The ceramics are actually iono-covalent compounds. Based on the major type of bonding, they are classified as either ionic compounds or covalent compounds. These bonding types are very strong compared to metallic and van der Waals bonding. The high modulus values and high melting points as well as the highly brittle nature of ceramic materials are due to the strong bonding.

Crystalline ceramics exhibit close-packed cubic or hexagonal structures. The major difference between metallic crystal structure and ceramic crystal structure is that the metal cations usually occupy the interstitial positions formed by the anions in ceramics. Any 3-D array of spheres forms voids between the spheres and they are called interstitial voids. Two types of interstitial voids are present in close-packed structures. They are tetrahedral and octahedral voids. A tetrahedral void is formed between three spheres in a plane and a fourth sphere on an adjacent plane, which occupies the valley formed by the three spheres. The centers of these four spheres form a tetrahedron and the void within the tetrahedron is called tetrahedral void. Two tetrahedral voids can form for every sphere in a 3-D array. For example, the effective number of atoms in an FCC unit cell is four and there are eight tetrahedral voids. An octahedral void is formed by three spheres in a plane with three more spheres in the adjacent plane. A tilted view of the six spheres looks like a square base with one sphere above and below in the valley formed by four spheres. It forms an octahedron, a shape with six corners and eight faces. The void at the center of this octahedron is octahedral void. Corresponding to every sphere in a 3-D array, there is one octahedral void. Hence, for the FCC unit cell, there are four octahedral voids. The size of tetrahedral voids formed by sphere of radius r is $0.225r$ and the size of octahedral voids is $0.414r$. Depending on the size of cations, they occupy either octahedral or tetrahedral voids formed by anions. When the cation valency is not more than two or at best three and the radius ratio is in the range of 0.414–0.732, the crystal structure can be described as an FCC or HCP packing of anions with cations in all or part of the octahedral voids. The fraction of octahedral voids occupied by cations depends on the cation to anion ratio according to the chemical formula.

The structure can be viewed as interpenetrating arrays. One of the common structures in ceramic materials is rock salt structure, in which oxygen ions (anions) make the proper FCC structure with the metal ions (cations) in the interstitial positions. A closer look at the ionic arrangement can reveal that both the ions form FCC structure on their own ions and they are interpenetrated. Ceramic materials having this structure include MgO, CaO, FeO, NiO, MnO, and BaO. Other variations of close-packed cubic structures are zinc blend (ZnS) and fluorite (CaF) structures. Ceramic materials having zinc blend structure are ZnO and β-SiC and fluorite structures are ThO_2, ZrO_2, and CeO_2. The hexagonal structure is also observed in ceramics, such as corundum (Al_2O_3) and α-SiC. Another polymorphic form of ZnS also crystallizes in hexagonal form.

Glass–ceramic materials are another important category of ceramics, in which 95%–98% by volume is crystalline phase and the rest is glassy phase. The grain size of crystalline phase is very small, often less than 1 μm. Such a fine-grain size is obtained by adding TiO_2 or ZrO_2 nucleating agents during melting. Important glass–ceramic materials are lithium aluminosilicate (LAS) and magnesium aluminosilicate (MAS).

Glasses are the amorphous inorganic materials, mainly based on silicates. These materials do not crystallize when cooled from the liquid state. The specific volume versus temperature curve is similar to amorphous organic polymers (Figure 3.24), and they also have a characteristic glass transition

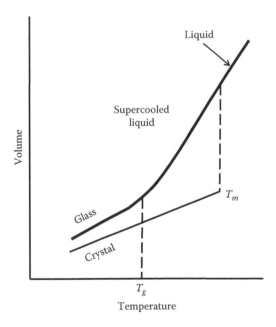

FIGURE 3.24
Volume changes in a glass and crystalline material during cooling from melt. (Reprinted from Kingery, W.D. et al., *Introduction to Ceramics*, John Wiley & Sons, New York, 1976, p. 92. With permission.)

temperature. The crystallization of glass under normal atmospheric conditions is a very slow process (it may take hundreds of years), but under certain conditions the crystallization can occur at a relatively faster rate, as in glass–ceramics.

Plastic deformation by slip in ceramic materials is very difficult because of strong chemical bonding. In addition to that, the ionic displacements during slip may lead to a position such that similar ions are placed one over other. This is an unfavorable condition, since the very high electrostatic repulsion between similar charged ions will not allow this to happen. Hence, the number of slip systems in ceramic materials is restricted. Moreover, the dislocations are narrow and relatively immobile in ceramic materials due to strong bonding. Ceramic materials are brittle because of these reasons. Hence, the main emphasis on ceramic matrix composites is to reduce the brittleness or to improve the toughness of ceramic matrix.

Oxides, such as alumina and mullite, and non-oxides, such as SiC and Si_3N_4, are the commonly used ceramic matrix materials. In certain applications, glasses and glass–ceramic materials are also used. The important characteristics of these common ceramic matrix materials are given in Table 3.10 and the possible reinforcements for these matrices are given in Table 3.11.

3.3.1 Crystalline Oxides

In principle, any oxide can be used as a matrix material for a ceramic matrix composite (CMC). However, it should be ascertained that the resulting composite is very useful for many applications and cost-effective. Because many oxides are refractory and stable in oxidizing atmosphere, they can be good choices for matrices in CMCs to be used at high temperatures under normal atmospheric conditions. However, the thermal expansion coefficients of many oxide ceramics are in the range 6–10×10^{-6} K^{-1}, and these values are rather higher than that of many typical reinforcements resulting in microcracking and poor properties. The main characteristics of those oxide matrices are described in the following sections.

3.3.1.1 Alumina

Alumina is perhaps the most versatile and cost-effective oxide ceramics. It has a range of useful properties including high hardness (up to 2000 H_v), high elastic modulus (up to 400 GPa), good strength (up to 500 MPa), and good electrical insulating properties. Alumina exists in various crystalline forms, such as γ, δ, θ, and α. Among these forms, α is the stable form with a hexagonal structure. The other forms are transformed into α-form on exposure at high temperature. Alumina ceramic components are usually made by compacting alumina powder and then sintering typically between 1400°C and 1800°C. The finer the starting α-Al_2O_3 powder, the lower the temperature required for sintering to achieve high density. Often

TABLE 3.10

Typical Properties of Common Ceramic Matrix Materials

Matrix	Crystal Structure	Melting Point (°C)	Theoretical Density (g cm^{-3})	Elastic Modulus (GPa)	Strength (MPa)	CTE (10^{-6} °C^{-1}) (25°C–1000°C)	Thermal Conductivity (W m^{-1} K^{-1}) (25°C)
α-Al$_2$O$_3$	Hexagonal	2060	3.95	380	200–310	8.5	~38
Y-TZP	Tetragonal	2300	6.10	210	800	10	~1.5
PSZ	Cubic/tetragonal	2300	6.05	200	500	10	~1.5
Mullite	Orthorhombic	1934	3.10	300	250	5	~10
α-SiC	Hexagonal	2300–2500	3.2	414		4.8	~80
Si$_3$N$_4$	Hexagonal	—	3.2	304	350–580	2.87	~30
MgO	Cubic	2800	3.6	210–300	97–130	13.8	45–60
YAG	Cubic	2000	—	280	250	8	~15
Spinel	Cubic	2135	3.59	280	250	8	~15
Borosilicate glass	—	—	2.3	60	100	3.5	—
LAS	—	—	2.0	100	100–150	1.5	—
MAS	—	—	2.6–2.8	120	110–170	2.5–5.5	—

Source: Adapted from Warren, R. (Vol. ed.), *Carbon/Carbon, Cement, and Ceramic Matrix Composites*, Vol. 4, in *Comprehensive Composite Materials*, eds. A. Kelly and C. Zweben, Elsevier Science Ltd., Oxford, U.K., 2000, p. 13.

Y-TZP, yttria stabilized tetragonal zirconia polycrystals; PSZ, partially stabilized zirconia; YAG, yttrium aluminum garnet; LAS, lithium aluminosilicate glass–ceramics; MAS, magnesium aluminosilicate glass–ceramics.

TABLE 3.11

Ceramic Matrices and Typical Reinforcements

Matrix	Particulates	Platelets	Whiskers	Fibers
Alumina	ZrO_2, SiC, TiC, TiN, TiB_2, ZrB_2, metals	SiC	SiC, B_4C	SiC, Al_2O_3, Mullite
Y-TZP	Al_2O_3	Al_2O_3	Al_2O_3	—
PSZ	Al_2O_3	SiC	—	—
Mullite	Al_2O_3, ZrO_2	Al_2O_3	SiC	SiC, Al_2O_3
SiC	SiC	SiC	B_4C, SiC	C, SiC
Si_3N_4	SiC, TiC, TiN, TiB_2	SiC, Si_3N_4	SiC, Si_3N_4	C, SiC, Al_2O_3
YAG	—	—	Al_2O_3	—
Spinel	Al_2O_3	—	—	—

Source: Adapted from Warren, R. (Vol. ed.), *Carbon/Carbon, Cement, and Ceramic Matrix Composites*, Vol. 4, in *Comprehensive Composite Materials*, eds. A. Kelly and C. Zweben, Elsevier Science Ltd., Oxford, U.K., 2000, p. 14.

some additives are added to alumina to modify its sintering characteristics and to control grain size or color. Small amount of MgO, ZrO_2, or Y_2O_2 is added to control grain size, and CaO or SiO_2 is added to reduce sintering temperatures by forming a liquid phase. MnO or TiO_2 can also be added to activate liquid-phase sintering, and transition metal oxides are mainly added to introduce color. In general, the greatest strength, hardness, and stiffness are obtained with fine-grained, high-purity products. Therefore, it is desirable to use the high-purity form as a matrix to achieve better properties in CMCs.

The properties of alumina can be improved by incorporating suitable reinforcements. For example, partially stabilized zirconia (PSZ) particulates and SiC whiskers improve toughness, non-oxide nanoparticles improve strength and wear resistance, and alumina or mullite fiber provides rigidity.

3.3.1.2 Mullite

Mullite is an aluminosilicate compound with a nominal Al_2O_3-to-SiO_2 ratio of 3:2. The principal advantages of mullite over alumina are a lower thermal expansion coefficient and better creep resistance in high-purity form. Traditionally this material was made by the reaction of a clay/alumina mixture at high temperature. Recently, many methods have been developed to produce sinterable high-purity mullite powders. The resulting ceramic products have good density, high strength, and a minimum of glassy or impurity phases. Although mullite ceramics are less widely used as structural ceramics than alumina, they offer some particular advantages when reinforced with mullite or alumina fibers. A modified matrix with improved toughness can be made by adding PSZ.

3.3.1.3 Spinel

Spinel is also a mixed oxide compound, formed with MgO and Al_2O_3 in the ratio of 1:1. It is usually made by the reaction sintering of MgO and Al_2O_3 mixture. The properties of this compound are similar to that of alumina, but this material is less stiff and hard. It is not widely used as a dense ceramic material, but the oxide–oxide composites made with this material have some special applications.

3.3.1.4 Zirconia

Zirconia has excellent refractoriness, but pure zirconia cannot be used as a refractory because of adverse phase transformations. Zirconia exists in three polymorphic forms, namely, cubic, tetragonal, and monoclinic forms. The high-temperature cubic form transforms to tetragonal and then to monoclinic form during cooling. The latter transformation produces a significant volume increase, which is very disruptive to the structure. In order to use this material for high-temperature applications, the phase transformations have to be suppressed or at least controlled by the addition of phase stabilizers.

Stabilization of cubic form to low temperatures is achieved by adding sufficient amount of MgO, CaO, or Y_2O_3 or a mixture of these oxides. This is known as fully stabilized zirconia and it can be used for refractory applications. The disadvantages of this form of zirconia are the relatively low strength and stiffness and high thermal expansion coefficient.

Use of insufficient stabilizer causes the formation of a cubic + tetragonal phase mixture during sintering and some monoclinic phase on cooling. The transformation to monoclinic phase causes microcracking, but such PSZ has thermal shock damage-resistant properties.

If a PSZ is sintered in the cubic phase field region and is then given an extended treatment in the cubic + tetragonal phase field region, then it forms a microstructure consisting of fine tetragonal zirconia particles inside the cubic grains. The small particles are restrained by the cubic phase to undergo phase transformation to monoclinic form on cooling. This zirconia is known as transformation-toughened PSZ (TTPSZ), which has enhanced strength and toughness. This material is also sometimes referred as inorganic steel because the stiffness and thermal expansion coefficient values are close to those of steel.

By using zirconia nanoparticles with 2.5–3 mol.% Y_2O_3, tetragonal zirconia with grain size less than 0.4 μm can be produced. In this material, the phase transformation to monoclinic form is suppressed by mutual grain interaction. This material is known as tetragonal zirconia polycrystals (TZP) and it has very high strength of the order of 1.2 GPa. TZP is now widely used to produce orthopedic parts, knives, guillotines, and fiber-optic ferrules.

Zirconia is seldom used as a matrix material for non-oxide whiskers or long fibers because of the risk of chemical reactions. Oxide particulate reinforcements are commonly used in zirconia. For example, the TZP reinforced with alumina particles has enhanced strength of 1.5 GPa. Oxide fibers or whiskers are not generally used with zirconia, because they need high consolidation temperatures or long durations of consolidation, which are not suitable for zirconia. However, the high-temperature properties of yttria-stabilized tetragonal zirconia polycrystals (Y-TZPs) have been improved by using alumina platelets, whiskers, or short alumina fibers.

3.3.1.5 Silica

Like zirconia, silica also undergoes phase transformations, which are highly disruptive. However, there are no effective means to stabilize a particular phase over a wide temperature range. Hence, silica is not generally used as a matrix material, but amorphous form of silica may be used.

3.3.1.6 Silicon Carbide

There are a variety of non-oxide ceramic materials with potential to be used as matrices for CMCs. However, they are generally hard and stiff refractory materials and it is very difficult to handle them. In particular, they are more difficult to sinter, and to achieve good density it is often necessary to use more complex processing routes.

SiC is a versatile non-oxide ceramic material, which is commonly produced in granular form for abrasive applications and solid form as refractory and wear- and chemical-resistant products. Despite the problem of sintering, many methods have been developed to prepare both porous and fully dense SiC ceramic products, and many of these methods can be adopted for the fabrication of CMCs using this material as the matrix.

Silicon carbide exists in two crystalline forms: α-SiC has hexagonal structure and β-SiC has cubic structure. The hexagonal form is stable at RT. SiC readily undergoes oxidation, and the protective SiO_2 layer formed on the surface prevents further oxidation.

3.3.1.7 Silicon Nitride

Silicon nitride is also a strong covalent ceramics. It is less refractory compared to SiC and has high vapor pressure at temperatures above 1700°C. It exists in two crystalline forms (α and β) and both are hexagonal. The main difference is that the c-axis of α-Si_3N_4 is about twice that of β-Si_3N_4. The α-phase is the harder phase, which is generally derived from the vapor phase and often has needle morphology, but is metastable at high temperatures. The β-phase is less hard and it is generally formed during solid- or liquid-phase reactions. It has a more blocky morphology. It is very difficult to process Si_3N_4, because of high vapor pressure and its reluctance to sinter. However, a number of methods such as hot pressing and reaction bonding have been developed to produce monolithic Si_3N_4 parts, which are also applicable to the formation of composites.

3.3.1.8 Boron Carbide

One of the most widely used boron carbides is B_4C. It is a very hard material and can be processed as the other conventional ceramic materials. Similar to other non-oxides, pressureless sintering is difficult for this material also, and normally hot-pressing is required to get optimum properties. As a consequence, its use as a matrix for composite materials is very much limited.

3.3.1.9 Boron Nitride

Boron nitride exists in two crystalline forms, hexagonal and cubic forms. The hexagonal BN (HBN) is a soft and platy material similar to graphite, whereas the cubic BN (CBN) is a very hard material similar to diamond. HBN is usually made by hot pressing. The hot-pressed material is slightly hygroscopic and machinable with conventional tooling. Because of hexagonal structure, the mechanical and thermal properties are somewhat anisotropic. It is an excellent electrical insulator and it finds widespread use as a refractory under inert or reducing conditions. The high-purity HBN is used as a crucible material for melting metals and semiconductor materials because of its non-wetting characteristics.

CBN is made from HBN by high-pressure, high-temperature technology. CBN finds applications primarily in cutting tools and wire drawing dies and these products are made by hot-pressing CBN with metallic binders.

Neither form of BN has achieved significant use as a matrix material. However, there is a scope for improving strength, stiffness, and hardness of HBN by introducing reinforcement. At present, HBN has been extensively used as a weak coating on fibers, as interlayer coating in layered composites, and as a solid lubricant.

3.3.1.10 Aluminum Nitride

The main characteristic of AlN is its high thermal conductivity. It also has high-temperature stability and good oxidation resistance. However, its mechanical properties are not comparable to silicon nitride. Although it has good oxidation resistance, it may form compounds with oxygen on long time exposure. An oxynitride (cubic AlON) has been used as a transparent ceramic material in hot-pressed form. AlN reacts with SiC, especially in the presence of liquid-phase sintering aids, and this causes problems in making SiC-reinforced composites. This material could be a choice of matrix material in applications, where there is a need of good thermal conductivity.

3.3.2 Glasses and Glass–Ceramics

Earlier there were some attempts to make CMCs using oxide-based glasses and glass–ceramics as matrix materials simply because at high temperatures glasses become viscous liquids, which can readily infiltrate/surround fibers and whiskers and on solidification form a dense matrix. Depending on the matrix formulations, either a glassy matrix is retained or a glass–ceramic matrix is formed by inducing crystallization during or after the solidification of matrix. A further advantage of a glass or glass–ceramic is that its composition can be suitably varied to achieve the desired thermal expansion coefficient for the matrix.

A glass is an amorphous material, in which the melt is cooled to a rigid state without forming a regular crystalline structure. The refractoriness/rigidity of a glass is determined by its glass transition temperature. This is the temperature at which the rigid, brittle solid is transformed into a flexible, rubbery solid. Another measure of glass transition temperature is based on stress relaxation. The minimum temperature at which stress relaxation can occur in a period of about 15 min is called glass transition temperature (T_g). The T_g varies depending on the composition of glass; for pure silica glass it is about 1000°C, for soda–lime glass it is about 600°C, and for lead oxide–based glass it can be as low as 350°C. Most of the glasses are based on silicates, and their properties depend on composition, particularly the extent to which silica network is broken up by modifying ions of lower valency.

Glasses have a key advantage over conventional ceramic materials in terms of composite processing; they transform to a highly viscous plastic material when heated sufficiently and flow readily under pressure. Thus, a mixture of glass powder and whisker/fiber can be easily hot pressed at a temperature typically 200°C–300°C above T_g to get a fully dense matrix. However, the advantage of easy processing is hampered by the restricted service temperature, which is little below the T_g. For many commercial glasses, the T_g lies in the range 500°C–800°C. On considering the service temperature, many metals and MMCs are better than glass matrix composites. Hence, glass matrix composites have only a few commercial applications.

Glass–ceramics are produced by the controlled crystallization of a suitable glass composition. Like glasses, composition can be varied to get materials of controlled properties. The main advantage of glass–ceramics is that they have better high-temperature properties than the precursor glass.

For making glass–ceramic-based composites, glass powder-based routes are more preferable than bulk glass-based routes. The essential difference is that nucleating agents are required for bulk glasses to promote fine-scale nucleation and growth of crystalline phases, whereas in powder-based methods, crystallization from powder particle surfaces is often more than adequate. To make a precursor powder, appropriate amounts of metal oxides or salts required for a suitable glass composition are mixed, melted, quenched as a glass, and ground to a fine powder. By using this powder, a composite can be processed like any other powder-based methods, but typically this is done by slurry impregnation of fiber tows, followed by lay-up and hot-pressing.

Depending on the behavior of the particular glass composition used as the matrix, two routes are available for densifying the glass–ceramic matrix. The glass powder/fiber lay-up is hot-pressed above the liquidus temperature of the glass composition and it densifies readily because of the formation of low-viscosity liquid. After hot-pressing, it should be allowed to undergo a controlled cooling schedule for the devitrification of glass to form the required glass–ceramics. On the other hand, if there is a large temperature interval between the T_g and devitrification temperature, then it may be possible to densify the matrix before devitrification halts the process, since the matrix becomes sufficiently fluid to permit densification. The advantages of the second method are that the fiber suffers less damage than the first method and the survival of graphite hot-pressing die system is obviously better.

The properties of glass–ceramics vary widely with composition, but in principle any glass–ceramic composition can be used as a matrix material. However, the potential matrices are those which have low thermal expansion coefficients suitable for matching to carbon or SiC fibers and a high-temperature capability but can still be made relatively easy as glasses. A number of additives are added most commonly to the glass–ceramics to control the devitrification process. Some of the more common glass–ceramic systems are outlined in the following texts.

3.3.2.1 Magnesium Aluminosilicates

The principal crystalline phase of interest for CMCs in this system is cordierite, for which the chemical formula is $2MgO \cdot 2Al_2O_3 \cdot 5SiO_2$. It has a very low thermal expansion coefficient and melting point of 1450°C. It is reasonably easy to produce this glass powder. However, in powder form, glass of cordierite composition does not sinter well because devitrification starts little above the T_g. Hence, the lower temperature window for hot-pressing is not available, and it should be hot-pressed above the liquidus. However, coarse grains are formed during the devitrification heat treatment and it becomes necessary to add a nucleating agent to control the grain growth. Typical nucleating agents for this glass–ceramics are TiO_2, ZrO_2, and P_2O_5. The low thermal expansion coefficient of this glass–ceramics is closely matching to that of carbon and SiC fibers. This material also has a higher Young's modulus than most of the other aluminosilicates, which is about 120 GPa.

3.3.2.2 Barium Magnesium Aluminosilicates

A glass–ceramic with similar thermal expansion and refractoriness as cordierite is barium osumilite, with a nominal composition of $BaMg_2Al_3$ $(Si_9Al_3O_{30})$. However, the sinterability of glass powder made with this material at lower temperature is good and also it undergoes a less complex series of phase transformations with increasing temperature.

3.3.2.3 Calcium Aluminosilicates

These compositions are also good glass formers. Refractoriness is similar to MAS and barium MAS (BMAS) systems, but they have relatively higher thermal expansion coefficients.

3.3.2.4 Lithium Aluminosilicates

LAS are a family of glass–ceramics with very low thermal expansion coefficients. Some of the common LAS are β-eucryptite (Li_2O Al_2O_3 $2SiO_2$), β-spodumene (Li_2O Al_2O_3 $4SiO_2$), and petalite (Li_2O Al_2O_3 $6SiO_2$). The thermal expansion coefficient and the degree of thermal expansion anisotropy decrease with increasing silica content. Hence, it is difficult to use β-eucryptite effectively because of its high anisotropy and the material survives only when the grain size is very fine. On the other hand, the thermal expansion coefficient of β-spodumene is close to zero, and it has good glass-forming and well-controlled nucleation/crystallization characteristics. Often, some MgO is added as a substitute for Li_2O, mainly to reduce the cost without much change in properties. Typically, TiO_2 or ZrO_2 is used as a nucleating agent.

3.3.3 Cement

Cement is one of the important construction materials, used to bind the other construction materials, such as bricks and aggregates. It differs from the other ceramic matrices in terms of processing. Unlike other ceramic matrices, composites based on cement can be processed at ambient conditions. Various types of cements have been developed from the prehistoric times, but the most important type is Portland cement. Portland cement consists of greater than two-thirds by mass of calcium silicates ($3CaO$ SiO_2 and $2CaO$ SiO_2) and the remainder is aluminum oxide, iron oxide, and other oxides. The main raw materials used for the production of Portland cement are limestone/chalk and clay/shale. These raw materials in correct proportions are intimately mixed, calcined at 800°C to drive off water and CO_2, and fired at 1300°C–1450°C in a rotary kiln. At this temperature, the material sinters into balls of diameter up to 25 mm and this is known as clinker. The cooled clinker is ground to a fine powder with the addition of about 3%–6% gypsum, which controls the setting time. The fine-powdered product is known as Portland cement. This cement is generally used in most building constructions. It is available in different grades based on a 28-day minimum strength in MPa and the main grades are 32, 42, 52, and 62 with strength levels 32.5, 42.5, 52.5, and 62.5 MPa, respectively. The grades 52 and 42 are having similar chemical composition, but the former grade is more finely ground leading to a more rapid strength gain.

The temperature of concrete during early stages of aging increases and that leads to expansion. During cooling it tries to contract, but the contraction is restrained at that stage leading to more cracking in the concrete. This problem can be minimized by adding a suitable inorganic material, which can control the hydration exotherm. This cement is known as blended cement. Typical materials are pulverized coal ash (commonly known as fly ash), ground-granulated blast-furnace slag (GGBS), and microsilica. Blended cements containing either fly ash or GGBS have considerable advantages over Portland cement apart from being cheaper. Due to the lower density of these materials, the volume of fine powder increases leading to increased cohesiveness and workability to the cement. Sulfate resistance and resistance to chloride ion penetration are also improved by these additions.

Microsilica is a by-product formed during the manufacture of silicon and ferrosilicon alloys from high-purity quartz and coal. The flue gases containing SiO oxidize and condense in the form of extremely fine particles (~0.1 μm in diameter) of amorphous silica. The microsilica particles have very high surface area and hence they are highly reactive with the calcium hydroxide formed during cement hydration. The advantages of this blended cement include very high strength (up to 150 MPa), reduced permeability, and good sulfate resistance. Sodium, calcium, and magnesium sulfates present in the groundwater can react with tricalcium aluminate hydrate in Portland cement and form calcium sulfoaluminate with a 227% increase in volume. Moreover, the sulfates can react with calcium hydroxide and form gypsum with a 124% increase in volume. To have better sulfate resistance, the cement should contain less tricalcium aluminate. For foundations or piles in sulfate-bearing soils, Portland cement with less tricalcium aluminate or other cements containing pulverized fly ash or slag should be recommended. Super-sulfated cement is highly resistant to seawater and sulfates. This cement can be made by grinding a mixture containing 80%–85% granulated blast-furnace slag, 10%–15% dead-burnt gypsum, and up to 5% Portland cement clinker.

A type of cement that undergoes rapid hardening is high-alumina cement. This cement is produced from bauxite and it gains 80% of its strength in 24 h. However, it is no longer used for the construction of buildings because it suffers from the conversion to a much weaker phase under warm damp conditions, which may lead to structural collapse. However, it is good cement for high-temperature refractory applications up to 1600°C with suitable refractory aggregates.

Questions

State whether the following statements are true or false and give reasons:

3.1   The matrix plays a minor role in the tensile load-carrying capacity of a composite.

3.2   Polymers are always brittle below their glass transition temperature.

3.3   Mechanical properties of polymers strongly depend upon the testing temperature.

3.4   Plastics are the toughest materials.

3.5   Recycling of thermoplastics is very difficult.

3.6   Thermosetting resins are widely used in composites.

3.7   Melting points of thermosetting resins are high.

3.8   Trace amount of inhibitor is added to polyester resin to control the curing reaction.

3.9   The HDT of polyester resin decreases with increasing saturated acid-to-unsaturated acid ratio.

3.10 High shrinkage levels of polyester resins are advantageous.

3.11 Curing at high temperature leads to better properties in polyester resin composites.

3.12 Epoxy resins are preferred for making prepregs.

3.13 It is better to have high cross-link density in polyester resin.

3.14 During the curing of epoxy resin, there is a continuous increase in viscosity.

3.15 The addition of CTBN rubber into epoxy resin improves the properties.

3.16 Epoxy resin is the most preferred thermosetting polymer in composite industries.

3.17 Vinyl ester resins possess good characteristics of epoxy resins.

3.18 Vinyl ester forms a good bond with glass fiber.

3.19 Titanium alloys are preferred over aluminum alloys for aerospace applications.

3.20 Cold working of magnesium is difficult.

3.21 Intermetallic matrix materials are preferred for high-temperature applications.

3.22 Ceramic materials are very useful for high-temperature applications, where there is large fluctuation in temperature.

Answer the following questions:

3.23 Explain the role of matrix materials.

3.24 Distinguish between thermoplastics and thermoset resin systems.

3.25 What is the main structural difference between thermoplastics and thermoset polymers?

3.26 Mention the advantages of thermoplastics.

3.27 Mention the disadvantages of thermoset polymers.

3.28 What are the desirable mechanical properties of a matrix for high-performance composites?

3.29 Explain the role of styrene in polyester resin.

3.30 What are the problems during the curing of thermosetting resins?

3.31 A thermoplastic polymer can be heat softened, melted, and reshaped as many times as desired. How is it possible?

3.32 Explain the unique characteristics of polymeric solids.

3.33 What is the effect of increasing cross-link density of epoxy resins?

3.34 How can the fracture toughness of epoxy resins be improved?

3.35 Point out the important differences between epoxy and polyester resins.

3.36 What is the effect of fillers in polyester resin?

3.37 What are the important characteristics of vinyl ester resin?

3.38 What is effect of the addition of thermoplastic polyimides into thermosetting polyimide?

3.39 What are the important characteristics of PEEK?

3.40 What are the important characteristics of titanium alloys?

3.41 What are the attempts for ductility enhancement in intermetallics?

3.42 What are the important characteristics of ceramic matrix materials?

3.43 Mention the important characteristics of LAS glass–ceramics.

References

Kingery, W.D. et al. 1976. *Introduction to Ceramics*. New York: John Wiley & Sons.

Kolczynski, J.R. et al. 1967. *Soc. Plast. Ind. Conf.* 22: 1A.

Mark, H.F. and J.I. Kroschwitz. 1988. *Encyclopedia of Polymer Science and Technology*, Vol. 12. New York: John Wiley & Sons.

Pandey, A.B., K.L. Kendig, and D.B. Miracle. 2001. Discontinuously reinforced alumi-
num for elevated temperature applications. Presented at *TMS Annual Meeting*,
New Orleans, LA, February 11–15.
Saito, T. 1995. *Adv. Perform. Mater.* 2: 121.
Scola, D.A. 2001. Polyimide resins. In *ASM Handbook*, Vol. 21, *Composites*. Materials
Park, OH: ASM International.
Unal, R. and K.U. Kainer. 1998. *Powder Metall.* 41: 119.

Bibliography

ASM Handbook. 2001. *Composites*, Vol. 21. Materials Park, OH: ASM International.
Callister, W.D. Jr. 2007. *Materials Science and Engineering.* New York: John Wiley & Sons.
Chawla, K.K. 1998. *Composite Materials: Science and Engineering*, 2nd edn. New York:
Springer-Verlag, Inc.
Lubin, G. 1982. *Handbook of Composites.* Berlin, Germany: Springer.
Mallick, P.K. 2008. *Fiber-Reinforced Composites*, 3rd edn. Boca Raton, FL: CRC Press.
Mazumdar, S.K. 2002. *Composites Manufacturing.* Boca Raton, FL: CRC Press.
Raghavan, V. 1998. *Materials Science and Engineering*, 4th edn. New Delhi, India:
Prentice Hall of India.
Warren, R. (Volume ed.). 2000. *Carbon/Carbon, Cement, and Ceramic Matrix Composites*,
Vol. 4. in *Comprehensive Composite Materials*, Kelly, A. and Zweben, C. (Eds.).
Oxford, U.K.: Elsevier Science Ltd.

4

Polymer Matrix Composites

Polymer matrix composites (PMCs) have become one of the important structural engineering materials. This is not only because of the development of high-performance fibers such as carbon, polyethylene, and aramid but also because of some new and improved polymer matrix materials. The mechanical properties of polymers in general are inadequate for many engineering applications. In particular, their strength and stiffness are lower than those of metals and ceramics. Hence, there is considerable scope for improving the mechanical properties of these materials by incorporating reinforcements. The resulting composites have the advantage of easy fabrication of polymers with improved properties. PMCs are very common compared to metal and ceramic matrix composites. In general, the processing of polymeric materials does not require high temperatures and pressures. Hence, the fabrication of complex-shaped components is easy with these materials. The problems associated with the degradation of reinforcement during the fabrication are insignificant for PMCs due to lower processing temperatures. Moreover, the equipment required for the fabrications of PMCs are also simpler. Hence, the development in PMCs happened rapidly, and they have become the accepted materials for many structural applications.

4.1 Processing of Polymer Matrix Composites

There is a variety of processing methods available to produce PMC parts. Depending on the quality, property, quantity, and cost of the product, a suitable processing method can be selected. A brief overview of the different processing methods is as follows.

Hand lay-up (contact molding) is a simple and cost-effective process. This process is often used to build large structures in small quantities. Reinforcements in the form of chopped strand mat (CSM) and woven rovings are generally used. The maximum fiber volume fraction that can be obtained by this process is 0.35. The quality of the product depends on the skill of the worker. Hence, it may not be possible to get consistent quality in the products. It is also a labor-intensive and time-consuming process. This process is more suited for the fabrication of custom-made parts in small numbers. It is not possible to produce high-performance composite parts by this method.

Spray-up is a more cost-effective process than hand lay-up, but the mechanical properties are much lower due to the presence of randomly oriented chopped fibers. In this process, continuous glass fiber roving is fed into a special gun that chops the fibers into short lengths and mixes them with resin, which is sprayed onto the mold. This process is similar to hand lay-up process in almost all aspects, but it is a slightly faster process. Both hand lay-up and spray-up processes produce products with good surface finish only on one side, which is the surface in contact with the mold.

Resin transfer molding (RTM) is a semiautomatic, closed mold process. The required amount of fiber reinforcement is placed within the mold cavity and a low-viscosity resin is injected under pressure to fill the closed mold. Since this is a closed mold process, it is possible to produce parts with good dimensional tolerances. Unlike the previous two processes, products with good surface finish on both the sides can be produced by this process. This process is well suited for the production of parts in moderate quantities with consistent quality. The mechanical properties of the parts produced by this process are better than those of the parts produced by hand lay-up process. Some capital investment is required for this process to acquire the resin injection system.

Compression molding is a matched-die process that uses molding compounds. The molding compounds are a mixture of short fibers, resin, and other ingredients. A predetermined quantity of molding compound is placed between the two die halves and then heat and pressure are applied. The molding compound flows to fill the die and then rapidly cures in 1–5 min. Short fibers are generally used in this process. The production rate is very high, and this process is suitable to produce parts in large quantities. Products with the best surface finish are obtained by this process. This process is more suitable for making small-to-medium-sized parts in large quantities.

Injection molding is another high-volume process, capable of producing small-to-medium-sized parts. The reinforcement (usually chopped glass fiber) is mixed with molten thermoplastic or liquid thermoset resin and then injected under high pressure into a matched metal die. The product is ejected after the thermoplastic part is cooled or the thermoset part is cured. It may not be possible to get significant improvement in mechanical properties by the reinforcement, since very short fibers (often less than the critical fiber length) are used in this process.

There are two processing methods that use continuous fibers. One is filament winding process and the other is pultrusion process. Very high fiber content (~70 vol.%) is achieved in these processes. The filament winding process is mainly used to produce axisymmetric shapes like cylinders and spheres. In this process, resin-impregnated fiber tows are wound on a mandrel at a specified angle and then cured. Very high mechanical properties could be realized in these composites, because of the usage of continuous fibers at high volume fractions. The pultrusion process is suitable for making long, constant cross-sectional parts. Fiber tows are pulled through a resin bath and then through a heated die, where curing occurs. The cured part is continuously pulled, and the desired length is cut from the continuous product. This process is suitable to produce long rods, tubes, angles, etc. As the filament wound products, these products also have very good mechanical properties. This process is not suitable to produce complex-shaped parts.

Vacuum impregnation (VI) method is used to produce medium-to-large-sized parts. The process is similar to RTM process, and the main difference is vacuum application that drives the mold filling by the resin instead of pressure used in RTM process. It is also possible to consolidate the composite in this process by using a flexible top mold. In that case, the process is known as vacuum bag molding. An extension of this process is autoclave molding. In autoclave molding, high temperature and pressure are applied over the vacuum bag setup, to activate curing and better consolidation. High-quality composite parts used in aerospace applications are mainly made by the autoclave molding process. Composites used in non-aerospace applications can also be made with VI method. High-pressure application during autoclave molding results in better compaction, high fiber contents, and less voids and porosity.

For different processing methods, different kinds of materials are used. Table 4.1 gives a summary of precursor materials and their processes. The details of the commonly used processing methods for polymer composites are given in the following sections.

TABLE 4.1

Precursor Materials and Processes

Material Form	Production Process					
	Pultrusion	RTM	Compression Molding	Filament Winding	Hand Lay-Up	Auto Tape-Laying
SMC			•			
BMC			•			
Swirl mat/neat resin	•	•	•	•	•	
Unidirectional tape			•		•	•
Woven prepreg			•		•	
Woven fabric/neat resin	•	•			•	•
Stitched material/ neat resin		•				
Prepreg roving				•		
Roving/neat resin	•			•	•	
Preform/neat resin		•				

Source: Data from Campbell, F.C., *Manufacturing Processes for Advanced Composites*, Elsevier, Oxford, U.K., 2004, p. 18.

4.1.1 Hand Lay-Up Process

Hand lay-up refers to the manufacturing process of fiber-reinforced plastics (FRPs) by laying fiber reinforcements with thermoset resins in or on a mold using hand or handheld tools. It is a nonmechanized open mold manufacturing process. This method is also known as contact molding or wet lay-up. Although hand lay-up process can be used for making products with any type of fibers, it is widely used for making glass fiber-reinforced polymer (GRP) products. Hand lay-up process is suitable for any thermoset resin system that is available in liquid form under room temperature. Polyester, epoxies, and certain grades of phenolics, furanes, silicones, and polyimides satisfy this requirement. The hand lay-up process explained in this section pertains to glass fiber-reinforced polyester resins. These processing details can be easily extended to other materials with suitable modifications.

Hand lay-up is mainly used for making FRP products in a mold. The mold has the shape of the product. The product will have smooth finish only on the side that is in contact with the mold. Only the male or the female half of the mold is generally used in hand lay-up process. The choice of male or female mold depends on which side of the product needs good finish. If the inside surface needs smooth finish, then the product is made over a male mold or vice versa. It is possible to make a product with glossy, mat, or textured finish on the surface by using a mold with the respective finish. The mold must be free from surface defects, because the imprint of such defects will form on the product. The typical cross-sectional structure (architecture) of layers in a hand lay-up product is shown schematically in Figure 4.1.

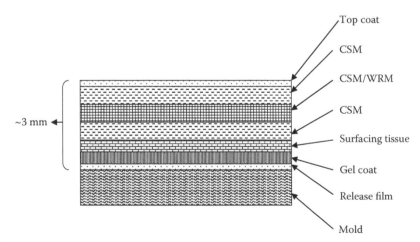

FIGURE 4.1
Exploded view of the layers in a typical laminate made by hand lay-up.

4.1.1.1 Molds

Molds for hand lay-up process are made out of plaster of Paris, wood, FRP, or metals. Plaster of Paris mold is good for making a few pieces, since the mold easily breaks during the release of product. Wooden mold requires finishing work after every cycle of molding. FRP and cast epoxy molds are ideal for making intricate shapes. For critical components and when heating and/or pressing is required, metallic molds are the best choice.

The thickness of the mold should be such that it should not deform excessively during lay-up and release of the product. For large-sized panels and relatively flat products, it is economical to provide closely spaced ribs or stiffeners rather than increasing the mold thickness. Steel or timber stiffeners can be provided for better stiffening. For deep drawn products, a taper has to be provided for easy release. A 1 in 1000 taper for polyester resin-based products is found to be adequate. For large-sized components, deep drawn moldings, and products with undercuts, split molds have to be used.

4.1.1.2 Release Film or Layer

A release film or layer should be placed over the mold surface for the easy release of product from the mold. For flat sheets and simple product shapes, a thin polyester film (Mylar) can be used. For complex shapes, the use of release layer of wax and polyvinyl alcohol (PVA) is recommended. PVA is a water-soluble material and 15% solution in water is applied with a sponge. A thin film of PVA forms on the mold after the evaporation of water. After the molding, PVA film can be washed-off from the mold and product.

4.1.1.3 Gel Coat

Gel coat is a thin layer of resin (~0.5 mm thickness) applied over the release film. Gel coat gives superior finish to the product. The required color of the product can be obtained by adding suitable pigment to the resin. The gel coat serves the following purposes:

1. It provides color, glossiness, and/or texture to the products.
2. It conceals the fiber pattern.
3. It also provides a resin-rich layer that protects the fiber from getting in contact with water and chemicals.

It is necessary to control the thickness of gel coat to an optimum level. If the gel coat is too thin, then the fiber pattern will become visible. If the gel coat is too thick, crazing and star cracks can appear on the gel coat.

The gel coat resin is generally of the same grade as the lay-up resin. Polyester gel coat resin is a thixotropic resin to which pigments, catalysts, and accelerator are added. Thixotropy means a decrease in viscosity of a liquid, when shear forces are applied. A thixotropic liquid ordinarily exists as a thick viscous fluid, but when a force (stirring with a brush or stick) is applied, it flows easily like a low viscous fluid. The gel coat resin becomes viscous immediately after the brushing is stopped because of thixotropic characteristics, and this prevents the flow of resin when it is applied on vertical or inclined surfaces. The required thixotropy can be achieved by adding about 2%–3% of aerosol powder to the resin with thorough mixing. The viscosity of the resin can also be increased by adding fine calcium carbonate or other fillers. The filler content should be as low as possible. If the filler loading is high, the resin will flow to the bottom due to more weight, and also the cured resin will become very brittle. Fillers are not generally added to the gel coat of products, which are in contact with water or corrosive chemicals.

The required amount of pigments and functional additives (conducting materials, UV stabilizers, etc.) can be mixed with the gel coat resin. Once the required ingredients are thoroughly mixed with the resin, a uniform coating is applied on the mold surface using a brush or spray gun. The lay-up of composite should be started only after the gel coat has gelled.

4.1.1.4 Surface Mat Layer

When there is a requirement of thick resin layer, as in chemical equipment for increased chemical resistance, a surface mat layer is used after the gel coat layer. In the surface mat, very fine fibers are bonded into a mat. The mat provides the required crack resistance and impact strength to the resin-rich layer. Surface mat layer is an optional layer and is used only in specific cases.

4.1.1.5 Laminates

Fiber layers wetted with resin are laid-up to the required thickness and the stacking is called laminate. The laminate provides strength and rigidity to the product. Glass fibers in the form of CSM are generally used. Woven roving mat (WRM) is used in between CSM layers to improve the strength of the laminates.

The first layer of laminate shall invariably be a CSM layer and the lamination should satisfy the following requirements:

1. The fibers should be uniformly placed, and they should fit correctly on the contours of the mold surface.
2. The fibers should not be damaged during lay-up.
3. The fiber-to-resin ratio should be correctly maintained.

The required quantities of filler, accelerator, and other additives are added to the resin and mixed thoroughly. Vigorous stirring can cause entrapment of air bubbles, and therefore slow stirring speed is recommended. The resin mix should be prepared one day earlier so that the entrapped air bubbles can partly escape. The resin mix container should be closed airtight and kept in a cool place.

The number of mat layers required for a certain thickness can be determined by taking into account the mat density and the fiber-to-resin ratio. The following points must be considered while preparing the mat:

1. The joints in a layer should be away from the corners.
2. Whenever there is a joint, a minimum overlap of 25 mm in the case of CSM and 50 mm in the case of WRM should be provided.
3. The joints in successive layers should be staggered in such a way that they are not placed one over the other.
4. Whenever there is a change in thickness of laminate, the thickness should not change abruptly. The mat for successive layers should be cut such that there is a slope of 1 in 6.

Weighing balances, rollers, brushes, mugs, and squeeze bottles are the tools and items required for hand lay-up, apart from the mold. Solvents are required for cleaning the rollers and brushes. Acetone or nitrocellulose thinner is generally used as a solvent.

4.1.1.6 Lamination

The required amount of catalyst is added from a squeeze bottle to a small quantity of resin mix taken in a mug. It is stirred well using a glass rod. A layer of resin is applied with a brush on the gel coat. One layer of CSM is then placed carefully over the resin. The resin is squeezed to the top surface by a stippling action using resin-wetted brush. The bristles of brush may come out and stick on the product during this action. Hence, for a high-quality product, the usage of brush should be avoided; instead a hand roller can be used. Care should be taken to ensure that the required fiber-to-resin ratio is uniformly maintained. The layer is consolidated and the air bubbles are removed by using the rollers (Figure 4.2). Subsequent layers are laid-up in a similar manner. More than four layers should not be laid at a time. If too many layers are applied at a stretch, excessive heat may build-up within the layers due to exothermic curing reaction. This may lead to the formation of cracks on the gel coat, debonding of layers and discoloration of the resin. However, the hand lay-up of successive layers should not be delayed too long, as it can lead to difficulty for the new layer to bond with the previous layer. When WRM is needed, it should be placed in between CSM layers in order to have better interlaminar shear strength. The lay-up procedure for WRM and CSM are identical except that less resin is used for WRM. All lay-up operations must be finished before the gelation of resin. The gel time depends on the amount of catalyst and accelerator added, but in a normal curing process, it is usually 15–30 min. Once the lay-up is completed, then the laminate is allowed to cure on the mold. The product can be released from the mold, when more than 90% of the curing is completed. Further curing

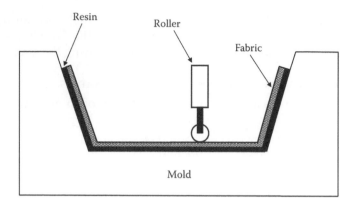

FIGURE 4.2
Schematic of hand lay-up process. (Reprinted from Mazumdar, S.K., *Composites Manufacturing*, CRC Press, Boca Raton, FL, 2002, p. 131. With permission.)

takes long time; normally, it may be 5–10 days. It is better to keep the product on a dummy mold during this period to avoid any warping. Better physical and mechanical properties are obtained for the product only after complete curing. The curing process can be accelerated by post-curing the product at an elevated temperature after releasing from the mold.

4.1.1.7 Advantages and Limitations

The large-scale use of hand lay-up process is attributed to the simplicity of the process. No expensive equipment is required, and only simple inexpensive tools like brushes and rollers are needed. Practically there is no restriction on the size of the product. It is possible to mold all shapes and incorporate inserts of any shape. Depending on the end user's taste, colors and decorative finishes can be incorporated in the product. This is the most suitable method for lining of tanks, ducts, and boats and repairing of concrete structures. There are some limitations in this process. The process is labor intensive, and the quality of the product depends largely on the skill of the person doing the fabrication. This process is not suitable if good surface finish is required on both the sides of the product. This process cannot compete with compression molding for the mass production of small items. Thickness control is not accurate and it is difficult to obtain uniform fiber-to-resin ratio.

Hand lay-up method requires comparatively a very low capital investment and is ideally suited for small fabrication units. It is a more suitable process to make tailor-made products in small numbers. The following conditions favor hand lay-up process as the choice for fabrication:

- Good surface finish is needed on only one side.
- Slight thickness variations are permissible.
- Labor charges are not prohibitively high.

- The product is large in size and is very complex in shape.
- Only a few numbers of products are required, and the number does not justify the use of costly metal dies and compression molding.

4.1.2 Spray-Up Process

Spray-up is another low-to-medium volume, open mold process suitable for producing medium-to-large-sized parts. Greater shape complexity is possible with spray-up than with hand lay-up. In the spray-up process, chopped fibers and resin are simultaneously deposited on an open mold. Usually glass fiber rovings are used in this process. The glass fiber rovings are fed through a chopper and propelled into the resin stream, which is directed to the mold. The spray gun consists of two nozzles: one nozzle sprays resin premixed with catalyst or catalyst alone, while another nozzle sprays resin premixed with accelerator. The resin mix pre-coats the strands of glass fiber, and the coated fibers with the resin is sprayed on the mold in an even pattern by the operator. After the deposition, it is rolled manually using hand rollers to remove air and compact the fibers (Figure 4.3).

In this process also, the quality of the final product ultimately depends on the operator skill. Additional control methods, independent of the operator, must be used to ensure quality. Laminate composition and thickness is a function of the fiber-to-resin ratio. Hence, proper control over any of the input material can control the resultant properties of the laminate. If the gun is set with a known resin to fiber ratio, spraying for a specified time will control the thickness of the laminate. In spray-up process also, fiber volume fraction is limited to 0.35.

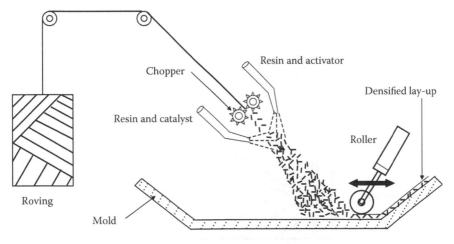

FIGURE 4.3
Schematic of the spray-up process. (Reprinted from Schewartz, M.M., *Composite Materials*, Vol. II, Prentice Hall, Upper Saddle River, NJ, 1997, p. 28. With permission.)

As in the hand lay-up process, superior surface finish can be achieved by spraying a gel coat onto the mold surface prior to spray-up of fiber–resin mix. Woven roving or woven cloth is occasionally incorporated in the laminate to provide high strength in certain locations. Core materials can also be easily incorporated. As with the hand lay-up process, the major advantages are low-cost tooling, simple processing, possibility of on-site fabrication (the spray equipment is portable), and virtually unlimited part size. An additional advantage of spray-up is that it is amenable to automation, thereby reducing labor costs and the risk of exposure of workers to potentially hazardous styrene vapor present in polyester resin.

4.1.3 Resin Transfer Molding

RTM is a closed mold semi-mechanized manufacturing method, generally used to produce fiber-reinforced thermoset polymer products. Unlike hand lay-up and spray-up processes, RTM process gives better control on product thickness and good surface finish on both sides. In this process, the fiber is packed to the required geometrical arrangement in the cavity of a closed mold, and a liquid resin of low viscosity is injected under pressure into the cavity (Figure 4.4). The resin wets the fiber completely and then cures. The RTM process gives faster production cycles than hand lay-up process, since fast curing polyester or epoxy resins are generally used.

The essential components for the RTM process are resin pumping/dispensing equipment and the mold of required shape. The pumping unit generates

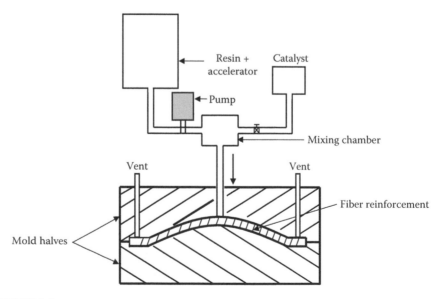

FIGURE 4.4
Schematic of RTM (catalyst injection system).

low-to-moderate pressure to inject the resin into the mold, which flows through the layers of reinforcement. There are three variants in the RTM equipment.

4.1.3.1 Two-Pot System

This system has two equal volume containers (pots). The resin is mixed with accelerator in one of the containers and mixed with catalyst in another container. Two pumps are used to pump resin from these containers to the mixing head, where they are mixed thoroughly. The main disadvantage of this system is that it needs two pumps. Another disadvantage is that there is a high probability of blockage by the gelled resin due to inadvertent interchange of pots.

4.1.3.2 Catalyst Injection System

The resin mixed with accelerator is pumped into the injection chamber. The catalyst is taken separately into the chamber by means of a control valve. A mixing head is an integral part of the injection chamber and that ensures adequate mixing of all the components of the resin system. Only a single pumping unit is needed for this system. The main advantage of this system is that the gel and cure times can be controlled easily by varying the amounts of catalyst mixing with the resin. The chance of resin curing in the container is eliminated. Most of the commercial RTM equipment are based on this design.

4.1.3.3 Premixing System

This is a simple system among the three variants. The required amount of resin, accelerator, and catalyst is mixed in a single vessel and then injected into the mold. A thick-walled airtight metallic cylinder is used in this system. The resin injection is carried out by means of compressed air. Precalculated quantity of resin required for filling a mold should be taken in the vessel, and the vessel should be cleaned immediately after the resin is injected into the mold. Otherwise the resin remaining in the vessel will gel and cure, and the cleaning of vessel will become a difficult task.

4.1.3.4 Selection of Resin Transfer Molding Equipment Capacity

The capacity of RTM equipment should be selected only after careful consideration of the size of product and the proposed volume of production. Machines capable of generating high pressures are needed for moderate-to-large-sized products. For small-size products, low-capacity machines should be used. The use of machine with higher capacities than required can lead to problems, such as washing away of reinforcement, bulging of mold, air entrapment, and wastage of resin. Use of low-capacity machines for large products warrants the use of resins with longer cure times to avoid gelling.

4.1.3.5 Mold

The sketch of a typical RTM mold section is shown in Figure 4.5. The mold essentially consists of clamped male and female halves, the guide pins, and the gasket along the parting line. The design of mold for RTM process is similar to that of hand lay-up mold. The inside surface of mold walls must have smooth surface finish, which determines the finish of the product. The mold wall thickness has to be designed such that it can withstand the injection pressure without excessive deformation. Any deformation during the injection will lead to change in product dimensions. If there is any deformation, the deformation must be such that it is within the allowable tolerances of the product dimensions. Mold wall should be sufficiently conductive to heat so that excessive heat buildup may not happen. By and large, FRP molds are suitable for most of the RTM processes. Fillers are added to the resin while fabricating the mold to get desirable properties, like dimensional stability, reduction in-mold shrinkage, and thermal conductivity.

Injection port is the nozzle through which resin is injected into the mold. As far as possible, the injection port must be located at the center of the mold so that the resin flows radially to the periphery. For large products, multiple injection ports may be needed.

Air vents are provided at suitable locations in the mold for allowing the volatiles and entrapped air to escape. Air vents are located at the points where the resin reaches at the end.

Guide pins are also generally provided in the mold for a perfect closure. Sealing gaskets are provided along the parting line to prevent the flow of resin through the parting line. The gasket thickness and the compression during tightening may alter the mold cavity volume. Hence, special care should be taken while designing the mold. A properly designed mold halves get seated directly on one another when the gasket is compressed.

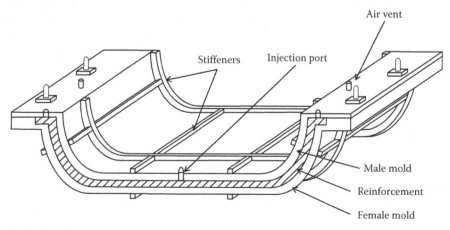

FIGURE 4.5
Schematic of a section of RTM mold with stiffeners.

4.1.3.6 Raw Materials

The raw materials used for making composites by RTM process include fibers, resins, catalysts, accelerators, additives, and fillers. Glass fiber is the most commonly used fiber in this process, but other fibers can also be used without much difficulty. The different forms of reinforcement products and their advantages and limitations are given in Table 4.2.

Polyester, vinyl ester, and epoxy resins are commonly used in the RTM process. The resin system should have the following characteristics:

1. Low viscosity to facilitate easy flow through the fiber packing (100–400 mPa s)
2. Low peak exotherm to avoid high heat build-up
3. Short gel time and cure time (the molding cycle time will be longer with slow curing resin and that makes the ineffective use of mold. At the same time, highly reactive resins are not suitable, since they do not provide enough time for mold filling)

Small quantities of fillers can be added to the resin, if necessary. However, the filler addition will increase the viscosity and hence higher injection pressure may be required.

4.1.3.7 Resin Injection Pressure

The resin is injected into the mold cavity under pressure to completely fill the mold. The pressure must be sufficient for the flow of resin into the entire space within the mold cavity, but the pressure should not be too high to cause fiber wash.

TABLE 4.2

Different Forms of Reinforcements Used in RTM

Reinforcement Form	Advantages and Limitations
Continuous strand mat (swirl mat)	Good formability and resistance to fiber wash, high bulk factor, moderate drapeability, high permeability
WRM	Gives high strength (biaxial), low bulk factor, good formability, poor drapeability, medium permeability
Unidirectional roving mat	High unidirectional strength, high stiffness, good formability and drapeability, high permeability
CSM	Low formability, low resistance to fiber wash, low cost, high bulk factor, good drapeability and permeability
Preforms	Highly complex shapes possible, little forming/handling necessary, high initial cost, low permeability
Veils/surface mat	Gives good surface quality and wear resistance

The flow behavior of resin in RTM can be approximated to fluid flow through a packed bed based on Darcy's law. Darcy's law can be expressed in differential form as given in Equation 4.1 (Greenkorn 1983):

$$\frac{dP}{dx} = \frac{\eta Q}{\kappa A}$$ (4.1)

where
  η is viscosity
  Q is flow rate
  A is the area of cross section
  κ is permeability constant

The κ value varies depending on the type of fiber and packing. The range of permeability values for CSM, WRM, and multiaxial preform are $1-4 \times 10^{-9}$, $0.05-0.30 \times 10^{-9}$, and $0.05-0.30 \times 10^{-9}$ m^2, respectively. The rate of change of pressure with distance of flow is directly proportional to the volumetric flow rate. For faster mold filling, the pressure difference and permeability should be high and the viscosity of resin should be low. The pressure required for injection is given by Equation 4.2:

$$P = \int_0^L \left(\frac{\eta Q}{\kappa A} \right) dx$$ (4.2)

where L is the total length that the resin covers while flowing through a strip of width dx.

4.1.3.8 Mold Filling Time

The total filling time t in minutes can be calculated by using Equation 4.3:

$$t = \frac{\left[V_{mo} \left(1 - V_f \right) \right]}{1000R}$$ (4.3)

where
  V_{mo} is the total volume of mold in cm^3
  V_f is the fiber volume fraction
  R is the rate of pumping of the resin in L min^{-1}

4.1.3.9 Molding Process

Inside surfaces of the two halves of the mold are cleaned thoroughly. The surfaces are polished using fine emery paper, wax polish is applied, and a thin film of PVA is applied over the layer of wax for easy release of the product. The disposable inlet port and air vents are then fitted to the mold.

A layer of gel coat resin with appropriate pigment is applied. The gel coat thickness should not exceed 0.5 mm. After the gelation of the gel coat, the calculated quantity of fiber reinforcement is placed in the cavity between the molds. The plies can be stitched together near the inlet port, since the injection pressure will be high at that place and fiber wash can occur. The inserts, if needed, must be correctly placed before the mold is closed. The clamping of mold should be tight enough to withstand the injection pressure, and it ensures that the product is molded to accurate thickness.

Once the mold is closed, the resin is injected into the mold using RTM equipment at the calculated pressure. The mixing head of the machine should be flushed with a solvent at 15 min interval to avoid curing of the resin within the mixing head. At the end of the work, the entire pipeline through which resin is being pumped must be flushed using a solvent.

The mold is left undisturbed until the resin is cured. Mold cooling is necessary for thick products, since the highly exothermic heat release may lead to degradation of the polymer. Mold cooling can be carried out by water circulation in the mold.

4.1.3.10 Advantages and Limitations

The molds used in RTM process need not be as sturdy as that used in compression or injection molding methods, because the resin injection is carried out at a relatively low pressure. It is possible to produce products with smooth finishes and desired colors on both sides by applying suitable gel coats on both sides with appropriate colors before the mold is closed. The molding within the cavity of a correctly profiled and accurately closed mold can help to make products with close dimensional tolerances. The other advantages are

1. Molding with inserts of wood, foam, or metal is possible.
2. The controlled usage of fiber and resin reduces material wastage.
3. A variety of products can be molded sequentially using a mobile RTM machine.
4. Styrene emission is practically eliminated.
5. Molding personnel requires less training to perform RTM operation.

There are a few limitations in the RTM process, which requires special attention. Some of the limitations are

1. There is a tendency for fiber wash, especially when high injection pressures are used during molding with tightly packed fibers or molding of large components.
2. Unlike compression molding, post trimming is required for this process.
3. The process is rather slow compared to compression molding or injection molding (the cycle time mainly depends on the mold filling time, which is longer for the large parts. The manual packing of reinforcements also extends the cycle time).

Over the years, some improvements have taken place in RTM process. Two of the improved versions of the RTM processes are vacuum-assisted resin transfer molding (VARTM) and Seemann composite resin infusion molding process (SCRIMP).

4.1.3.11 Vacuum-Assisted Resin Transfer Molding Process

VARTM is an adapted version of RTM process, in which the application of vacuum assists the better impregnation of fiber packing. It is a very cost-effective process to make large structures such as boat hulls. In this process, fibers are placed in a female mold and covered with a flexible material to form a vacuum-tight seal. Hence, tooling costs are very much reduced because only one half of the rigid mold is used to make the part. An application of vacuum facilitates the removal of entrapped air and better flow of resin through the fiber packing. Since it is a closed mold process, styrene emission is close to zero. Moreover, a high fiber volume fraction can be achieved, and therefore the structural performance of the part is generally high.

4.1.3.12 Seemann Composite Resin Infusion Molding Process

SCRIMP is a patented technology of the Seemann, TPI, and Hardcore composites companies. This process is suitable to produce medium-to-large-sized parts, and it is similar to the VARTM process. In this process, a steady vacuum is applied to first compact the fiber layers and then to draw resin into the layers. In this way, fiber layers are compacted before the resin started flowing, thus eliminating void formation and resin flow between mold surface and fiber pack. Since the fiber pack is compacted by vacuum application, the fiber volume fraction in the part is increased. In addition, a patented resin distribution system consisting of a special resin flow medium combined with simple mechanical devices is used. The arrangement of different layers in this process is schematically shown in Figure 4.6. This process is mainly

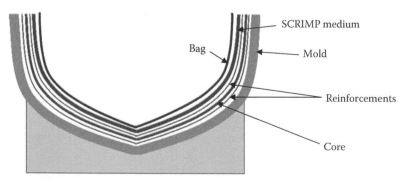

FIGURE 4.6
Schematic of SCRIMP lay-up. (Reprinted from Mazumdar, S.K., *Composites Manufacturing*, CRC Press, Boca Raton, FL, 2002, p. 176. With permission.)

used in marine industry. The other applications of this process include the fabrication of marine docking fenders for seaports, windmill blades, antenna dishes, railroad cars, bus bodies, and amusement park rides.

4.1.4 Vacuum Impregnation Methods

In this process, fiber wetting by resin is assisted by vacuum. Because of vacuum application, the resin penetrates through the reinforcement easily without air entrapment. Thus, the defects in the product can be minimized. The materials used for the molds in the VI process are comparatively cheaper than used in compression molding.

VI process is useful for the production of large-sized parts requiring closer thickness control than the hand lay-up process, and parts requiring a precise fiber-to-resin ratio. This process is also suitable to make pipes used in electrical insulation applications, since there is some air entrapment when the pipes are made by filament winding or centrifugal casting process. The presence of air bubbles can lead to the breakdown of insulation under high-moisture conditions. The VI process is the basis for many FRP fabrication processes like autoclave molding, vacuum-assisted compression molding, and VARTM.

The advantages of VI methods are cheaper molds, faster mold turn around, less labor requirement, smooth finish on both sides, large-sized molding, reduced styrene emission, and good dimensional control. The disadvantages are costlier polyester resin, because diallylphthalate should be used instead of styrene, and higher investment compared to hand lay-up process.

There are three variants in the vacuum-assisted process. They are

1. VI
2. Vacuum injection molding
3. Vacuum bag molding

4.1.4.1 Vacuum Impregnation

VI is used for the manufacture of parts, which need controlled mechanical properties and good dimensional control. Some of the examples are nose cones, radomes, etc. This process can also be used for producing circuit breakers tubes that need void-free structure with good mechanical properties and high electrical resistivity. The schematic of VI process is shown in Figure 4.7.

The mold is made with FRP, cast aluminum, or cast iron depending on the number of products to be made. Inside surfaces of the mold are highly polished or chrome plated. An appropriate release agent is then applied. The mold is closed after placing the reinforcements. Special care should be taken to ensure airtight mold closure.

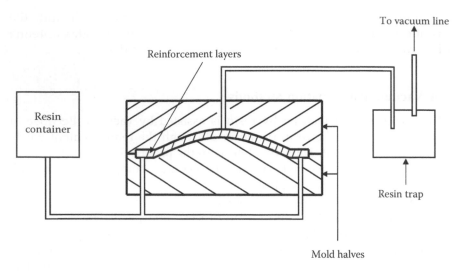

FIGURE 4.7
Schematic of VI process.

The topmost part of the mold is connected to the vacuum line through a resin trap. The resin trap prevents any accidental sucking of resin to the vacuum lines. The bottom parts of the mold are connected to the resin container. During the application of vacuum, the resin flows from the bottom to the top of the mold. Once the mold is completely filled with resin, the resin starts flowing into the resin trap. At this stage, the vacuum line valve is closed and the resin is allowed to cure. The product is taken out from the mold after curing.

Different types and forms of fibers reinforcements can be used in the VI depending on the application. The advantages and limitations of different forms of fiber reinforcements are the same as that in the RTM process. Pre-impregnated fabrics and roving can also be used to produce a part with a controlled fiber-to-resin ratio.

The curing conditions, initial viscosity, gelation time, and wettability are the important properties of resins to be considered for VI. Other factors to be considered for the selection of resin are service temperature of the part, mechanical properties, chemical resistance, electrical resistance, etc. Commercially available impregnating grades of polyester, epoxy, and phenolic resins are suitable for this process. The only condition is that the volatile content should be as low as possible. Because of this condition, diallylphthalate is used in polyester resin instead of styrene. The resin used in this process should have a long pot life and low viscosity (less than 0.1 Pa s). A long pot life allows the resin to completely fill the large and complex parts with high reinforcement content. Lower viscosity permits the resin to fill all the areas of mold quickly.

4.1.4.2 Vacuum Injection Molding

This process is a combination of VI and resin injection. It is known as Hochst process. FRP can be used for making the mold. The lower half of the mold is rigid (6–8 mm thick) and the upper half is more flexible. A vacuum channel is built around the periphery of the mold for better mold closure. The molding process is almost the same as described for VI. Catalyzed resin is injected under pressure (0.05–0.3 MPa). More than one injection point should be used for large products. The air present in the mold is sucked out by applying vacuum and then the resin starts flowing into the mold. The flexible top half forces the resin to flow through the reinforcement, instead of through the space between reinforcement and mold surface. The resin is injected until the reinforcement is thoroughly impregnated. Vacuum is maintained till the resin is cured.

4.1.4.3 Vacuum Bag Molding

This process is used for producing complex components in small numbers. Both large and small parts can be made by this process. This process is useful to make aircraft parts, since the quality and reliability of the products made by this process are very good. Actually, it is a consolidation process, in which prefabricated parts are consolidated. In one case, the products are fabricated by hand lay-up process and then consolidated using vacuum bag molding. In another case, the products are fabricated using prepregs, and the final consolidation is carried out by vacuum bag molding. At present, vacuum bag molding technology is advanced and sophisticated. Precise control of temperature, pressure, and other parameters is possible.

The schematic of vacuum bag molding setup is shown in Figure 4.8. The mold is made of aluminum or steel coated with polytetrafluoroethylene (PTFE). Polyamide or polyester film can be used as release layer, if the mold is not coated with permanent release coating. A peel ply is placed on top and bottom of the product to be molded, which is an optional layer. It is usually a woven fabric made of polyamide, polyester or PTFE coated glass fabric. Peel ply renders the cured laminate surface paintable or bondable to other components.

A separator is placed on both the sides of the laminate after peel ply. It allows the volatiles to escape and excess resin to bleed through. The separator can be a porous or perforated fiber layer, usually made of fluorocarbon polymers. The degree of porosity or spacing of perforations determines the amount of resin flow out of the laminate.

A bleeder layer is used to absorb excess resin from the lay-up. Fabric made out of glass fiber or other absorbent materials can be used as bleeder. The amount of bleeder material needed is determined based on its absorbency and the desired fiber volume fraction.

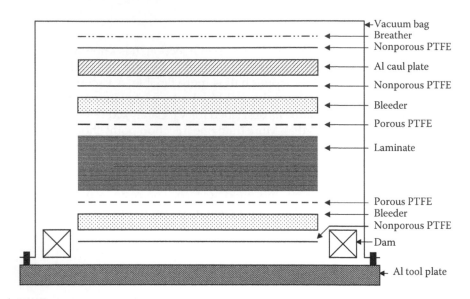

FIGURE 4.8
Schematic of the vacuum bagging lay-up. (Reprinted from Mallick, P.K., *Fiber-Reinforced Composites*, 3rd edn., CRC Press, Boca Raton, FL, 2008, p. 391. With permission.)

Barrier film is a nonstick film. It is placed after the bleeder to control the resin flow. Polyamide or cellophane films can be used as barrier. Some perforations in the barrier film can help to prevent the clogging of excess resin in the bleeder.

A breather film placed above the barrier film is used for the uniform application of vacuum all over the area of the lay-up and removal of volatiles formed during curing. A loosely woven material made of glass, cotton, or jute fiber can be used as a breather film.

The vacuum bag is an impermeable layer made out of polyamide, neoprene rubberized cloth, or silicone rubberized fabric. It applies pressure on the lay-up, when vacuum is generated by the vacuum pump. Autoclaves are used for the resin systems requiring high temperature and pressure for curing. In that case, the bag material should be resistant to the temperature and pressure applied during the autoclave process.

The primary purpose of vacuum application is to reduce void content in the product. Vacuum lines of capacity 4.45×10^{-4} m^3 s^{-1} should be located approximately for every 0.23 m^2 of bag area. Care should be taken so that the volatiles formed during suction or curing reaction do not enter the vacuum pump or the control valves.

4.1.5 Autoclave Process

Autoclave is a pressure vessel with a provision for heating. When the molding is carried out in an autoclave, the molding process is called autoclave molding.

It is actually an extension of vacuum bag molding process. An autoclave is a closed vessel (spherical or cylindrical) in which curing process occurs under simultaneous application of high temperature and pressure. The quality of the products made in autoclave is very good because of the application of pressure and temperature. The temperature initiates the curing reaction and the pressure aids the better consolidation of the product during curing. High quality of the products is essential for the aerospace industry. Hence, autoclave molding is a preferred method in the aerospace industry. The production rate of autoclave molding is low. This is another reason for the suitability of this process for aerospace industry. Prepregs are generally used in the autoclave process. Composites made from prepregs provide superior mechanical properties, which can meet the stringent requirements of the aerospace industry. Instead of prepregs, chopped fibers mixed with resin can also be used. During the molding process, the viscosity of resin decreases. The low-viscosity resin flows and fills the voids created due to the removal of entrapped air and residual solvents.

4.1.5.1 Prepregs

Prepreg is the short form of pre-impregnated fiber reinforcements. A prepreg is a thin sheet or lamina of unidirectional or woven fiber-reinforced polymer composite. It is protected on both sides with easily removable separators. Prepregs are thus considered as an intermediate product in the fabrication of certain polymer composites. Prepregs have higher fiber content (up to 65% by volume). They are available in tape and cloth forms. Usually woven cloths are pre-impregnated, but woven rovings and CSMs can also be pre-impregnated. Prepregs have uniform fiber-to-resin ratio and the variation is only at ±2%. Resin flow and volatile content can also be controlled to close tolerances. Prepregs are extensively used in aerospace, electrical, and sport goods industries.

Prepregs are generally made with carbon fibers, but other fibers can also be used. An important characteristic of a prepreg made with a thermoset polymer matrix is that the resin is in a partially cured state with a moderately self-adhesive tack. Epoxy resin is the preferred resin system for prepregs, since better control of curing is possible with this resin. Epoxy resins cure in two stages and it is easy to stop the curing after the first stage. Other resins, such as polyester, phenolics, silicones, and polyimides, can also be used. When the resin system is not undergoing curing in two stages, the prepreg tack can be achieved by the addition of liquid rubbers or resins. Typically, a unidirectional prepreg can be in the form of a long roll of length 50–250 m. The width of prepreg varies from 75 to 1500 mm. The thickness is usually 0.125 mm.

Prepregs can be made in two ways. In the first method, the fiber cloth is dipped in the resin or solution of the resin, and then the resin is allowed to cure up to B-stage by applying heat. The temperature is then reduced to

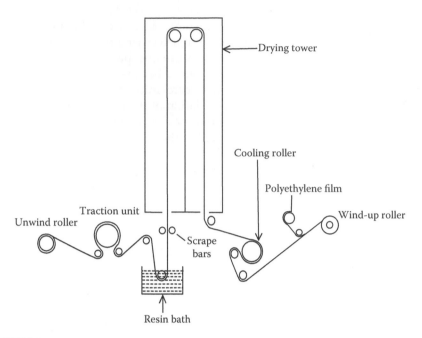

FIGURE 4.9
Tower for prepreg production. (Reprinted from Talreja, R. and Manson, J.E. (volume eds.), in *Comprehensive Composite Materials*, Vol. 2, eds. A. Kelly and C. Zweben, Elsevier Science Ltd., Oxford, U.K., 2000, p. 705. With permission.)

arrest further curing. Very long and continuous prepregs are produced by using mechanized process. One such type of machine is known as tower (Figure 4.9). Fibers under tension pass through a resin bath and are impregnated with resin. Resins impregnate the fiber as liquid, hot melt, solution, or liquid monomer reactants. The impregnated fibers pass through a set of scrape bars to squeeze out excess resin and obtain uniform fiber/resin ratio. The wetted fibers then pass through a drying oven, where the temperature gradually increases. Volatiles are removed and the resin starts curing up to the B-stage. Once the resin reaches the B-stage curing, further curing is stopped by reducing the temperature of the prepreg. Time and temperature necessary for reaching the B-stage depends on the resin formulation. The prepregs are protected by polyethylene films and the complete roll is covered in aluminum foils.

In the second process, the fiber layer is coated with resin powder. The powder melts at high temperature and the viscosity is low enough to impregnate the fibers. The molten resin then starts curing and the curing is stopped at the B-stage.

The prepregs are stored in refrigerated chambers. The temperature maintained in the chamber is an important factor and that determines the shelf life of prepreg. The shelf life of prepreg is 6–8 months when stored at −18°C.

Moisture condensation within the chamber must be completely avoided to preserve the quality of prepregs.

The amorphous thermoplastics can be dissolved in solvents, and prepregs can be made using these solutions similar to thermoset resins. However, the solvent must be completely removed from the prepreg before the component is made. Even trace amounts of solvent can reduce the glass transition temperature (T_g) and mechanical properties of the composite due to the plasticizing effect of the solvent. Melt impregnation of thermoplastics is a difficult process. Much higher temperatures are required to melt the high molecular weight polymer, and even after melting, their viscosities are several orders of magnitude higher than that of the thermoset resin. This makes uniform fiber impregnation very difficult. Hence, melt impregnation with viscous polymers requires high pressures, slow speeds, and thin fiber beds.

An alternate method developed for impregnating fiber tows with thermoplastics is powder coating method. A fluidized bed is used to coat thermoplastic powder onto the fiber surface and the polymer is fused on the fibers in an oven. Other processes such as electrostatic coating and powder slurry coating have also been developed. Another method of making thermoplastic prepreg is by commingling the reinforcement fibers and fine thermoplastic fibers. Both the powder-coated and the commingled tows are normally woven into a tackless, but very drapable, prepreg. The fiber distribution in powder-coated and commingled prepregs will not be as uniform as the prepregs formed by solution impregnation and melt impregnation. In addition, wrinkling and buckling of fibers can happen during material placement and consolidation of powder-coated and commingled fiber prepregs due to their extensive drapeability.

Thermoplastic prepregs are also made by film calendering process. The fiber layer is sandwiched between two thermoplastic films and calendered using hot rollers so that the film melts and coats the fiber. Thermoplastic prepregs need not be stored under refrigerated conditions. They are tack free and boardy. They do not stick to each other and so protective films are also not needed.

Tack is an important characteristic of prepregs which should be controlled to facilitate lay-up operations. Tack is a measure of the adhesion quality of prepregs. It is affected by the resin and inert volatile contents, the extent of resin curing, and the lay-up ambient temperature and humidity. Tack is increased by increasing the resin and volatile contents, reducing the extent of resin curing or by slightly increasing the ambient temperature. Tack can also be increased by adding some special additives, but the additives may alter the final properties of laminate.

It is necessary to have adequate tack in the prepreg. The prepregs with adequate tack adhere well to the mold surface or the preceding plies during lay-up. The required tack should be such that the prepreg remain adhered to the backing material until a predetermined force is applied to separate it. The level of tack is comparable to the tack that exists in the normally used self-adhesive tape. Prepregs with excessively high tack cannot be handled easily. It may lead to disorientation of fiber or roping (fiber bundling) of the reinforcements. It is also very difficult to control the fiber/resin ratio, since undetermined amounts of resin are removed during processing. The processing of composites with high tack prepreg is almost similar in many respects to the wet lay-up process. Thermosetting resin prepregs with no tack either are excessively cured or have exceeded their storage life. Such prepregs are useless, since they cannot attain adequate properties on curing. Exceptions are silicone and some polyimide-based prepregs. These prepregs are prepared without tack. Composites based on these prepregs are mainly used for electrical or high-temperature applications, where the mechanical properties are not very important.

Sometimes the tack can be modified by changing the ambient conditions during lay-up. The lay-up should be carried out in an air-conditioned work space, to minimize the effect of temperature and humidity variations. The prepreg tackiness can be increased by the judicious use of hot-air guns.

Tack is not very important with thermoplastic prepregs. They are thermoformed above their glass transition temperature and fused at their melting temperatures. Tack is imparted to the prepregs during fabrication. Hence, the thermoplastic prepregs can be stored indefinitely at room temperature. Another advantage of thermoplastic prepregs is that the magnitude of moisture degradation is very less compared to thermosetting resin-based prepregs.

4.1.5.1.1 Characteristics of a Good Prepreg

Prepregs should have certain basic characteristics that make them suitable for lay-up and fabrication of high quality composites. The basic requirements of a good prepreg are the following:

- The fiber-to-resin ratio should be high and should not vary from place to place.
- Volatile content should be a minimum.
- The prepreg should have adequate tack and flexibility.
- It should have fairly long storage life.

4.1.5.2 Automatic Tape-Laying

Automatic tape-laying (Figure 4.10) is more suitable for the fabrication of large structures. The manual lay-up process is limited by the extent to which the workers can reach. The other advantage of automatic tape-laying is faster lay-up and accurate placement of tapes. Using computer-aided design (CAD), the component to be manufactured is developed mathematically onto a flat surface. This is discretized into layers to be fabricated by tape strips. A software program is used to translate strips in each layer to the final product shape

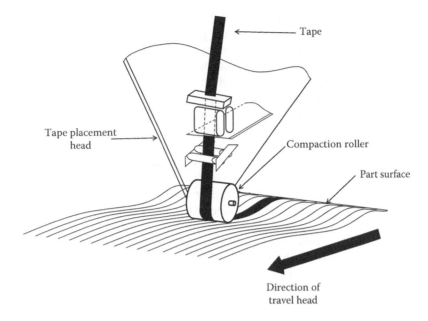

FIGURE 4.10
Automatic tape lay-up. (Courtesy of Cincinnati Milacron, Batavia, OH.)

by a series of numerical control steps. Plies are oriented to within ±1° with the use of tape-laying heads. The lay-up machines can lay 76 mm and wider tapes directly onto the molds, but their use is restricted to simpler shapes.

Unidirectional fiber prepregs are wound on a mandrel or laid-up on a mold. Based on the design conditions, the prepregs are wound at ±θ angles. The plies are laid at optimum angles combined with hoop wraps to develop the desired directional properties for the laminate.

Lay-up is carried out at 0°, ±45°, 90° or 0°, ±θ, 90° sequences. To optimize the properties of a π/4 sequence, additional plies are added in the preferred direction, and the numbers of plies are reduced in the other directions. To maintain symmetry about the principal directions, plies in the ±45° directions are added or reduced in pairs. It is always desired to have the plying sequence symmetrical about the laminate midplane to prevent bending–stretching coupling effects. Anisotropy of fabric-reinforced prepregs in one ply is corrected with equal but opposite anisotropy in adjacent plies. The corrections to achieve symmetry are important to avoid distortions to the cured laminates.

Templates are normally used to expedite lay-up operations efficiently. Indexed Mylar templates are used to reduce the lay-up time on the mold and for perfect lay-up. The prepregs are cut to the size and first laid-up on the templates. They are then transferred to the mold by referring the indexes on the templates. Once the plies are transferred from the template, the bleed-out systems are placed and the assembly is bagged and cured.

The drape characteristics of woven fabric prepregs depend on the weaving patterns. Long-shaft satin fabric-based prepregs are the most complicated, while prepregs made with square weave or basket weave fabrics attain the least drape. Woven fabric prepregs may be darted to comply with convoluted shapes. Darting is the practice of slitting the prepregs at locations where folds would normally occur in a lay-up. The excess material at those locations may be removed completely or the whole of excess material may be allowed to overlap, provided that it is wrinkle free. When the excess material is completely removed, an additional ply is required to compensate the strength at the weak butt-joints. However, it should be ensured that the joints do not coincide in any of the successive plies.

Thermoplastic prepregs contain either collimated fiber reinforcements or woven fabrics. The required number of plies is stacked on the mold to form the lay-up. The lay-up is then vacuum bag or autoclave molded at the fusion temperature of the thermoplastic matrix. The dwell time at the fusion temperature depends on the lay-up thickness. Approximately 30 min dwell time is required for eight-ply graphite fiber-reinforced laminate. Some of the problems with thermoplastic prepregs are degraded prepreg surfaces, contamination, and nonuniform fusion. Surface contaminations result in delamination and increased voids content. Nonuniform fusion results in poor properties for laminate and the laminates are prone to easy delamination.

Generally, the thermoplastic prepregs are consolidated by compression molding. The shaped part made by filament winding of thermoset polymer prepreg is cured in a hot chamber and the part made by tape-laying is usually cured in an autoclave.

4.1.5.3 Autoclave Molding

The schematic of autoclave molding process setup is shown in Figure 4.11. The mold surface is covered with a PTFE-coated glass fabric separator. Plies are cut from the prepregs roll into the desired shape, size, and orientation. The backup release film is removed from each ply before lay-up. The prepreg plies are laid-up in the desired fiber orientation and sequence on the separator.

A porous release cloth and a few layers of bleeder fabric are placed on the prepreg stack. The complete lay-up is covered with a thin heat-resistant vacuum bag after placing another sheet of PTFE-coated glass fabric separator and a caul plate. The entire assembly is transferred into a preheated autoclave where a combination of external pressure and vacuum is applied to consolidate the plies.

The cure cycle of an epoxy prepreg consists of two stages. In the first stage, the temperature is increased up to 130°C and this temperature is maintained for 60 min. During this period, the minimum viscosity is reached. The external pressure is applied on the prepreg stack, which causes the excess resin to flow into the bleeder fabric. The resin flow is critical since it allows

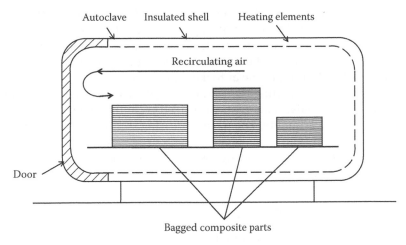

FIGURE 4.11
Schematic of autoclave molding. (Reprinted from Schewartz, M.M., *Composite Materials*, Vol. II, Prentice Hall, Upper Saddle River, NJ, 1997, p. 25. With permission.)

the removal of entrapped air and volatiles from the prepreg, which reduces the void content in the cured laminate.

At the end of the dwell, the autoclave temperature is increased to the curing temperature of the resin. The temperature and pressure are maintained for 2 h or more till the resin is completely cured. At the end of the curing, the temperature is slowly reduced without reducing the pressure. After reaching the room temperature, the pressure and vacuum are released and the laminate is removed from the vacuum bag. If necessary, the laminate is postcured at an elevated temperature in a hot-air oven.

Face bleeding (normally to the laminate face) is preferred over edge bleeding when thin structures are made by autoclave molding process. Face bleeding is more effective since the resin path before gelation is shorter in the thickness direction than parallel to the surface.

Tooling is less expensive for the autoclave molding than compression molding. Molds are required to withstand only the curing conditions without any distortion or degradation. The thermal expansion coefficients of composites are generally low compared to the metallic tooling materials. While heating, the tool expands more compared to the lay-up, and the uncured lay-up can deform to that extent. During cooling, the tool shrinks more than the part and the cured laminate is rigid, but not rigid enough to completely resist shrinkage. As a result, the final dimension of the part will be slightly larger than the expected dimension. The recommended empirical correction for steel or nickel tools is to make the tool at 0.999 of the engineering dimension and for aluminum tools at 0.998. These corrections are needed to ensure an acceptable fit of mating composite parts. To avoid the problem of thermal expansion coefficient mismatch, graphite–epoxy molds can be used.

The pressure exerted on the vacuum bag is transmitted to the lay-up through the caul plate. At the initial stage of curing, the viscosity of resin is low. The caul plate exerts pressure on the lay-up and the excess resin squeezes out from the lay-up. Hence, the metal caul plate must have high rigidity so that it does not deflect under the autoclave pressure.

Although autoclave molding is an expensive process, the higher cost can be amortized by taking advantage of its improved capability to produce complex geometries with high quality. The parts requiring secondary bonding can be economically produced in autoclave by a co-curing process. Sometimes the operating cost of autoclave molding can be considerably reduced by curing many parts at a time.

4.1.6 Filament Winding Process

Filament winding is a semiautomatic process for manufacturing axisymmetric fiber-reinforced composite parts. Continuous reinforcements in the form of rovings or yarns coated with matrix materials are wound using predesigned machines, in which the fiber feeder traversing at speeds synchronized with the mandrel rotation controls the winding angles and the placement of the reinforcements. The fiber tows may be wrapped in adjacent bands or in repeating patterns, which ultimately cover the mandrel surface. Successive layers are wound at the same angle or different angles until the desired thickness is reached. The winding of fiber tows can be polar or helical winding patterns. Depending on the design criteria and shape of the product, the winding angle is selected to wind the fibers in specific directions and to pack them well. The fiber tension produced when wound on a positively curved (convex) surface provides the required compaction. In filament winding process, it is possible to align fibers in the direction of principal stresses. Moreover, very high fiber contents can be reached in filament wound products.

Although filament winding was initially developed for thermoset resins, winding with thermoplastic materials have also been practiced. Processes have also been developed for winding ceramic slurry-impregnated fibers. Parts made with ceramic materials are subsequently sintered at high temperature to achieve the desired properties.

Filament winding process can be used to produce any product with positively curved surface. Cylindrical, spherical, conical, and geodesic shapes are within winding capabilities. Surface with reverse curvature (concave) are not suitable since the fibers cannot take the contour of the reverse curve. If needed, surface with reverse curvature can be made by internal pressurization of an uncured winding in a closed mold. Although products with axisymmetric surfaces of revolution are ideal, other profiles and tapering sections can also be wound with modern filament winding machines. Combination winding is another possibility, in which filament winding is carried out on thermoplastic pipes or metal pressure vessels to improve the performance. Filament winding machines are available with various levels of sophistication. The machines range from simple mechanically controlled machines to more sophisticated computer-controlled machines with three or four axes of motion.

The basic raw materials are the reinforcement fibers and the polymer matrices. Fillers can be added in the resin, but they are not generally used because they increase the viscosity of resins.

4.1.6.1 Reinforcements

Nearly all filament winding is carried out with E-glass roving. It is a common reinforcement for commercial applications including chemical tanks, petroleum tanks, and pipelines, which are not subjected to excessive bending or buckling loads. When the loading is more conducive to buckling, wall thickness can be increased, or stiffener can be incorporated in the regions prone to buckling.

An alternative method for imparting greater rigidity is to use higher-modulus fibers, such as aramid, polyethylene, or carbon fibers. These reinforcements have much higher specific moduli as compared to glass fibers. Aramid fibers are used for producing aerospace structures, rocket motor casing, etc., where specific stiffness and strength are very critical. However, aramid fibers have poor compressive and shear properties. Polyethylene fibers have properties similar to aramid fibers. These two types of fibers are not suitable for high temperature curing resins, since the melting temperatures of these fibers are low. Carbon fiber is another versatile fiber because of its high modulus, strength, and temperature resistance. However, fiber breakage during winding can lead to practical winding difficulties, and hence special precautions are required while winding with carbon fibers. The price of carbon and aramid fibers, though greatly reduced in recent years, still remains prohibitive for widespread commercial applications. A cost-effective alternative is the usage of hybrid composites, in which the incorporation of certain amount of high-modulus fibers improves the rigidity of glass fiber-reinforced composites.

For noncritical applications, jute fibers can also be used. Modular houses, silos, etc., are some of the possible products. Jute fibers provide strength for such applications, but further research is needed to improve their durability.

Fibers in different forms, such as rovings, yarn, tape, and cloth, can be used in filament winding process. There are some problems encountered during the filament winding process and special care should be taken to overcome them:

1. *Fuzziness*: Fuzziness indicates fiber breakage. The fiber breakage during winding can lead to a problem of keeping the filament integrity. Moreover, the broken fibers tend to fall off, and the broken fibers accumulated in the resin bath can increase the viscosity of resin, making resin pickup uneven. Hence, fuzziness must be controlled below the acceptable levels.

2. *Catenary formation*: Catenary means unequal tension of filaments within a strand or roving. A roving is a collection of strands, which themselves are a collection of filaments. If there is slackness in few filaments, then they form catenary. The fiber that forms catenary may remain wavy in the wound product. This phenomenon will affect the product quality, since the wavy fibers may not take the load. Roving must be tested to ensure that the catenary formation is absent.

3. *Wettability*: The time available for fiber wetting during filament winding process is very short. Hence, a suitable resin with short wetting time should be selected.

4.1.6.2 Matrix Materials

Most of the filament winding operations are carried out using thermoset resins. Polyester, vinyl ester, and epoxies are the commonly used thermoset resins. Isophthalic and bisphenol polyester resins are used in majority of the chemical plant, petroleum tank, and pipeline applications. Vinyl ester also finds applications in chemical-resistant products. Epoxy resins are mostly used in high-performance products because of their superior mechanical properties, heat resistance, strong bonding to the reinforcement, and lower shrinkage during curing. Epoxies have been used in aircraft, space, and military applications, where the cost is not the main criterion. Polyimides, silicones, phenolics, and furan resins find applications where some special requirements such as high temperature resistance or fire retardance have to be met. However, phenolic resins are not generally recommended for filament winding process since they require pressure application during curing to remove the reaction by-products (volatiles).

4.1.6.2.1 Epoxy Resins

Epoxy resins used for filament winding process are essentially the same as the laminating resins. Minor modifications are made to meet specific processing conditions. The viscosity and pot life are major considerations, while gel time and resin flow under winding operation and during cure are other factors. A low viscosity is essential for good wetting of the fibers and the removal of entrapped air and volatile materials. At the same time, the viscosity should not be too low. At low viscosities, it is difficult to control the resin content and uniformity, and some fibers, such as carbon fibers, will not pick up enough resin. It may also lead to the migration of resin to the outer layer, leaving inner layers dry. Too high viscosity causes fiber fuzzing in the resin bath and feeder, uneven fiber coating, excessive tension, and air entrapment. Reactive diluents are used to bring down the viscosity. Typical diluents used for this purpose are butyl glycidyl ether, phenyl glycidyl ether, and diglycidyl diol. Nonreactive diluents should be avoided, since they produce voids in the product. If possible, the resin viscosity can be reduced by heating the resin. Suitable viscosities for filament winding fall in the range from 0.35 to 1.5 Pa s at 25°C.

A pot life of several hours is required to prevent gelation before the completion of winding. It is not advisable to wind over a gelled on partially gelled layer. Continuous metering of the resin/curing agent mixture, or a controlled batch mixing will reduce the problem of premature gelation. The shrinkage of epoxy resins during curing is lower than that of the other filament winding resins. A low shrinkage is an advantage in reducing internal stresses and at the same time facilitates the easy removal of product from the mandrel. The shrinkage value depends on the resin type, curing agent, rate of heating, and cure temperature.

4.1.6.2.2 Polyester and Vinyl Ester Resins

These resins are widely used in commercial applications because of their lower cost. The physical and chemical properties of these resins are good for these applications. Processing characteristics of these resins are comparatively easy to control for filament winding process. Processing viscosity is controlled by the amount of reactive monomer. In some cases, the viscosity needs to be increased since the commercially available resins may have lower viscosity. The viscosity can be increased by the addition of suitable thixotropic agents. Impregnating viscosities of 0.25–1.0 Pa s are typical for these resins. This range is somewhat lower compared to epoxy resins, which reflects in higher winding speeds. Cure shrinkage of the polyester resins is in the order of 4%–6%, which is higher than epoxy resins.

4.1.6.2.3 Thermoplastics

Filament winding with thermoplastics is a recent development. Almost all thermoplastics can in principle be used for filament winding. Usually thermoplastic prepreg tapes are used. In some cases, commingled yarns can be used. The prepreg tapes are made by dip coating of fibers in solution or melt of a thermoplastic material or by powder coating on the fibers. These prepreg tapes are wound using a tape winding machine. After winding, the product is heated to the appropriate temperature to fuse and form bonding between the adjacent layers.

4.1.6.3 Mandrels

The mandrel is the tool around which the impregnated fibers are wound. Its profile controls the inner profile or geometry of the product. Hence, the mandrel must maintain dimensional accuracy and integrity throughout its operational life. The inner surface finish of the product depends on the surface finish of the mandrel. Hence, it is necessary to maintain good surface finish on the mandrel. In addition, the mandrel should have the capacity to withstand the pressure exerted due to winding tension and stability at the curing temperature. The mandrel must be easily removable after the product is cured.

The mandrels used in filament winding process can be broadly divided into two categories. One category is open-ended non-collapsible mandrels and the other is collapsible mandrels. Open-ended mandrels are relatively easy to design. Either steel or aluminum alloys with appropriate thickness is suitable for making the mandrels. Careful consideration must be given to mandrel design and selection of a suitable material for the collapsible mandrels.

4.1.6.3.1 Non-collapsible Mandrels

These mandrels are used for open-ended structures such as cylinders or conical shapes. They should have smooth surface finish with an axial taper of at least 1:200 for the easy release of product. When the matrix material shrinks on the mold during curing, the removal of product becomes difficult. Screw or hydraulic extractors can be used to overcome the high frictional forces.

There are metallic or thermoplastic mandrels, which form part of the product. They are usually called liners. The products with liners have good impermeability to gases. These products are ideally suited for storing gases.

4.1.6.3.2 Collapsible Mandrels

These mandrels are made of different materials depending on the size, quantity, and quality of the products. Each material has its own advantages and disadvantages. Segmented metallic mandrels are suitable for the products with diameter in the range of 1–1.5 m, and the minimum number of products is 25. The mandrel segments have the required product shape after assembly. The segments can be dismantled during the release of the product. The removal of mandrel may be difficult with small polar (end) openings.

A water-soluble polymer-based mandrel can be made by casting over a central axis and polar fittings, using sand and PVA slurry. This mandrel is suitable for products up to 1.5 m diameter in limited quantity. The mandrel can be removed easily by flushing with hot water, since the polymer binder readily dissolves in hot water.

Frangible mandrels can be made with plaster of Paris. These are suitable for large-diameter products. The mandrel is made by casting over a removable or collapsible tooling. After winding, the central tool is removed and the plaster is chopped off. Sometimes the breakout is difficult and can cause damage to the product. Chains can be embedded in the plaster for easy breakup and removal.

Inflatable mandrels are made by inflating a bag. However, it is very difficult to maintain close tolerances in the products while using these mandrels. One technique to avoid dimensional change of mandrel is by filling with sand and applying vacuum. The inflatable mandrel can be used wisely to apply pressure also for better consolidation of the product. The uncured product is transferred to a closed mold and then cured with pressurization through the mandrel.

4.1.6.4 Winding Machines

The most essential item for the filament process is the winding machine. The winding machine has the facilities for wetting the fiber, tensioning the tows, and winding the fibers or tapes at the required angle in a uniformly spaced pattern. Winding machines can be broadly divided into three groups:

1. Helical winding machines
2. Polar winding machines
3. Advanced winding machines

4.1.6.4.1 Helical Winding Machines

The helical winding machine lays the fiber tows on a rotating mandrel at winding angles varying from 0° to 90° with respect to the axis of rotation. The basic movements of this machine are mandrel rotation and the feed traverse motion. It is possible to vary the winding angle by varying the speed of the two movements. The feed eye moves to and fro from one end of the mandrel to another end creating an angle ply or netting structure on the mandrel. Figure 4.12 shows the layout of a typical helical winding machine. In addition to normal movements, a cross slide perpendicular to the mandrel axis and a fourth axis of motion and rotation of feed eye can be added. These permit more accurate fiber placement over the ends.

The fiber tows pass through a resin bath in the wet winding process. The resin bath is not needed for the dry or prepreg winding. In the wet winding process, the resin bath also moves with the feeder. The fibers or tapes come from a creel stand. The creel stand can be stationed away from the machine as a stationery stand or can be mounted on the resin carriage. The later

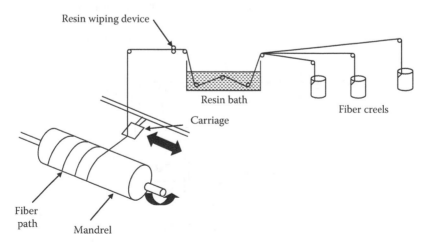

FIGURE 4.12
Schematic of a filament winding process. (Reprinted from Mallick, P.K., *Fiber-Reinforced Composites*, 3rd edn., CRC Press, Boca Raton, FL, 2008, p. 409. With permission.)

arrangement is preferred for very long products, since the fibers need not bend too much while winding from one end to the other end of the mandrel.

Two types of helical winding machines are available. In one type, helical winding can be made with constant helix angles along the length of the mandrel. In the other type, the angle of winding can be varied from very close to 0° to 90° along the length of mandrel. All the commercial machines available today belong to the latter type. The variation in angle is achieved by varying the mandrel rotation speed and feeder speed. The speed variations can be achieved by gear systems or by using numerically controlled step motors. These machines have the added advantage of changing winding angle by preprogramming using computers.

4.1.6.4.2 Polar Winding Machines

Polar winding is generally used for spherical, ellipsoidal, or other closed axisymmetric shapes. The fiber tows are wound from one pole to the other in this process. Schematic of polar winding is shown in Figure 4.13. In order to facilitate easy winding, end bosses are provided and fiber is taken around the end bosses. After one revolution of fiber around the mandrel, the mandrel is rotated in such a way that the next roving is placed side by side.

The polar winding machines can be made in two different types:

1. The feeder revolves around the mandrel, while the mandrel rotates on a fixed axis. This arrangement needs the resin bath and fiber spool to travel with the feed eye. Hence, this is more suitable for prepreg winding than wet winding.

2. The feeder is fixed while the mandrel rotates about its axis and about one of the mounting supports. The rotation of the mandrel is carried out using a cantilever arrangement, and the support system must

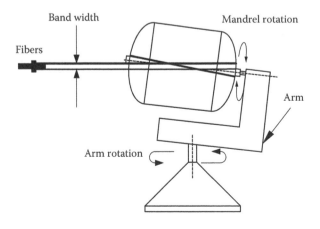

FIGURE 4.13
Schematic of a polar winding process. (Reprinted from Joselin, R. and Chelladurai, T., *J. Fail. Anal. Prev.*, 11, 344, 2011. With permission.)

be rigid enough to carry the load of mandrel without causing any deflections. It will be a difficult task to wind on heavy mandrels used for larger products. This method is suitable for wet winding of smaller products.

Cantilever type support helps the fiber to be easily taken around the mandrel without any obstruction. In such machines, the mandrel can be mounted with its axis horizontally or vertically placed. The vertical mounting is preferable for bigger products, since the load of the mandrel can be supported in a better way.

4.1.6.4.3 Advanced Winding Machines

Many products like pressure vessels and mobile tankers have cylindrical shell with end domes in spherical or ellipsoidal shape. End closures for pressure vessels can be made either by mechanically fastening to the cylindrical portions or by winding integrally. The former method is used for many commercial applications. In this case, the joints will be weak compared to the rest of the positions. A better way is to wind the shell with the end domes in one operation. This is possible by using a combined helical and polar winding machine. In this machine, the winding in the cylindrical region is helical winding and in the end dome region is polar winding. It is possible to wind more precisely with computer-controlled machines, which change from helical to polar mode according to the program in the computer (Figure 4.14).

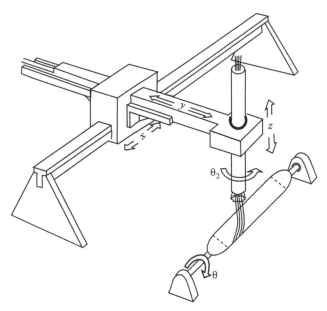

FIGURE 4.14
Basic motions of a five-axis filament winding machine. (Reprinted from Bowen, D.H., *Concise Encyclopedia of Composite Materials*, ed. Kelly, A., Pergamon Press, Oxford, U.K., 1998, p. 181. With permission.)

For high-performance vessels, such as rocket motor cases, integral end closures are essential. End closures, which deviate from spherical shape, are less efficient. The fiber path should have a balance of meridional and circumferential forces under the winding conditions to avoid slippage.

4.1.6.5 Advantages of Filament Winding

Filament winding has several advantages over the other processing methods. Some of the advantages are

1. Filament winding can be done more effectively with a few operators, since it is a semiautomated process.
2. Precise control of fiber-to-matrix ratio, geometrical arrangement of fibers, and fiber tensioning is possible, which gives products of high strength and consistent quality.
3. In filament winding, the fiber content can be as high as 70% by volume.
4. It is possible to produce a part having high strength in any particular direction by suitably varying the angle of winding. The structures can be made for any specific loading conditions, such as internal pressure, external pressure, torsional loading, or compressive loading.
5. Much larger products can be made with less expensive machinery compared to autoclave process. Filament winding process has been used for making storage tanks up to 15 m diameter.
6. Wastage of materials is very less compared to many other polymer composite processing methods.

4.1.6.6 Limitations

Some of the limitations of filament winding process are the following:

1. Products with surfaces having complicated profiles and reverse curvature are difficult to make by the filament winding process. This process may not be suitable to make a product with very small end opening, since it will be difficult to release the mandrel.
2. Although the process is not as expensive as autoclave molding, it is a capital intensive process.
3. The interlaminar shear strength and compressive strength of the filament wound parts are low.
4. The external surface finish is poor unless external molds or compaction devices are used.
5. The quality of filament wound components is generally lower than that of the components made by autoclave molding. Filament wound products have higher void content than autoclave-molded products.

4.1.7 Pultrusion

The pultrusion process involves pulling continuous rovings and/or mats through a resin bath and heated dies. This process is useful to produce long products with constant cross section, such as rods, tubes, and angles. The schematic of this process is shown in Figure 4.15. This technique is used for the manufacture of continuous composite structural profiles. It is similar to aluminum or thermoplastic extrusion; the main difference being that the product is pulled from the die, whereas it is pushed through the die in the extrusion process.

This processing has the additional advantages of providing dependable control of filament orientation and tensioning. It is also possible to produce composite products with high fiber contents (>70 vol.%). Normal production rates are of the order of 0.6–0.9 m min^{-1}. Production rates of as high as 4.5–6.0 m min^{-1} are possible now.

Unidirectional fiber composites with high fiber content can be made by this process. It is also possible to incorporate surface mat/woven mat to modify the properties of products. A surface mat improves the product's surface appearance, chemical resistance, and weather resistance. Woven mats can improve the mechanical properties in the transverse direction. The construction should be balanced (same sequence of reinforcements above and below the horizontal centerline) to prevent the part from twisting and warping during cure. Unbalanced profiles are sometimes made for specific design conditions. Many pultruded products are made only with rovings.

Lower void content and uniform fiber content are the other advantages. Interlaminar shear properties are also good for the pultruded products. The current development in pultrusion process is property optimization for specific products by incorporating fiber reinforcements in exact orientation. Various types of fibers and different orientations impart different properties

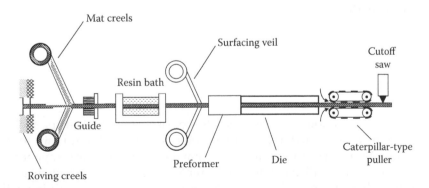

FIGURE 4.15
Schematic of pultrusion process. (Reprinted from Mallick, P.K., *Fiber-Reinforced Composites*, 3rd edn., CRC Press, Boca Raton, FL, 2008, p. 405. With permission.)

to the product. For example, polyester veil imparts surface lubricity, continuous strand mat imparts omnidirectional properties, longitudinal roving improves bending strength, circumferential roving imparts burst strength, and glass veil imparts corrosion resistance.

4.1.7.1 Raw Materials

The reinforcements normally used are glass fibers in the form of continuous strand mats and continuous rovings. Carbon, aramid, and S-glass fibers are being used to produce structural shapes with superior properties. Hybrid composites to satisfy both engineering and economic parameters have become a commercial reality now.

The laminating resin can be unsaturated polyester, vinyl ester, or epoxy resin, but majority of the products are made with polyester resin. A recent development is the reaction injection pultrusion, in which thermoplastics are used as the matrix. The other ingredients are curing agents to cure the resin, pigments to impart color, internal release agents, inert fillers, etc.

The resin mix should have a viscosity of less than 1 Pa s for rod stock and between 1 and 3 Pa s for mat/roving profiles. If the resin mix tank is surrounded by a warm water bath (50°C–60°C), the viscosity of resin decreases; hence, a slightly higher viscous resin can also be used.

4.1.7.2 Pultrusion Process

In the pultrusion process, continuous rovings and/or mats are pulled through a resin bath and then into preforming fixtures, where the section is partially shaped and excess resin and entrapped air are removed. The impregnated fibers then pass through heated dies, where the section is cured. The basic pultrusion machine consists of the following elements:

1. Creel
2. Resin bath
3. Preforming fixtures
4. Matched metal die with heating arrangement
5. Puller or driving mechanism
6. Cutoff saw

The creel for continuous rovings generally consists of bookcase-type shelves with ceramic eyes located at convenient intervals to lead the rovings to the resin bath. One must be careful to ensure that the rovings do not scrape across one another as this will generate considerable static charge and fuzziness.

The resin bath is generally a steel or aluminum trough containing rolls that force the reinforcement to pass through the resin for complete wetting. Most of the wet-out tanks contain a set of rollers or slots at the exit to remove excess resin from the reinforcement. A comb or grid plate is generally provided at the entrance and exit ends of the resin bath to keep the rovings in horizontal alignment. In another variant of pultrusion process, the oriented fiber package is consolidated dry and then impregnated by pumping resin through the packing.

Preforming fixtures consolidate the reinforcements closer to the final shape. They may be made with fluorocarbon or ultrahigh molecular weight polyethylene, since they are easy to fabricate and clean. For longer life and high production runs, chrome-plated steel fixtures can be used.

Dies are usually made with chrome-plated steel. The metal die is heated using electric cartridges, strip heaters, or hot oil. Conductive heat through the die walls is sufficient to cure thin sections. The cure of thick sections can be speeded up and made more uniform by using conductive heat with radio-frequency (RF) radiation. The use of RF heating in conjunction with conventional heating not only increases production speeds but permits the manufacture of massive profiles and those that have widely varying masses in the profile. For RF cure, it is necessary to have a short section of the mold constructed of a RF transparent material such as PTFE or to support wet rovings on each side with grills or guide members.

The puller can be either a pair of continuous caterpillar belts containing pads that engage pultruded sections or a pair of cylinders with pad pullers. Cutoff saw is a conventional cutoff wheel or a continuous rim diamond wheel, which is sometimes used with a coolant spray. In addition to cooling the cutoff wheel and product, the coolant spray minimizes dust. The saw carriage is clamped to the pultrusion product during the cutting operation.

4.1.7.3 Pultrusion Die Design Criteria

Any good tool steel with a reasonably high hardness can be used to make a pultrusion die. Dies should be heat treated to a high hardness after machining. Stainless steel dies and ceramic-coated steel dies can also be used. In order to facilitate the wet reinforcement to enter the die without much difficulty, a bell mouth is machined around the die entrance. For very large structural shapes, the bell mouth should have large width and area.

The die must be fastened to the heating platens with clamps or bolts. A less expensive and reliable method is to fasten the dies with platens using bolts and angle clamps. All internal surfaces of the die should have a good surface finish. Final polishing should be carried out in the longitudinal direction. The internal surface of the die can be chrome plated to a thickness of 0.0375–0.05 mm for a long working life.

The die platens can be heated with electrical strip heaters, electrical cartridge heaters, or hot oil. The use of electrical cartridge heaters with thermocouple controls in the platens has now become a standard practice. With this arrangement it is possible to have several zones with different temperatures and a different temperature at the start-up.

A cold junction is provided on the portion of the die that extends outside the heated platen area. Cooling water enters from the bottom port and flows to the top plate.

4.1.7.4 Pultrusion Products

Pultruded products find a variety of applications from structural to electrical. Electrical applications include transformers, air duct spacer sticks, ladders, bus bar supports, cable support trays, switch actuators, and fuse tubes. Sports applications include fishing rods, sail battens, tent poles, skateboards, ski poles, hockey sticks, paddle shafts, bows and arrows, golf shafts, flagpoles, and pole vault poles. Corrosion-resistant products include bridges and platforms, floor grating, handrails, pipe supports, pipes and tubes, and tank supports.

Products in transportation applications include landing bars in trucks and railcars, kickplates, bus luggage racks, seating, and flat sheets for refrigeration trucks. Construction applications include portable work platforms, signposts, lampposts, roof gutters, greenhouse structures, and panel sections. Miscellaneous applications include heat shields in photo copiers, farm wagons, and pallets in food processing plants. Figure 4.16 shows the different cross-sectional shapes produced by pultrusion process.

Continuous production of pipes in different sizes ranging from 25 mm to 3 m is now a well-proven pultrusion process. Pipes manufactured by this process compete well with those produced from traditional materials, such as concrete and steel. As there is development in fittings manufacture, bedding and trenching technology, and mechanical joining techniques, composite pipes have captured a substantial share of the overall market.

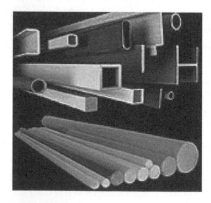

FIGURE 4.16
Various cross-sectional shapes formed by pultrusion process. (Reprinted from the Product profile of Fiber Tech Composite Pvt. Ltd., India. 2012. http://www.fibretech.co.in.)

4.1.8 Compression Molding

Compression molding is one of the fastest polymer composite processing methods. It is a closed mold processing method, in which the product is made within the cavity of a die. A predetermined quantity of molding charge is placed in the die cavity. The molding charge can be bulk molding compound (BMC), sheet molding compound (SMC), prepreg, or wet lay-up. The die is closed and the required pressure is applied by means of a press. The die remains closed till the charge cures and becomes rigid enough to maintain its shape. After the completion of curing, the pressure is released and the die is opened to remove the component. There are two types of compression molding processes:

1. Hot pressing, in which the molding charge is heated while shaping
2. Cold pressing, in which the product is cured without the application of heat

Molding compounds and prepregs are generally hot pressed. Parts made by hand lay-up process are cold pressed, for better consolidation and better surface finish on both the sides.

The advantages of compression molding process are

- Good surface finish on both sides
- Faster production
- Uniform product quality
- Less labor requirement
- Requirement of very little finishing operation

The disadvantages are

- Not suitable for low volume production because of high investment needed for dies and press
- Not suitable to produce very large-sized products

As mentioned earlier, the charge for compression molding process can be BMC, SMC, prepreg, or wet lay-up. The processing of prepregs and wet lay-up is already described. Here, the main emphasis will be given for the BMC and SMC.

4.1.8.1 Molding Compounds

The BMC is also known as "premix" or dough molding compound. It is just a mixture of reinforcements, resin, fillers, etc., in right proportions. SMCs are made in the form of sheets with the same ingredients but with slightly different composition. Molding compounds eliminate the need for separately weighing and mixing the ingredients at the time of molding. The resin remains as it is, without any advancement in curing. This is due to the usage of high temperature curing resin system. To prolong the life of molding compounds, they are normally stored in air-conditioned rooms at 18°C.

Another peculiar property of molding compound is thickening. While mixing the resin with other ingredients, the resin viscosity is low enough to wet all the ingredients. However, the viscosity is high enough to prevent the segregation of ingredients during molding. This transition of viscosity from a low value to high value is achieved by chemical thickening of the resin. Chemical thickening converts the low-viscosity resin mixture into a high-viscosity molding compound with a viscosity of around 1×10^5 Pa s.

This high viscosity makes the compound tack free and leather hard for easy handling and also prevents the separation of ingredients from the resin while molding. Chemical thickening is achieved by adding oxide or hydroxide of magnesium or calcium. It is believed that the difunctional oxide reacts with the terminal carboxyl group of polyester resin resulting in doubling or tripling the average molecular weight of the resin. This bond is not permanent and diminishes slightly when subjected to high temperature and pressure during molding. The rate of thickening and induction time is important parameters for a molding compound. Induction time is the time at which the rate of thickening is slow. If the rate of thickening is too rapid, then the penetration of resin will be incomplete and may result in dry spots. Short induction time leads to difficulties in the application of resin by impregnation.

The viscosity should remain constant during molding and it should lie within the range of $0.75–1.5 \times 10^5$ Pa s. Viscosity below 0.75×10^5 Pa s may lead to the separation of resin from fiber, and viscosities higher than 1.5×10^5 Pa s are likely to give short molding.

4.1.8.1.1 Raw Materials for Molding Compounds

A typical molding compound contains reinforcement, resin, filler, release agent, curing agent, thickener, and sometimes a low profile additive. The discussion on raw materials will be limited to material requirements specific to BMC and SMC. The selection of raw materials for molding compounds is very important. Wide ranges of materials are used and their formulation is also diverse. For making consistent quality molding compounds, all raw materials should be subjected to stringent quality control. The following categories of raw materials are used for making molding compounds:

1. Fiber reinforcement
2. Resin
3. Curing agent
4. Fillers
5. Chemical thickener

6. Release agent
7. Low profile additives
8. Miscellaneous additives

The fiber gives mechanical strength to the product. For the BMC the fiber content varies from 10% to 25% and the length of fiber varies between 6 and 12 mm. Reinforcements for BMC include glass, carbon, aramid, sisal, and other natural organic fibers. For the SMC, about 10%–35% fibers of length between 25 and 50 mm are used. SMCs are commonly made with glass fiber roving chopped to 25 mm length, although it is possible to use fiber lengths between 12 and 75 mm. Originally SMC was made from CSM having fibers of length 50 mm. Recently, continuous fibers of glass, carbon, and aramid are being used to obtain higher strength.

Increase in fiber length improves the strength and rigidity of the products. Short fibers provide better flow of compound, increased strength along weld lines, and reduced sink marks over ribs and bosses. Fiber length beyond 12 mm cannot be used in BMC because the longer fibers are more prone to entanglement and degradation in the mixer. Moreover, wet-out becomes difficult with long fibers. Mostly chopped fibers are used in molding compounds. However, some SMCs are made with CSMs. Molding compounds are generally made using glass fibers as the fiber reinforcement. Other reinforcing fibers include sisal, carbon, aramid, chopped polyamide rag, and wood fibers.

Glass fibers generally give good mechanical properties. Sisal fibers make an easy flowing compound suitable for large and moderately complex parts not requiring high water resistance. A very low cost product with excellent water, stain, and electrical resistance can be made with diced polyamide tricot fabric. The color and weave structure of fabric are difficult to mask, and the mechanical properties of the product made with this fiber are also low. Some molding compounds have been made with carbon and aramid fibers. Although they increase the mechanical properties, the increase is not proportional to their basic mechanical properties.

The resin should ideally have a low viscosity for easy mixing, but high enough to prevent the separation from other ingredients during molding. It should cure rapidly and have high hot strength to permit the removal of part without damage under hot conditions. Unsaturated polyester resin is the most widely used matrix in molding compounds. Highly reactive grades of polyester resin are used to reduce the processing time. The polyester resin can be suitably modified for different applications. The water content and molecular weight of the polyester resin affect its thickening behavior. Polyester resins mostly used for the molding compounds have viscosity of about 2.5 Pa s. However, resins with viscosity from 1 to 250 Pa s are being used. Resins of viscosity up to 60 Pa s can be used in conventional mixing equipment without using viscosity-reducing solvents.

Phenolic, vinyl ester, or epoxy resin can also be used as the matrix material in BMC. Epoxy resin offers better strength and chemical resistance. However, its high cost and slow cure rate have restricted its use in molding compounds. Vinyl ester offers toughness, chemical resistance, and flexibility in compounding. In spite of its higher price than polyester resin, high-performance products are commercially viable with this resin. On the other hand, SMC must be made with a resin system that will thicken to proper viscosity after the impregnation of the reinforcement.

Fillers perform more functions than mere filling. They reduce the shrinkage, increase the modulus and hardness, and modify the electrical properties. Fillers also alter the rheology of molding compounds and prevent fiber–resin segregation during molding.

Four groups of fillers are commonly used in molding compounds. They are silicates, including silica, carbonates, sulfates, and oxides. The particle size of fillers should be in the range of 0.5–50 μm. Most of the fillers are produced from natural minerals by appropriate grinding.

Calcium carbonate and clays are the principal fillers used in molding compounds. Calcium carbonate has less oil absorption characteristics, but it affects the flow during molding. Clay-filled compounds have better flow and the molded products have better properties, except color. A combination of clay and calcium carbonate provides good flow and better molded properties. Addition of small quantities of talc, a highly oil-absorbing material will also improve the flow without affecting the color much. Other fillers are calcite, carbon black, and mica. Filler loading up to 150 phr (parts per 100 parts of resin) is common in molding compounds.

The following list of desirable properties can be used to select the fillers for molding compounds:

- Low specific gravity
- Low cost
- Low oil absorption
- Nontoxic
- Nonabrasive
- Ready dispersibility
- Chemical purity and whiteness

A fairly wide particle size distribution will provide a compact arrangement of particles. The ideal situation warranted is a thin coating of resin to bond the particles and not the resin to fill the gaps formed by particles. This will ensure high filler loading and at the same time good flow during molding and better mechanical properties.

Curing agents used in polyester molding compounds are high temperature curing catalysts, such as *t*-Butyl perbenzoate (TBPB), *t*-butyl peroctoate (TBPO), and benzoyl peroxide (BPO). The curing reaction is initiated by heat, which decomposes the peroxide catalysts into free radicals. These radicals activate the cross-linking reaction. The recommended curing temperature with TBPB is 150°C and that of TBPO is 130°C. BPO is a good, economical curing agent for BMC. However, the molding compound should be used shortly after mixing. TBPB requires higher molding temperatures but permits high temperature mixing. It is the standard curing agent for BMC and SMC, since the molding compound can be stored for longer duration. Peroxyesters and peroxyketals give shelf life equal to TBPB with slightly faster curing rate.

Thickeners are the materials that increase the viscosity of the compound without curing. Commonly used thickening agents are magnesium oxide, magnesium hydroxide, calcium oxide, and calcium hydroxide. It is possible to control the rate of thickening and the degree of final thickening by selectively combining the thickeners. The thickening agent should be uniformly distributed to get good result. High-speed and high-shear mixers are effective for uniform distribution. However, at high-speed mixing, the temperature rises and consequently the viscosity increases very rapidly due to thickening. To overcome this problem, the thickener can be pre-dispersed in a non-active medium such as styrene or thermoplastic solution.

A relatively new thickening system is based on the in situ formation of polyurethane rubber network distributed throughout the polyester matrix. The ideal thickeners would be the one that does not start its thickening reaction until the resin thoroughly wets all the ingredients and then rapidly thickens to the desired viscosity and maintains that viscosity until the compound is molded. However, the metal oxide/hydroxide thickener starts the thickening process immediately after mixing with the resin and never stops thickening. The process slows down after sometime and there is a period of time, at which compounds will have satisfactory molding viscosity. If the viscosity is not in the molding range, then the molding parameters have to be altered.

The polyurethane rubber system involves a finite chemical reaction. When the reaction between an isocyanate and polyol is completed, the thickening process stops and the viscosity will remain constant. However, the maturation time may be very short. Hence, a better control of reaction is needed, for complete wetting of ingredients.

Release agents are used during molding for the better release of product. Internal release agents are more effective than external release agents. Internal release agents are mixed in the molding compound and they are usually long-chain fatty acids and their salts. While heating, they migrate to the surface and prevent the bonding of resin with the tool. These release agents are chosen in such a way that their melting point is just below the molding temperature. This avoids marring of the surface because of premature melting. The commonly used release agents are stearic acid ($T_m = 70°C$), zinc stearate ($T_m = 122°C$), magnesium stearate ($T_m = 130°C$), and calcium stearate ($T_m = 155°C$).

Low profile additives impart smooth surface to the product. A smooth surface is indicated by shallow hills and valleys in a profilometer. The use of low profile additives has been a major factor for the increased replacement of metal parts by BMC and SMC composites. All the low profile additives are based on thermoplastics, and they also impart dimensional stability to the molded part. Another advantage of these additives is that the molded parts can be painted with little or no abrasive treatment. Molded in color is also excellent with some of these additives. Warpage is a serious problem with the conventional resin systems. Warpage can be practically eliminated by the use of low profile additives. Hence, moldings can be made with close dimensional tolerances, equal to or better than precision cast metals. Low profile additives usually reduce strength and modulus but improve impact strength.

The amount of these additives added in the compound range from 5 to 20 phr. Commonly used additives for polyester resin are polystyrene, polyethylene, thermoplastic polyester, polyvinyl acetate, polymethyl methacrylate, etc. The additives compatible with the polyester resin are usually mixed with the resin itself. Those additives soluble in styrene monomer are added to the formulation as a solution.

4.1.8.1.2 Compound Formulation

The ideal molding compound should flow easily and fill extremities and details of the mold. In addition, the compound must, to the maximum degree, remain homogeneous while flowing in the mold. If the resin/filler/reinforcement separates, variations in properties at different locations in the part will occur. The flow properties mainly depend on the degree to which the resin is absorbed by, or adsorbed on, the filler and the reinforcement. Each of the dry ingredients has its own resin absorption characteristics. For example, china clay has more than double absorptive power compared to calcium carbonate. Longer glass fibers are less absorbing than short fibers. The dryer the compound (more absorption of resin), the lower will be the plasticity or flow. The resin absorption will also change with resin properties. The ability to wet the filler and fibers depends on the viscosity, basic chemical nature, type and quantity of monomer, etc.

The type and amount of ingredients are selected based on the desired final properties, such as strength, stiffness, toughness, electrical insulation, corrosion resistance, and fire resistance. However, the first and foremost condition for the selection of ingredients is moldability. Typical compositions of BMC and SMC are given in Table 4.3.

Resin content for the molding compound may vary from 18% to 50% by weight. Usually it is about 30%. The resin content depends on the absorption characteristics of the other ingredients. When very low absorption fillers such as calcium carbonate are used, the resin content can be low. When high absorption fillers are needed, the resin requirement will be higher. The catalyst concentration is in proportion to the amount of resin. Lower or higher concentration can be used, depending on shelf life and curing temperature requirements.

TABLE 4.3

Typical Molding Compound Formulations

Ingredients (wt.%)	BMC		SMC	
	GP	Electrical	Structural	Automotive
Polyester resin				
Isophthalic	20	25	25	
High fumarate				18
Polyvinyl acetate				3
Styrene				2
Filler				
$CaCO_3$	60.8	—	39.75	45.75
Alumina trihydrate	—	52.75	—	—
Catalyst (BPO/TBPB)	0.2	0.25	0.25	0.25
Thickening agent (MgO)	—	—	1.0	1.0
Release agent (zinc stearate)	2.0	2.0	2.0	2.0
Pigment (TiO_2)	2.0	$1.0 + 1.0$[a]	2.0	2.0
Glass fiber				
(6 mm length)	15	18	—	—
(25 mm length)	—	—	30	25

Source: Data from Mark, H.F. and Kroschwitz, J.I., *Encyclopedia of Polymer Science and Technology*, Vol. 12, John Wiley & Sons, New York, 1988, p. 287.

[a] Iron black pigment.

GP, general purpose.

Fiber contents in BMC may vary from as little as 5% to about 50%. Higher glass fiber content makes the molding compound fluffy, whereas compounds high in resin content are very wet and sticky. Up to 20% fiber provides compounds that can be handled easily. Commonly SMC formulations contain about 30% glass fibers of length 25 mm. Compounds can be made with up to 65% short fibers and 75% continuous fibers. The filler content is inversely proportional to fiber content. Depending on the required properties of the product, either more fiber content or more filler content is used. Better mechanical properties are obtained with more fibers, and better electrical resistance or fire retardance is obtained with more fillers. Usually fillers can replace the reinforcement without significantly affecting the moldability.

Internal release agents such as stearates are used at the 1%–3% level. Some properties may be affected by the use of release agents; hence, their quantity should be as minimum as possible. The most common thickeners, MgO and $Mg(OH)_2$, are used in the range of 1%–1.5% and 3%–5% of the resin content, respectively. The MgO thickens the molding compound faster than the $Mg(OH)_2$, but the compound remains in a moldable range for a longer time when $Mg(OH)_2$ is used. Most of the low profile additives are pre-dissolved in styrene to the solid contents of about 30%–40%. This solution is used at a level of about 50% of the basic resin. Fine powders of low molecular weight polyethylene are usually added directly and the concentration can be 3%–5% of the resin.

4.1.8.1.3 Molding Compound Preparation

BMC is made in batches, while SMC is usually made in continuous process. Both the compounds are made in two distinct stages. At the first stage, all ingredients, except reinforcement, are combined, and at the second stage, the materials are combined with the reinforcement. The first stage of mixing should be intense and long duration, but it should not be too intense or too long. The second stage of mixing should be in such a way that it is just sufficient to ensure wetting and uniform distribution of the reinforcements. Intense mixing of fibers will lead to fiber breakage and degradation. Fast-acting thickeners such as MgO should be added just prior to the addition of reinforcement.

For making BMC, two types of mixers are generally used. The first-stage mixer can be a simple propeller-type mixer. The prime concerns are mixing efficiency and speed. The second-stage mixer must be a heavy-duty mixer. Sigma mixer is used for mixing all ingredients with the resin. The sigma mixer has two Z- or E-shaped blades, which rotate at different speeds. Mixing with sigma mixer is efficient, fast, and thorough, but over mixing can cause some degradation of fiber. The drum and blades of the sigma mixer should be made preferably with stainless steel.

A typical mixing schedule for BMC would be as follows:

1. Blending resin, catalyst, release agent, and pigment for 10 min
2. Mixing for 5 min after adding 50% filler
3. Transferring to sigma mixer
4. Mixing for 10 min after adding remaining filler
5. Reinforcement addition and mixing for 3 min

The first stage of SMC preparation is essentially similar to that of BMC. SMC always has a thickener. For producing SMC, resin slurry is made with all ingredients except glass fiber. It is preferable to prepare the resin paste in small quantities so that the resin paste is completely utilized before thickening. The slurry is mixed thoroughly and fed to an impregnator. The impregnator spreads the slurry uniformly on two carrier films by knife (doctor blade). The reinforcement is applied over a slurry film. The reinforcement can be CSM or chopped strands produced from the rovings during the impregnation itself. The reinforcement layer is then covered with the second slurry film. The reinforcement sandwiched between the resin spread is compacted by rollers. The SMC sheet with carrier film is wound on a mandrel, covered with aluminum foil, and kept for maturing. It takes 1 to 2 days to reach the required thickening. Aluminum foil covering prevents styrene loss from polyester resin. They should be stored in an air-conditioned room for longtime storage. The molding compounds can be stored up to 6 months at 5°C–10°C.

4.1.8.2 Compression Molding Press

The function of press in compression molding process is to apply the required pressure for molding the products. The three types of presses available are mechanical, pneumatic, and hydraulic. Pressure required for composite molding can be up to 10 MPa. Mechanical and pneumatic presses are not suitable since they can apply pressure only up to 1.8 MPa. Hydraulic presses are the most suitable for compression molding. There are various types of hydraulic presses available, but the four-column upward or downward moving platen, direct-acting press is widely used for molding of composites. Schematic of a typical hydraulic press used for compression molding is shown in Figure 4.17.

Various parameters are considered for the selection of a press. The first and most important parameter is press capacity. It is the maximum force exerted on the molding compound during molding by the press platens. It is determined by the maximum molding pressure required and the maximum size of the product. For example, if the maximum pressure required is 4 MPa and the maximum planar surface area of the product is 0.25 m², the press capacity required will be 1×10^6 N (100 ton). Molding temperature mainly depends on the resin type and part thickness. The molding temperature range for polyester resin is 80°C–150°C and epoxy resin is 145°C–200°C.

Another parameter is breakaway or return force. It is the force required to separate the molds after molding. It is generally 25%–30% of the press capacity. It is provided by an auxiliary cylinder or by a double-acting main cylinder.

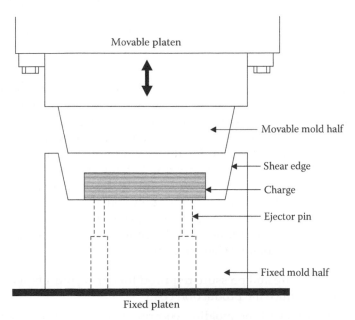

FIGURE 4.17
Schematic of compression molding press. (Reprinted from Mallick, P.K., *Fiber-Reinforced Composites*, 3rd edn., CRC Press, Boca Raton, FL, 2008, p. 395. With permission.)

Daylight is the maximum distance between the upper and lower platens of the press. It determines the depth of the part that can be molded. It must be three times of the maximum depth of the largest part. The stroke (movement) should be at least twice the depth of the largest part.

4.1.8.3 Molds (Dies)

Dies give shape to the molding compound. Dimensional accuracy and surface finish of the product depends mainly on the dimensional accuracy and surface finish of the dies. Inner surfaces of the die should have high-class surface finish and high abrasion resistance, since several thousand parts are usually made using the same die.

Alloy steel AISI 4140 or its equivalent (IS 40 or EN 19C) pre-hardened to Rockwell C hardness of 30–32 is used for high-class dies. AISI 1045 steel or its equivalent (IS-C14 or EN43B) can be used for nonappearance or low production run dies.

The standard dies used for compression molding are classified into two types. They are flash-type die and fully positive die. The flash-type die is not generally recommended for high-quality products. However, it may be used for producing large parts using BMC and SMC. Fully positive dies are used for large deep drawn parts, where maximum density is required. Appropriate draft angle should be provided in the die to facilitate easy removal of product from the die.

The shrinkage of the compound after molding must be taken into account while designing the die. For a molding temperature of 150°C, a shrinkage value of 0.012 mm mm^{-1} should be used for low-profile compounds, and 0.02–0.03 mm mm^{-1} should be used for general grade compounds.

4.1.8.4 Molding Process

Molding temperatures for FRP molding lie in the range 120°C–200°C. Heating should be uniform across the mold faces. Both the dies should be heated to the same temperature, and the temperature difference should be within ±5°C. Sometimes, die (female mold) is heated to a slightly higher temperature than the punch (male mold) for the easy removal of product. Heating of the die can be carried out by conduction heating from the platen, electric heating through cartridge heater inserted in the die or circulating hot oil or steam. Electric cartridge heaters are extremely convenient, since they can be easily accommodated in the die and are inexpensive. However, they tend to give uneven heating, resulting in faulty moldings. The essential requirements for an efficient compression molding process are

- The temperature and pressure required for curing must be as low as possible to reduce the production cost.
- The consistency of the molding compounds must be such that the material flows into all parts and cavities of the die without blow holes and other defects.

- The molding operation should not cause any fiber degradation or fiber/matrix segregation.
- The compression-molded product should have a fairly smooth surface finish and must be free from crazing, cracks, fractures, edge chips, and porosity.
- The fiber pattern should not be visible on the surface of the product.
- Shrinkage after compression molding should be a minimum.

4.1.8.4.1 Molding Temperature

Lower temperatures are used for deep drawn and complicated shapes, since it takes longer time to completely fill the die. Simple shapes and components with uniform wall thickness can be molded at higher temperatures. Thick section moldings should be cured for longer times at lower temperatures.

4.1.8.4.2 Molding Pressure

Comparatively low molding pressures can be used for molding compounds, from which no volatiles are liberated during molding. Generally, SMC requires pressure in the range of 3.5–10.5 MPa, while BMC requires pressure in the range of 1.4–5.6 MPa. Lower pressures are sufficient for simple shapes, while higher pressures are needed for complicated shapes.

The charge (molding compound) should be placed in such a way that it covers 60%–80% of the die surface. SMC components are normally made to have wall thickness in the range of 2–6 mm. However, wall thicknesses as low as 1.5 mm and as high as 25 mm are possible. Special initiator systems should be used when molding thick sections. BMC can be molded to little thinner than SMC. The wall thickness of BMC products can range from 1.3 to 50 mm and the variation in wall thickness should not exceed 2:1. However, wall thickness variations as high as 6:1 can be tolerated in the BMC moldings.

4.1.9 Thermoplastic Composite Processing

The main reason for the lack of widespread acceptance of thermoplastic matrices is that they are more difficult to process into a composite than thermoset polymer matrices. In fact, a number of techniques have been developed to process thermoplastic matrix composites, but they are of limited use. Hence, intensive research and development efforts are underway to develop new techniques and produce amenable thermoplastic composites.

Among the many factors, the important factors that influence processability are matrix viscosity and processing parameters (i.e., temperature, pressure, time). Low viscosity facilitates easy flow and the better impregnation of reinforcements so that each reinforcing fiber is ideally surrounded by matrix without any voids. Fully polymerized thermoplastics have very high molecular weights; hence, their melt viscosities are at least two orders of magnitude higher than thermoset polymer viscosities, which make impregnation more difficult. Since the viscosity of molten thermoplastics is higher, pressure is applied during processing to achieve the same degree of material flow as with thermosets, even then the flow is not significant in many cases. However, there are some advantages with thermoplastic matrices. Thermoplastics only need to be melted, shaped, and then cooled to get the product and the time needed is very short. Hence, thermoplastic composites can be manufactured more rapidly than thermoset composites, which usually require a few hours. Chemical reactions involving volatiles and potentially toxic substances occur during the curing of thermoset polymers, whereas fully polymerized thermoplastics do not involve any hazardous substance during processing.

Molding compounds and prepregs are mostly used in the manufacturing of thermoplastic composites, and use of reinforcement and matrix separately is very rare due to impregnation difficulties. However, it is possible to impregnate the reinforcement with a liquid monomer, which after impregnation undergoes polymerization to form high molecular weight thermoplastics. This concept is not new, but there has been no significant commercial success until recently, due to intolerant processing requirements. Such in situ polymerizable thermoplastics can be used in many thermoset composite manufacturing techniques, thus opening new opportunities for thermoplastic composite. However, the good work environment facilitated by an inert thermoplastic matrix may not be available with this approach, since the monomer undergoes chemical reactions in the presence of catalysts and initiators.

Thermoplastic composite processing is carried out above the melting temperature or at least near the melting temperature of the matrix. In many cases, shaping and consolidation take place simultaneously. Some of the commonly used processing methods are described in the following sections.

4.1.9.1 Film Stacking

Film-stacking process is used to produce thermoplastic composites, in which alternate layers of thermoplastic film and woven cloth (reinforcement) are laid up and consolidated. The time required for the successful consolidation of lay-up is longer, since the high-viscosity polymer should flow longer distances. A typical processing cycle time for a film-stacking laminate would be 1 h at 1 MPa applied pressure.

4.1.9.2 Thermoplastic Tape-Laying

Thermoplastic tape-laying machines are also available, but they are not as common as thermosetting tape-laying machines. In this process, only the area that is being consolidated is heated above the melt temperature. Heating can be carried out using hot shoes or a focused laser beam. The heated tape is consolidated by a cold roller. The consolidation can occur in less than 0.5 s. Figure 4.18 shows the schematic of the setup. A controllable tape head has

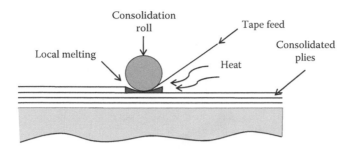

FIGURE 4.18
Thermoplastic tape-laying. (Courtesy of Cincinnati Milacron, Batavia, OH.)

the tape-dispensing reels and heating shoes. There are three heating and two cooling/compaction shoes. The hot shoes heat the tape to molten state, and the cold shoes compact and cool the tape instantly to a solid state.

A potential problem with hot tape-laying is lack of consolidation due to insufficient diffusion at the short processing time. If there are intraply voids, then these voids may not be healed and consolidated during this process. A post consolidation cycle will be required to achieve full densification. The interlaminar shear strength of the composite will be reduced by about 7% for each 1% of voids up to a void content of 4%. It is preferable to produce a composite with void content of less than 0.5%. Usually the hot tape-laying process results in only 80%–90% consolidation, hence a secondary processing is necessary to get full consolidation.

4.1.9.3 Commingled Fibers

The thermoplastic matrix can be used in the form of a fiber. The matrix fiber and the reinforcement fiber are commingled to produce a yarn. Such a commingled yarn can be woven, knit, or filament wound to form fabrics or some shapes. The yarn formed into the appropriate shape is then subjected to heat and pressure. During this time, the thermoplastic fibers melt and wet out the reinforcement fiber, and on cooling the composite part is obtained.

4.1.9.4 Other Processing Methods

Many other processes have been evaluated for fabricating thermoplastic composites, which includes roll forming, pultrusion, and RTM. The schematic of roll-forming process is shown in Figure 4.19. A pre-consolidated blank is heated above the melt temperature of the matrix and then passed through a series of rollers to form the desired shape gradually. In some operations, only the regions of the blank that undergo deformation are heated. The important condition in roll forming is to maintain uniform pressure on all portions of part during heat-up, forming, and cooldown. If uniform pressure is not maintained on the molten portions, deconsolidation will occur due to relaxation of the fiber bed.

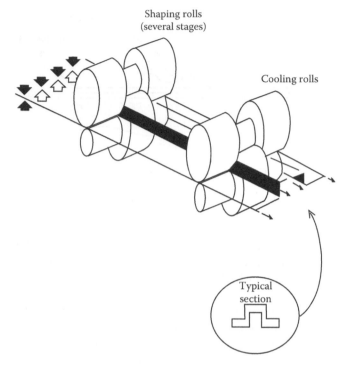

Shaping rolls
(several stages)

Cooling rolls

Typical
section

FIGURE 4.19
Roll-forming process. (Reprinted from Campbell, F.C., *Manufacturing Processes for Advanced Composites*, Elsevier, Oxford, U.K., 2004, p. 389. With permission.)

Thermoplastic composites have also been successfully pultruded, but the process is very expensive and difficult due to the high melt temperatures and viscosities of thermoplastics. The melt viscosity of thermoplastics is too high for the long flow paths required for RTM; hence, total wet-out of a reinforced fiber bed is rarely achieved.

4.1.9.5 Consolidation of Thermoplastic Composites

There are several methods available to consolidate thermoplastic composites. In the two press process, collated plies are preheated in an oven and then immediately transferred into a press for consolidation. The press should have provisions for heating if the material requires time for flow or crystallization. The collation of ply layers is usually carried out by hand lay-up. Soldering irons heated to 425°C–650°C are frequently used to join the edges to prevent the layers from slipping. Subsequently, it is consolidated by various methods.

For the complex parts, autoclave consolidation is the best option. However, there are several difficulties with the autoclave consolidation. First, it may be difficult to find an autoclave that can go up to a temperature of 350°C–400°C and pressure of 0.7–1.4 MPa. These temperature and pressure levels are required for some advanced thermoplastics. Second, the tooling used at these conditions will be more expensive. Third, the coefficient of thermal expansion (CTE) of the tool should match with the part since the processing temperatures are high. Monolithic graphite, cast ceramic, and Invar 42 are the tool materials normally used for carbon fiber-reinforced thermoplastics. Fourth, the bagging materials must be capable of withstanding the high temperatures and pressures. The typical bagging materials are polyimide and silicone rubber. The polyimide bagging materials are not flexible as that of polyamide, a commonly used bagging material for thermoset polymers. The high-temperature silicone rubber has minimal tack at room temperature; hence, sealing may not be effective. Clamped bars are often placed around the periphery for better sealing.

Consolidation of prepregs made with fusible thermoplastics consists of heating, forming, and cooling, as depicted schematically in Figure 4.20. The main processing variables are temperature, pressure, and time. Heating can be accomplished with infrared radiation, conventional ovens, heated platen presses, or autoclaves. The time required to reach consolidation temperature depends on the heating method and the mass of tooling. The consolidation temperature depends on the specific thermoplastic material used. Generally, the temperature should be above T_g ($\sim$200°C above T_g) for amorphous thermoplastics and above T_m ($\sim$100°C above T_m) for semicrystalline thermoplastics. The time at the consolidation temperature is primarily a function of the nature of the prepreg. For example, melt-impregnated tape can be successfully consolidated in very short times, while powder-coated or commingled prepregs require longer times since the polymer should flow and impregnate the fibers during consolidation. The pressure application during consolidation forces the layers to intimate contact and helps better impregnation of fibers.

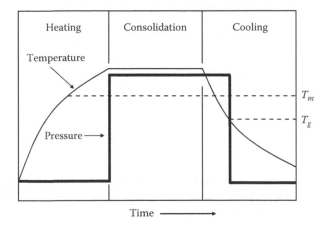

FIGURE 4.20
A typical consolidation cycle for thermoplastic composites. (Reprinted from Muzzy, J. et al., *SAMPE J.*, 25, 23, 1989. With permission.)

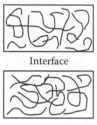
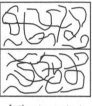

Interface

Before contact   Intimate contact   Partially diffused   Fully diffused

FIGURE 4.21
Schematic of autohesion. (Reprinted from Campbell, F.C., *Manufacturing Processes for Advanced Composites*, Elsevier, Oxford, U.K., 2004, p. 373. With permission.)

The cooling rate from the consolidation temperature depends on the mass of the tooling. Semicrystalline thermoplastic-based composites should not be cooled at a faster rate (i.e., quenching), since they fail to form the desired semicrystalline structure. The pressure should be maintained until the temperature falls well below the T_g of the polymer. This restricts the nucleation of voids, suppresses the elastic recovery of fiber, and helps to maintain the desired dimensions. The properties of laminates produced using solution impregnated prepreg, powder-coated, commingled, and film-stacked fibers are not good as those made from melt-impregnated prepreg, since there is better fiber–matrix bond formation during melt impregnation.

The consolidation of thermoplastics occurs by a process called autohesion as depicted in Figure 4.21. When two surfaces come together, they must have intimate contact for better diffusion of polymer chains across the interface. Due to the low flow nature of thermoplastics, the surfaces must be physically deformed under heat and pressure to obtain the required intimate contact. To obtain intimate contact and autohesion, the amorphous and semicrystalline thermoplastic-based materials should be heated above T_g and T_m of the polymers, respectively. In general, higher pressures and higher temperatures lead to faster consolidation. As the contact time increases, the diffusion of polymer chains across the interface increases and the extent of polymer entanglement increases and that results in the formation of strong bond at the interface. Consolidation times are usually longer for amorphous thermoplastics since they have higher viscosities at the processing temperature. However, the consolidation time can be shortened by increasing the applied pressure, because the pressure aids flow and intimate contact.

4.1.9.5.1 Thermoforming

One of the main advantages of thermoplastic composites is their ability to be rapidly transformed into structural shapes by thermoforming process. This process uses heat and pressure to transform a flat sheet into a structural shape. The blank is heated close to the melting temperature of the thermoplastic matrix and then quickly transferred to a press containing dies of the desired shape (Figure 4.22). The part must be held under pressure until it cools below its T_g to avoid warpage.

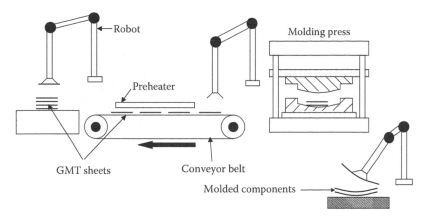

FIGURE 4.22
Schematic of glass mat thermoplastic (GMT) composite thermoforming. (Reprinted from Mazumdar, S.K., *Composites Manufacturing*, CRC Press, Boca Raton, FL, 2002, p. 214. With permission.)

The main preheating equipments used for thermoforming are infrared heaters, conventional ovens, and heated platen presses. The heating time is short (1–2 min) in IR heating, but temperature gradients can form within the thick stacks. Since the surface heats faster than the center, there is a danger of overheating at the surface unless the temperature is carefully controlled. In addition, it is difficult to obtain uniform heating on complex contours. Still, IR heating is a good choice for thin pre-consolidated blanks having moderate contour. On the other hand, conventional heating in hot oven takes longer time (5–10 min) but is more uniform through the thickness. It is the preferred method for unconsolidated blanks and blanks having complex contour. The forming dies can also be made with internal heating arrangements.

Although matched metal dies can be used for thermoforming, they are expensive. If they are not made to precise dimensions, the pressure application may not be uniform across the part that will result in the formation of defective parts. This problem can be avoided by using a facing made of silicone rubber on any one of the die halves. Instead, one of the die halves can be made entirely with rubber, either as a flat block or as a block that is having the shape of the product. Although the flat block is cheaper and easier to fabricate, the shaped block provides a more uniform pressure distribution and better part consolidation. Silicone rubber with hardness of 60–70 Shore-A is commonly used for the dies. For deep drawn parts, it is better to make the male tool half with metal and female tool half with rubber. For moderate draws, the female half is usually made with metal by incorporating draft angles of 2°–3° to facilitate easy removal of the part.

The thermoforming process seems to be simple, but the process is quite complicated because of the presence of continuous fiber reinforcement. The fibers restrict the easy flow and deformation of the material during

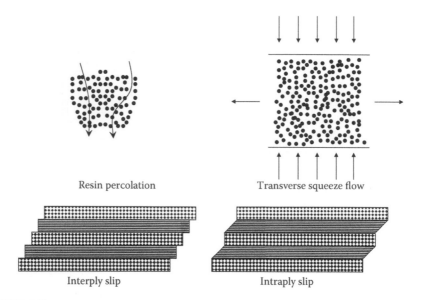

FIGURE 4.23
Polymer flow phenomena in thermoforming. (Reprinted with permission from Campbell, F.C., *Manufacturing Processes for Advanced Composites*, Elsevier, Oxford, U.K., 2004, p. 384. With permission.)

thermoforming. There are four primary polymer flow phenomena that must be dealt during thermoforming. They are (1) polymer percolation, (2) transverse squeeze flow, (3) interply slip, and (4) intraply slip (Figure 4.23). Polymer percolation and transverse squeeze flow normally occur during consolidation, but they are also important factors during thermoforming. Polymer percolation is the flow of the viscous polymer through the fiber bed that allows the plies to bond together, while transverse squeeze flow eliminates slight variation in blank thickness by allowing the excess polymer to flow latterly due to applied pressure. The polymer matrix tends to flow parallel to the fiber axis, since flow through the fiber bed is more difficult. When flow occurs off-axis to the fiber direction, the fibers tend to move with the polymer matrix. Interply slip is the slip between plies at the ply interfaces and intraply slip is the slip within the ply. These slip mechanisms are encountered in the thermoforming process by a combination of transverse and axial shear. If these slip mechanisms are not operating during forming, then the reinforcing fibers will either break or buckle or it is not possible to form the part.

Fiber reinforcements are usually strong in tension but will buckle and wrinkle readily under compression especially when the matrix is in the viscous state. Therefore, the part shape on the die has to be designed to keep the fibers in tension throughout the forming process. A special holding/clamping fixture can be used during the forming process, if neither the part shape nor the die design is amenable to prevent compression buckling.

The transfer time from the heating station to the press is critical in any thermoforming process. The heated part must be transferred to the press and formed before it cools below its T_g for amorphous thermoplastics or below its T_m for semicrystalline plastics. This warrants a transfer time of 15 s or less. Presses with fast-closing speeds of the order of 5–13 m min^{-1} are preferred for optimum forming and the typical pressure required is 1.5–3.5 MPa.

4.1.9.5.2 Hydroforming

It is another method of applying pressure during forming process, in which an elastomeric bladder presses the part in the lower die half using fluid pressure (Figure 4.24). Typical thermoforming pressures are 0.7–3.5 MPa; however, some hydroforming presses are capable of applying pressures as high as 70 MPa.

4.1.9.5.3 Diaphragm Forming

It is a rather unique thermoforming process that is capable of making parts with a wider range of configurations, especially parts with double curvatures. Pressure is applied from one side, which deforms the diaphragms and makes them to take the shape of the tool. The laminate is freely floating and very flexible above the melting point of the matrix. It readily conforms to the tool shape. After the completion of the forming process, the tool is

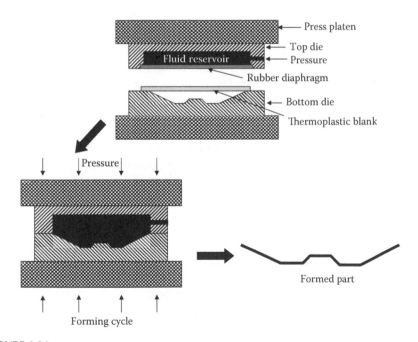

FIGURE 4.24
Hydroforming. (Reprinted from Campbell, F.C., *Manufacturing Processes for Advanced Composites*, Elsevier, Oxford, U.K., 2004, p. 383. With permission.)

cooled and the diaphragms are stripped off from the composite part. The diaphragms are the important element to this forming process and their stiffness is a very critical parameter. Compliant diaphragms can be used for simple components, whereas stiff diaphragms are needed for very complex shapes requiring high forming pressures. A significant transverse flow of the melt can happen at high pressures, and this can lead to undesirable thickness variations in the final composite part.

Diaphragms are usually made with superplastic aluminum (Supral) and high temperature polyimide films (Upilex-R and S). Superplastic aluminum sheet is more expensive than polyimide films but is less susceptible to rupturing. Polyimide films work well for thin parts with moderate complexity, while superplastic aluminum sheet is preferred for thicker parts with complex geometries. Typical diaphragm-forming temperatures are 400°C and 300°C–400°C for aluminum sheet and polyimide films, respectively. Even with diaphragm forming, thinning occurs in the male radii and thickening occurs in the female radii. The main disadvantage of this process is that the materials can be formed only at the forming temperatures of the available diaphragm materials. In addition, the process is expensive because of the high cost and non-reusable nature of the diaphragm materials.

A schematic of diaphragm forming setup for a thermoplastic composite part is shown in Figure 4.25. In this process, unconsolidated ply stacks are placed between two flexible diaphragms. The unconsolidated ply stack is preferred because it readily promotes slippage compared to pre-consolidated blanks. A vacuum pump is connected to the diaphragms to remove air within the diaphragms and to apply pressure on the stack. The evacuated diaphragm is then placed on a tool and heated above the melt temperature. Gas pressure is used to deform the pack over the tool surface. During forming, the plies slip within the diaphragms and the tensile stresses created can reduce the tendency for wrinkling. The gas pressure forms the part according to the tool contour and at the same time consolidates the part. Pressures applied during this process range from 0.35 to 1.0 MPa and cycle times are usually 20–100 min; however, for massive parts, cycle times of 4–6 h are not unusual. Slow pressurization rates are recommended to avoid out-of-phase buckling.

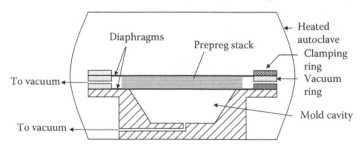

FIGURE 4.25
Schematic of diaphragm forming process. (Reprinted from Mallick, P.K., *Fiber-Reinforced Composites*, 3rd edn., CRC Press, Boca Raton, FL, 2008, p. 425. With permission.)

4.1.9.6 Injection Molding

Injection molding is a very common processing method used to form different-shaped products with thermoplastics (Figure 4.26). The same method can also be used to form thermoplastic composites with short fibers, but the incorporation of fiber will increase the melt viscosity. Another method has been developed to overcome the problem of high melt viscosity and the method is called as reinforced reaction injection molding (RRIM). Actually this is an extension of reaction injection molding (RIM) of polymers. In RIM, two liquid precursors of the polymer are pumped at high speed and pressure into a mixing head and then into a mold where the two components react and polymerize rapidly. In RRIM, short fibers are added to one or both of the components. The RRIM equipment must be able to handle rather abrasive slurries. The fibers used in RRIM are generally short fibers, owing to viscosity limitations. For effective fiber reinforcement, the fiber length should be above the critical length, but more often fibers used in RRIM process have shorter lengths; hence, the fibers act as fillers rather than reinforcements. Both RIM and RRIM methods are mostly used in automotive industry.

In RRIM, milled glass fibers are generally used. The fiber length is restricted to less than 0.5 mm in order to keep the viscosity low. Although RRIM parts are stiffer and more damage tolerant than RIM parts, their strength is not high enough for many structural applications. To overcome this problem, structural reaction injection molding (SRIM) process was developed. In this process, two liquid reactive components of a polymer are mixed in a chamber at a high velocity before injected into a mold containing the reinforcement. The reactants flow at a speed of 100–200 m s^{-1}, and the pressure generated during collision is in the range of 10–40 MPa. However, the polymer is injected into the mold at a pressure of less than 1 MPa. This pressure is not high enough for the washout of fibers near the injection port. The schematic of this process is shown in Figure 4.27.

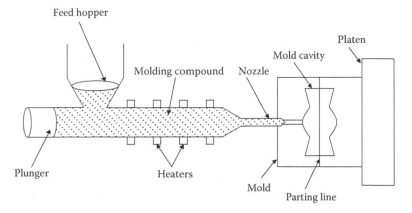

FIGURE 4.26
Schematic of the injection molding process. (Reprinted from Mazumdar, S.K., *Composites Manufacturing*, CRC Press, Boca Raton, FL, 2002, p. 199. With permission.)

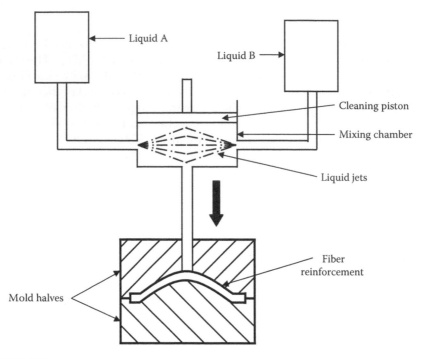

FIGURE 4.27
Schematic of SRIM process. (Reprinted from Mazumdar, S.K., *Composites Manufacturing*, CRC Press, Boca Raton, FL, 2002, p. 176. With permission.)

4.2 Advances in Polymer Matrix Composites

Over the years, there are many developments in PMCs to increase the performance levels. Some of the recent developments include advanced curing processes, smart composites, and metal-composite laminates.

4.2.1 Advanced Curing Processes

One of the problems with the thermoset resin is long curing time and this restricts the production rate. Curing with electron beam or microwave is a suitable option to achieve faster curing rate.

An electron beam accelerated using 1–10 MeV potential is a good source of ionizing radiation, which can be used to initiate the curing reaction of a thermoset resin. The resin formulation needs to be modified for electron beam curing by adding 1–3 phr of a cationic photo-initiator in place of the conventional curing agents. Parts having thickness up to 25 mm can be cured with this electron beam, whereas high energy beams are required for higher thickness. The major advantages of electron beam curing are

- Shorter curing time
- Possibility to cure at a desired temperature
- Lower energy consumption than in thermal curing
- Elimination of undesirable volatile components, since the hardener is not used

However, electron beam curing is not widely practiced because of the following limitations:

- Electron beam curing facilities are very expensive and the cost will increase with the size of the products.
- On-site curing is not possible, since it is problematic to shift the equipment.
- Long-term performance of electron beam cured products is not well established.

Electromagnetic radiation processing, based on microwave curing, of thermoset composites has generated much interest in recent years. This is because microwave curing of thermoset composites has the advantage of significantly shorter cure time and greater consistency compared to conventional thermal curing. Conventional curing is time-consuming as it relies on heat conduction and convection. In contrast, electromagnetic radiation in microwave processing penetrates the matrix and induces fast curing through rapid dielectric-related heat generation. There are many reports on the microwave curing of composites based on epoxy resin system (Nightingale and Day 2002, Zhou et al. 2003, Prasad and Hsu 2004).

In conventional thermal processing, energy is transferred to the core through convection, conduction, and radiation of heat from the surfaces of the material. In contrast, microwave energy interacts directly to the whole material through molecular interaction. In heat transfer, energy is transferred due to thermal gradients. In microwave heating, electromagnetic energy is converted to thermal energy throughout the body of material.

Microwaves belong to the portion of the electromagnetic spectrum with wavelengths from 1 mm to 1 m and the corresponding frequencies lie between 300 MHz and 300 GHz. However, only two frequencies (0.915 and 2.45 GHz) reserved by the Federal Communications Commission (FCC) for industrial, scientific, and medical purposes are used in microwave heating.

Energy is transferred to materials by the interaction of the electromagnetic field with the molecules and the dielectric properties of the material determine the extent of interaction. The interaction of microwaves with molecular dipoles results in alignment of the dipoles, and energy is dissipated as heat from internal resistance to the alignment. Maxwell–Wagner polarization, which results from the accumulation of charge at the material interface, is also an important heating mechanism in composite materials. Since the microwave energy is absorbed within the material, the electric field decreases as a function of the distance from the surface. Hence, the penetration depth of the microwaves must be at least equal to the thickness of the material.

Since the microwaves penetrate the materials, heat is generated throughout the volume of the material. Hence, it is possible to achieve rapid and uniform heating of thick materials and this can result in significantly reduced processing times. In traditional heating, the cycle time is often decided by slow heating rates to be used to minimize steep thermal gradients.

Moreover, in thermal curing, the exothermic chemical reactions generate internal heat that is slowly dissipated through conduction. It results in a complex temperature distribution within the laminate, which may lead to nonuniform cure, uncontrolled spatial solidification, thermal degradation, and process-induced stresses. The volumetric heating due to microwaves can potentially reduce some of these problems.

Microwaves can be utilized for selective heating of materials. The molecular structure of materials is responsible for the interaction with microwaves. When the materials exposed to microwaves have different dielectric properties, microwaves will selectively couple with the constituent material having higher dielectric loss. In fiber-reinforced composites, the dielectric heating of the composite is dominated by the dielectric properties of the polymer matrix.

As the application of microwaves can affect the molecular orientation of the polymer, the resulting mechanical properties may also be affected. An improvement of the mechanical properties, such as the modulus of elasticity, has been reported for the microwave-cured polymers. It could be due to molecular alignment in the electric field, because the alignment of molecules may produce higher molecular packing with lower free volume, resulting in a higher modulus. It has been found (Belle 2007) that the bonding between the glass fiber and polyester resin is increased on microwave post-curing of glass fiber-reinforced polyester composite (Figure 4.28).

4.2.2 Metal-Composite Laminates

Metal-composite laminates consist of alternating layers of a metal sheet (~0.3 mm thick) and a PMC. The metal can be aluminum alloys, steel, titanium and its alloys, magnesium alloys, etc. In principle, the fiber used in the composite can be continuous or discontinuous and it can be glass, aramid, or carbon and the matrix can be any polymer, but it is commonly epoxy resin because of its good adhesive properties. Hence, a variety

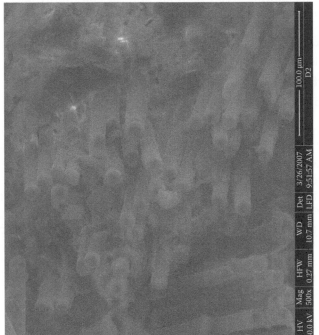

Microwave postcured

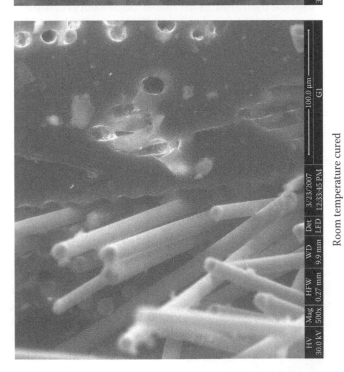

Room temperature cured

FIGURE 4.28
SEM micrographs of glass fiber-reinforced polyester composites with and without microwave post-curing.

of metal-composite laminates can be made and the commonly used metal-composite laminates are

- ARALL (aramid fiber-reinforced aluminum laminates)
- GLARE (glass fiber-reinforced laminates)
- CARE (carbon fiber-reinforced laminates)
- TIGR (titanium/graphite–epoxy laminates)

The ARALL is made with epoxy adhesive prepregs containing 50 vol.% high-modulus aramid fiber and high-strength aluminum alloy sheets. GLARE laminates are made with aluminum alloy sheets and adhesive prepreg containing 60 vol.% high-strength glass fiber oriented in single direction or two directions. Since the glass fiber is stronger than aramid fiber, the GLARE laminates are stronger than ARALL. The main advantage of these laminates is the ability to impede fatigue cracks. In monolithic aluminum alloys, fatigue cracks can grow until the failure of the component. However, in the composite laminates, the cracks developed in the aluminum alloy layer are arrested in the composite layer because of its high strength and toughness. Moreover, the fiber bridging the cracks carries higher portion of the load, thus decreasing the stress intensity at the crack tip and consequently arresting further crack growth. In addition to the superior crack growth resistance, ARALL and GLARE laminates are lighter and more damage tolerant than aluminum alloys. Such hybrid composite laminates combine the advantages of metals and PMCs, such as (Chawla 2012)

- Excellent fatigue and impact properties
- High strength and stiffness
- Flame resistance
- Easy machinability and formability
- Capability for mechanical fastening

In the Airbus A380 aircraft, the panels of the upper fuselage section are made with GLARE.

4.2.3 Smart Composites

Most engineering materials and structures developed until recently have been "dumb." That is, they are produced and/or designed to offer only a limited set of responses to external stimuli. Similarly, advanced materials such as glass and carbon FRPs can be tailored to suit the requirements of their end applications but to a single combination of properties. These dumb materials and structures differ significantly from the nature, where animals and plants have the clear ability to adapt to their environment in real time. They are actually made up of smart materials and structures. Thus, a smart structure is a flexible structure to adapt its form in real time to counteract the external influence. For example, leaf

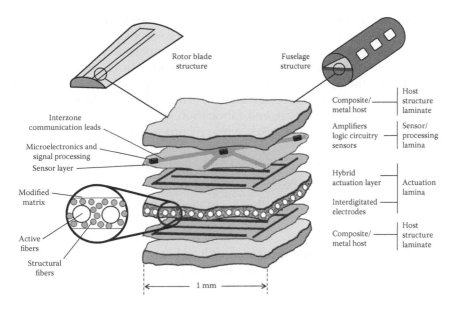

FIGURE 4.29
Smart composite actuator concept and army applications. (Reprinted from the Federation of American Scientists website. 2012. http://www.fas.org/man/dod-101/army/docs/astmp98/sec5b8.htm.)

surfaces of plants follow the direction of sunlight and animals limp to avoid overload of a damaged region. Thus, the materials and structures of natural systems have the capability to sense their environment, process the data, and respond. They are the truly "smart systems" capable of integrating information technology with structural engineering and actuation or locomotion. The composite structures having this capability are called smart composites. Smart composites are used in real-time health monitoring, damping, and adaptive and self-healing structures. Specific potential military applications of smart composites include shock isolation, vibration control, and stability augmentation systems in rotary wing aircraft to extend structural fatigue life and reliability, structural damage detection and health-monitoring systems, more accurate weapon systems, fire control and battle damage identification, and assessment and control of active, conformal, and load-bearing antenna structures (Figure 4.29).

4.2.3.1 Structural Health Monitoring

There are many unique problems with FRPs. The material is essentially inhomogeneous, since it is made up of at least two different types of materials. Its microstructural details, especially in thick sections are often totally unknown. It is difficult to avoid the formation of voids and resin-rich/resin-depleted regions. The properties of polymer matrix are usually dependent on the curing conditions. The in-service fatigue, wear, and damage characteristics are unknown. It undergoes delamination during impact loading, which is also not visible to the surface. The properties of polymer are affected by moisture, so as the properties of composites. Many of these problems could be solved by introducing optical fiber sensors into the composites with appropriate instrumentation. The optical fibers are compatible with the fiber reinforcements, having the ability to withstand strain cycling, embedded within the composite without loss in mechanical properties of the composite, and they are useful to monitor the cure and fabrication processes. However, the following precautions should be taken while incorporating optical fibers:

• They should be oriented appropriately with the reinforcing fibers.
• Their diameter should be less than 10% of the thickness of the composite.
• They should be coated with a resin-compatible material, typically polyimide.

A combination of composite structures for load bearing and sensors and actuators for information acquisition and control has led to the development of integrated composite structures that have shown capabilities in process control and health monitoring. The main objectives for the development of smart composite structures is improvement of quality control, detection of impact-induced damage during service, identification of damage size and location, reduction in maintenance costs, and enhancement of safety and reliability. The system uses light to continuously monitor optical fiber sensors located at multiple locations in the structure. It consists of three major components, namely, the optical fiber sensor, the optical interrogation unit, and a computer for data acquisition, display, and storage. All three components work together to provide a multisensor real-time interrogation system for structural monitoring.

The advantages of optical fiber-based structural monitoring system are

• Placement of multiple sensors on each optical fiber
• Availability optical fiber in long cable lengths
• Small size of cabling and sensors
• No interference from electrical and magnetic fields
• No requirement of electrical power outside the interrogation unit
• Linear variation of sensor signal with respect to strain/temperature

This type of structural health monitoring is very useful to monitor composite bridges and aircraft structures.

4.2.3.2 Structural Damping

Structural damping is carried out using piezoelectric material patches as actuators/sensors. The internal resonance and saturation phenomena are used to suppress transient and steady-state vibrations of flexible isotropic and composite structures. This technique uses the model interactions and saturation phenomenon arising from internal resonance for transferring energy from a vibrating structure to one or more electronic circuits. Once the structure is excited to resonance, the response of excited modes saturates and the oscillatory energy is channeled into the circuits.

Large amplitude vibrations often impede the effective operation of various military systems, including missiles, land vehicles, and weapon systems. Hence, it is desirable to introduce structural damping to achieve a more satisfactory performance and to delay fatigue damage.

New generations of faster, lighter, and lethal multifunctional air, land, and sea systems can only be realized if heterogeneous materials can be successfully integrated with arrays of microelectronic devices, sensors, and actuators. These devices, which are fabricated from active materials, such as piezo- and ferroelectrics, are self-contained data processors and embedded components of control devices. They can potentially be used in a number of military applications, such as control of vibration and noise in land and air vehicles, submarines, and ships for the detection and active mitigation of structural damage, the control of precision weapon systems, and stability augmentation systems in aircrafts.

4.2.3.3 Smart Electromagnetic Antenna Structures

Current aircrafts carry a proliferation of antennae to support communication, navigation, and weapon system electronics. For that, they have numerous antenna apertures located at myriad of sites, which are harmful for structural integrity. In the design of future aircrafts, it is desirable to integrate antennae as conformal, load-bearing structural elements. These multifunctional antennae will satisfy the requirements of advanced weapon system electronics and provide benefits such as lower weight, reduced drag, and the associated cost savings. The preferable locations for smart skin panels are dorsal deck, centerline, weapon bay door, outer wing, radome, trailing edge flaps, and vertical tail.

4.2.3.4 Self-Healing Composites

Structural composites are susceptible to damage, and cracks are generally formed deep within the structure, where detection is difficult and repair is almost impossible. The presence of cracks leads to reduction in mechanical properties of structural composites, as well as electrical failure in microelectronic substrates.

In nature, damage to an organism (part) elicits a healing response. This concept has been used to develop self-healing composites. Tiny capsules containing a healing agent are incorporated in the polymer matrix. When the material is damaged, the healing agent is released from the capsule, which repairs the cracks. The concept of self-healing in composites is schematically illustrated in Figure 4.30. The materials involved for the self-healing are a healing agent, microcapsules, and a catalyst. The healing agent should be stable and having a long shelf life. The healing agent suitable for this application is

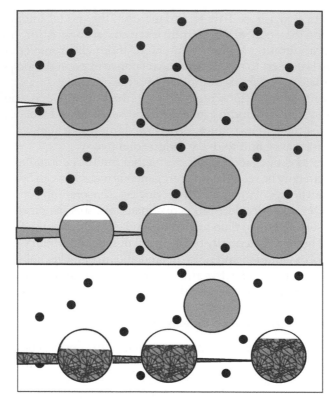

FIGURE 4.30
Conceptual illustration of self-healing in composites. (Reprinted from White, S.R. et al., *Nature*, 409, 794, 2001. With permission.)

dicyclopentadiene. This healing agent is incorporated in microcapsules and the microcapsules are dispersed in the polymer matrix. The microcapsule should rupture in the presence of a crack, but not during the manufacture of composite. Urea formaldehyde meets this condition, and with this material microcapsules of 100 μm can be made. A suitable catalyst to cure the healing agent is also dispersed in the matrix. For cyclopentadiene, ruthenium-based Grubbs' catalyst is used. This catalyst remains active, even after triggering the curing reaction. The potential applications of self-healing composites are deep-sea exploration, satellites, rocket motors, prosthetic organs, space stations, and composite bridges.

4.3 Structure and Properties of PMCs

The properties of a composite depend on the type of fiber and matrix, volume fraction of constituents, fabrication process, and the fiber orientation. Continuous fiber-reinforced polymer composites show anisotropic properties, whereas randomly oriented short fiber composites and particulate composites show isotropic properties.

4.3.1 Structural Defects in PMCs

The final stage in any PMC fabrication is called debulking (compacting a thick laminate under moderate heat and pressure and/or vacuum to remove the entrapped air), which serves to reduce the number of defects. Nevertheless, there are some common structural defects present in PMCs, which are

- Resin-rich (fiber-depleted) regions
- Voids, mainly present at roving crossovers in filament winding and in between layers having different fiber orientations (this is a very serious problem and low void content is necessary for better mechanical properties)
- Microcracks (may form due to stresses developed during curing or moisture absorption during processing)
- Debonded and delaminated regions
- Change in fiber alignment

4.3.2 Mechanical Properties of PMCs

Continuous fiber-reinforced composites invariably show anisotropic properties, that is, the mechanical properties are very high in the fiber direction and very poor in the direction perpendicular to the fiber. In a randomly oriented short fiber-reinforced composite, it is expected that the properties will be isotropic, but in certain cases, the properties are anisotropic because of the processing-induced orientation of fibers. For example, in an injection-molded thermoplastic composite, it has been found that the fibers are parallel to the mold fill direction in the two surface layers (Friedrich 1985).

Laminates of PMCs made by stacking of appropriately oriented plies also show highly anisotropic behavior. In particular, the mechanical properties of continuous fiber-reinforced polymers are very high in the longitudinal direction than in the other directions. Very often, the longitudinal properties of composites are being quoted in the literature without mentioning the direction. One must bear this discrepancy in mind when comparing such data of composite materials with the data of isotropic materials such as common metallic materials. Besides that, the composites made with aramid and polyethylene fibers will not have attractive compressive properties in the fiber direction, even though the tensile properties are very good. A summary of important properties of some PMCs is given in Table 4.4.

TABLE 4.4

Properties of Epoxy Matrix Composites

Reinforcement Materials	Specific Gravity	Tensile Modulus (GPa)		Tensile Strength (MPa)		In-Plane Shear		Compressive Strength (MPa) L	Flexural Modulus (GPa)	Flexural Strength (MPa)	ILSS (MPa)
		L	T	L	T	Strength (MPa)	Modulus (GPa)				
Unidirectional E-glass	1.8	39	4.8	1130	96.5	83	4.8	620	36.5	1135	69
E-glass cloth	1.7	16.5	16.5	280	280	—	3	100	15	220	60
E-glass CSM	1.4	7	7	100	100	—	2.8	120	7	140	69
Unidirectional S-glass	1.82	43	—	1215	—	83	—	758	41	1170	72
Boron	2.0	207	19	1585	63	131	6.4	2480	—	—	110
Kevlar 29	1.38	50	5	1350	—	—	3	238	51.7	53.5	44
Kevlar 49	1.38	76	5.5	1380	28	60	2.1	276	70	—	48
Carbon T-300	1.55	138	10	1450	45	62	6.5	1450	138	1790	96.5
Carbon HMS	1.63	207	13.8	827	86	72	5.9	620	193	1035	72
Carbon GY-70	1.69	276	8.3	585	41	97	4.1	517	262	930	52

Source: Data from Chamis, C.C., *Hybrid and Metal Matrix Composites*, American Institute of Aeronautics and Astronautics, New York, 1977. L, longitudinal direction; T, transverse direction; ILSS, interlaminar shear strength.

4.3.2.1 Tensile Properties

Most of the thermoset PMCs exhibit elastic behavior right up to fracture, that is, there is no yielding or plasticity. The strain-to-failure values of these composites are also very low, typically less than 0.5%. Consequently the work of fracture is also very low. This has some practical implications for the design engineer, since there is no local yielding to take care of stress concentrations.

In general, the continuous fiber-reinforced composites are stiff and strong along the fiber axis, but these properties fall rather sharply in the off-angles. The stiffness and strength of composites can be increased by increasing the fiber content. However, the maximum amount of fiber that can be incorporated in a composite depends on the fiber alignment. In a unidirectional fiber-reinforced PMC, the typical fiber volume fraction is about 0.65. In composites made with bidirectional fiber reinforcement, this value falls to 0.5, while in a composite containing in-plane random distribution of fibers, the volume fraction will rarely be more than 0.3.

By using the rule-of-mixture relationship, one can easily estimate the Young's modulus of a composite in the longitudinal direction from the data provided in Chapters 2 and 3 on the mechanical properties of fibers and matrix materials, respectively. The modulus values of composites made with different fibers indicate that glass fiber will only give a modest increase in modulus value, while aramid fiber will provide a significant increase. Aramid/polymer composites, however, will show a higher creep rate than glass/polymer composites because of the organic nature of aramid fiber. Aramid fiber has superior impact properties; therefore, aramid fiber-based polymer composites, in general, show better ballistic resistance against impact. In general, the long-term fatigue characteristics of aramid/epoxy composites are better than those of glass/epoxy composites, but inferior to that of carbon/epoxy composites. Hence, it reveals that the composite properties are highly dependent on the fiber properties.

4.3.2.2 Compressive Strength of Unidirectional Fiber Composites

Fiber-reinforced composites under compressive loading can be considered as elastic columns under compression. Thus, the main failure modes in a composite under compression are the ones that occur in the buckling of columns. Buckling occurs when a slender column under compression becomes unstable against lateral movement. The critical stress corresponding to failure of a column by buckling is given by

$$\sigma_{crit} = \frac{\pi^2 E}{16}\left(\frac{d}{l}\right)^2 \tag{4.4}$$

where d and l are the diameter and length of column. When the aspect ratio of a column is high, it will result in a low σ_{crit}. Of course, the fiber is not loaded directly in a fiber-reinforced composite and the matrix provides some stability in the lateral direction. By means of photoelasticity, Rosen (1965) showed

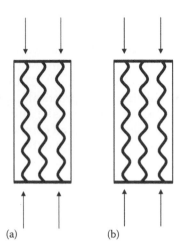

FIGURE 4.31
Fiber-reinforced composites subjected to compression in longitudinal direction: (a) in-phase buckling and (b) out-of-phase buckling. (Reprinted from Kelly, A., *Strong Solids*, Clarendon Press, Oxford, U.K., 1966, p. 149. With permission.)

that fiber-reinforced composites fail by periodic buckling of fibers and the buckling wavelength is proportional to the fiber diameter. Figure 4.31 shows schematically the two situations of a fiber-reinforced composite under compression. They are in-phase buckling and out-of-phase buckling. The in-phase buckling of fibers involves shear deformation of the matrix, and the composite strength in compression is proportional to the matrix shear modulus, G_m:

$$\sigma_c = \frac{G_m}{V_m} = \frac{G_m}{1-V_f} \tag{4.5}$$

For an isotropic matrix, G_m is given as

$$G_m = \frac{E_m}{2(1+v_m)} \tag{4.6}$$

where E_m and v_m are the matrix Young's modulus and Poisson's ratio, respectively. Hence,

$$\sigma_c = \frac{E_m}{2(1+v_m)V_m} \tag{4.7}$$

Therefore, the compressive strength of a composite can be increased by using a high-modulus matrix. Out-of-phase buckling of fibers involves transverse compressive and tensile strains. The compressive strength is proportional to the geometric mean of the fiber and matrix Young's moduli (Rosen 1965):

$$\sigma_c = 2V_f \left(\frac{V_f E_m E_f}{3V_m}\right)^{1/2} = 2V_f \left(\frac{V_f E_m E_f}{3(1-V_f)}\right)^{1/2} \tag{4.8}$$

These equations imply that, when the same fiber is incorporated in two different matrices having different moduli, the compressive failure modes differ. The volume fraction of fiber also influences the failure mode. An out-of-phase buckling mode predominates at low fiber volume fractions, while in-phase buckling mode prevails at high fiber volume fractions because fibers exert more influence on each other. The above compressive strength equations are very approximate, since they imply that as $V_m \to 0$, $\sigma_c \to \infty$, that is, the fibers are infinitely strong. In fact, no fibers are infinitely strong.

Interface bonding and matrix yielding also affect the compressive strength of fiber-reinforced composites. Poor interlayer bonding in the laminated composite also lead to easy buckling of fibers. Like tensile strength, compressive strength can also be affected by heat and moisture, as well as the ply stacking sequence.

4.3.2.3 Fracture

Fracture in PMCs, as well as in other composites, is associated with the characteristics of fibers, matrix, and interface. Specifically, fiber/matrix debonding, fiber fracture, fiber pullout, matrix crazing, and fracture of the matrix are the energy-absorbing phenomena that can contribute to the overall failure process of the composite. Of course, debonding and pullout processes depend on the interface bonding. At low temperatures, fracture of a thermoplastic PMC involves brittle failure of the polymer matrix accompanied by pullout of the broken fibers transverse to the crack plane. Such type of fracture was observed in a short glass fiber/PET composite at $-80°C$ (Friedrich 1985). The same polymer matrix is found to deform locally in a plastic manner at room temperature and showed crazing. Generally, the stiffness and strength of a PMC increase with the amount of stiff and strong fibers introduced in the polymer matrix. The same may not be applicable for the fracture toughness. The toughness of the matrix and several microstructural features related to the fibers, matrix, and the fiber/matrix interface has a strong influence on the fracture toughness of the composite. Friedrich (1985) described the fracture toughness of short fiber-reinforced thermoplastic composite in an empirical manner by the following relationship:

$$K_{cc} = MK_{cm} \tag{4.9}$$

where
  K_{cm} is the fracture toughness of the matrix
  M is a microstructural efficiency factor

M can be larger than 1 and depends on fiber content, fiber orientation, and the fiber orientation distribution with respect to the fracture plane, as well as the relative effectiveness of all the energy-absorbing mechanisms.

4.3.2.3.1 Fracture Modes in Composites

The failure of a composite occurs by a great variety of deformation modes. The operating failure mode mainly depends on the loading conditions and the microstructure of a particular composite. The important microstructural features include fiber diameter, volume fraction and distribution, and the damage/defects resulting from thermal stresses that may develop during fabrication and/or service. Since there are many factors contributing to the fracture of composites, it is not surprising that a multiplicity of failure modes is observed in composites.

Generally, the fiber and matrix have different strain-to-failure values. When the component with the lower strain-to-failure breaks (e.g., a brittle fiber or a brittle matrix), the load carried by that component should be taken by the other component. If the component with a higher strain-to-failure can bear this additional load, the composite will show multiple fractures in the brittle component. Eventually, a particular transverse section of the composite cannot take the load any further, and the composite fails.

Consider a fiber-reinforced composite, in which the failure strain of fiber is less than that of the matrix, for example, a ceramic fiber in a metallic matrix. The composite will show a single fracture, when (Hancox 1975):

$$\sigma_f^u V_f > \sigma_m^u V_m - \sigma_m^* V_m \tag{4.10}$$

where

σ_m^* is the matrix stress at fiber fracture strain

σ_f^u and σ_m^u are the ultimate tensile strengths of the fiber and matrix, respectively

This implies that when the fibers break, the matrix will not be able to bear the additional load and the composite will fail at the weakest region. This is commonly observed in composites containing a large quantity of brittle fibers and a ductile matrix. All the fibers break in more or less in a single plane (if the interface bonding is strong) and the composite also fails in that plane.

However, when the matrix can take the additional load after the failure of fiber, then the fibers will be broken into small segments until the matrix failure strain is reached. For this to happen,

$$\sigma_f^u V_f < \sigma_m^u V_m - \sigma_m^* V_m \tag{4.11}$$

When fibers with higher strain-to-failure values are incorporated in a brittle matrix, then multiple fractures in the matrix will result. According to Hale and Kelly (1972), the situation can be expressed as

$$\sigma_f^u V_f < \sigma_m^u V_m - \sigma_f^* V_f \tag{4.12}$$

where σ_f^* is the stress in the fiber at the matrix fracture strain.

Debonding, fiber pullout, and delamination are some of the failure modes commonly observed in fiber-reinforced composites and those are not observed in homogeneous, monolithic materials. Consider a fiber-reinforced composite, in which a crack in the matrix grows and approaches the fiber/matrix interface. In a short fiber-reinforced composite having fibers longer than the critical length (l_c), the fibers with extremities within a distance $l_c/2$ from the crack plane may not break but debond and pullout of the matrix. Continuous fibers ($l \gg l_c$) invariably have flaws distributed along their length. Thus, in a continuous fiber-reinforced composite, some of them fracture in the crack plane while others fracture away from the crack plane. The final fracture of the composite may involve some fiber pullout. When a fiber is pulled out from the matrix, the adhesion between the matrix and the fiber will produce a shear stress τ parallel to the fiber surface. Hence, the shear force acting on the fiber is given by $2\pi r_f \tau l$, where r_f is the fiber radius and l is the length of fiber below/above crack plane (distance from the fiber end to crack plane). Let τ_i be the maximum shear stress possible at the interface and σ_f^u be the fiber fracture stress in tension, the maximum tensile force acting on the fiber is $\pi r_f^2 \sigma_f^u$. For maximum fiber strengthening, fiber break at the crack plane is more desirable than fiber pullout, whereas for toughening, fiber pull-out may be more desirable.

Hence, the condition for fiber breakage is

$$\pi r_f^2 \sigma_f^u < 2\pi r_f \tau_i l \quad \text{or} \quad \frac{\sigma_f^u}{4\tau_i} < \frac{l}{2r_f} = \frac{l}{d_f} \tag{4.13}$$

where d_f is the fiber diameter.

On the other hand, the condition for fiber pullout is

$$\pi r_f^2 \sigma_f^u > 2\pi r_f \tau_i l \quad \text{or} \quad \frac{l}{d_f} \leq \frac{\sigma_f^u}{4\tau_i} \tag{4.14}$$

These equations imply that a weak interface is preferable for fiber pullout and a strong interface favors fiber breakage.

It is more likely that a fiber would break away from the main fracture plane because of nonuniform properties along the length of fibers. Debonding then occurs around the fiber break point. On further straining, the broken fiber parts are pulled out from the matrix. Work is done for the debonding process as well as fiber pullout against frictional resistance at the interface. Outwater and Murphy (1969) showed that the maximum energy required for debonding is

$$W_d = \left(\frac{\pi d^2}{24} \right) \left(\frac{\left(\sigma_f^u \right)^2}{E_f} \right) x \tag{4.15}$$

where x is the debonded fiber length.

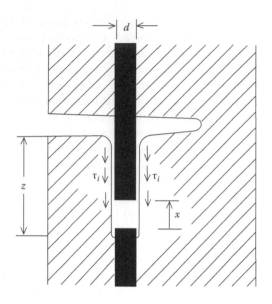

FIGURE 4.32
Schematic of an isolated fiber pullout against interfacial shear stress. (Reprinted from Chawla, K.K., *Composite Materials: Science and Engineering*, 2nd edn., Springer-Verlag Inc., New York, 1998, p. 385. With permission.)

Cottrel (1964) has pointed out the importance of fiber pullout on the toughness of composites. The debonded length, l, should be large but less than l_c for maximizing the fiber pullout work. It should be noted that for $l < l_c$, the fiber will not be loaded to its maximum possible strength level, and thus full fiber strengthening potential will not be realized. The work done during pullout of an isolated fiber can be theoretically estimated. Consider a fiber break at a distance z below the principal crack plane, such that $0 < z < l_c/2$ (Figure 4.32). Let the fiber be pulled out through a distance x against the interfacial frictional shear stress, τ_i. Hence, the force opposing the pullout is $\tau_i \pi d (z - x)$. When the fiber is further pulled out to a distance dx, the work done by this force is $\tau_i \pi d (z - x)dx$. The total work done in pulling out the fiber over a distance z can be obtained by integrating that force:

$$\text{Fiber pullout work} = \int_0^z \pi d\tau_i (z - x)\, dx = \frac{\pi d\tau_i z^2}{2} \qquad (4.16)$$

The pullout length can vary between a minimum 0 and maximum $l_c/2$. The average pullout work per fiber is then

$$W_{pf} = \frac{1}{l_c/2} \int_0^{l_c/2} \frac{\pi d\tau_i z^2}{2}\, dz = \frac{\pi d\tau_i l_c^2}{24} \qquad (4.17)$$

The earlier estimation is based on the assumption that all fibers are pulled out. In a discontinuous fiber-reinforced composite, it has been observed that fibers with ends within a distance $l_c/2$ from the main fracture plane suffer pullout (Kelly 1970). Thus, it is more likely that a fraction l_c/l of fibers will be pulled out, where l is the average length of short fibers. Hence, the average work done to pull out a fiber can be written as

$$W_{pf} = \left(\frac{l_c}{l}\right)\frac{\pi d \tau_i l_c^2}{24} \tag{4.18}$$

In general, fiber pullout contributes more significantly to composite fracture toughness than fiber/matrix debonding. However, the debonding must precede the fiber pullout.

One of the attractive characteristics of composites is the possibility of obtaining high strength with improved fracture toughness. Fracture toughness depends on the resistance to crack propagation. The crack propagation resistance in a fibrous composite can be increased by one of the following means:

1. Plastic deformation of the matrix (usually in metal/thermoplastic matrix)
2. Fiber pullout
3. Debonding and crack deflection

Yet another failure mode, which is observed in laminated composites, is delamination failure associated with the separation of plies and fiber/matrix. This is very important in structural applications involving long-term use, for example, under fatigue conditions. Highly oriented molecular structures such as that present in aramid fiber can also contribute to the work of fracture.

4.3.2.4 Damping Characteristics

High damping or the ability to reduce vibrations is very useful in many applications. For example, in mechanical equipment subjected to variable speeds, resonance problems lead to unacceptable levels of noise. In sporting goods such as tennis rackets and golf clubs also, it is desirable to have high damping. In general, aramid fiber/polymer composites exhibit good damping characteristics, because of the superior damping properties of aramid fiber.

4.3.2.5 Fatigue Properties

The degradation of mechanical properties of a material under cyclic loading is called fatigue. Understanding the fatigue behavior of composites is important, since many composite components are subjected to cyclic loading.

However, the study of fatigue behavior of composites has lagged behind the studies on stiffness and strength. The conventional approaches to fatigue are not directly applicable to the fatigue of composites. This is mainly due to the inherent heterogeneity and anisotropic nature of the composites. These result in entirely different damage mechanisms in composites compared to conventional, homogeneous, monolithic materials. Unlike the conventional materials, the failure of composites is characterized by a multiplicity of damage modes, such as matrix crazing, matrix cracking, fiber fracture, delamination, debonding, and multidirectional crack propagation, which appear rather early in the fatigue life. Progressive loss of stiffness during fatigue of a composite is another important characteristic. In the case of homogeneous material, a single crack propagates in a direction perpendicular to the cyclic loading axis, whereas a variety of subcritical damage mechanisms lead to a highly diffused damage zone in the fiber-reinforced composites.

$S-N$ curves are commonly used with monolithic materials to assess the fatigue behavior, where S is the stress amplitude $[(\sigma_{max} - \sigma_{min})/2]$ and N is the number of cycles to failure. For ferrous metals, there is a fatigue limit or endurance limit and at a stress level below this limit; theoretically, the material can be subjected to infinite number of cycles without failure. For the materials where there is no endurance limit, an arbitrary value of 10^6 cycles is assigned. In a unidirectional fiber-reinforced composite, elastic modulus and strength improve in the fiber direction. This also influences the fatigue behavior, that is, the strength improvement generally improves the fatigue resistance. As a rule of thumb, the ratio of fatigue strength/tensile strength is almost a constant. Hence, a better way to improve fatigue strength is to improve the tensile strength. The maximum efficiency in terms of strength gains in fiber-reinforced composites occurs when the fibers are continuous and uniaxially aligned to the loading direction. At off-angles, the strength drops sharply, and the role of matrix becomes more important in the deformation and failure processes. One major drawback of this $S-N$ approach is that there is no distinction made between the crack initiation and crack propagation.

A variety of operating mechanisms during fatigue failure of composites has been reported. Owen and Dukes (1967) have reported the following sequence of events in the fatigue failure of CSM glass/polyester composite: (1) debonding, generally at fiber-oriented transverse to the loading direction, (2) matrix cracking, and (3) final separation or fracture. It was found that the debonding and cracking phenomena set in early fatigue life. Lavengood and Gulbransen (1969) investigated the effect fiber aspect ratio has on the fatigue behavior of B/epoxy composites. The fiber content was 50–55 vol.%, and a low frequency (3 Hz) was used to minimize the hysteretic heating effect. The fatigue life was found to increase with increasing aspect ratio up to 200, beyond which there was little effect. Irrespective of the aspect ratio of fibers, the failure of the composites consisted of a combination of interfacial fracture and brittle failure of the matrix at 45° to the fiber axis. Composites containing fibers aligned along the stress axis and in large volume fractions show high monotonic strength values, which are translated into high fatigue strength values. This has been proved in a glass/ polysulfone composites containing different amount of glass fibers (Mandell et al. 1983). Izuka et al. (1986) studied the fatigue behavior of carbon/epoxy composites containing different strength (4.5 and 3.5 GPa) carbon fibers and found that the fatigue strength of the composite containing high-strength carbon fiber was better than the composite made with low-strength fiber.

The polymer matrices show a viscoelastic behavior and generally they are poor conductors of heat. Owing to the viscoelastic nature of the polymer matrix, there will be a phase lag between the stress and strain during cyclic loading and that excess energy is stored in the form of heat. The heat generated may not be dissipated completely because of the low thermal conductivity of the polymer matrix. This can cause a significant increase in temperature at the interior, and hence the fatigue behavior of PMCs becomes even more complex. The internal heating phenomenon, of course, depends on the frequency of cyclic loading, and it is generally negligible at frequencies less than 20 Hz. Carbon fibers are much more effective in improving the fatigue strength of a given polymer matrix than glass fibers. This is mainly due to the high stiffness and thermal conductivity of carbon fibers.

4.3.2.5.1 Fatigue of Hybrid Composites

There is a possibility of obtaining a synergistic effect in the fatigue behavior of hybrid composites. Phillips (1976) observed enhanced fatigue strength in a carbon–glass hybrid composite compared to glass fiber as well as carbon fiber-reinforced composites. However, such synergistic effects are not universal. In the case of 100% unidirectional carbon fiber/polyester, 100% chopped glass fiber/polyester, and polyester composite made with chopped glass fiber at the core and unidirectional carbon fiber at the sheath, the hybrid composite fatigue strength was intermediate between the 100% carbon and 100% glass fiber composites (Riggs 1985).

An interesting type of hybrid composite is aramid aluminum laminate (ARALL), which consists of alternating layers of high-strength aluminum alloy sheets and unidirectional aramid fiber/epoxy composites. The fatigue resistance of ARALL was found to be better than that of monolithic aluminum alloy structures, since the cracks can grow only a short distance before being blocked by the aramid fibers. The potential applications of ARALL are in tension-dominated fatigue structures such as aircraft fuselage, lower wing, and tail skins. In addition to better fatigue resistance, the use of ARALL will result in 15%–30% weight savings over conventional construction (Froes 1987).

4.3.2.6 Creep of PMCs

One can observe creep even at room temperature in many polymers and PMCs such as aramid/epoxy. At a given temperature, thermoset polymers undergo less creep than thermoplastics. Creep is usually measured by applying a constant stress and observing the strain as a function of time. There is another related phenomenon called stress relaxation, which is also important in polymers and PMCs. In the stress relaxation test, a constant strain is imposed on the specimen, and the drop in stress as a function of time is observed. To have better creep resistance in PMCs, it is necessary to use creep-resistant fibers such as alumina, SiC, or even glass instead of polymeric fibers such as aramid or polyethylene.

4.4 Environmental Effects on PMCs

Moisture present in the atmosphere can penetrate polymeric materials by diffusion process. The absorbed moisture works as a plasticizer for polymer, resulting in a decrease in stiffness, strength, and glass transition temperature. Hence, the problem of moisture absorption in PMCs is critical. It may lead to the degradation of matrix, fiber, and interface, resulting in early failure of the PMC. Hence, it is essential to select the polymer matrix, which absorbs very less moisture. Among the thermosetting resins, epoxy resin has the low moisture absorption characteristics. The epoxy resin is also impervious to a range of fluids commonly encountered in the aerospace industry, such as jet fuel, hydraulic fluids, and lubricants.

When the moisture absorption is combined with temperature rise, the degradation becomes more aggressive, leading to significant reduction in mechanical properties. Therefore, the combined effect of these two, namely, temperature and humidity, must be taken into account when designing components made of PMCs. This is especially important for the high performance composites such as those used in the aerospace industry.

Degradation of polymer composites due to ultraviolet radiation is another important environmental effect. Ultraviolet radiation breaks the covalent bonds of organic polymers, resulting in a reduction in molecular weight and hence properties. This may also lead to the formation of fine cracks on the surface. Sometimes the exposure of epoxy laminates to ultraviolet radiation can slightly increase the strength initially, due to post-curing of the resin, but there will be gradual loss of strength afterwards as a result of surface degradation.

4.5 Applications of PMCs

The commercial applications of PMCs are innumerous and it is impossible to list them all. There are many books that give a detailed account of applications (Chung 2004, Gay and Hoa 2007, Jain and Lee 2012, Mallick 2008). More information on the applications can also be found in the following websites:

www.acmanet.org/resources/07papers/wu179.pdf
http://www.eng.buffalo.edu/Classes/cie500a/oconnor%20-%20
    9-21-09%20class%20on%20FRP%20Composites.pdf
www.fibergrate.com/applications.aspx

GFRPs are widely used in several industries, from sporting goods to construction industries. Tanks and reaction vessels (with and without pressure) in the chemical industry, as well as process and effluent pipelines, are frequently made with glass fiber-reinforced polyester resin. Walk ways and many other offshore structures are fabricated using a wide variety of glass fiber/resin structural shapes made by the pultrusion technique. Glass fiber and aramid fiber-reinforced PMCs are used in the storage bins and floorings of civilian aircraft. Other PMC aircraft parts include doors, fairings, and radomes. Aramid fiber-reinforced PMCs are also used in some critical parts of helicopters and small planes. Aramid fibers can be substituted for glass fibers in most applications without much difficulty, but at a higher cost. Racing yachts and private boats are examples of composites, where glass fibers are replaced with aramid fibers, since performance is more important than cost in these applications. Currently drumsticks are being made with a pultruded core containing aramid fibers and thermoplastic injection-molded cover. These drumsticks are lightweight, will not warp, are more consistent, and last longer than the wooden ones. Military applications of PMCs vary from helmets to rocket engine cases. One has to be careful in using aramid fiber reinforcements in components involving compressive, shear, or transverse tensile loadings. In such components, the use of hybrid composites consisting of more than one type of fiber reinforcements (aramid with some other fiber) is recommended. PMCs are extensively used in military and commercial helicopters. The main driving force in such applications is weight reduction.

Some of the important structural applications are highlighted in this section, which include aerospace, automotive, sporting goods, marine, and infrastructure applications. The selection of PMCs for any application requires careful design practice and appropriate process development based on the understanding of their unique mechanical, physical, and thermal characteristics.

4.5.1 Aerospace Applications

The major structural applications of FRPs are in the aerospace industry, for which weight reduction is critical for high speeds and increased payloads. The use of FRPs has experienced a steady growth in the aircraft industry since the development of boron fiber-reinforced epoxy skins in 1969 for F-14 horizontal stabilizers. After the development of carbon fibers in 1970s, carbon fiber-reinforced epoxy has become the primary material for many wing, fuselage, and empennage components. The structural integrity and performance of these early components have built-up confidence on these materials and that prompted the development of other aircraft components, resulting in an increasing use of FRPs in military aircrafts. For example, the airframes of a vertical and short takeoff and landing (VSTOL) aircraft, AV-8B and F-22 fighter aircraft, contain nearly 25% by weight of carbon fiber-reinforced polymers. The outer skin of B-2 and other stealth aircrafts is almost made of CFRPs. The use of carbon fibers, special coatings, and other design features reduces radar reflection and heat radiation from these aircrafts.

The use of FRPs in commercial aircrafts began with a few selective secondary structural components. Later many components were replaced with FRPs. By 1987, about 350 components were placed in service in various commercial aircrafts. There is a steady increase of FRP components in commercial aircrafts as shown in Figure 4.33. Boeing 757 and 767 were the first large commercial aircrafts to make use of PMC structural components. About 95% of the visible interior parts in 757 and 767 cabins are made from nonconventional materials, including polymer composites. Airbus was the first commercial aircraft manufacturer to make extensive use of FRPs in aircrafts, and they used about 10% of the A310 aircraft's weight with FRPs. Some of the major components include lower access panels and top panels of the wing leading edge, outer deflector doors, elevators, fin box, nose radome, and rudder. A recently introduced aircraft, A380, contains about 25% by weight of FRPs. The major FRP components are central torsion box, rear-pressure bulkhead, tail, and flight control surfaces, such as flaps, spoilers, and ailerons.

Boeing has started making use of FRPs from 1995 in its 777 aircraft. About 10% of Boeing 777's structural weight is made of carbon fiber-reinforced epoxy. The composite components include horizontal stabilizer, vertical stabilizer, elevator, rudder, engine cowlings, and fuselage floor beams. About 50% of the structural weight of Boeing 787 Dreamliner is made of carbon FRPs. Two of the major composite components in 787 are fuselage and the forward section.

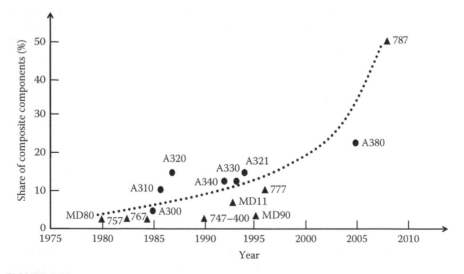

FIGURE 4.33
Usage of composites in commercial aircrafts over the years. (Courtesy of L. Herbeck and H. Voggenreiter, DLR Center of Excellence in Composite Structures, Braunschweig, Germany.)

Nowadays many aircraft structural parts are made with carbon fiber-reinforced PMCs. One of the main reasons for using this material is steadily reducing price of carbon fibers. Similarly, there has been an increasing use of PMCs in military aircraft and helicopters. Weight and cost savings are the driving forces for the use of composites. For example, the fabrication of the fuselage of Sikorsky H-69 helicopter is very labor intensive with conventional metal construction. It involves 856 parts, 75 assemblies, and 13,600 fasteners. However, a fuselage made with composites (carbon, aramid, and glass fiber/epoxy) has only 104 parts, 10 assemblies, and 1700 fasteners. The fuselage made with composites is 17% lighter and 38% cheaper than the metal fuselage.

The weight reduction by the use of composites in aircraft structures results in energy savings, in terms of fuel consumption. For a given aerodynamic configuration, there is a direct correlation between aircraft weight and fuel consumption. Boeing has brought out some estimates of the resultant fuel savings from weight reduction. For a typical operating schedule of a Boeing 757 or 767 plane, it burns 120–150 L of fuel per kilogram of weight per year. Assuming a weight savings of about 1500 kg by the use of composite materials, about 4 million liters of fuel will be saved per airplane for a 20-year economic life.

FRPs are used in many military and commercial helicopters for the fabrication of baggage doors, fairings, vertical fins, tail rotor spars, etc. One of the important applications of composites in helicopter is the rotor blades. With FRPs, the vibration characteristics of blades can be controlled by varying the type, concentration, distribution, and orientation of fibers along the blade's chord length. Another advantage of FRP rotor blade is the manufacturing flexibility. Near-net-shaped composite blades can be made by filament winding or molding techniques with little or no secondary manufacturing costs.

The main limiting factors in using CFRPs in aircraft structures are relatively low-impact damage tolerance and susceptibility to lightning damage. Another problem is that they can induce galvanic corrosion in the contacting metal components. This problem can be avoided by giving a coating on the contacting surfaces but with an additional cost.

The usage of FRPs is also increasing in many space vehicles, mainly due to the high specific properties. Mid-fuselage truss structure, payload bay doors, remote manipulator arm, and pressure vessels are the important structures of space shuttle made with FRPs. They are also used for the support structures of solar arrays, antennas, and optical platforms. Carbon fiber-reinforced polymers are mainly used in these applications because of their dimensional stability (CTE is close to zero) and high stiffness–weight ratio.

4.5.2 Automotive Applications

FRPs are used in automobiles as body and chassis components. Exterior body components require high stiffness and damage tolerance (dent resistance) as well as a "Class A" surface finish. The components are generally made by the compression molding of SMC. Although "Class A" surface is not easily obtainable in compression molding, it is achieved by in-mold coating of the exterior molded surface with a flexible resin. However, the external appearance is not critical for many underbody and under-the-hood components, such as radiator supports, bumper beams, roof frames, and oil pans. Hence, these components are generally made by compression molding of SMC. The major advantages of using SMC instead of steel in these components are weight reduction, corrosion resistance, lower tooling cost, and better parts integration.

Another manufacturing process used for making body panels of automobiles is the SRIM. In the composites made by SRIM process, randomly oriented discontinuous E-glass fibers are distributed in a polyurethane or polyurea matrix. A cargo box molded into one piece by SRIM replaces four steel panels that are joined together using welding.

Another important application of PMCs is natural gas cylinders used in automobiles. The use of compressed natural gas as a fuel in automobiles requires onboard storage of gas at a high pressure of about 20 MPa. Earlier, steel cylinders were used in automobiles to store natural gas, but they are quite heavy and thus results in a reduced pay load. Much lighter, filament-wound PMC cylinders were developed to replace those steel cylinders. Such composite cylinders are made with hoop-wrapped glass fiber/polyester on steel or aluminum liner or hoop- and polar-wound glass or carbon fiber-reinforced polymer on thermoplastic liner.

The first major application of FRPs as the chassis components is the corvette rear leaf spring. E-glass fiber-reinforced epoxy springs have been used to replace steel springs with as much as 80% weight reduction. Other chassis components such as drive shafts and wheels are proven to be useful and have been used in limited quantities in vehicles.

Carbon FRPs have become the material of choice in race cars where lightweight structure is used for gaining the advantage of higher speed. Major body, chassis, interior, and suspension components in today's Formula 1 race cars are made with CFRPs.

4.5.3 Sporting Goods

Fiber-reinforced polymers are extensively used to make sporting goods, such as tennis rackets, golf club shafts, fishing rods, bicycle frames, snow and water skis, pole vault poles, hockey sticks, sail boats and kayaks, bows and arrows, and helmets. The advantages of using FRPs for these applications are lightweight, good vibration damping, and design flexibility.

Weight reduction leads to higher speeds and quick maneuvering in competitive sports, such as bicycle races and canoe races. The frames of racing bicycles are mostly made of carbon fiber-reinforced epoxy composites.

In snow skis, sandwich construction of carbon fiber-reinforced epoxy skin and urethane foam core produces a higher weight reduction without sacrificing stiffness. Faster damping of vibrations provided by the FRPs reduces the shock transmitted to the player's arm in tennis and badminton games, when using FRP rackets. In archery bows and pole vault poles, the high stiffness–weight ratio of FRPs is used to store high elastic energy per unit weight, which helps in propelling the arrow over a longer distance or the pole vaulter to jump a greater height.

Aramid, carbon, and S-glass fibers are used in skis, boots, poles, and gloves and as threads to stitch many sports items. The main advantages of using composites in sporting goods industry are higher safety, lighter weight, and higher strength than conventional materials. For example, polymer composite ski poles are lighter and stiffer than aluminum poles. The skate shells are made with composites of aramid and carbon fibers and fabrics. PMCs are also used in rifle stocks to carry the rifles over a distance of up to 20 km during biathlons. Both lightweight and strength are important factors for the rifle stocks.

4.5.4 Marine Applications

Glass fiber-reinforced polyesters have been used in different types of boats, such as sail boats, fishing boats, dinghies, life boats, and yachts, ever since the introduction of FRPs as a commercial material. Hulls, decks, and various interior components are made with GFRPs. In recent years, aramid fiber is replacing glass fibers in some of the components because of its high strength–weight and modulus–weight ratios. The principal advantage of using FRPs is weight reduction, which translates into higher cruising speed, acceleration, maneuverability, and fuel efficiency.

The use of FRPs in naval ships started in the 1950s and has grown steadily since then. They are used in hulls, decks, masts, propulsion shafts, and rudders of mine hunters, frigates, destroyers, and aircraft carriers. FRPs are extensively used in Royal Swedish Navy's 72 m long, 10.4 m wide Visby-class corvette, and this ship is the largest composite ship in the world today. Recently, the U.S. navy has commissioned a 24 m long combat ship, called Stiletto, in which carbon fiber-reinforced epoxy is the primary material of construction. The high strength–weight ratio of CFRPs leads to high speed, maneuverability, range, and payload capacity. In addition, they impart stealth characteristics by minimizing radar reflection.

4.5.5 Infrastructure

FRPs have a greater potential for replacing reinforced concrete and steel in bridges, buildings, and other civil structures. The principal advantage of using composites in these applications is their corrosion resistance, which leads to

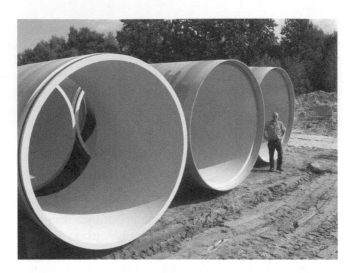

FIGURE 4.34
HOBAS 3 m FRP pipes. (Courtesy of HOBAS Engineering GmbH, Klagenfurt, Austria.)

longer life and lower maintenance and repair costs. Polymer composite pipes are widely used in gas and sewage lines (Figure 4.34). Reinforced concrete bridges tend to deteriorate after certain years mainly due to the corrosion of steel reinforcing rods (rebars) used in their construction. The corrosion problem becomes more severe in many parts of the world because of deicing salt spread on the bridge surface in winter months. The concrete surrounding the steel rebars start cracking due to the expansion of steel rods on corrosion and ultimately fall off. Another advantage of using FRPs in large bridge structures is their light weight, which means lower deadweight and easier transportation and installation. It is also possible to design FRP bridges with longer span between the supports because of their high stiffness–weight and strength–weight ratios.

A number of composite bridge decks have been constructed in the United States and Canada. An example of one such construction is the Wickwire Run Bridge located in West Virginia, United States. Glass fiber-reinforced polyester or vinyl ester resin composites are generally used for this application.

Polymer composites are being used in civilian infrastructures, such as bridges and highway overpasses. Many damaged structural beams are strengthened using PMCs to strength levels exceeding the original values. Although glass fiber-reinforced PMCs are widely used for such applications, the use of carbon fiber-reinforced PMCs is also becoming popular.

FRPs are also used for upgrading, retrofitting, and strengthening the damaged, deteriorating, or substandard concrete or steel structures. FRP strips and plates are attached to the cracked or damaged areas of the concrete structures using adhesive, or they are laid by wet lay-up or resin infusion.

(a)

(b)

FIGURE 4.35
Repairing of corroded concrete column using FRP: (a) before repair and (b) after repair.
(Courtesy of Dr. Jerome O'Connor, University at Buffalo, Buffalo, NY.)

Retrofitting of steel girders is accomplished by attaching FRP plates to their flanges. The strengthening of reinforced concrete columns is carried out by wrapping them with fiber-reinforced composite jackets (Figure 4.35). Composite jackets are much easier to install and they do not corrode like steel.

4.5.6 Armor

Spectra shield is a product of Allied Signal that is made by using woven fabrics of polyethylene fiber. It is used to make helmets, hard armor for vehicles, and soft body armor. These spectra shield helmets were used by the U.N. peacekeeping force from France in 1993 and then introduced to police forces in the United States and Europe. Body armor is a high-performance, soft armor system used for protection against rifle bullets.

4.6 Recycling of PMCs

One of the major concerns of PMCs is the recycling or reclamation of the components after their useful life. The advantageous factors during the fabrication/service of a part become disadvantageous during recycling. For example, paint with good adhesive characteristics is selected for better life, but the paint removal is a difficult task during recycling. Another example is the selection of thermosetting polymers for the fabrication of composites. Since these polymers are available as low-viscosity liquids, the fabrication of composites is very easy because they are amenable to many processing methods. After the fabrication, these polymers cure by forming permanent cross-links. Although this cross-linking phenomenon imparts thermal stability to the part, the same makes recycling very difficult, since these polymers cannot be melted to separate the constituents.

About 14 million vehicles are disposed of in Europe alone every year. Out of the total weight of a car, some 75% can be recycled (mainly metal), and the remaining 25%, including thermoset composites, is generally dumped on waste sites or used as landfill (Fisher 1991). In order to curtail the mounting of waste materials, stringent environmental regulations came into existent. Impending federal legislation in Germany, effective from 1995, has forced the automobile manufacturers to take back or arrange for the disposal of materials from discarded cars. The Netherlands had already banned land filling with FRP scrap from late 1997. The international landfill taxes are on an upward trend. In California, a state in the United States is said to produce more waste than the whole of China, a Green party aim is to ban composites and other materials that cannot be recycled. The increasingly stringent environmental regulations may squeeze out composites usage, unless recycling methods are developed. Hence, a growing number of research projects and practical initiatives have been undertaken for the recycling of PMCs (Marsh 2001).

It is recognized that a single solution appropriate for all scrap reinforced plastics is difficult to find and hence a range of options has been tried. The search for viable recycling has gone in several directions, and each avenue has produced some opportunities, but not the total solution for all reinforced plastics. Recycling techniques under investigation include grinding the composite to particulate matter for use as filler, pyrolysis to get organic compounds, reclamation of reinforcement by selective degradation of polymer matrix, and incineration as a source of heat energy.

4.6.1 Grading

A viable recycling infrastructure requires some means, preferably automated, to identify the materials in the incoming scrap and grade it according to the material type. Various methods have been tried for identifying used polymers that include ultrasonic, density differentiation, differential solution, electrostatic, and infrared spectroscopic methods. The identification of materials is a cumbersome process, since various materials in unknown quantities are used in polymer composites. Many times a combination of methods has to be tried to identify all the materials present in the composites. A simple machine has been devised in the United Kingdom by Southampton University's faculty of engineering and christened PolyAna®, which works on the principle of infrared spectroscopy. An infrared beam is directed onto the composite material surface through an optical cell, and the reflected beam is read by a spectrometer, which identifies the materials present in the composite. It could identify all the usual fiber and matrix materials, as well as fillers and additives present in the composites (Marsh 2001).

4.6.2 Shredding

Once the materials present in the scrap are identified, the next step is the size reduction. All thermoset polymer composites go through an initial size reduction called shredding. It can be a true shredding or crushing or milling. The scrap composite must be reduced to a size and shape for easy handling/packing and transportation. Some recycling strategies include separating the glass fiber, since the glass fiber will provide a higher value if it still has an aspect ratio.

4.6.3 Recycling of Thermoset Composites

There exist a number of barriers, which make the recycling of thermoset resin-based composites difficult. They are

- Thermosetting polymers cannot be remolded, since they do not melt.
- A wide range of reinforcements and fillers are used and also they are present in varying proportions.
- The major constituents in many thermoset polymer composites are inert glass fiber and mineral fillers.
- Dissolution of these polymers is also not possible, since they are resistant to most of the common solvents.

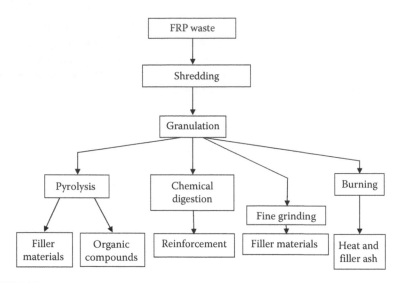

FIGURE 4.36
Recycling options for reinforced thermoset polymers.

Even with these barriers, many recycling methods for the reinforced thermoset plastics have been tried over a period of time and they are summarized in Figure 4.36. A detailed account of various recycling options can found in the book by Scheirs (1998).

4.6.3.1 Fine Grinding

One simple method of recycling thermoset resin-based composites is just grind them into fine powder and use the fine powder as filler. However, glass fiber is difficult to grind and it becomes more difficult in the presence of cured resin and mineral fillers. Hence, the glass fiber remains long that requires extra resin for wet-out. In many cases, the filler cost may not even match the grinding cost. Even then it is a more suitable option for the composites fabricated from molding compounds. Although this option offers acceptable economics for recycling composites, it may not produce a product value and process economics (Simmons 1999).

Many German automobile manufacturers are using this option of recycling. It has been found that the addition of less than 10 wt.% of ground SMC filler does not result in significant loss in strength of the final product (Petterson and Nilsson 1994). An added advantage of using SMC recyclate is that it is possible to produce lighter and stronger parts because of its lower specific gravity and the presence of glass fibers with some aspect ratio.

4.6.3.2 Advanced Grinding Processes

In the advanced grinding processes, the composite scrap is not ground to a powder, instead shredded in a controlled manner using a multistage shredding process, which allows significant preservation of the fiber length. Thus, the recyclate can be an excellent reinforcing agent.

The use of ground composite scraps as a filler in new SMC parts is a simple and inexpensive recycling option. However, much potential value of composites is not recovered, since it is not retaining the valuable properties of resins or fibers. Moreover, the grinding of used composites for reuse may be considered a satisfactory approach for SMCs used in automotive applications (which contain high levels of inorganic filler); it is not an acceptable option for the recycling of advanced composites used in aircraft and other high-performance applications, since this method does not recover the full value of the expensive fibers and resins used in such applications.

4.6.3.3 Selective Chemical Degradation

Another recycling option for thermoset polymer composites is the selective chemical degradation of polymer by using an appropriate chemical agent. It is useful to recover some compounds from the polymer matrix. It has been shown that cured polyester resin powder reacted with nucleophiles (e.g., potassium hydroxide/benzyl alcohol, hydrazine or benzylamine) and produced soluble products such as copolymers of styrene and fumaric acid derivatives (Winter et al. 1995). This processing option has certain problems, such as reduced quality of glass fibers and the requirement of neutralization step for potassium hydroxide. Hence, chemical reagents, which are less aggressive toward glass fibers, such as ethanolamines, have been evaluated (Winter et al. 1995). The ethanolamine allows the recovery of good-quality glass fibers. An additional advantage of this reagent is that it can be recovered by distillation.

The selective dissolution of epoxy from aramid/epoxy composite has been tried with 100% dimethyl sulfoxide solvent at 105°C. Although better fiber–matrix separation is achieved, the quality of aramid fiber is reduced in this process (Buggy et al. 1995).

4.6.3.4 Pyrolysis of Composites

Tertiary or chemical recycling is another option for the thermoset resin-based composites, especially for the continuous fiber-reinforced composites. This process originates from oil cracking. The heating of high molecular weight hydrocarbon (oil) in the absence of oxygen breaks the large organic molecules into smaller hydrocarbons. The resulting liquids and gases can often be used as fuels. The pyrolysis of composites involves decomposing the polymer matrix at temperatures between 700°C and 1000°C under inert atmosphere. Organic monomers are produced in this process, which can be used for the synthesis of polymers. Significant quantities of useful gases such as methane, ethane, propane, propylene, and isobutylene are also produced during this process. The solid residue comprising carbon, calcium carbonate, and glass fiber can be ground and used as filler (Anon 1991).

4.6.3.4.1 Low-Temperature Pyrolysis

The high temperatures used in pyrolysis process may degrade the quality of fibers; hence, a low-temperature pyrolysis process using a catalyst will be a more suitable option. It involves the conversion of polymer into a gaseous mixture of low molecular weight hydrocarbons using a low-temperature catalytic process. Adherent Technologies (Albuquerque, New Mexico) have developed a catalytic process to recover fibers from polymer-based composites (Allred et al. 1996). In this process, a catalyst is employed to trigger pyrolytic degradation of the composite at temperatures below 200°C. It shows excellent potential for the recovery of expensive fibers from advanced composites. It has been found that the fibers recovered by low-temperature pyrolysis retain almost all of their mechanical properties. These hydrocarbon fractions can be used as monomers, chemicals, or fuels. The reclaimed fibers and fillers can also be reused. It was found that although the reclaimed fibers are not continuous, their structural and mechanical characteristics are not altered much by the chemical recycling. It should be noted that the recycled materials are more suitable for use in less-demanding applications because of considerably reduced intrinsic properties of the materials. The processing cost is also a major concern in this recycling option.

A concept of applying high-pressure steam to cause thermal decomposition of the polymer is being evaluated in the United States for the recovery of carbon fiber from composites. This process is already practiced in Japan for the recycling glass fiber-reinforced composites. The advantage over pyrolysis is that the glass fiber and filler come out of the process are relatively clean (Simmons 1999).

4.6.3.5 Energy Recovery

Another option, which may not be considered as recycling, but a useful option for disposal, is incineration. The heat generated by burning the polymer composites can be utilized in many ways. However, this type of recycling is not generally preferred due to the following reasons:

- Most composites have relatively low fuel values because of the high level of filler contents.
- The cost of collection and transportation of composite scrap is relatively high.

Hence, this method of recycling is selected only when disposal is prohibitive. The ash after burning has a large quantity of inert reinforcement and filler. This process may become attractive when the inorganic material as well the energy content of the polymer is utilized. The two main methods that have been identified for energy recovery and utilization of inorganic material from SMC waste are

- Co-combustion of SMC scrap with coal in thermal power plant
- As a fuel and additive in cement manufacture

Co-firing with coal not only supplements the energy from the coal but the calcium carbonate fillers present in the SMC absorb sulfur oxides produced during coal combustion (Pickering and Benson 1993). However, the glass fibers are disintegrated during combustion, and the total ash can be used only as particulate filler in composites or in agriculture to reduce pH of acidic soils.

Waste SMC is ideally suited in cement manufacturing, since the chemical composition of reinforcements and fillers in SMC are almost similar to the raw materials used in the manufacture of cement. The SMC scrap is burnt in cement kilns to utilize the energy content of the polymer and to enable the incorporation of mineral fillers and reinforcements into the cement clinker (Pickering and Benson 1991). However, the quantity of SMC scrap should be restricted; otherwise the presence of boron oxide in the cement arising from the glass fiber can affect the early strength during setting of the cement. To prevent this potentially deleterious effect, it is necessary to limit the use of scrap composites such that boron oxide levels do not exceed 0.2%. Considering the huge quantity of cement produced, this seems to be a better option for the recycling of SMC scrap.

4.6.4 Recycling of Thermoplastic Composites

Although recycling of thermoplastics is easier than thermoset polymers, the presence of fibers and fillers poses problems. A summary of recycling options available for thermoplastic composites is shown in Figure 4.37.

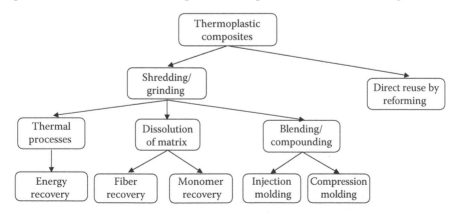

FIGURE 4.37
Recycling options for fiber-reinforced thermoplastics. (Reprinted from Henshaw, J.M., in *ASM Handbook*, Vol. 21, *Composites*, ASM International, Materials Park, OH, 2001, p. 1010. With permission.)

Thermoplastic composites can be reused several times, even though there is some degradation in mechanical properties after each cycle (Anon 1991). The following methods can be used for the reuse of thermoplastics:

- Parts can be ground to chips and used as raw material for producing sheets.
- Several smaller parts can be heated and remolded into large-sized part.
- Parts can be ground and used in place of short glass fiber for injection molding or extrusion purposes.

Sulfuric acid dissolution has been found to be a successful method to dissolve polyether ether ketone (PEEK) from carbon/PEEK composite. The PEEK polymer is recovered from acid solution by adding drops of ice water to the solution at 0°C. The other methods are similar to those used for thermoset polymer composites.

Questions

State whether the following statements are true or false and give reasons:

4.1 Hand lay-up process is the most preferred process for making FRPs.

4.2 Good surface finish on both the sides can be obtained by the hand lay-up process.

4.3 Tooling cost is low for the RTM process.

4.4 It is very easy to make composites by RTM process using thermoplastics.

4.5 Thermal expansion coefficient of mold material is an important parameter while designing the mold.

4.6 Viscosity of the resin decides the flowability within the RTM.

4.7 Structural composites can be made by the RRIM process.

4.8 Generally thin sections are produced by compression molding process.

4.9 Components with double curvatures can be produced by the diaphragm forming process.

4.10 Void-free components are produced by filament winding.

4.11 Filament winding process can be used to produce complex geometries.

4.12 Strength of PMC can be accurately predicted by the rule of mixtures.

4.13 Matrix cracking resulting from thermal mismatch is a more serious problem in short fiber composites than in continuous fiber composites.

4.14 Fatigue behavior of GFRP composite is better than carbon fiber-reinforced polymer composite.

4.15 High strength values translated into high fatigue strength values in composites.

4.16 Composites generally show lower fatigue resistance in compression than in tension.

Answer the following questions:

4.17 Explain the criteria for selecting a process to manufacture FRP products.

4.18 What are the advantages and limitations of spray-up process compared to hand lay-up?

4.19 What are the advantages and disadvantages of hand lay-up process?

4.20 What are the common structural defects in PMCs?

4.21 Find out the modulus of Kevlar–epoxy composite containing 60 wt.% of Kevlar fiber ($\rho_{Kevlar} = 1.45$; $\rho_{epoxy} = 1.2$; $E_{epoxy} = 3$ GPa; and $E_{Kevlar} = 125$ GPa).

4.22 Compare hand lay-up and RTM processes.

4.23 Explain the draft and corner radii requirements of a mold.

4.24 What are the limitations of the filament winding process?

4.25 What are the fracture modes in fiber-reinforced polymer composites?

4.26 What are the common structural defects in fiber-reinforced polymer composites?

4.27 Explain the damping characteristics of polymer composites.

4.28 How does the fatigue behavior of composites differ from the homogeneous isotropic material?

4.29 Unidirectional composites may not be useful or cost-effective in some applications. Explain. What are the ways to overcome those problems?

4.30 What are the factors responsible for the degradation of composite materials?

4.31 Mention the advantages of using optical fiber sensors in composites.

References

Allred, R.E., A.B. Coons, and R.J. Simonson. 1996. In *Proc. 28th Int. SAMPE Tech. Conf.*, November 4–7, Seattle, WA.

Anon. 1991. *Reinf. Plast.* 35 (7–8): 20.

Belle, A. 2007. Influence of microwave post-curing on the mechanical and physical properties of a glass/polyester composite. MTech dissertation, IIT Madras, Chennai, India.

Bowen, D.H. 1998. *Concise Encyclopedia of Composite Materials*, ed. Kelly, A. Oxford, U.K.: Pergamon Press, p. 181.

Buggy, M., L. Farragher, and W. Madden. 1995. *J. Mater. Process. Technol.* 55: 448.

Chamis, C.C. 1977. *Hybrid and Metal Matrix Composites*. New York: American Institute of Aeronautics and Astronautics.

Chawla, K.K. 2012. *Composite Materials: Science and Engineering*, 3rd edn. New York: Springer-Verlag Inc.

Cottrel, A.H. 1964. *Proc. R. Soc. Lond.* A228: 508.

Fisher, G. 1991. *Reinf. Plast.* 35 (9): 34.

Friedrich, K. 1985. *Compos. Sci. Technol.* 22: 43.

Froes, F.H. 1987. *J. Metals* 39 (11): 28.

Greenkorn, R.A. 1983. *Flow Phenomena in Porous Media-Fundamentals and Applications in Petroleum, Water and Food Production*. New York: Dekker.

Hale, D.K. and A. Kelly. 1972. *Annu. Rev. Mater. Sci.* 2: 405.

Hancox, N.L. 1975. *J. Mater. Sci.* 10: 234.

Henshaw, J.M. 2001. In *ASM Handbook*, Vol. 21, *Composites*. Materials Park, OH: ASM International, p. 1010.

Izuka, Y., T. Norita, T. Nishimura, and K. Fujisawa. 1986. In *Carbon Fibers*. Park Ridge, NJ: Noyes Publications, p. 14.

Jain, R. and L. Lee. (Eds.). 2012. *Fiber Reinforced Polymer (FRP) Composites for Infrastructure Applications*. New York: Springer.

Joselin, R. and T. Chelladurai. 2011. *J. Fail. Anal. Prev.* 11: 344.

Kelly, A. 1966. *Strong Solids*. Oxford, U.K.: Clarendon Press, p. 149.

Kelly, A. 1970. *Proc. R. Soc. Lond.* A319: 95.

Lavengood, R.E. and L.B. Gulbransen. 1969. *Mater. Sci. Eng.* 9: 365–369.

Mandell, J.F., F.J. McGarry, D.D. Huang, and C.G. Li. 1983. *Polym. Compos.* 4: 32.

Mark, H.F. and J.I. Kroschwitz. 1988. *Encyclopedia of Polymer Science and Technology*, Vol. 12. New York: Wiley, p. 287.

Marsh, G. 2001. *Reinf. Plast.* 45 (6): 22.

Muzzy, J. et al. 1989. *SAMPE J.* 25: 23.

Nightingale, C. and R.J. Day. 2002. *Composites* A33: 1021.

Outwater, J.O. and M.C. Murphy. 1969. Fracture energy of unidirectional laminates. In 24th *Annu. Tech. Conf., SPI (Soc. Plast. Ind.) Reinf. Plast./Compos. Div., Proc.*, 11-C-1-8. New York: Society of Plastics Industry.

Owen, M.J. and R. Dukes. 1967. *J. Strain Anal. Eng.* 2: 272.

Petterson, J. and P. Nilsson. 1994. *J. Thermoplast. Compos.* 7: 56.

Phillips, L.N. 1976. *Composites*, 7: 7.

Pickering, S.J. and M. Benson. 1991. In *Proc. 2nd Int. Conf. Plast. Recycl.*, March 13–14, London, U.K., p. 23/1–10.

Pickering, S.J. and M. Benson. 1993. In *Proc. ECCM 6 – Recycl. Concepts Procedures*, September, Bordeaux, France, p. 41.

Prasad, K.D.V. and S.H. Hsu. 2004. *J. Mater. Process. Technol.* 155–156: 1532.

Riggs, J.P. 1985. In *Encyclopedia of Polymer Science and Engineering*, 2nd edn., Vol. 2. New York: John Wiley & Sons, p. 640.

Rosen, B.W. 1965. In *Fiber Composite Materials*. Metals Park, OH: American Society for Metals, p. 58.

Scheirs, J. 1998. *Polymer Recycling*. Chichester, U.K.: John Wiley & Sons.

Schewartz, M.M. 1997. *Composite Materials*, Vol. II. Upper Saddle River, NJ: Prentice Hall, p. 28.

Simmons, J. 1999. *Reinf. Plast.* 43 (10): 64–65.
Talreja, R. and J.E. Manson. (volume eds.). 2000. In *Comprehensive Composite Materials*, Vol. 2, eds. Kelly, A. and Zweben, C. Oxford, U.K.: Elsevier, p. 705.
White, S.R. et al. 2001. *Nature* 409: 794.
Winter, H., H.A.M. Mostert, P.J.H.M. Smeets, and G. Paas. 1995. *J. Appl. Polym. Sci.* 57: 1409.
Zhou, J., C. Shi, B. Mei, R. Yuan, and Z. Fu. 2003. *J. Mater. Process. Technol.* 137: 156.

_____

Bibliography

ASM Handbook. 2001. *Composites*, Vol. 21. Materials Park, OH: ASM International.
Astrom, B.T. 2002. *Manufacturing of Polymer Composites*. Cheltenham, U.K.: Nelson Thornes.
Campbell, F.C. 2004. *Manufacturing Processes for Advanced Composites*. Oxford, U.K.: Elsevier.
Chawla, K.K. 1998. *Composite Materials: Science and Engineering*, 2nd edn. New York: Springer-Verlag Inc.
Lubin, G. 1982. *Handbook of Composites*. New York: Springer.
Mallick, P.K. 2008. *Fiber-Reinforced Composites*, 3rd edn. Boca Raton, FL: CRC Press.
Mazumdar, S.K. 2002. *Composites Manufacturing*. Boca Raton, FL: CRC Press.
Schwartz, M.M. 1997. *Composite Materials*, Vol. II. Upper Saddle River, NJ: Prentice Hall.
Wang, R.M., S.R. Zheng, and Y.P. Zheng. 2011. *Polymer Matrix Composites and Technology*. Cambridge, U.K.: Woodhead Publishing Limited.

_____

Further Reading

Chung, D.D.L. 2004. *Composite Materials: Science and Applications*. London, U.K.: Springer-Verlag.
Gay, D. and S.V. Hoa. 2007. *Composite Materials: Design and Applications*, 2nd edn. Boca Raton, FL: CRC Press.
Zhang, S. and D. Zhao. 2013. *Aerospace Materials Handbook*. Boca Raton, FL: CRC Press.

5

Metal Matrix Composites

Metal matrix composites (MMCs) are a kind of composite material in which rigid reinforcements are embedded in a ductile metal or alloy matrix. The reinforcements can be in the form of particle, short fiber, or continuous fiber. MMCs combine metallic properties like ductility and toughness with ceramic characteristics like hardness and modulus, leading to greater strength in shear and compression and to higher service temperature capabilities. Interest in MMCs for use in the aerospace and automotive industries and other structural applications has increased over the past 30 years because of these attractive physical and mechanical properties. Moreover, the availability of relatively inexpensive reinforcements and the development of various processing routes to produce MMCs with reproducible microstructure and properties increase the interest on MMCs further.

MMCs have long been of technological significance. For example, layered structures of metallic and nonmetallic constituents, made by repeated lamination and hammering, were produced in several ancient civilizations. The development of aluminum or copper reinforced with 30%–70% of continuous tungsten or boron fibers was initiated in the 1960s. However, commercial usage of long-fiber-reinforced MMC has not yet reached the level that could be considered industrially significant.

The development of high-strength B and SiC monofilaments enabled significant efforts on fiber-reinforced MMCs throughout the 1960s and early 1970s. Problems associated with processing, fiber damage, and fiber–matrix interactions were identified to produce quality MMC materials. Although they were very expensive and had marginal reproducibility, important applications were established, mainly in aerospace sector. In the late 1970s, efforts on MMCs were renewed with the development of SiC whisker-reinforced composites. The high cost of whiskers and difficulty in avoiding whisker damage during consolidation led to the development of particulate MMCs. These particulate MMCs provided nearly equivalent strength and stiffness, but with much lower cost and easier processing.

Discontinuous fiber-reinforced and particulate composites based on metal matrix were developed during the 1980s, especially Al-based matrices with SiC particles, or Al_2O_3 particles or fibers. A combination of good properties, low cost, and high workability has made them attractive for many applications, and they have now become commercially important MMCs. There has been progressive development in MMCs on a number of fronts over the years. Most of the aspects of MMCs are now fairly understood, and the limitations and attractions of their processing and performance characteristics are reasonably clear. The volume of MMCs used in industrial applications remains relatively low compared to PMCs, but this is changing, at least for certain types of discontinuous fiber-reinforced and particulate MMCs.

When selecting a MMC, the objective might be a combination of the attractive properties of matrix, such as high ductility or thermal conductivity, and the properties of reinforcement, such as high stiffness or low thermal expansion coefficient. Unlike PMCs, MMCs are nonflammable, do not outgas in a vacuum, and suffer minimal attack by organic fluids such as fuels and solvents.

Simultaneously new applications for MMCs were explored, and MMCs of aluminum, magnesium, iron, and copper find applications in automotive, thermal management, tribological, and aerospace industries. In addition, titanium MMCs were developed for high-temperature components in aeronautical systems, including high-mach craft frames and critical rotating components for advanced gas turbine engines. In 1990s, MMC insertions in the surface transportation and thermal management/electronic packaging industries far exceeded the growth in the aerospace industry.

There are three types of MMCs based on the dispersed phase, and they are as follows:

1. Particulate MMCs
2. Short-fiber-reinforced MMCs
3. Continuous fiber-reinforced MMC

Particulate or short-fiber-reinforced MMCs are very common compared to continuous fiber-reinforced MMCs, because they are relatively easy to manufacture. They are less expensive and mostly have isotropic properties. More recently, the discontinuously reinforced MMCs have attracted considerable attention as a result of (1) the availability of various types of reinforcement at competitive prices, (2) the successful development of manufacturing processes to produce MMCs with reproducible microstructure and properties, and (3) the availability of standard or near-standard metalworking methods, which can be utilized to fabricate these composites. The family of discontinuously reinforced MMCs includes both particulate and whisker- or short-fiber-reinforced MMCs. These MMCs have been produced by several processing techniques such as powder metallurgy, spray deposition, mechanical alloying, and various casting techniques.

Particulate composites have been the subject of intensive study for a range of industrial applications because of the processing advantage compared to continuous fiber-reinforced MMCs. Most of the composites are made with aluminum alloy matrices, and some composites are made with titanium-, steel-, cobalt-, and magnesium-based matrices. The important attributes of particulate MMCs are high stiffness, low density, high hardness, adequate toughness, and the availability of low-cost powders of suitable size.

5.1 Selection of Reinforcements

Many processing methods are available now, by which reinforcement can be successfully incorporated into a metal matrix. It is important to note that the right choice of processing method is just as important in terms of the microstructure and performance of a component as it is for its commercial viability. However, before looking into the details of various processing options, it is worthwhile dwelling for a moment on the selection of reinforcement. Clearly, the size, shape, and mechanical properties of the reinforcing particles or fibers are of prime importance. A list of some important dispersed phases used in MMCs and their aspect ratio and diameter is given in Table 5.1.

There are many issues, which may need consideration while selecting the particulate material. Some of them are density, cost, stiffness, thermal expansion, thermal/electrical conductivity, chemical compatibility with the matrix during processing, and the formation of strong interfacial bond. SiC and alumina particles are the better choices, in terms of availability and cost, and they are also quite attractive in terms of relevant mechanical properties. While ceramics are in general electrical insulators, the thermal conductivity of some of the ceramics is comparable or even greater than metals. The thermal conductivities of SiC, aluminum nitride, and diamond are reasonably high. In these materials, heat can be conducted by phonons as well as electrons, and phonon transport is favored by light and stiff crystal lattice.

Chemical reaction during processing can occur with some particulate materials used in composites. Several reports concerning the reaction between SiC and aluminum melts are available. There is also evidence for the reaction between SiC and Ti during solid-state consolidation. Reactions also occur between SiC and ferrous alloy matrices. Alumina is usually less reactive than SiC in aluminum, but reacts quite strongly with titanium at elevated temperatures. Magnesium is rather different from Al and Ti, that is, it does not react readily with SiC, but it does react with alumina. Surface treatments or coatings are normally given to fiber reinforcements to prevent reaction, whereas particulates are usually used in the virgin state. Surface oxidation of nonoxide

TABLE 5.1

Typical Reinforcements Used in Metal Matrix Composites

Type	Aspect Ratio	Diameter (μm)	Examples
Particle	~1–4	1–25	SiC, Al_2O_3, WC, B_4C_3, TiB_2, SiO_2, TiC, BN, ZrO_2, fly ash
Short fiber	~10–1000	0.1–25	SiC, SiC_w, Al_2O_3, Al_2O_3–SiO_2, C
Continuous fiber	>1000	3–150	SiC, Al_2O_3, B, C, W

Source: Adapted from Chawla, K.K., *Composite Materials: Science and Engineering*, 2nd edn., Springer-Verlag, Inc., New York, 1998, p. 165.

particulates is a viable option, which has a beneficial effect on interfacial bonding and other characteristics.

For many years, alumina has been recognized as a promising reinforcement for MMCs used at high temperatures and chemically aggressive environments. It retains its modulus, strength, and abrasion resistance up to 900°C. It is also inert to most metals.

Particulate MMCs are usually manufactured on a commercial scale by melt infiltration, by casting, or by powder metallurgy (P/M) technique. More specialized processes, such as reactive processing and spray co-deposition, are sometimes used. Quality control objectives include the prevention of excessive interfacial reaction during processing and also the avoidance of microstructural defects such as poor interfacial bonding, internal voids, and segregation of the particulates. Typically, particulate MMCs are made with the particles of diameter range from 10 to 20 μm and content in the range of 10%–30% by volume. MMCs with finer particles or higher particle contents are also made commercially.

The range of applications for particulate MMCs is wide, but more applications are in the automobile industry, especially wear-resistant components are being developed. Al–MMCs give superior wear performance and at the same time the components weigh less compared to conventional cast-iron components. However, the temperature rise should be controlled to avoid excessive wear above a critical temperature. The good thermal conductivity of SiC–Al combination favors heat dissipation, and the active cooling by proper design of the component ensures the better control of temperature rise.

Short-fiber-reinforced MMCs attracted much attention in the mid-1980s with the development of short alumina fiber-reinforced aluminum diesel engine pistons. Other similar fibers have also been used for this type of applications. This type of components is normally made by melt infiltration process. Other processing methods used for short-fiber-reinforced MMCs are squeeze infiltration and P/M techniques. Typical fiber diameters are a few microns and lengths are several hundred microns. However, the fibers are often broken during processing, and the final aspect ratio of fibers lies in the range of 3–100. The formability of these composites is generally inferior to that of particulate composites. However, there are some property advantages over particulate MMCs, particularly the resistance to creep and, to a lesser extent, wear.

Whiskers are usually submicron in diameter, with aspect ratios up to several hundreds. Their tensile strengths are often very high and elastic moduli are also very high. However, they are not widely used in commercial applications. The main barrier initially was the difficulty in producing them at economically attractive cost. In the late 1980s, many methods were developed to produce SiC whiskers from various cheap starting materials, such as rice husks. Even then, the commercial exploitation of whisker-reinforced MMCs didn't take off largely due to handling difficulties. There are inherent problems with very fine fiber like materials, which tend to form tenacious ball-like structures, and it is difficult to disperse them uniformly and orient them in a controlled manner. Another major problem of whiskers concerns perceived health hazards. Whiskers in the submicron size range can become airborne very easily and are likely to reach the lungs. Once they reach lungs, they have similar effect as asbestos fibers. Hence, rigorous safety measures would be necessary during handling, and this led to strong discouragement to further research and development activities.

A number of long-fiber-reinforced MMCs have been developed, and some of these have been used in certain industrial and military applications. However, as a consequence of processing difficulties and limited ductility and toughness of these composites, the usage has in general remained rather specialized and limited. The diameter of the fibers is ranging from 5 to 30 μm. These fibers are flexible enough to produce complex-shaped parts because of their small diameter. The fibers commonly used in MMCs are C, SiC, and various alumina-based fibers. Glass fibers and organic fibers are not suitable for MMCs, since these fibers cannot survive at the elevated temperatures used for MMC production.

Melt infiltration may be a suitable process for this type of MMCs, although some problems arise with unidirectionally aligned fibers. The pressure applied to the melt in the transverse direction to the fiber axis tends to bring the fibers in close contact and reduce the channels between them, and hence the melt is unable to penetrate the fiber bed completely. This problem can be overcome to some extent by the introduction of particulates or fibers in the transverse direction (melt flow direction). Other processing techniques, including P/M techniques, are often unsatisfactory.

Carbon fibers are usually not preferred for MMCs, mainly because of the interfacial reaction during processing and galvanic corrosion during service. Chemical reaction problems are less pronounced with magnesium, since it does not form a stable carbide. Attempts have been made to prevent the reaction by giving a protective coating, such as titanium nitride on carbon fibers, but it is a difficult and expensive process. There is a peculiar problem with aluminum alloys reinforced with carbon fibers; that is, the formation of hygroscopic interfacial reaction product, Al_3C_4. Hence, this type of composites undergoes rapid corrosion in aqueous environments.

SiC continuous fibers are not very common as SiC particulates and whiskers. Commercial SiC fibers are available under trade name Nicalon which actually contain considerable amounts of free carbon and silica. The presence of free carbon leads to excessive reaction with most metallic matrices during processing. There are a number of oxide fibers, which are fairly resistant to molten metals. Notable among these fibers are polycrystalline alumina fibers, which have been primarily used in aluminum alloy matrices and to a lesser extent in titanium alloy matrices. Commercial interest in continuous fiber-reinforced MMCs has remained low because of the difficulties mentioned earlier, and they are used mainly in some specialized applications.

Even though they can significantly improve the creep resistance and undergo minimum interfacial reactions, the composites made with this type of fibers are prohibitively expensive.

Often the choice between continuous and discontinuous reinforcements is relatively straightforward, both in terms of performance and processing cost. However, within each category there exist wide variations in reinforcement size and morphology. For example, consider SiC particulates; the most convenient form is the particles of 3–30 μm diameter, which is cheap and relatively easy to handle. Extensive studies carried out on the influence of particle size on properties suggest a narrow size distribution for better properties. Furthermore, a change from the common angular particles to spherical particles may lead to improved performance for the composites and superior flow behavior during the handling of particles. It is also presumed that platelets or ribbons may have certain advantages over equiaxed particles. Alumina is available in the form of platelets of aspect ratio 5–25 and diameter of the order of 10 μm. These platelets can be readily assembled into preforms for melt infiltration, often with better control over volume fraction. However, the benefits of changing the type or nature of the reinforcement must be clearly established, if such a change is accompanied by increased processing or raw materials costs.

5.2 Processing of Metal Matrix Composites

Over the years different types of processing techniques have evolved in an attempt to optimize the microstructure and mechanical properties of MMCs. MMCs can be produced by solid-, liquid-, or gaseous-state processes. Some of the solid-state processing methods are compaction and sintering, hot-pressing, roll bonding, and diffusion bonding. The most common solid-state processes are based on P/M techniques. These mainly involve particulates or discontinuous fiber reinforcements because of easy mixing and consolidation. The ceramic reinforcements and metal powders are mixed thoroughly, compacted, and sintered. Since the solid-state processing is carried out at a relatively lower temperature compared to liquid-state processing, the adverse reactions taking place between the constituents are minimized. (Normally sintering is carried out at 0.75–0.8 T_m of the metal matrix.) Another important advantage is the possibility of incorporating very high reinforcement content (up to 90 vol.%). However, there are some problems with the solid-state processing methods. Since chemically dissimilar materials are used for the MMCs, the diffusion processes during sintering are very much affected and result in incomplete densification. In many cases, hot pressing (simultaneous application of temperature and pressure) is required to achieve full densification by enhancing diffusion processes. The hot pressing process increases the cost of product, reduces the rate of production, and also may damage the fiber reinforcement. As an alternative, secondary processing methods such as extrusion or forging can be used to improve the density.

For the MMCs, liquid-state processing methods are relatively easy compared to solid-state processing, since most of the MMCs are made with aluminum alloys, which have low melting point. In these processes, the reinforcements are incorporated into the molten metal matrix. There are several advantages in using liquid-state processing. These include near-net-shaped formation (even complex geometries), faster rate of production, uniform wetting of reinforcements, and dense matrix. The main drawbacks of liquid-state processing are the difficulty in controlling reinforcement distribution and obtaining a uniform matrix microstructure. Furthermore, adverse reactions at the interface between matrix and reinforcement are more likely to occur at the high temperatures used in liquid-state processing. These reactions lead to the formation of brittle compound, which will adversely affect the mechanical properties of the composite. Another disadvantage of these processing methods is the restriction on the quantity of reinforcement added. Although the initial viscosity of molten metal is less, the viscosity will increase drastically by the addition of reinforcements (this effect increases with decrease in particle size). The segregation of reinforcement due to the difference in density between the matrix and the reinforcement is another major problem. Even though there are many disadvantages, the liquid-state processing methods are widely used for the fabrication of MMCs.

Most of the commercial applications of MMCs have focused on aluminum and its alloys as the matrix materials. The combination of lightweight, corrosion resistance, and useful mechanical properties has made aluminum alloys very popular. The melting point of aluminum is high enough for many applications, yet low enough for reasonably convenient composite processing.

5.2.1 Liquid-State Processes

Various liquid-state processing techniques have been developed that involve the matrix becoming at least partially molten when it is brought into contact with the reinforcement. This generally favors intimate contact and hence a stronger bond, but it may also lead to the formation of a brittle interfacial layer. These features are sensitive to a number of factors such as the type of reinforcement and matrix, melt/reinforcement contact time, temperature, and pressure, which vary widely between the different processes.

In the liquid-state processes, the ceramic reinforcements are incorporated into molten metallic matrices. This is followed by mixing and casting into components of different shapes or billets for further fabrication. The process involves a careful selection of the ceramic reinforcement depending on the matrix alloy; mainly the compatibility with the matrix is considered. Since most ceramic materials are not wetted by the molten metals or alloys, wetting agents should be added to the melt or an appropriate coating should be given to the ceramic reinforcements.

Several approaches have been utilized to introduce reinforcement particles into molten matrix materials. These include (1) injection of powders entrained in an inert carrier gas into the melt using an injection gun, (2) incorporation of particulates into the molten stream as it fills the mold, (3) addition of particulates to the melt through the vortex introduced by mechanical stirring, (4) addition of small briquettes made by mixing and pressing matrix alloy powder and particulates into the melt followed by stirring, (5) dispersion of the particulates in the melt using centrifugal acceleration, and (6) injection of particulates into the melt while the melt is subjected to ultrasonic vibration.

A strong bond between the matrix and the reinforcement is achieved by utilizing high processing temperatures and alloying the matrix with an element that can improve wetting between the matrix and reinforcement. Some kind of agitation during processing is necessary to facilitate interfacial bonding by disrupting contamination films and adsorbed layers.

Liquid-state processes have reached an advanced stage of development, and SiC or Al_2O_3 particulates of size 3–150 μm are generally added to a variety of aluminum alloy matrices. It is possible to incorporate up to 20 vol.% of either SiC or Al_2O_3 particulate in various aluminum alloys to obtain MMCs with attractive combinations of properties.

Despite many MMCs have been produced successfully with liquid-state processes, some difficulties still exist in this process. These include agglomeration and settling of reinforcement particulates, segregation of reinforcement/secondary phases in the metallic matrix, adverse interface reactions, and reinforcement fracture during mechanical agitation.

5.2.1.1 Stir Casting

The simplest and most economical method to produce MMC is simply mixing the solid reinforcement in the liquid metal and then allowing the mixture to solidify in a suitable mold. The mixture can be continuously agitated while the reinforcement is progressively added. In principle, this can be carried out using conventional processing equipment on a continuous or semicontinuous basis. This type of processing is now commercially used for the production of Al–SiC_p composites. However, the details of the conditions employed during the stir casting process have not been published. Though the process looks simple, there are many difficulties in this process. Some of the problems are increased viscosity due to the addition of reinforcement, inhomogeneous microstructure, and adverse chemical reactions.

Typically particulate reinforcements are used in this process because of the difficulty in casting with fibrous forms. A mixture of particles and molten matrix material is cast into ingots and a secondary mechanical process, such as extrusion or rolling is usually applied to the composite to remove casting defects.

Casting of MMCs requires some modifications to the existing metal casting process, such as

1. *The viscosity should be kept within the allowable limit.* The addition of particulates increases the viscosity of the melt. For a molten metal with non-metallic particles, the increase in viscosity is given by (Thomas 1965):

$$\eta_c = \eta_m \left(1 + 2.5V_p + 10.05V_p^2\right) \qquad (5.1)$$

   where
   η_m is the viscosity of molten metal without reinforcement particles
   V_p is the volume fraction of particles

   To keep the viscosity within the allowable limit, the temperature of the composite melt should be increased (~745°C for Al–Si/SiC casting).
2. *Alloys with minimum reactivity to the reinforcement must be used.* Al–Si alloys are typically used with SiC particulates. The amount of Si required to prevent the formation of Al_4C_3 increases with increase in melt temperature. Hence, the appropriate amount of Si should be available in the matrix alloy.
3. *Covering the melt with an inert gas atmosphere.* The oxidation of metal will also increase with increasing temperature. Covering the melt with an inert gas atmosphere reduces oxidation.
4. *Stirring of the melt.* To minimize the settling of particles due to density difference, continuous stirring is often required. For example, the density of SiC is 3.2 g cm^{-3} and Al is 2.7 g cm^{-3}. The dense particles will try to settle at the bottom of vessel because of this density difference, unless the melt is stirred continuously.

The schematic of stir casting of MMC is shown in Figure 5.1. The following procedure is generally used to prepare SiC$_p$/aluminum alloy composites. The required amount of aluminum alloy is placed in a graphite crucible and heated to 800°C. Simultaneously, SiC$_p$ is preheated to 900°C in a separate muffle furnace for a soaking period of 1 h. Prior to particle addition, 1 wt.% of Mg powder is added in small packets (wrapped in an aluminum foil) into the melt in order to improve the wettability. The SiC particles are then added manually in small quantities using a feeder through the vortex created by stirring the molten metal. The stirrer speed is generally 500 rpm. After the completion of SiC$_p$ addition, the stirring is continued for 20 min. The composite melt is finally poured into cast-iron die preheated to 250°C.

The possible sources of microstructural inhomogeneity in stir cast MMCs are many; those include particle agglomeration and sedimentation in the melt, gas bubble entrapment, porosity due to inadequate liquid feeding during casting, and particle clustering as a result of particle pushing by an

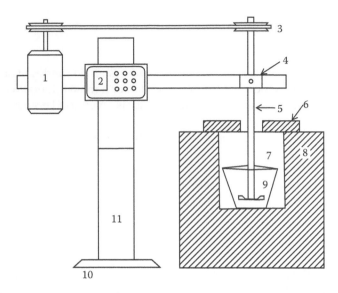

FIGURE 5.1
Schematic of stir casting process. 1. Variable speed motor, 2. speed controller, 3. pulley drive, 4. bearing assembly, 5. stirrer shaft, 6. insulating cover, 7. melting crucible, 8. furnace, 9. impeller, 10. base, 11. pedestal.

advancing solidification front. The last of these is the most difficult to eliminate. This effect can be aggravated by the post-casting heat treatment given in the semisolid state. Rapid solidification is a possible solution to this problem, since it can lead to the formation of fine microstructure and the solidification velocity is fast, which is above the critical growth velocity needed for enveloping the solid particles rather than pushing.

Stir casting involves prolonged melt/reinforcement contact, which can cause excessive interfacial reaction. In the Al–SiC system, the formation of Al_4C_3 can be extensive due to the interfacial reaction between Al and SiC. This reaction product degrades the final properties of the composite, and also the reaction raises the viscosity of the slurry substantially, making subsequent casting difficult. The rate of reaction is reduced, if the melt is Si rich, either by prior alloying or naturally by the reaction. Hence, the stir casting of Al–SiC$_p$ is generally suitable for the conventional casting alloys (high Si content), but not to most wrought alloys. If the conventional casting aluminum alloys are not suitable for any particular application, then the suitable alloy can be reinforced with Al_2O_3 particulates.

5.2.1.2 Squeeze Casting

As mentioned earlier, it is difficult to use the fibrous form of reinforcement in the stir casting process. Various liquid-state processing methods have been developed to overcome this problem, and one such processing method is squeeze casting. In this process, a reheated composite mixture or stirred molten matrix material with reinforcement is used for the fabrication of MMCs. The term squeeze casting is used to various processes in which pressure is imposed on a solidifying system, usually through a single hydraulically activated ram. This process has certain characteristic features, such as relatively fine microstructures (as a result of the rapid cooling induced by contact with a massive mold) and low porosity levels (due to efficient liquid feeding). The main difference between this process and conventional pressure die casting is that the ram continues to move during solidification, hence deforming the growing dendrite and compensating the freezing contraction. In addition, the ram movement is slower and the applied pressure is often greater than that used in typical die casting.

5.2.1.3 Slurry Casting (Compocasting)

The increase in viscosity of liquid metal on the addition of solid reinforcement is well documented, and the problem is severe with fibrous reinforcements. Hence, MMCs containing more than a few percentages of ceramic fibers cannot be made in this way. The idea of overcoming the problem of high viscosity increase at higher volume fraction of particulates came from the rheocasting process. Many metallic alloys behave like low-viscosity slurry, when subjected to agitation during solidification. This behavior has been observed for solid fractions as high as 0.5. During stirring the solid dendrites are broken into spheroidal solid particles, which are suspended in the liquid melt as fine particulates. This unique characteristic of numerous alloys, known as thixotropy, can be regained by raising the temperature even after complete solidification. A similar operation can be carried out with a suspension of ceramic reinforcements, and the process is called compocasting. In this process, agitation is used to encourage initial particle wetting and discourage particle agglomeration and sedimentation. During this process, the viscosity may not increase by more than a factor of two with up to about 25 vol.% particulate, provided that the particles remain well dispersed in the melt. This viscosity is sufficiently low for the conventional casting operations. However, it is difficult to ensure a good dispersion, and some problems may be encountered during the initial incorporation of the particles into the melt.

To produce MMC, fine ceramic particulates are introduced in the low-viscosity slurry. The ceramic particulates are mechanically entrapped and prevented from agglomeration by the presence of solid alloy particles. Subsequently, they are bonded with the matrix during the solidification of surrounding liquid matrix. Moreover, continuous deformation and breakdown of solidified matrix during agitation prevent ceramic particulate agglomeration and settling. This method has been successfully utilized to make Al_2O_3/SiC particulate-reinforced Al–5Si–2Fe alloy composites containing up to 30 wt.% ceramic particulates.

Compocasting is one of the most economical methods of fabricating discontinuously reinforced MMCs. This process has several advantages and some of them are listed here. The pouring temperature can be lower than those conventionally employed for stir casting, resulting in reduced interfacial reaction. It offers forming at low pressure in the semisolid state, resulting in a defect-free product. Conventional foundry methods are applicable to this process. There are some disadvantages also with this process:

- Residual pores between fibers cannot be eliminated completely.
- This method is not suitable for making continuous fiber-reinforced MMCs.

5.2.1.4 Centrifugal Casting

One of the disadvantages of MMCs produced by the previously mentioned methods is that they are more difficult to machine than unreinforced alloys. For many tribological components, high hardness is needed only on the contact surface; that means it is sufficient to incorporate the reinforcements in the surface layer. For example, in brake rotors, good wear resistance is needed on the rotor face, but not in the hub area. Centrifugal casting would be an ideal process for this type of components, since the centrifugal force drives all the particles toward the periphery from the central region. Hence, the concentration of particles will be high near the face and less in the middle region (hub). If there is a need for machining in the hub region, then it will be easier because of the presence of less number of particles.

In centrifugal casting, the optimal placement of reinforcement can be achieved by inducing a centrifugal force immediately after transferring the molten mix to the mold (Figure 5.2). Figure 5.3 shows the microstructure of a

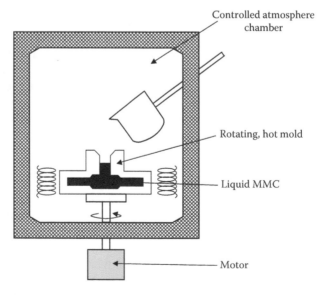

FIGURE 5.2
Schematic of centrifugal casting process. (Reprinted from Chawla, N. and Chawla, K.K., *Metal Matrix Composites*, Springer-Verlag, Inc., New York, 2006, p. 68. With permission.)

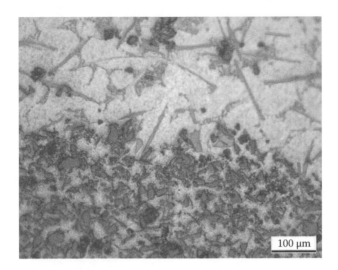

FIGURE 5.3
Microstructure of a centrifugally cast MMC. (Reprinted from Rajan, T.P.D. et al., *Mater. Char.*, 61, 923, 2010. With permission.)

centrifugally cast MMC having different concentration of particles at different regions. This process is relatively inexpensive, but it is restricted to only certain geometries.

5.2.1.5 Microstructure of Cast MMCs

During casting, the reinforcement is generally mobile and the movement or the pushing of the reinforcement by the liquid controls the microstructure of cast MMCs. Usually particulates or short fibers are used as reinforcement in the casting process. If the particles are pushed out by the solidifying liquid, the particles are concentrated in the remaining liquid, resulting in the segregation/nonuniform distribution of particles after complete solidification. The extent of pushing depends on the cooling rate. At relatively slow cooling rates, enough time is available for pushing out, resulting in the formation of large dendritic structure with several areas of high particle clustering. At faster cooling rates, the degree of particle pushing is diminished significantly, resulting in a microstructure with much more homogeneous distribution of particles and smaller dendrites.

The following observations have been made during the casting of aluminum alloy-based MMCs (Mortensen and Jin 1992): (1) in eutectic and hypereutectic Al–Si alloy-based matrices, segregation of particles is not taking place for all growth conditions; (2) in hypoeutectic alloy systems, segregation takes place because of the presence of a nonplanar solid/liquid interface; and (3) in several systems, the segregation is not taking place when the particles move above a critical velocity during solidification. The critical velocity depends on the size, shape, and density of particles and the type of matrix alloy. For a planar solid/liquid interface, the critical velocity, S_c, can be related to the particle diameter, d, by the following relationship:

$$S_c d^n = C \tag{5.2}$$

where
  C is a system-dependent constant
  The exponent "n" varies between 0.5 and 3

A more detailed model to predict the critical velocity has also been proposed (Cisse and Bolling 1971). According to that,

$$S_c^2 = \frac{4kT\gamma_{sl}a_o\alpha(1-\alpha)^3}{9\pi\eta^2R^3(1-3\alpha)} \tag{5.3}$$

where
  η is the viscosity of the melt
  T is temperature
  γ_{sl} is free energy of solid/liquid interface
  R is the radius of the particle
  a_o is the distance between the particle and the solid/liquid interface
  k is the curvature of the solid/liquid interface
  α is the ratio of particle radius to interface radius

5.2.1.6 Melt Infiltration

Casting methods are not suitable beyond certain amount of fibrous reinforcement. Indeed the infiltration process is best suited for the production of composites containing relatively high volume fractions of reinforcement. For many MMCs, infiltration is the most efficient and versatile processing route. In general, infiltration is the process whereby a fluid invades open space within a porous solid. A composite can be produced by this process using a mechanically self-sustaining porous preform of a reinforcing phase, which is infiltrated by a fluid matrix phase. The fluid need not be always a liquid; in some processes, such as chemical vapor infiltration, the fluid can be in the gaseous form. Since many common metals used in MMCs can be easily wetted and handled in liquid form for MMC production, infiltration is generally carried out using molten matrix material.

  Melt infiltration as a MMC fabrication process has many advantages. Since most molten metals have low viscosities, this process can be quite fast in the production of MMCs. For example, when cold-die pressure infiltration is used, the infiltration and solidification of the matrix can be completed in a few seconds. This process is suitable for the production of MMC components to their final or near-final shape. It is definitely a great advantage since it is very difficult to machine the hard MMCs. Another advantage is that it is possible to produce components with reinforcements placed at any selective region by using appropriate preform fabrication and placement procedures. A classical example is diesel engine piston, in which reinforcement is incorporated selectively in the region of high wear. Another significant advantage is that this process is more effective to avoid the formation of internal defects in the composites, especially in the pressure infiltration processes. Infiltration processing is suitable for many, but not all, of the matrix/reinforcement combinations. Its main limitations (though it is common for any melt-based processes) are the greater tendency for interfacial reaction between reinforcement and matrix than in the solid-state processes. Hence, for the reactive systems, the relatively lower-temperature solid-state powder metallurgical or diffusion bonding processes are often preferred. This consideration coupled with the general difficulty in handling molten titanium nearly excludes this metal and its alloys from the melt infiltration process. With other metals used in MMCs, this requirement of chemical inertness is generally less stringent. Melt infiltration has nonetheless been demonstrated for the preparation of composites with nearly all metal matrix systems.

The melt infiltration process stands out as one of the most economical and versatile processes for the production of MMCs with relatively high volume fractions of reinforcement. This process is extensively used for the production of MMCs based on low to medium melting point metal matrices, such as aluminum and copper. Examples of currently mass-produced MMCs by infiltration include automotive engine components such as diesel engine pistons, engine block cylinder liners, and crankshaft pulleys, electronic substrates made of SiC–Al composites, and tungsten fiber-reinforced copper electrical contacts. Scientific understanding and the level of technological development are now sufficiently advanced for this process; hence 3-D models similar to those developed for conventional casting processes can be developed for better optimization of process parameters.

In melt infiltration process, the wetting of the preform by the melt depends on alloy composition, preform material, reinforcement surface treatment, surface geometry, interfacial reactions, atmosphere, and melt temperature. Some of the problems in this process include reinforcement damage, preform compression, inhomogeneous microstructure, coarse grain size, reinforcement fiber or particulate contacts, and undesirable interfacial reactions.

The requirement of chemical inertness also plays a major role in the selection of a suitable reinforcement for a given matrix. However, the extent of this limitation varies significantly with the process also, since some infiltration processes are carried out at lower temperatures and short durations than others. For example, aluminum matrix composites reinforced with SiC can be produced without any interfacial reaction by squeeze infiltration process, whereas for the pressureless infiltration processes, the matrix must be alloyed with silicon to prevent aluminum carbide formation at the interface.

Some degree of internal connectivity between reinforcements is a common feature in composites produced by infiltration. Since the reinforcement must form a self-sustaining preform that does not collapse, either under its own weight before infiltration or under forces exerted by the liquid during infiltration, sufficient connectivity is needed. The extent of connectivity increases with increasing volume fraction of reinforcement. However, after certain volume fraction of reinforcement, the connectivity between the voids is closed, and infiltration may not be complete at this condition, that is, isolated voids will not be filled by the matrix.

Generally, the infiltration process consists of three stages, namely, preform preparation, infiltration, and solidification. At the first stage, an appropriate preform of required shape and reinforcement volume fraction is made. During the infiltration stage, the molten matrix invades open porosity within the preform. At the final stage, the matrix melt transforms into a solid and microstructure develops. This last stage is similar to metal casting processes; however, there are certain solidification phenomena that are specific to infiltrated composites.

5.2.1.6.1 Preform Preparation

Before the reinforcement is infiltrated, it must be assembled into a preform of similar shape as the product. The preform should have sufficient strength for handling and to withstand any stress it may experience during infiltration. It determines the lower limit on the volume fraction of reinforcement within infiltrated composites. Equiaxed particulate reinforcements of about a micrometer or more in diameter pack naturally to a volume fraction of 0.4 or higher. The packing density of preforms can be increased by many ways, which include (1) the use of particulates with multimodal size distribution, (2) cold pressing, (3) conventional sintering, and (4) oxidative sintering or reaction bonding. Parallel continuous fibers of diameter approximately 10 μm can be packed to about 40–60 vol.%. Randomly oriented non-equiaxed reinforcements, such as short fibers or platelets, pack to very low volume fractions. The minimum volume fraction of reinforcement in the preform can be as low as 0.05, depending on the reinforcement shape and aspect ratio. Volume fractions above this minimum are easily produced by controlled compaction.

It is also possible to reduce the volume fraction of reinforcement in the preform below the natural packing volume fraction. One method is to mix matrix powder with the reinforcement during the preparation of preform. However, the matrix powder may melt during the infiltration, resulting in the collapse of the preform. Another method is to combine reinforcements of different size and shape in the preform. Generally continuous fibers are combined with finer particles, and finer chopped fibers are combined with coarser particles. There is also an upper limit on the volume fraction of reinforcement, which is decided by the requirement of open porosity for effective infiltration. For sintered or pressed equiaxed particulates, the pore closure begins at a volume fraction of about 0.9. This is not a serious limitation, since composites are not generally made with this high volume fraction of reinforcement.

Binders are often used in the preforms to ensure their cohesion during handling and/or to prevent disintegration during infiltration. Either fugitive binders, which decompose and leave the preform at high temperatures, or permanent binders, which are stable at high temperatures, can be used. When the preform covers the entire die, the die is capable of holding reinforcement together during infiltration; hence it is sufficient to ensure preform cohesion before infiltration. In this case, fugitive binders, which evaporate or decompose during infiltration, can be used. Often, it is necessary to hold the preform intact during infiltration, and in such cases refractory binders should be used. Various silica- and alumina-based compounds have been popular as high-temperature binders. Colloidal silica confers sufficient strength to the preforms, but there is a problem with this binder. It reacts with aluminum and magnesium matrices during infiltration. Alterative refractive binders are being developed to overcome this problem. The binding agent is normally introduced in the suspension liquid, and it preferentially deposits at fiber contact points, where it serves to lock the fibers into a strong network. Typical binder content in the preform is of the order of 5%–10% by weight. A different approach to improve the bonding of reinforcement is to sinter the reinforcing elements. However, this is generally practiced with particulate preforms, since most fibers would degrade during sintering.

Several processes are used to prepare preforms. The reinforcement can be simply filled in the mold, or it can be pressed or injection molded to produce a preform of well-defined shape. There are two processes commonly used for making short-fiber preforms. They are (1) press forming and (2) suction forming (Figure 5.4). In the press forming process, aqueous slurry made with short fibers is agitated and poured into a mold containing very fine pores at the bottom. Pressure is applied to the slurry to squeeze out water, and then the preform is dried. The applied pressure will determine the fiber volume fraction (usually 0.05–0.3) and also influence the degree of fiber fracture. In the suction forming process, the mold contains very fine openings. Suction is applied through the mold and then the mold is immersed in the slurry tank. The fibers are attracted toward the mold and get deposited on the surface of the mold because of suction. Once the required thickness is built up, the mold is taken out from the slurry tank. The preform is removed from the mold and then dried.

Continuous fibers can be wound using filament winding techniques or woven into 2- or 3-D preforms. Preform production techniques are also developed to prepare open-celled foamlike preforms, in which the reinforcing phase is arranged along a 3-D connected network. This type of preforms can be easily made with particulate reinforcements. Organic polymer based foam is infiltrated with ceramic powder suspension, dried, pyrolyzed to decompose the organic polymer, and then sintered, resulting in the formation of foam like preform.

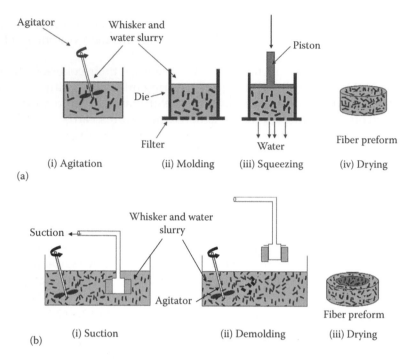

FIGURE 5.4
Processes for preparing fiber preforms: (a) press forming and (b) suction forming. (Reprinted from Chawla, K.K., *Composite Materials: Science and Engineering*, 2nd edn., Springer-Verlag, Inc., New York, 1998, p. 171. With permission.)

Squeeze infiltration is a preferred method for very fine fibers and whiskers, since it is difficult to blend these materials with metallic powders because of the size difference and the tendency to form tenacious agglomerates. It is also impractical to stir mix these materials in a melt. Furthermore, during handling these fine whiskers are becoming airborne, thus creating a risk of health hazardousness. It is advisable to prepare the preform within a glove box when the whiskers are selected as reinforcement.

Once the preform is prepared, then it is infiltrated with molten metal or alloy. In some cases, the alloy is formed during the process from the elemental metals. The melt infiltration processes can be broadly classified into two groups. They are (1) pressureless or spontaneous infiltration and (2) pressure or forced infiltration. In the pressureless infiltration process, the reinforcement preform is spontaneously wet by the molten metal without the application of external force. The pressureless infiltration process should be conducted under inert atmosphere to minimize interfacial reactions, since it involves long infiltration times at high temperatures. Typically, the infiltration temperatures of pure aluminum is in the range of 700°C–800°C, while the infiltration of Al–Mg is conducted between 700°C and 1000°C and the infiltration rates are less than 0.25 m h^{-1}.

In the forced infiltration process, external forces are used to drive the molten matrix into the preform. Such forcing is mainly used to overcome capillary forces, which resist the infiltration. In addition to that the forced infiltration speeds up the process and prevents matrix solidification shrinkage. The force is applied to the molten metal by various means. The principal method is to apply pressure on the molten metal using a solid piston or a pressurized gas. There are a few alternative processes that force the metal into the preform using a directional body force instead of applying pressure to the melt. These include Lorenz force infiltration, ultrasonic infiltration, centrifugal infiltration, and reactive infiltration.

5.2.1.6.2 Pressureless Infiltration (Capillary-Driven Infiltration)

Most molten engineering metals and alloys do not wet reinforcements. However, a large number of composite systems capable of undergoing spontaneous infiltration have been developed. These consist of specifically engineered chemical process alterations that produce wetting conditions in inherently non-wetting systems. Such methods can be divided into four broad categories:

1. Tailoring the matrix alloy chemistry
2. Modification of reinforcement surface
3. Infiltration at very high temperatures
4. Use of reactions between matrix and gases in the preform

In the first process, alloying elements that react with the reinforcement are added in the matrix. For example, lithium is added to aluminum for the infiltration of oxide preforms, and carbide-forming elements are added for the infiltration of carbon fiber preforms. In some cases, wetting can be promoted by elements that are absorbed along the matrix–reinforcement interface instead of reacting. An example is oxygen addition to copper to promote wetting of oxides.

The second process involves the use of coating on the reinforcement that is well wetted by the molten matrix. The most classical example is nickel coating on reinforcements to improve wettability. Other examples are titanium boride coatings for carbon fiber–aluminum composite, SiO_2 coatings for carbon fiber–magnesium composite, and K_2ZrF_6 crystal coatings for wetting of various reinforcements by aluminum alloys.

Infiltration at temperatures significantly above matrix melting range often concomitant with chemical interaction between the matrix and the reinforcement and/or between the matrix and the gas initially present in the preform. Examples include infiltration of boron carbide, silicon carbide, or aluminum nitride by aluminum. Rapid infrared heating is a suitable option to avoid extensive reactions taking place at these high temperatures.

Wetting can also be improved due to the reactions between the elements present in the matrix and the gas in the preform. An example is the Lanxide

"Primex" process, in which wetting is aided by chemical interaction between Mg and Al in the matrix and nitrogen gas in the pores. Other examples are those that use reactions between the gases present in air and the matrix (oxygen and magnesium) or between air and an element present in the reinforcement (air and Ti). Because of the consumption of gas in the preform, a pressure differential of up to 1 atm is created and that drives the molten metal mechanically into the preform while aiding wetting chemically.

The spontaneous infiltration (pressureless infiltration) technique is relatively simple and attractive. A schematic of this process is shown in Figure 5.5. A preform and a solid piece of matrix material are placed in contact within a controlled atmosphere furnace, and the molten matrix material simply flows spontaneously into the preform. Sometimes a vacuum is applied in the mold, which provides a small pressure differential to aid infiltration.

The relatively small investment for tooling is the main advantage of this technique. However, there are some disadvantages with this technique. The fact that chemical constraints imposed by the need for good wetting prevent the optimization of composite properties and in some cases result in poor composite properties. Reactions that occur in chemically induced spontaneous infiltration are generally detrimental to composite properties, either because of the degradation of reinforcement or due to the formation of brittle secondary phases. Coatings on reinforcement to improve wetting and control reactions have been developed, which can produce impressive results. However, the fiber coatings should not be exposed to air prior to infiltration

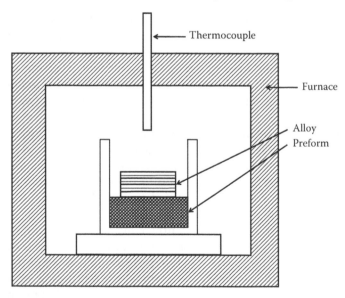

FIGURE 5.5
Schematic of spontaneous melt infiltration process. (Adapted from Agarwal, A. et al., *Carbon Nanotubes: Reinforced Metal Matrix Composites*, CRC Press, Boca Raton, FL, 2010, p. 33. With permission.)

because surface oxidation alters the positive effects of coating. The infiltration will be quite slow with fine reinforcements, since the permeability varies inversely to the square of the reinforcement size. Some porosity remains in the composite due to insufficient infiltration or matrix solidification shrinkage. Most of the spontaneous infiltration processes suffer from any one of the problems listed earlier, and the rate of production in these processes is lower than the pressure-assisted infiltration processes. Even though there are many problems in this process, there is a scope for the development of this attractive spontaneous infiltration process for MMC production. However, the majority of infiltrated composites produced at industrial scale currently use pressure-assisted infiltration.

5.2.1.6.3 *Squeeze Infiltration (Pressure Infiltration)*

Pressure-assisted infiltration is generally used, mainly because of the difficulties with reinforcement wetting by the molten metal in spontaneous melt infiltration. Squeeze infiltration or pressure infiltration involves forcing the melt into a porous preform by utilizing either inert gas pressure or pressure applied through a mechanical device. The pressure is maintained until solidification is complete. This method obviates the requirement of good wettability of the reinforcement by the molten metal, because the pressure applied on the molten metal can force the flow of melt into small pores. Pressure infiltration is better than pressureless infiltration and stir casting in many aspects. The main advantages are short processing time and minimal residual porosity, shrinkage cavities, and interfacial reaction. Ready to use components (net-shaped components) can be produced inexpensively by this process. Complex-shaped composites with a rather high fiber volume fraction are also obtainable.

The most classical example of piston-driven infiltration is squeeze infiltration. Many aluminum-based composite parts for automobile applications are made by this process. The schematic of squeeze infiltration process is shown in Figure 5.6. The preheated preform is placed in a die and the die is heated to a preselected temperature. The required amount of melt is poured into the die, and then pressure is applied through the upper punch. When the contact angle between reinforcement and melt is large, the pressure required to force molten metal through a channel of width "w" is roughly equal to the surface tension of the metal divided by "w." For example, when the surface tension of metal is ~1 J m^{-2} and w is ~1 µm, a pressure of the order of 1 MPa (14 psi) is at least required to drive the molten metal. This is somewhat high pressure and hence the infiltration setup may not be simple or inexpensive. The pressing capacity needed in the machine will also increase with the increasing size of product. Applied pressures on the order of 70–100 MPa (10,150–14,500 psi) are typically used. This high pressure provides good composite soundness and dimensional accuracy and enables the infiltration of preforms containing reinforcements as small as 25 nm in diameter. On the other hand, this high applied pressure can cause preform deformation, if the

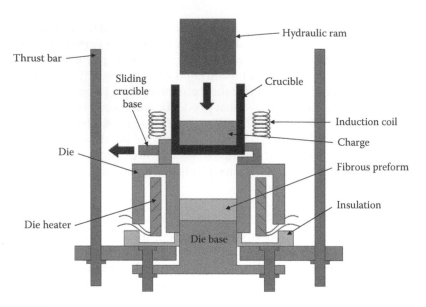

FIGURE 5.6
Schematic of squeeze infiltration process. (Reprinted from Clyne, T.W. and Withers, P.J., *An Introduction to Metal Matrix Composites*, Cambridge University Press, Cambridge, U.K., 1993, p. 325. With permission.)

rate of pressure application is not well controlled. It is desirable to keep the preform at a lower temperature than that of the matrix liquidus temperature to minimize interfacial reaction and to obtain a fine-grain matrix. The process control parameters are piston movement speed and initial melt, die, and preform temperatures.

Alternatively, the piston can push the melt poured in a shot chamber located upstream of the preform-containing die. This setup is similar to a pressure die-casting machine. The advantages of this setup over squeeze infiltration setup are that the melt need not be metered with much precision and that more complex part geometries can be produced.

In the gas-driven pressure infiltration process, gas pressure is used to push the melt into the preform instead of a piston. A schematic of this process is shown in Figure 5.7. A preform is placed in a nonporous mold kept in a pressure vessel. Above the mold is a crucible containing the molten metal with a bottom-pouring device to transfer the melt into the mold. There can be separate heating arrangements for heating the crucible containing the matrix metal and the mold containing the preform. The pressure vessel is first evacuated and then the melt is poured over the preheated preform. The melt seals the preform from the surrounding atmosphere, and then pressurized gas is let into the vessel. The melt is pushed into the preform by the gas pressure and infiltration takes place. Once infiltration is completed, the melt is allowed to solidify. The solidification can be accelerated by using a cooling setup at the bottom of the vessel.

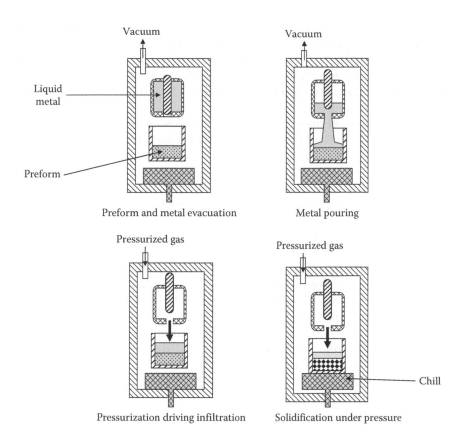

FIGURE 5.7
Schematic of a gas-driven pressure infiltration process. (Reprinted from Clyne, T.W., in *Comprehensive Composite Materials*, Vol. 3, eds. A. Kelly and C. Zweben, Elsevier Science Ltd., Oxford, U.K., 2000, p. 545. With permission.)

When the bottom of the mold is in contact with the cooling setup, solidification proceeds from the bottom to top of the mold. Hence, solidification shrinkage is compensated by feeding metal under the dual action of gravity and the pressurized gas. Typical gas pressure levels used in this process vary from 1 to 10 MPa.

There are some drawbacks in the gas-driven process compared to piston-driven pressure infiltration process: (1) production rates are very much lower, (2) the usage of inert gas adds up to the product cost, and (3) the use of gas pressurized above 1 MPa requires special care in equipment design and operation due to safety concerns.

On the other hand, there is greater flexibility in selecting mold materials. Unlike piston-driven infiltration, the mold is subjected to gas pressure from all the sides, and hence, the strength requirement of mold is not very critical. Many high-temperature ceramic materials can be used for making molds. With these molds, intermetallic and superalloy matrix composites can also be produced.

More complex and precise geometries can be produced by this process. Other advantages of this process include the greater ease of evacuating the preform prior to infiltration and also the possibility of infiltration without the problem of solidification, since both the preform and the mold can be heated to temperatures above the matrix solidification range. This also lowers the pressures required for infiltration.

These two classes of pressure infiltration processes have their own advantages. Depending on the production rate, cost, matrix melting point, and the complexity of product, the appropriate process can be selected. Aluminum matrix composite automotive parts, such as diesel engine pistons or engine blocks, are produced by squeeze infiltration. High production rates, cost of the product, and the use of low-melting-point metal more than compensate the high cost of tooling. On the other hand, composites with complex and precise geometries, such as SiC/Al electronic substrates, are often produced by gas-driven pressure infiltration process.

The squeeze infiltration process can be used either to form simple shapes, such as cylinders, or to form the final product in the as-cast form. It can be used to prepare composites with relatively high reinforcement volume fractions (>0.4). Achieving homogeneous reinforcement distribution at high volume fractions is difficult in stir casting. It is also possible to incorporate the reinforcement at a selected region in the part by this process. A typical example is diesel engine piston, in which the reinforcement is incorporated in the region where high wear resistance is needed. Squeeze infiltration can also be used to fabricate continuous fiber-reinforced MMCs. Although this technique has been primarily used for infiltrating low-melting-point metals, continuous fiber-reinforced intermetallic matrix composites such as alumina fiber-reinforced TiAl, Ni_3Al, and Fe_3Al have also been prepared.

The fabrication of composites takes place within a confined vessel/die in the pressure infiltration processes, and this limits the size of the products that can be fabricated. Moreover, it is very difficult to use this process for the continuous production of long-fiber composites. Even then, there are many attempts made for the continuous production of composites by suitably modifying the equipment.

5.2.1.6.4 *Infiltration Mechanics*

The infiltration of molten metal into the reinforcement preform is a complex phenomenon. Various factors are controlling the infiltration in a synergistic manner. The infiltration predicted by considering a single factor at a time may not be very useful. However, it can give some idea about the factors controlling the infiltration; based on that, the process can be controlled in a better way to get a quality composite. A glimpse of the important factors governing the infiltration is discussed here. More detailed discussion on this subject can be found in the book on *Comprehensive Composite Materials* by Clyne (2000).

The infiltration of molten metal is governed by two important classes of phenomena, which are capillary phenomena and transport phenomena. The capillary phenomena are pronounced in infiltration processing due to the fineness of pores within reinforcement preforms. Among the transport phenomena, fluid flow is the important one. The infiltration of molten metal through a particulate or fibrous preform can be considered as liquid infiltration through a porous medium. The capillarity of the melt and its permeability in the preform will affect the infiltration characteristics.

Consider the fabrication of a composite by infiltration, and assume that there are no interfacial reactions between molten matrix and reinforcement. If this process is conducted quasistatically, the change in energy per unit volume of composite infiltration is

$$\Delta G = A_i(\gamma_{fm} - \gamma_{fa}) \tag{5.4}$$

where

A_i is the area of matrix–reinforcement interface per unit volume of composite

γ_{fm} is the reinforcement–matrix interface surface energy

γ_{fa} is the surface energy of the reinforcement in contact with the initial atmosphere

If it is assumed that infiltration takes place along a well-defined surface, across which essentially all pores in the preform are filled with molten metal, the energy change ΔG manifests itself as a capillary pressure difference ΔP_{ma} between the metal and the atmosphere. This pressure must be applied to the metal free surface along the infiltration front:

$$\Delta P_{ma} = (1 - V_f)^{-1}\Delta G = (1 - V_f)^{-1}A_i(\gamma_{fm} - \gamma_{fa}) \tag{5.5}$$

where $(1 - V_f)^{-1}A_i$ is the reinforcement–matrix surface area per unit volume of matrix.

In some systems, $\gamma_{fm} - \gamma_{fa}$ is negative, and hence the infiltration is spontaneous. The molten metal in contact with the preform will flow by itself. However, in most of the MMC systems, this term is positive; hence, external force is necessary for infiltration.

The term $(\gamma_{fm} - \gamma_{fa})$ can be determined using Young's equation:

$$\gamma_{fm} - \gamma_{fa} = -\cos(\theta)\gamma_{ma} \tag{5.6}$$

where

θ is the contact angle of liquid metal with a flat substrate of the solid (should be same as the reinforcement material) in the relevant atmosphere

γ_{ma} is the surface tension of liquid metal in that atmosphere

For spontaneous infiltration, the wetting angle should be less than 90°. When θ > 90°, it is generally termed "poor wetting," which implies that the infiltration of melt requires the use of applied force. However, for a wetting fluid to infiltrate the porous medium spontaneously, it is not sufficient that the wetting angle be less than 90°, but it should be well below 90°, in fact, very near to zero.

A further complication to the infiltration mechanism is the influence exerted by chemical interaction between the matrix and the reinforcement. In practice, spontaneous infiltration is more easily observed with reactive systems. However, the reaction poses several problems and is therefore controversial at present. The chemical interaction between the matrix and the reinforcement can influence the infiltration in another way; the heat released by the exothermic chemical reactions behind the infiltration front can significantly raise the local temperature. In many systems, an increased temperature results in a marked improvement in wetting of the preform by the molten matrix.

Capillary phenomena indicate whether infiltration can be spontaneous or not. Several other phenomena intervene the infiltration and control the rate of the process and the structure of the composite. The rate of infiltration into the preform is measured by the rate of flow of the liquid metal. This rate of flow is generally described by the superficial velocity, v, which is defined as the average velocity of the fluid multiplied by the volume fraction of pores $(1 - V_f)$. The superficial velocity gives the quantity of melt flows per second through unit area of composite, perpendicular to the flow direction. If the porous medium is assumed as transversely isotropic with respect to the infiltration direction, then the rate of flow of the melt is governed by the Forscheimer equation, which is given as

$$F_x = \frac{\partial P}{\partial x} = \frac{\eta}{k} v + B\rho_m v^2 \qquad (5.7)$$

where
  F_x is the local volumetric value of all forces
  P is the local average pressure in the melt
  ρ_m is the metal density
  η is melt viscosity
  k and B are constants, the first one is called the permeability of the porous medium

The first term on the right-hand side accounts for viscous friction in the flowing liquid, while the second term accounts for inertial losses.

When the relevant Reynolds number falls below a critical value, that is, ≈1, the second team on the right-hand side of Equation 5.7 can be neglected, and the body forces are often negligible; the Forscheimer equation then reduces to Darcy's law (because the diameter of reinforcement is generally below 100 μm), which is given as

$$v = -\frac{k\partial P}{\eta \partial x} \tag{5.8}$$

The simplification is most often valid in infiltration processing. The permeability constant varies proportional to the square of the average pore diameter and is also a strong function of pore structure and volume fraction. The most universally applicable equation to calculate this parameter is Blake–Kozeny equation.

Infiltration process is often nonisothermal. The preform temperature is different from the melt temperature, and there is an exchange of heat between melt and reinforcement during infiltration. In some cases, heat exchange between preform and melt or heat flow to the surroundings or the die wall can cause solidification of the melt. During infiltration, there can be liquid, semisolid, and fully solid regions depending on the extent of heat exchange. The formation of solid metal within the preform during infiltration significantly lowers the local preform permeability. Such solidification also strongly influences the formation of microstructure in the composite. Matrix solidification due to preform-matrix heat exchange results in fine equiaxed matrix grains in pure metal matrix and significant macrosegregation across the composite in alloy matrix.

When the melt is superheated (i.e., heated well above its melting point), it remelts at least a certain portion of the metal that has solidified by heat exchange with the preform. This results in the formation of a region of fully liquid metal. This region in the infiltrated composites can be often clearly identified from the large-grained and columnar microstructure. The partial remelting of solid metal by a superheated matrix accelerates the infiltration kinetics to some extent. However, the effect is small in practice because the superheated melt seldom remelts all solid metal formed by contact with the preform. The bottleneck regions formed by the solidification of melt near the reinforcement contact points remain unchanged and still control the rate of flow.

Infiltration is usually carried out with the equipment, schematically shown in Figure 5.6. The important features are the onset and progression of infiltration and the heat flow and solidification characteristics.

The pressure necessary for the onset of melt entry can be calculated using the Kelvin equation:

$$P_i = \gamma\left(\frac{1}{r_1} + \frac{1}{r_2}\right) \tag{5.9}$$

where
γ is the surface tension of the melt
r_1 and r_2 are the principal radii of curvature at the front

A lower limit of these radii is half of the interfiber spacing, thus leading to

$$P_i = \frac{4\gamma}{d\left[\left(\pi / 2\xi V_f(1 - f_{CL})\right)^{1/2} - 1\right]}$$ (5.10)

where
  d is the fiber diameter
  ξ is the ratio of the interplanar spacing to the in-plane fiber spacing
  V_f is the volume fraction of fibers in the preform
  f_{CL} is the fraction of fibers forming cross-links between square arrays of
     in-plane fibers

This model is over simplistic and there is a difficulty in specifying appropriate parameters. However, a number of broad features can be explored using this equation. This equation indicates that vacuum infiltration is possible with relatively coarse fibers ($d \geq 20$ µm), whereas whiskers may need high pressures for infiltration. The most important parameter neglected in the previously mentioned model is wetting. Wetting will lower P_i by allowing reduced curvature. There is considerable interest in wetting characteristics, especially adding wetting agents that allow infiltration under reduced pressure or under normal atmospheric pressure. An example of the system, which shows spontaneous infiltration, is Al alloy melt containing Mg infiltrated under nitrogen atmosphere into various ceramic preforms. This approach is promising for the preparation of MMCs with very high reinforcement contents, since pressurized infiltration tends to be difficult for these kinds of preforms.

Another area of major interest is the heat flow and solidification during infiltration. Since the preforms are usually made with relatively fine fibers, local heat exchange with the melt is effectively instantaneous. Hence, it is easy to estimate the minimum melt superheat temperature needed to avoid premature solidification based on global heat balance

$$\Delta T = \frac{(T_{mm} - T_p)C_f V_f}{C_m(1 - V_f)}$$ (5.11)

where
  T_{mm} is the fusion temperature of matrix material
  T_p is the initial preform temperature
  C_f and C_m are the specific heats of fiber and matrix, respectively
  V_f is the volume fraction of fiber in the preform

Often it is necessary to superheat the melt at a higher temperature than that indicated by the previously mentioned equation to avoid excessive solidification immediately behind the advancing melt front. The excessive solidification is a common cause of incomplete infiltration in many composites.

This problem can be overcome by the use fiber coatings, which release heat on reacting with the melt, or by preheating the preform to a very high temperature and then placing in the die just before infiltration. Although the second approach controls the solidification, it can increase the danger of melt penetration into air escape paths. Hence, the process parameters should be optimized to avoid incomplete infiltration and excessive reinforcement/matrix reaction.

5.2.1.6.5 Microstructure Development

Once the infiltration of the preform is completed, bulk solidification starts. Matrix solidification in a composite is governed by the same physical processes as in unreinforced metals. It starts with nucleation and is followed by growth, and these processes are governed by heat transfer, diffusion, and convection in the liquid. The presence of reinforcement can exert a significant influence on all of these processes. In infiltrated composites, matrix solidification can actually begin during the flow of matrix itself. Such solidification is caused by heat extraction from the melt to a preform and mold kept at sufficiently lower temperatures. Of particular interest is solidification caused by a low initial fiber temperature. This type of solidification is relatively unique to the infiltrated composites, and it results in the formation of a fine-grained zone, in pure metals and significant macrosegregation region in alloy matrices. These regions are often distinct from the large-grained region formed due to remelting.

Often, the preform and mold temperatures are kept sufficiently high during infiltration so as to avoid solidification. After the infiltration, solidification begins with the nucleation of solid matrix. The matrix nucleation can be catalyzed by the reinforcement, which depends on the type of reinforcement–matrix system. It occurs when the contact angle of liquid matrix on reinforcement material is less than 180° or more precisely smaller than on heterogeneous nucleation sites, usually present in the unreinforced matrix. As a consequence of this, fine-grained microstructure is formed in the matrix. This has indeed been observed in many MMC systems: hypereutectic Al–Si alloys reinforced with graphite, SiC, SiO_2, and Al_2O_3 (Mortensen and Jin 1992, Wang et al. 1994), Al–Cu (4.5 wt.%) alloys with TiC (Mortensen et al. 1986), Ti–51.5Al–1.4Mn with TiB_2 (Bryant et al. 1990), and AZ91 magnesium alloy with SiC (Luo 1994). On the other hand, most of the aluminum-rich alloys in the presence of commonly used reinforcements and also pure or slightly alloyed copper in the presence of W or Al_2O_3 don't exhibit such grain refinement because of high contact angle (Gonzalez and Trumble 1996, Bystricky et al. 1999). However, most of the MMCs produced by infiltration are based on these alloys.

The microstructure development in alloyed matrices is also influenced by the presence of reinforcement. Depending on the solidification conditions, the nature, and distribution of reinforcement, the microstructural features can be altered in a composite. If it is assumed that the contact angle is large, then the nucleation of matrix primary phase begins away from the reinforcement. The growth of that phase is locally governed by solute transport to the liquid–solid interface. Since the reinforcement does not allow the diffusion of solute through it, a solute-depleted region will form near the reinforcement. This implies that the reinforcement–matrix interface also tends to have the same microstructural elements as that formed at the end of solidification, such as eutectic phases.

When infiltration is carried out through continuous fiber reinforcements, the matrix is almost entirely surrounded by the fibers with a few narrow passages connecting it. Hence, the matrix is constrained to solidify in a large number of tiny, elongated cavities. Since such interstitial spaces are typically only a few micrometers wide and often finer than solidification microstructural features (dendrite arm spacing, grain size, etc.), some modifications in the solidification of matrix are expected. Some of the microstructural observations made in these types of composites are (Mortensen 1993, Mortensen and Flemings 1996)

1. When matrix solidification is dendritic, the dendrite arm coalescence can be significantly accelerated in the composite, and this causes the disappearance of dendritic microstructure.
2. Since the diffusion distances are limited by the interstitial space between the reinforcements, microsegregation is restricted to the available space. This causes a reduction in microsegregation during solidification.
3. The cell to dendrite transitions and also the undercooling of cell tips are strongly modified by the geometrical constraint imposed by the reinforcement on diffusion.

In essence, the presence of reinforcement significantly influences the microstructural development in the matrix. However, the effect varies with the type of matrix and reinforcement, the nature, and size and distribution of reinforcement.

5.2.1.7 Spray-Forming Technique

It is a relatively new and promising technique for making particulate MMCs. It involves the use of atomization techniques that have been employed for some time to produce metallic powders. This process uses a spray gun to atomize a molten alloy (matrix material) into which particulates are injected. An optimum particle size is required for efficient incorporation. Whiskers are very difficult to be incorporated, because of their fine nature. A preform deposit produced by the atomized molten metal with ceramic particulates can have density up to 97%. This MMC preform is subjected to scalping, consolidation, and finishing processes to get the product. The formation of reaction products is minimized in this process because of the extremely short time of contact between reinforcement and molten metal.

This process has the great flexibility in selecting different types of reinforcement particulates. Layered MMCs consisting of different reinforcement in alternate layers can be made using two particle sprayers. However, this process needs high capital investment for the processing equipment.

Spray forming is characterized by its inherent simplicity and nonequilibrium solidification phenomenon. It consists of two stages: spray atomization and spray deposition. In the atomization stage, a stream of molten metallic material is impacted by highly energetic gas jets, and the molten stream is disintegrated into small, irregular ligaments and then into spherical droplets. These droplets deposit on the substrates in the deposition stage. For the formation of MMCs, the reinforcement is incorporated into the metal matrix at certain stage.

Since the mid-1980s, spray forming has emerged as a promising process for the production of particulate/discontinuous fiber-reinforced metal/intermetallic matrix composites. A wide variety of MMC material systems have been investigated over the years. The chief among the matrix materials are Al and Al-based alloys, Mg-based alloys, Cu-based alloys, and intermetallics. Particulates are the generally used reinforcements, which include Al_2O_3, graphite, SiC, TiB_2, TiC, B_4C, WC, and Cr_3C_2.

Several approaches have been utilized for producing MMCs by this process. Although the principle remains same in all these approaches, each one has some unique features. Some of the methods are described in detail as follows.

5.2.1.7.1 Spray Forming of Premixed MMCs

This method is similar to the spray forming of alloys. Instead of spraying molten alloy, a mixture of the molten matrix alloy and the solid reinforcement is sprayed to get a deposit of MMC. This method is used to produce SiC-reinforced Al–Si alloy-based composites and TiB_2-reinforced Ti–Al alloy-based composites. A composite melt stream for this method may be obtained by any one of the following two routes:

1. Melting of a ready-made MMC
2. Mixing of the reinforcement with the molten matrix alloys

The first route is not generally preferred because it involves the formation of composites two times. This route is used only when there is a necessity to alter the microstructure of already formed MMC.

Sufficient mixing of the reinforcement and the matrix is crucial for getting high-quality MMCs. Good mixing may be achieved by inserting an agitator in the melt. Alternatively, RF induction heating can be used instead of resistance heating. The strong electromagnetic interactions between the induction source current and the eddy current in the melt may provide sufficient agitation.

A major advantage of spray forming using premixed MMCs is the uniform distribution of reinforcement in the deposit. This is achieved by the combined effect of rather uniform distribution of reinforcement in the molten matrix, the disintegration of melt stream into micron-sized droplets, often containing reinforcement surrounded by matrix, and the high cooling rate. However, there are some drawbacks in this method; the matrix–reinforcement reaction can be severe, since the reinforcement is in contact with the molten matrix for long duration and low fluidity of the melt due to the dispersion of reinforcement. The chemical reaction at the matrix–reinforcement interface may severely affect the properties of composite, and the decreased melt fluidity may cause great difficulty in atomizing the melt. Hence, this method is selected for the production of particulate MMCs only when the chemical reaction at the interface is not strong and the volume fraction of reinforcement is at low to intermediate levels (1–20 vol.%).

5.2.1.7.2 Osprey Deposition

In this process, the reinforcement particulates are injected into a stream of molten alloy, which is subsequently atomized by jets of inert gas. The atomized droplets with reinforcement particulates are deposited on a substrate in the form of a MMC billet. This process combines the blending and consolidation steps of P/M process and promises major cost savings in the production of MMCs.

5.2.1.7.3 Spray Forming by Co-Injection

Spray atomization has been used commercially to prepare metal and alloy powders. The metal or alloy is melted and the liquid stream is atomized with water or an inert gas. The fine liquid droplets formed during atomization rapidly solidify into fine powder. This technique can be modified, by injecting reinforcement particles into the melt spray and co-depositing the reinforcement particles with the matrix alloy. Spray forming by co-injection is one of the most widely used methods for the fabrication of MMCs. In this method, a reinforcement injection unit is incorporated into a normal spray-forming facility used for depositing alloys. A schematic of this setup is shown in Figure 5.8. The particulates are carried by a high-pressure protective gas in injector tubes. The injector tubes are positioned at a certain distance below the atomizer. In order to obtain uniform distribution of the reinforcement in the matrix, these injector tubes are placed symmetrically around the central line of the spray cone. The molten matrix alloy is released through a nozzle and subsequently atomized, forming the spray cone. As the droplets travel toward the water-cooled substrate, the injector tubes release the particulates at high velocity. The reinforcement particulates penetrate into the matrix droplets, thus forming composite droplets. The micron-sized composite droplets deposit and solidify on the substrate as a dense composite preform.

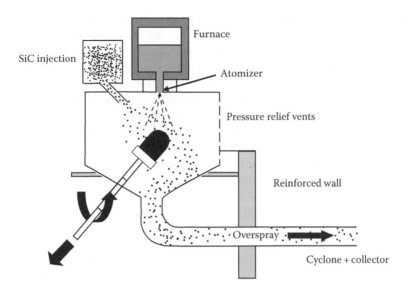

FIGURE 5.8
Schematic of a spray deposition arrangement for MMC manufacture. (Reprinted from Willis, T.C., *Metals Mater.*, 1988, 4, 485. With permission.)

The advantages of this process are the high rate of production (6–10 kg min^{-1}) and the fast solidification rate, which minimizes any reaction between reinforcement and matrix. The deposit in the as-produced condition is not fully dense; hence secondary processing is often needed to fully densify and homogenize the composite. The distribution of particles in the deposit is very much dependent on the size of the reinforcement and at what point the reinforcement is joining the matrix. For example, it is very difficult to get uniform distribution of whiskers since they are too fine to be optimally injected. When the reinforcement particles are injected into the matrix stream near the atomization nozzle, the particles are able to envelope themselves into the liquid droplet, and a relatively homogeneous distribution is obtained. Injection of the particles away from the nozzle, that is, when the matrix droplet is semisolid, does not allow the easy ingression of the particles into the matrix and causes the particles to reside along the periphery of the matrix particle. If these particles are consolidated to form a composite, then the composite will have inferior properties as a result of the high degree of particle clustering. This process is quite flexible in that the metal and reinforcement sprayers can be adjusted to obtain laminates with different amount of reinforcement in each layer, that is, functionally gradient materials can be prepared. However, the process is quite expensive owing to the high cost of capital equipment.

This method can also be used for the production of MMCs, because of the following reasons. By the time the atomized droplets reach the particulate injection point, the droplets have undergone sufficient cooling and the temperature of droplets is much lower than the melt temperature. While coarse droplets may still remain in a liquid state, fine droplets normally have been partially solidified. Under such conditions, the opportunity for detrimental chemical reactions at the interface is substantially reduced. After the penetration of reinforcement into the matrix droplets, the high cooling rate experienced by the droplets further limits the interfacial reaction. Hence, a wide variety of MMC systems, whether reactive or nonreactive, can be produced by this method. Since the reinforcement particulates may penetrate the matrix droplets, uniform coverage of matrix over the reinforcement will no longer be limited by the droplet sizes, whereas this is a major problem in P/M techniques in which smaller matrix particulates may not cover the larger reinforcement particulates. If the matrix droplets are allowed to partially solidify prior to the contact of reinforcement, then there is a high probability for nonequilibrium solidification. It is possible to incorporate higher volume fraction of reinforcements in this method compared to spray forming of premixed MMCs.

To produce layered MMCs, the injection is carried out in an intermittent pattern, that is, the injection process is divided into a series of injection pulses. The injection time and interval and the mass flow rate in each injection pulse determine many of the layered structural characteristics, such as the layer thickness and the reinforcement content in each layer. Unlike the functionally gradient composites produced by other methods, the layered composites produced by spray forming by co-injection are characterized by a continuous variation of reinforcement across each layer. The layered structures produced by co-extrusion, diffusion bonding, and P/M techniques have a very sharp variation of reinforcement across each layer. A diffused distribution across the layers is beneficial to the toughness and fatigue properties, since it will generate lower stress concentrations.

5.2.1.7.4 VCM Process

In this process, the matrix material is atomized using a high-velocity gas jet. Simultaneously, one or more jets of a reinforcement material are injected into the atomized spray at a location where the droplets contain a limited amount of liquid. Contact time and thermal exposure of the particulates with the matrix are minimum in this process, and hence interfacial reactions can be controlled. In addition, better control of gases in the spray chamber minimizes oxidation.

5.2.1.7.5 Reactive Spray Forming

In the spray-forming methods previously discussed, readily available reinforcements are incorporated into matrices to form composites, whereas in reactive spray forming of MMCs, the reinforcements are formed during spray forming. The reinforcement formation can take place prior to atomization, during atomization, and/or after atomization. The reinforcement can be formed prior to atomization by the injection of reactive gases in the molten matrix alloy. In situ formation of reinforcements during and after atomization may adopt the following approaches:

1. *Liquid–gas interactions*: In this approach, the atomization gas is selected to be a mixture of protective (He, Ar, and N_2) and reactive (O_2, CO, CO_2, H_2, and N_2) gases. During atomization, chemical reaction between the matrix droplets and the reactive gas is triggered and may continue even after the droplets are deposited. The selection of the appropriate combination of protective and reactive gases is necessary to promote and at the same time to control the chemical reactions between the atomization gas and melt droplets. Examples include

$$NiAl\ (l) + O_2(g) \rightarrow NiAl(l) + Al_2O_3(s)$$

$$NiAlY(l) + O_2(g) \rightarrow NiAl(l) + Y_2O_3(s)$$

$$CuAl(l) + O_2(g) \rightarrow Cu(l) + Al_2O_3(s)$$

$$FeAl(l) + O_2(g) \rightarrow Fe(l) + Al_2O_3(s)$$

$$Al(l) + O_2(g) \rightarrow Al(l) + Al_2O_3(s)$$

where g, l, and s denote gas, liquid, and solid phases, respectively.

2. *Liquid–liquid interactions*: In this approach two different materials are atomized individually and merged at a certain distance below the atomization zone. Micron-sized droplets from the two materials collide and react with each other, forming matrix and reinforcement materials. An example for this approach is

$$CuTi(l) + CuB(l) \rightarrow Cu(l) + TiB_2(s)$$

3. *Liquid–solid interactions*: This approach is similar to spray forming by co-injection. However, the injected particulates react with the liquid droplets to form the required reinforcements. Examples for this approach include

$$CuAl(l) + CuO(s) \rightarrow Cu(l) + Al_2O_3(s)$$

$$FeTi(l) + FeC\ (s) \rightarrow Fe(l) + TiC(s)$$

When the reinforcement is formed during and/or after atomization, the high-temperature exposure time for the reactants is limited. Accordingly, even though the chemical reaction between reactants is promoted, the amount and size of the resultant reinforcements are still small. Hence, the reactive spray forming of MMCs is more suitable for the preparation of dispersion-strengthened MMCs. Often the liquid/liquid and liquid/solid reactions may not complete at the deposition stage, but will be completed during post-processing heat treatments. However, it is important that the reinforcement to be formed should be thermodynamically stable. Otherwise, brittle intermetallic compounds may be formed at the interface during heat treatment, and those compounds severely degrade the mechanical properties of the spray-formed MMCs.

5.2.1.7.6 Spray Forming of Continuous Fiber-Reinforced MMCs

Spray forming of continuous fiber-reinforced MMCs is a challenging task. MMCs with single layer of fiber reinforcement are relatively easy to make compared to multilayered fiber composites. By the time the droplets reach the fiber substrate, they are almost solidified. The solidified/partially solidified metal matrix droplets cannot penetrate the fiber stack. However, it is possible to produce bulk fiber-reinforced composites by stacking and consolidating spray-formed single-layer MMC sheets. Such a processing route will severely compromise the simplicity and cost-effectiveness of spray-forming process.

5.2.1.7.7 Microstructure of Spray-Formed MMCs

The microstructures of spray-formed materials are almost always characterized by equiaxed grains. These grains are finer than those encountered in conventionally cast products. Another feature concomitant with the fine equiaxed grains is the absence of macrosegregation and the minimal microsegregation. The equiaxed grains are even finer in spray-formed MMCs compared to monolithic alloys because of the presence of reinforcements.

The volume percentage of pores in spray-deposited materials is normally in the range of 1%–10%, and the pore size distribution is frequently bimodal, with small pores in the range of 1–2 μm and large pores in the range of 10–100 μm. By optimizing the processing conditions and incorporating novel designs, the porosity can be significantly reduced in as-spray-formed materials. The spray-formed MMCs tend to contain a higher percentage of pores than spray-formed alloys. Since the cooling rate of melt droplet in contact with the reinforcement is higher, less liquid is available to fill the irregular interstices formed during deposition, leading to higher percentage of porosity in the spray-formed composite.

5.2.1.8 In Situ Processes

Most of the processing techniques of MMCs are based on the addition of readily available reinforcements to the matrix materials, which may be in molten or powder form. That is, the reinforcing phases are prepared separately prior to the composite fabrication. Hence, these MMCs can be viewed as ex situ MMCs. In this case, the scale of the reinforcing phase is limited by the starting material dimensions, which is typically on the order of micrometers to tens of micrometers and rarely below 1 μm. Other main drawbacks of ex situ MMCs are interfacial reactions between the reinforcements and the matrix and poor wettability of the reinforcement by the matrix as well as contamination during processing.

In the in situ techniques, the reinforcement phase is formed in situ during the fabrication of composite. The in situ formation avoids the problems of nonuniform distribution of reinforcement and the incompatibility of reinforcement and matrix. A classical example of this process is the controlled unidirectional solidification of a eutectic alloy. This process can result in one phase being distributed in the other, in the form of fibers or ribbons. One can control the fineness of the reinforcement phase by simply controlling the solidification rate. However, the solidification rate in practice is limited to a range of 1–5 cm h^{-1}, because of the need to maintain a stable growth front.

Efforts have been made to the development of in situ MMCs, in which the reinforcements are formed in a metallic matrix by chemical reactions between elements or between an element and compound during the fabrication of the composite. With this approach, a wide variety of MMCs with aluminum, titanium, copper, nickel, and iron matrices containing borides, carbides, nitrides, oxides, or their mixtures have been produced. Compared to conventional MMCs (ex situ MMCs), the in situ MMCs exhibit the following advantages: (1) formation of reinforcements that are thermodynamically stable, leading to less degradation in elevated-temperature service; (2) clean reinforcement/matrix interfaces, resulting in strong interfacial bonding; and (3) the formation of fine reinforcements with a more uniform distribution in the matrix, which yields better mechanical properties.

The in situ processes fall into two major categories, namely, reactive and nonreactive processes. In the reactive processes two components react and form the reinforcement phase. Generally, a rather high volume fraction of reinforcement is formed in the matrix alloy, and this master alloy is mixed with the matrix alloy to obtain a composite of desired reinforcement volume fraction. Typical examples are TiB_2- and TiC-reinforced aluminum alloy composites. Processing variables such as reaction temperature and cooling rate can be used to control the reinforcement size. The size of reinforcement is usually in the 0.25–1.5 µm range. The common matrix materials are Al and Ni and their alloys and intermetallics.

Nonreactive in situ processes take the advantage of phase transformation in two-phase systems, such as eutectic alloys, to form the reinforcement and matrix in situ. Directional solidification has long been used to produce anisotropic materials, which are often having a high degree of microstructural regularity and perfection. This procedure can be used to produce castings, in which the reinforcement and the matrix are formed during solidification. A binary alloy melt of eutectic composition normally forms an aligned two-phase structure on freezing. If the volume fraction of the minor phase is sufficiently low, then it can solidify in the form of fibers rather than the usual lamella structure because of the favorable energy conditions for the formation of fibers. Continuous fibers can form when the solidification is directional, that is, the solidification front is moved slowly in a particular direction. The critical volume fraction, below which fiber formation occurs, depends on interfacial energy and solute diffusion characteristics, and it is typically about 0.05. The fibers are usually intermetallic compounds with high strength and stiffness. Since the fibers are grown from the parent alloy, the fiber/matrix interface will be strong. By selecting the appropriate growth conditions, the fiber diameter and spacing can be controlled to some extent. A considerable amount of research has been carried out for the production of such in situ composites, particularly based on cobalt and nickel alloys. Promising materials with good creep resistance have been made in this way. The level of interest has dropped over the years, mainly because of inherent limitations in the nature and volume fraction of the reinforcement and also due to very slow growth/solidification rate. Although there is no problem of interfacial reaction, the microstructures are often found to be unstable, particularly when there is a thermal gradient. The schematic of controlled directional solidification setup to form the continuous reinforcement in the matrix is shown in Figure 5.9. The eutectic composition is melted in a graphite crucible, which is contained in a quartz tube under vacuum or inert gas atmosphere. Induction heating is normally used and temperature gradient is obtained by the chilling arrangement around the crucible. Electron beam heating can also be used, particularly when reactive metals such as titanium are used. It is possible to control the size and spacing of the reinforcement by controlling the solidification rate. Typical solidification rates lie in the range of 10–50 mm h^{-1}. This relatively slow rate is due to the necessity to maintain a stable solidification growth front. Figure 5.10 shows the microstructure of an in situ processed composite at various solidification rates.

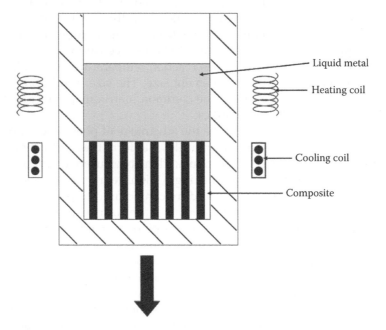

FIGURE 5.9
Schematic of directional solidification process to form in situ MMC. (Reprinted from Chawla, N. and Chawla, K.K., *Metal Matrix Composites*, Springer-Verlag, Inc., New York, 2006, p. 86. With permission.)

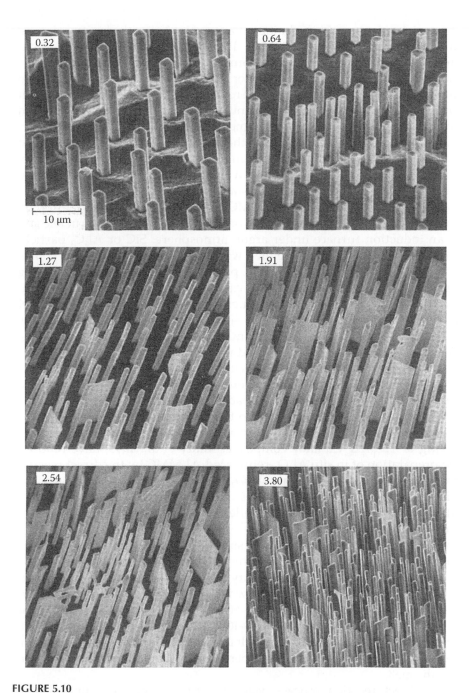

FIGURE 5.10
Transverse section of in situ TaC/Ni alloy composites obtained from a eutectic at different solidification rates indicated in left-hand top corners (cm h^{-1}) after etching the matrix from the top surface. (Reprinted from Walter, J.L., *In Situ Composites IV*, Elsevier, Amsterdam, the Netherlands, 1982, p. 85. With permission.)

A variety of processing techniques have evolved over the last two decades to prepare in situ particulate MMCs. Two of the commonly used processes are direct reaction and flux-assisted processes. In the direct reaction process, the powders or the compacts of powders to form the reinforcement phase are directly added into molten metal. The reinforcing particles are formed in situ through the reaction between reactants in the powder or between reactant powder and some component of the melt.

Nakata et al. (1995) have reported the preparation of carbide particulate-reinforced Al-based MMCs by this in situ process. The reaction between liquid aluminum alloys containing a thermodynamically stable carbide-forming element and relatively unstable carbide such as SiC or Al_4C_3 is used to form aluminum-based MMC with a stable carbide reinforcing phase. Aluminum containing Ti, Zr, or Ta is melted in a MgO crucible kept in an induction furnace under argon atmosphere. SiC or Al_4C_3 particles are then incorporated into the melt while stirring, and stirring is continued for some time. TiC, ZrC, or TaC particles are formed during stirring, and the melt with these particles is cast in a permanent mold. It is found that the size of SiC particles has an effect on the rate of reaction as well as on the dispersion behavior of in situ formed TiC particles. When SiC particles of size between 0.6 and 3 μm are added, TiC particles of size less than 1 μm are formed and dispersed uniformly in the aluminum matrix. However, spherical TiC aggregates with unreacted SiC particles, and large Al_3Ti intermetallic compounds are found when SiC particles of 14 μm size are added. In another process, instead of adding Ti into Al, Ti powder is mixed with Al_4C_3 and then added to the Al melt. After the completion of in situ reaction, Mg or Cu is added at a lower temperature to get TiC_p/Al–Mg or TiC_p/Al–Cu alloy composite. Maity et al. (1993) have reported the preparation of in situ Al_2O_3-reinforced Al-MMC by the direct reaction process. The required amount of TiO_2 particles to form Al_2O_3 on reaction with Al is added in small quantities into the vortex developed in molten Al by stirring with a graphite impeller. Al_2O_3 particles of ~3 μm size are found in the cast composite.

Aluminum-based composites with submicron size TiC or TiB_2 particulates have been prepared by a modified direct reaction process (Zhang et al. 1999). In this process, stoichiometric quantity of reactant powders are mixed with an appropriate amount of matrix powder and uniaxially pressed into compacts to a theoretical density of 50%–60%. These compacts are placed in an aluminum melt maintained at 800°C–1000°C. After allowing sufficient time for the completion of reaction, the melt is stirred and then cast in a mold. The processing time and temperature depend on the quantity of mixed powders and the melt.

The direct reaction process was also used to prepare copper-based composites. C powder or a mixture of B_2O_3 and C powders was added to a Cu–Ti alloy melt at a temperature of 1400°C–1600°C. The following reactions that took place in melt led to the formation of reinforcement phase (Chrysanthou and Erbaccio 1995):

$$Ti + C \rightarrow TiC$$

$$Ti + B_2O_3 + 3C \rightarrow TiB_2 + 3CO$$

By this process, TiC/Cu composites containing up to 55 wt.% TiC particles with a size of 1–10 μm and TiB_2/Cu composites containing up to 18 wt.% TiB_2 particles with a size of 3–5 μm have been prepared.

The flux-assisted synthesis was developed by the London & Scandinavian Metallurgical Company, to produce aluminum matrix composites through in situ process. A salt mixture containing potassium hexafluorotitanate (K_2TiF_6) and potassium tetrafluoroborate (KBF_4) in the stoichiometric ratio to form TiB_2 is introduced into a stirred aluminum melt. The salts undergo reactions with aluminum according to the following sequence (Kellie and Wood 1995):

$$3K_2TiF_6 + 13Al \rightarrow 3TiAl_3 + 3KAlF_4 + K_3AlF_6$$

$$2KBF_4 + 3Al \rightarrow AlB_2 + 2KAlF_4$$

$$AlB_2 + TiAl_3 \rightarrow TiB_2 + 4Al$$

After the reaction, stirring is stopped, the slag containing $KAlF_4$ and K_3AlF_6 is removed, and then the molten material is cast into a mold. The viscosity of the melt limits the maximum TiB_2 content, and TiB_2 content up to 12% has been reported. In situ MMC-based Al–4Cu and A356 aluminum alloy with TiB_2 have also been successfully prepared (Wood et al. 1993, Lu et al. 1997). Chen and Chung (1996) have prepared TiB (TiB_2) particle-reinforced 2024 aluminum alloy composite using a powder mixture containing TiO_2, Na_3AlF_6, and KBF_4 by the in situ process.

Since the flux-assisted synthesis is a well-established technology, it could be amenable for the commercial production of MMCs. However, it has been found that the TiB_2 particles formed in the process are coated with some undesirable reaction products due to the presence of salts in the melt (Koczak and Premkumar 1993). A comparative study has been made by Chen et al. (1999) using mixed salt (K_2TiF_6–KBF_4–Na_3AlF_6) and mixed oxide (TiO_2–H_3BO_3–Na_3AlF_6). It was found that the mechanical properties of $TiB_2/$ Al–4Cu composite fabricated using mixed salt are lower than that of the composite made using mixed oxide. The reaction rate of mixed salt system was also found to be markedly lower than that of the mixed oxide system. It has also been reported that it is rather difficult to produce composites containing more than 8 vol.% TiB_2 using mixed salt system (Chen et al. 1999).

5.2.1.8.1 XD™ (Exothermic Dispersion) Process

The XD™ process is another type of in situ process, in which an exothermic reaction between two components produces the reinforcement phase. This process is a proprietary method developed by Martin Marietta Corporation for producing particle-reinforced MMCs. It involves heating various compounds to high temperatures, such that a self-propagating exothermic reaction takes place. Generally, a master composite containing a rather high volume fraction of reinforcement is produced by the reaction. This composite is mixed and remelted with the base alloy to produce the final composite with the desirable amount of reinforcement particulates. Typical composites produced by this method are SiC, TiC, or TiB_2 particle-reinforced aluminum, nickel, or intermetallics.

This process has been extensively studied for the development reinforced intermetallic composites, such as TiB_2/Ti_3Al_4 and TiB_2/Ni_3Al_2. The as-formed composites may contain pores, which can be eliminated during the subsequent hot isostatic pressing (HIPing) operation. Moreover, there is a scope for controlling the porosity by the choice of reactions and conditions. Although the products of these in situ reactions are expected to be thermodynamically stable, they may suffer from morphological instability during service at very high temperatures.

5.2.1.8.2 Exo-Melt™ Process

This process provides a method for melting intermetallics utilizing the exothermic heat generated during the formation of intermetallic compound from elemental materials. For example, the heat released due to reactions between nickel and aluminum is utilized in a novel way to melt the stock of Ni_3Al intermetallics. Similarly, the reaction between iron and aluminum leads to the melting of Fe_3Al and FeAl intermetallics. The Exo-Melt™ process has revolutionized the interest in intermetallics and thus paved the way for commercial use of them.

Heat release and temperature rise due to the exothermic reaction are strongly dependent on the constituent elements of the intermetallic compound, their relative amount, and the initial temperature of the components. Thermodynamic principles can be used to calculate the maximum attainable temperatures during the formation of intermetallics. Several other parameters such as interfacial contact area between the reacting elements and the particle size of elements also control the reaction.

A schematic for the preparation of Ni_3Al is shown in Figure 5.11. The total amount of nickel required to form Ni_3Al is divided into two parts. The nickel content in one part is equivalent to the formation of NiAl compound. This part is placed at the top, surrounded by the required quantity of aluminum. The remaining nickel is loaded at the bottom with alloying elements placed between the two layers. At a crucible temperature of 700°C or above the melting point of aluminum, the molten aluminum reacts with nickel and forms

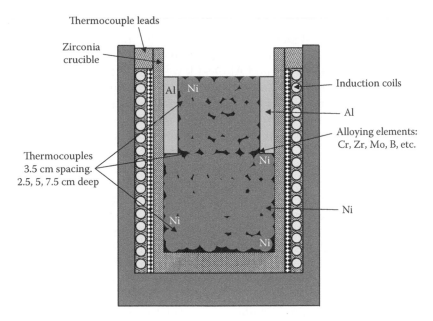

Thermocouple leads

Zirconia crucible

Induction coils

Al

Alloying elements: Cr, Zr, Mo, B, etc.

Thermocouples 3.5 cm spacing. 2.5, 5, 7.5 cm deep

Ni

FIGURE 5.11

A schematic of furnace loading sequence employed for the Exo-Melt™ process to prepare nickel aluminide. (Reprinted from Deevi, S.C. and Sikka, V.K., *Intermetallics*, 5, 19, 1997. With permission.)

NiAl liquid. This liquid NiAl dissolves alloying elements, drips down to the bottom, and forms Ni₃Al on reaction with the remaining nickel. This division of nickel into two parts allowed to raise the temperature to a significantly higher value than what could have been accomplished with the direct formation of Ni_3Al. The entire stock can be melted within 15 min.

Different melting techniques, such as vacuum-induction melting, air-induction melting, vacuum-arc melting, electroslag remelting, and electron beam melting, can be used for this process. Castings without solidification porosity and with low levels of impurities are obtained in the vacuum-induction melting, electroslag remelting, and electron beam melting techniques. However, the melting cost associated with these processes is prohibitively expensive for commercial applications. Air-induction melting technique is the most economical melting route for these kinds of materials and can be used to obtain castings with low levels of impurities. A protective slag is formed during the melting of aluminide intermetallics in air, and the surface can be covered by argon gas to further limit the diffusion of oxygen, nitrogen, hydrogen, and sulfur impurities.

Crucibles used for melting intermetallics should be chemically inert with high melting points and have reasonable strength and excellent thermal shock resistance. Generally, crucibles made with MgO, BeO, or ZrO_2 are more expensive than that made from zircon, cordierite, mullite, and forsterite.

The stability of oxides varies depending on the nature of the oxide and temperature. MgO is more stable than Al_2O_3 up to 1600°C, but Al_2O_3 is much more stable above 1600°C even though it has a relatively lower melting point. ZrO_2 has lower stability than MgO and Al_2O_3 at lower temperatures, but it is more stable at high temperatures. Among the oxides, BeO is the most stable oxide, but crucible made out of BeO is not easily available and more expensive.

The energy requirement for the Exo-Melt process is significantly lower compared to the conventional process. In the conventional melt process to form Ni_3Al, the crucible must be heated above the melting point of nickel (>1453°C), whereas in the Exo-Melt process, only melting of aluminum (>667°C) is needed to initiate the reaction. The other advantages of this process are reduced melting time, minimal oxidation of alloying elements, and easy vacuum melting, since all alloying elements can be loaded at the beginning itself. Another important advantage is that it eliminates the safety concerns in melting nickel and iron aluminides. The Exo-Melt process should in principle be applicable to several aluminides, and the methodology can be adapted to reduce violent reaction and maximize energy savings. Based on the heat of formation, specific heat, and adiabatic combustion temperatures of the particular intermetallic compound, the furnace loading pattern should be altered.

A variant of this process has already been used for the preparation of MMCs, which is called combustion-assisted casting (Tjong and Ma 2000). In this process combustion synthesis and traditional ingot metallurgy are combined to produce MMCs. A stoichiometric quantity of reactant powders is mixed thoroughly and compacted into pellets. These pellets and the required amount of matrix material are heated. Once the reaction starts in the pellets, the exothermic heat released from the reaction helps to melt the matrix material. After thorough mixing, the molten slurry is cast into a mold.

This combustion-assisted casting process has been used to prepare (TiB + TiC)/Ti composites (Ranganath et al. 1992). Pellets made from Ti and B_4C powder mixture were melted with the required amount of titanium sponge in a water-cooled copper crucible using nonconsumable vacuum-arc melting. It has been suggested that the following reaction takes place between B_4C and Ti:

$$5Ti + B_4C \rightarrow 4TiB + TiC$$

The heat released from this exothermic reaction is utilized for the propagation of this reaction as well as for the melting of reinforcements. In another study (Zhang et al. 1999), stoichiometric amounts of sponge Ti, B_4C, and graphite powder were used to prepare the pellets, and the composite was made by using the previously mentioned procedure. It was found that the TiC content was higher because of the introduction of graphite.

Lin et al. (1991) have reported the preparation of Ti matrix composite with 45 vol.% of TiC reinforcement by this process. A mixture of Ti, C, and Al powders was dried in a graphite crucible at 200°C for 12 h in an oven and then heated in an induction furnace. An ultrasonic vibration was used to enhance mixing and reduce entrained gas in the melt. The molten material was cast in a mold.

5.2.1.9 Melt Oxidation Process (Lanxide Process)

In this process, a ceramic reinforcement preform formed into the final product shape is continuously infiltrated by a molten alloy as it undergoes an oxidation reaction with a gas. Basically a chemical reaction is used to cause spontaneous infiltration of melt into the preform. An aluminum alloy ingot containing 3–10 wt.% Mg is placed on top of a preform made of Al_2O_3 or SiC. This assembly is heated under nitrogen atmosphere at temperatures between 800°C and 1000°C. The molten alloy spontaneously infiltrates into the preform under this condition. Aluminum nitride (AlN) is also formed in the composite because of the presence of nitrogen. The amount of AlN formed inversely depends on infiltration rate, that is, more AlN forming at slower infiltration rates. An increase in the amount of AlN increases the modulus of the composite and reduces coefficient of thermal expansion (CTE). Infiltration rate is slower at lower temperatures and at lower nitrogen content in the infiltration atmosphere. Therefore, modulus and CTE can be tailored by controlling the process temperature and the amount of nitrogen in the infiltration atmosphere. Similarly, the high-temperature oxidation of the molten alloy in the interstices of preform produces a matrix material composed of a mixture of oxidation reaction products and unreacted metal alloy. The primary advantage of this process is its ability to form fully dense, complex-shaped composites.

5.2.2 Solid-State Processes

The main drawbacks of liquid-state processing techniques are the difficulty in controlling reinforcement distribution and obtaining a uniform matrix microstructure. Furthermore, adverse interfacial reactions between matrix and reinforcement are more likely to occur at the high temperatures involved in liquid-state processing. These reactions can have a detrimental effect on the mechanical properties of the composite. These problems can be well controlled in the solid-state processing techniques.

It is possible to produce MMCs without the melting of matrix, although this may result in less intimate interfacial contact. One of the main problems with this approach is that it needs heavy processing equipment and hence expensive. However, mixing of metallic powder and fiber or particulate reinforcement is a convenient and versatile technique for MMC production. It offers excellent control over the reinforcement content across the complete range. The blending can be carried out in dry on wet condition. This is usually followed by cold compaction, (wet mixed powders should be dried before cold compaction), canning, evacuation (degassing), and high-temperature consolidation, such as HIPing. HIPing results in higher interfacial bond strength and hence improved properties for the composites. It is difficult to achieve a homogeneous mixing during blending, particularly with fibrous materials, which tend to form tangled agglomerates with interstitial spaces too small for the penetration of matrix particles. The relative size of matrix and reinforcement particles has a significant influence on the uniformity of mixing. Another notable feature of most metallic powders is the presence of oxide layers of a few tens of nm thick, constituting about 0.05–0.5 vol.%, depending on the powder history and processing conditions.

5.2.2.1 Powder Metallurgy Techniques

P/M techniques are the most common solid-state processes. In these techniques, discontinuous reinforcements and matrix material powders are generally used, due to the ease of mixing and the effectiveness to densification. The reinforcement and metal matrix powders are mixed, cold compacted, and sintered or directly hot-pressed after mixing. The near-full density compact is then subjected to secondary operation such as extrusion or forging. The sinter forging is a novel low-cost technique that can eliminate the expensive hot-pressing step, to get a quality composite. P/M processing is mainly used to fabricate particle- or whisker-reinforced MMCs. The matrix powder and the reinforcement are blended to produce a homogeneous distribution. This mixture is cold-pressed in a die by applying pressure. This cold-pressed green body is about 80% dense and can be easily handled. P/M components are usually sintered without the application of pressure. However, pressureless sintering may not be effective for the composites. Sintering is a diffusion-assisted process. Since there are chemically dissimilar materials present in the composite, the second material acts as a barrier for the diffusion. Furthermore, fibers and whiskers can form a network that may inhibit the sintering process. Therefore, the densification rate of matrix during sintering will, in general, be retarded by the presence of reinforcements. Another reason for poor sintering of MMCs is the presence of oxide skin on the metallic powders, particularly on aluminum powders, which retard the diffusion process. Hence, it is very difficult to achieve full densification under pressureless sintering.

Some form of hot pressing is frequently resorted to achieve full densification in MMCs. The simultaneous application of pressure and high temperature can accelerate the diffusion processes, and a pore-free and fine-grained microstructure can be obtained. The pressure is applied uniaxially or isostatically during hot pressing depending on the complexity of the component.

The P/M processes have been successfully used to produce a variety of MMCs from a large number of metal–ceramic combinations. Overall strength values of Al–SiC MMCs produced by P/M processes are higher than that of MMCs produced by liquid-state processes; the elongation values, however, are lower. P/M process allows the development of novel matrix materials outside the compositional limits dictated by equilibrium thermodynamics. Major limitations and constraints of P/M technique are limited availability of suitable prealloyed metal powders and high capital investment (particularly for hot pressing/HIPing operations) and may not be cost-effective, if produced in small numbers.

Continuous fiber-reinforced MMCs can also be produced by the P/M technique. Fiber tows are impregnated by slurry made out of metallic powders in a suitable liquid, dried, arranged to the desired shape, and then subjected to hot pressing/HIPing.

The ratio of reinforcement size to matrix particle size is an important parameter in controlling a homogeneous distribution of reinforcement in the matrix. When the matrix particles are larger than the reinforcement particles, the reinforcement particles are pushed and packed in the interstices between larger matrix particles, yielding a more clustered microstructure. A particle size ratio close to unity is preferable for a more homogeneous distribution. Several techniques have been used to quantify reinforcement particle clustering in MMCs. The coefficient of variance (COV) of the mean near-neighbor distance is sensitive and effective in characterizing particle clustering. This parameter is also relatively insensitive to reinforcement volume fraction, size, and morphology. The coefficient, COV, can be found by Equation 5.12 (Yang et al. 2001):

$$COV = \frac{\sigma_d^2}{d} \tag{5.12}$$

where
σ_d is the variance in the mean near-neighbor distance
d is the average of the mean near-neighbor distance of the particles sampled

The degree of clustering tends to a minimum at a particle size ratio close to unity.

In some composite systems, simple cold pressing and sintering can be used to successfully prepare the dense composites. These composite systems are sintered in a temperature range to obtain some amount of liquid phase. The liquid phase flows through the pores in the compact, and also diffusion is fast through the liquid phase, thus resulting in the better densification of the composite. A classical example of these composite systems is WC/Co composite, popularly known as cemented carbide. In this composite, very high volume fraction of WC particles is distributed in a relatively soft cobalt matrix. This composite is widely used in machining of hard metals and rock drilling operations. During sintering, molten cobalt wets WC very well with little or no interfacial reaction.

The P/M route has the advantage of reducing the interaction between the reinforcement and the matrix, since the processing is carried out in the semi-solid or solid state. Other attractive features P/M route includes

1. Ability to combine various types of reinforcements within the same composite
2. Ability form nonequilibrium matrix alloys, as produced by rapid solidification
3. Ability to form composites with high reinforcement volume fractions

However, the P/M route does have certain disadvantages. These include

1. Higher capital investment compared to melt processing
2. Relatively complex process
3. Requirement of high safety precautions, since many powders are highly reactive

5.2.2.1.1 Reinforcements

Mainly discontinuous fibers or particulates are used as reinforcements in this process. The selection of appropriate reinforcement is application specific. For example, elastic modulus, tensile strength, and density are considered as important properties for structural applications. The thermal conductivity of reinforcement material is important for heat exchanger applications. For high-temperature applications, the thermal expansion coefficient of reinforcement is considered. Thermal expansion mismatch between the reinforcement and the matrix should be minimum, if the composite is to be subjected to repeated thermal cycling. The reactivity of reinforcement with matrix is considered, while selecting the reinforcement for reactive metallic matrices.

5.2.2.1.2 Matrices

Originally wrought alloy compositions (6061, ZK 60H, and IN 718) were selected as matrices. However, it was found that minor alloying elements commonly included in wrought alloys (Mn and Cr) were detrimental to the composites' mechanical properties. This is due to the formation of Mn- and Cr-based intermetallic compounds during densification and subsequent processing. Moreover, it was shown that the alloy compositions, lying at the lower end of standard alloy specifications, had a better combination of strength, ductility, and fracture toughness. Normally prealloyed atomized powders of mean size 15 µm are utilized in P/M technique.

5.2.2.1.3 Blending

Uniform mixing of reinforcement with matrix powder is of prime importance in P/M processing. There are many variables that can control the

TABLE 5.2

Variables Controlling the Effective Blending of the Constituents

1	Type of particulate
2	Type and geometry of the mixer
3	Mixing time
4	Volume ratio of mixer to particulate
5	Volume ratio of component particulate and their characteristics
6	Inner surface area of the mixer
7	Construction material and inner surface finish of the mixer
8	Rotational speed of the mixer
9	Mixing medium, that is, gaseous or liquid
10	Existence of lubricant

Source: Data from Clyne, T.W., in *Comprehensive Composite Materials*, Vol. 3, eds. A. Kelly and C. Zweben, Elsevier Science Ltd., Oxford, U.K., 2000, p. 690.

effective blending of the constituents. These variables are listed in Table 5.2. Apart from those variables, the size, shape, and density of reinforcement and matrix particles also influence the extent of mixing. A homogeneous distribution can be achieved easily when spherical particles are used. Uniform mixing is more difficult to achieve when whiskers or fibers are present in the mixture. Mixing problems are generally encountered with whiskers or fibers, since they easily form agglomerates. It has also been shown that particulates with aspect ratio greater than 30 have a strong tendency to cluster. As a result, ball milling is often used to obtain uniform mixtures, recognizing the fact that the final length of the reinforcement may be reduced by excessive mixing. Difference in the constituents' densities is another major cause of particulate segregation. Unless adequate precautions are taken, this can result in settling of heavier particles at the bottom of the mixture.

Wet and ultrasonic blending is another alternative process for deagglomeration. Wet blending is not only effective for achieving a better degree of mixing but also reduces whisker and fiber damage considerably. Some additives can be incorporated during blending for efficient mixing by reducing interparticle friction and interparticle attraction. The use of nonpolar solvents is particularly effective with smaller matrix powders and reinforcements.

Often a homogeneous mixture is obtained after a relatively short period of mixing. Indeed prolonged mixing can cause a decrease in a mixing uniformity as electrostatic charges accumulate with time. The degree of mixing uniformity is also considerably reduced when the mixing vessel is insufficiently filled or overfilled. As a rule of thumb, the volume occupied by the balls and powders should be approximately 50% of the volume of container. Among the wide variety of blenders, the tumble-type blenders are more suitable for mixing reinforcements with matrix powders. Drum, double-cone, twin-shell, and cross-flow blenders can also be used to achieve a good degree of mixing. When the fragmentation of reinforcements is not a major constraint, ball and Attritor mixers represent a good choice. High-energy mills have been used to prepare MMC powders by mechanical alloying.

5.2.2.1.4 Consolidation

Blending is normally followed by cold compaction, outgassing, and sintering. Cold compaction to approximately 50% of composite density can be achieved by either single- or double-action pressing. In any compaction procedure, it is essential to control local densities to ensure an open interconnecting pore structure. This is necessary, since it is through these open channels that the various gaseous products liberated during outgassing and heating must flow and come out of the compact. Outgassing involves the removal of remnant liquid from the wet-blending operation (if used) and adsorbed and chemisorbed water from reinforcement and matrix powder. It is normally carried out by heating the compact under vacuum.

Once outgassing is completed, the partially densified compact is heated to the final densification temperature. The selection of this temperature is dictated by the melting point of matrix material (often 0.75 T_m) and the need to minimize the reactions between the reinforcement and matrix.

5.2.2.1.5 Sintering

Sintering is the consolidation of a green product at elevated temperature (usually at 0.7–0.8 of T_m), producing a dense body with a microstructure giving the required physical properties. Most solids, if sufficiently divided, but in contact, will densify at high temperature. The driving force for sintering is the reduction of total free energy of the compact associated with a reduction in the total surface energy. The reduction in surface area occurs by the growth of contact area (neck) between particles. Mass is transported to the neck region by surface diffusion, grain boundary diffusion, and bulk diffusion. As the neck grows, the centers of particles come closure, and as a result the compact shrinks. During this process, the porosity between the particles is also eliminated.

For a compact containing spherical particles of a substance with molar volume V_{mol} and radius r, the molar energy that is released by full densification is

$$E_d = \frac{3\gamma_{sv}V_{mol}}{r} \tag{5.13}$$

where γ_{sv} is the solid–vapor surface energy. The surface energy per unit volume depends on the inverse of the particle diameter, since the surface energy is directly related to surface area of particles. Thus smaller particles with high specific surface area (surface area per unit weight) have more surface energy and sinter faster. The driving factor for diffusion is provided by the curvature of the particles' free surfaces and an externally applied pressure, if used.

Diffusion is thermally activated; hence specific energy is necessary for atomic movement. The energy should be equal or above the activation energy for an atom to break from its current site and move into a vacant site.

The population of vacant atomic sites and the number of atoms with sufficient energy to move into those sites vary with the temperature. An Arrhenius-type relation exists for this:

$$\frac{N}{N_o} = \exp\left(\frac{-Q}{RT}\right) \tag{5.14}$$

where
  N is the number of available vacant sites or active atoms
  N_o is the total number of atoms
  Q is activation energy
  R is gas constant
  T is absolute temperature

Diffusion, hence sintering, is faster at higher temperatures, because of the increased number of active atoms and available vacant sites.

In powder compacts, there are many particles in contact with each particle. The contacting areas between the particles enlarge by replacing solid–vapor interface, and the particles merge as sintering progresses. Consider two spherical particles in contact, as shown in Figure 5.12.

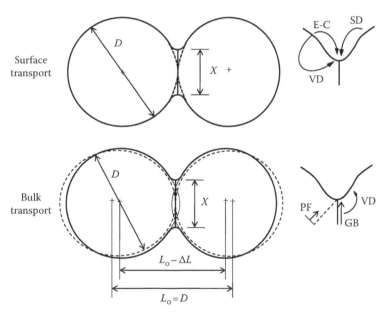

FIGURE 5.12
Two classes of sintering mechanisms as applied to the two-sphere sintering model. Surface transport mechanisms provide for neck growth by moving mass from surface, and bulk transport processes provide for neck growth using internal mass. (Reprinted from German, R.M., *Powder Metallurgy Science*, Metal Powder Industries Federation, Princeton, NJ, 1984, p. 250. With permission.)

The Laplace equation gives the stress associated with a curved surface as

$$\sigma = \gamma \left(\frac{1}{R_1} + \frac{1}{R_2} \right) \tag{5.15}$$

where
   γ is the surface tension
   R_1 and R_2 are the principal radii of curvature for the surface

The sign convention gives a positive value for the convex surface and negative value for the concave surface. Consider the initial stage of sintering, at which particles are having point contacts. At a distance from the neck, the curvature is constant and both R_1 and R_2 are equal to the sphere radius $(d/2)$; hence Equation 5.15 becomes

$$\sigma = \frac{4\gamma}{d} \tag{5.16}$$

Using a circle approximation to the neck shape with a radius of R_n, where R_n is approximately equal to X^2/d (X is the diameter of neck), the stress at the neck region is given as follows:

$$\sigma = \left(\frac{2}{X} - \frac{d}{X^2} \right) \tag{5.17}$$

It is clear from Equations 5.16 and 5.17 that there is a large stress gradient in the neck region. Thus, there is a strong driving force for mass flow to the neck. As the neck grows, the curvature gradient is reduced and the process slows down.

Transport mechanisms determine the method of mass flow in response to the driving force. The transport of mass can occur either by surface transport or by bulk transport or by both. Surface diffusion and evaporation–condensation are the two most important contributors during surface transport. The surface diffusion dominates the low-temperature sintering of many materials, whereas the evaporation–condensation is not as important and it dominates only during the sintering of low-stability materials such as lead.

In the bulk transport mechanism, the mass originates from the interior of the particle and deposits at the neck. Bulk transport mechanisms of crystalline materials include volume diffusion, grain boundary diffusion, and plastic flow. Plastic flow is usually important for compacted powders, where the initial dislocation density is high. However, the role of plastic flow declines as the dislocations are annealed out during sintering. Grain boundary diffusion is fairly important for the densification of most crystalline materials. While both surface and bulk transport processes lead to neck growth, the main difference is in shrinkage during sintering. Surface transport involves neck growth without a change in particle spacing and hence no shrinkage on densification, whereas bulk transport causes shrinkage. Although bulk transport processes can operate at lower temperatures, they are more active at high temperatures.

The green density of the component will also affect the sintered density. Generally, higher green densities will lead to higher sintered densities and lower shrinkage. Therefore it is important during green body formation to produce a compact with good packing of particles and minimum binder content. Differential densification/shrinkage in the compact can occur if the reinforcement phase is nonuniformly distributed in the matrix. Therefore, it is essential to properly mix the raw materials to achieve the uniform distribution of reinforcement phase.

5.2.2.1.6 Hot Pressing

Densification of powder-based composites occurs by sintering or pressure-assisted sintering methods such as hot pressing and HIPing. Conventional pressureless sintering is the most economical process for densification. It has the advantage of being able to produce near-net-shaped components in large numbers at a time. However, the addition of a large volume fraction of reinforcement phase tends to prevent densification. This is due to the steric hindrance of secondary phases (reinforcements), which may prevent the matrix from densification during sintering. Hence, in many cases, pressure-assisted densification methods are used to produce dense composites. Hot pressing, which leads to the achievement of low to zero porosity levels, is applicable in principle to all powdered materials.

The material used for hot pressing can simply be a mixture of discontinuous reinforcement and matrix powder or a preform. During this process, temperature and pressure are applied simultaneously. The application of pressure accelerates densification, which is attributed to the increase in contact stress between the powder particles. During hot pressing, the matrix particles rearrange and sinter or flow viscously between the reinforcements and fill voids. The application of pressure becomes more important for fiber-reinforced composites, since the densification rate is significantly retarded by the presence of fibers, especially at high fiber loading.

A schematic of hot pressing setup is shown in Figure 5.13. It consists of a furnace surrounding a high-temperature die and a hydraulic press. The die is perhaps the most important part of the hot press, and a number of refractory metal and ceramic dies are available. Refractory metal dies are generally very expensive, though molybdenum and tantalum dies have been used in certain cases. Refractory metals tend to be reactive, have high thermal expansion, and deform easily at the temperatures used to process most ceramic composites (>1500°C). On the other hand, ceramic die materials, especially Al_2O_3 and SiC, are nonreactive and have low thermal expansion, but limited by their use temperature (Al_2O_3 can only be used up to 1200°C and SiC up to 1400°C). Graphite is the most widely used die material in hot

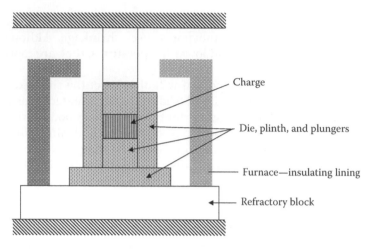

Charge

Die, plinth, and plungers

Furnace—insulating lining

Refractory block

FIGURE 5.13
Schematic of a hot pressing equipment. (Reprinted from Warren, R., in *Comprehensive Composite Materials*, Vol. 4, eds. A. Kelly and C. Zweben, Elsevier Science Ltd., Oxford, U.K., 2000, p. 658. With permission.)

press due to its ability to withstand high temperatures and transient thermal stresses. It does not react with most materials and can be coated with BN to avoid any reaction. Low thermal expansion of graphite is another added advantage. If the thermal expansion of the die material is higher than that of the material being hot pressed, the die will shrink more and fit tightly around the material during cooling. This will make the ejection of the material extremely difficult.

The heating in hot press is generally carried out using RF induction through water-cooled copper coils and a graphite susceptor that is capable of producing temperatures greater than 1800°C. The furnace should be evacuated or backfilled with argon gas during processing to prevent the oxidation of graphite. Furnaces with resistant heating elements have also been used, for hot pressing, but they are very rare.

The pressure during hot pressing is usually applied through a hydraulic press with water-cooled platens. The size of the sample and pressure requirement determine the capacity of the press. Most hot pressing is carried out in the pressure range of 10–40 MPa.

Although hot pressing process is more expensive than normal sintering process, it has several advantages. The simultaneous application of pressure allows the reduction of densification temperatures, some 300°C–500°C lower than that used in the pressureless sintering process and at the same time increases the densification rate. The reduced temperature and time suppress grain growth, resulting in fine-grain structure in the matrix. It is possible to achieve near-theoretical density by this process. Since the sintering aids are either not used or used in small quantities, the resulting composite will have very good high-temperature properties, such as creep.

Hot pressing can cause preferred orientation of grains especially when matrix powders with large aspect ratio are used. As a consequence, the matrix material will have different properties in different directions. Such an anisotropic behavior can also occur due to flattening of aggregates during hot pressing. The main disadvantage of hot pressing is that it is restricted to produce only simple shapes. It may not be possible to produce nonuniform cross sections and intricate or contoured shapes by this process. Nevertheless, composites produced by hot pressing have superior properties compared to those produced by other routes.

5.2.2.1.7 Hot Isostatic Pressing

HIPing is a process in which a uniform pressure is applied from all directions to the composite at high temperatures. In recent years, considerable progress has been made in using this process to fabricate ceramics and ceramic matrix composite parts. Modern HIPing units are well equipped to allow computer-controlled programming of HIPing cycles for optimizing process parameters such as temperature, pressure, and time. A typical HIPing equipment consists of a pressure vessel, a high-temperature furnace, and a gas-handling unit. The process allows fairly high pressures (up to 400 MPa) and temperatures as high as 2000°C. The pressure is applied by pressurizing inert gas such as argon.

Unlike hot pressing, HIPing does not impose any constraint on the shape of the part being produced. Hence, complex-shaped components to near-net-shaped can be made by this process. This method offers the possibility to substantially reduce or even eliminate the use of sintering additives. This in turn can minimize the grain boundary glassy phases and their deleterious effect on high-temperature properties. It is important to keep the HIPing cycle duration (i.e., time to reach the required temperature and pressure, dwell time, and cooldown time) as short as possible to get products with better properties and at the same time to reduce the processing cost. The main disadvantage of HIPing process is the prohibitively high cost of the equipment. However, the extra cost can be justified if the properties obtained are unique to the process and required for a particular application.

There are three main classes of HIPing techniques. Sinter-plus-HIPing is one of the HIPing processes, in which a green body is sintered by pressureless sintering in a separate furnace to a state that is not permeable with the pressurized gas. The partially sintered body with closed surface pores is then fully densified by HIPing. Sinter HIPing is another HIPing process, which involves pressureless sintering to closed porosity in the HIPing furnace itself, followed by the HIPing treatment. The combined sinter-HIPing process eliminates the extra heating-up, cooling, and handling required in the sinter-plus-HIPing technique. Both these techniques do not need encapsulation of the part during HIPing. The third HIPing technique is capsule HIPing that uses encapsulation of the green body before HIPing. The encapsulation material can be a glass or metal, which softens at the HIPing temperature and acts as a flexible material. This encapsulation material prevents the pressurizing gas from entering into the pores. The encapsulation material should be chemically inert with respect to the composite material during HIPing and easily be removed after HIPing without damaging the HIPed material. The main advantage of the capsule-HIPing technique over other techniques is that the green body does not require pre-sintering, which eliminates the need for high levels of sintering additives. This HIPing process is ideally suitable for ceramic matrices such as nitrides that decompose under ambient pressures at high temperatures.

Densified billets, generally >98% theoretical density, can be deformed into different shapes with semistandard metalworking equipment. The processing conditions should be adjusted for each composite system, since the forming temperatures, deformation rates, and flow conditions vary for different composite systems.

5.2.2.2 Diffusion Bonding

Diffusion bonding is a common solid-state joining technique used for joining similar and dissimilar metals. Interdiffusion of atoms at elevated temperature through the clean intimately contacting surfaces leads to bonding. This process can also be used for the fabrication of MMCs. The main advantages of this process are the ability to fabricate composites with a wide variety of metallic matrices and better control of fiber orientation and volume fraction. However, there are some drawbacks with this process, and they are long processing times, the requirement of high processing temperatures and pressures, and limitation on the complexity of product shapes. There are many variants in the basic diffusion bonding process; in one of the processes, matrix alloy foil and fiber arrays, or matrix-coated fiber, or monolayer laminate are stacked in a predetermined order and then consolidated.

5.2.2.2.1 Foil–Fiber–Foil Process

Matrix alloy foils or thin sheets made from matrix powder and a fugitive organic binder and fiber arrays are stacked alternatively to a predetermined thickness, as shown in Figure 5.14. The stacked layers are vacuum hot pressed by applying the required pressure and temperature. The atomic diffusion at these conditions leads to bonding between the layers. Instead of uniaxial hot pressing, HIPing can also be used. Before HIPing, the stacked layers are transferred to a can, evacuated by applying vacuum, and then sealed. The gas pressure applied from all sides of the can consolidates the composite part. With HIP, it is relatively easy to apply high pressures at elevated temperatures on complex geometries. Hot pressing is mainly suitable for flat products, although corrugated structures have also been produced.

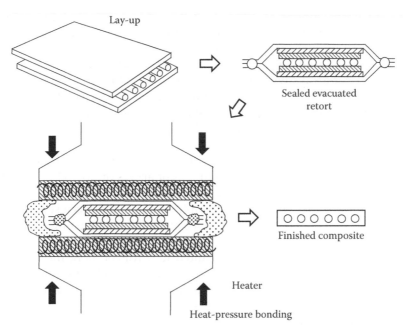

FIGURE 5.14
Foil–fiber–foil diffusion bonding process. (Reprinted from Lynch, C.T. and Kershaw, J.P., *Metal Matrix Composites*, CRC Press, Boca Raton, FL, 1972. With permission.)

5.2.2.2.2 Matrix-Coated Fiber Process

In this process, the fibers are first coated with the matrix material by plasma spraying or physical vapor deposition (PVD). It should be noted that the fiber may react or be damaged by the high impact velocity of liquid droplets during plasma spraying. The fiber spacing in the composite is easier to control, and the coated fibers can be handled more easily. Since the thickness of coating decides the matrix volume fraction, composites with very high fiber volume fraction (>0.75) can be obtained by using fibers with very thin matrix coating. The pre-coated fibers are arranged in a predetermined order and then hot pressed or hot isostatically pressed to form the MMCs. If the coated fibers are flexible, then axisymmetric shapes can be fabricated by the filament winding process.

In another variant, C, SiC, or Al_2O_3, fiber bundles are wound on a drum made of a metal having good thermal conduction. The composite preforms are then made by applying the matrix material on reinforcements using various coating methods, such as plasma spraying, chemical coating, electrochemical plating, chemical vapor deposition (CVD), and PVD. These preforms are subjected to hot pressing or HIPing, to enhance the density of the composites by removing voids.

Fiber coating treatment to prevent interfacial reactions may not be needed during diffusion bonding, because of the relatively low temperatures involved, but the pressure applied to enhance diffusion may cause damage to the fibers. The applied pressure and temperature, as well as the time required for diffusion bonding, vary with the composite system. Of the several processing methods used for fiber-reinforced superalloy composites, diffusion bonding of fiber arrays with matrix alloy powder or matrix foil is the most effective process. Hot- roll diffusion bonding is another variant of the diffusion bonding process, which can be used to produce a composite consisting of different metals in alternate layers. Such a composite is called sheet-laminated MMC.

A technique in commercial use for the production of continuous fiber-reinforced titanium involves the placement of arrays of fibers between thin metallic foils, followed by a hot pressing operation. This process is attractive for titanium MMCs because of the following: (1) there is a scope for significant improvements in creep resistance and stiffness by using continuous fiber reinforcement, (2) routes involving Ti melt suffer from rapid interfacial chemical reaction, and (3) Ti is well suited to diffusion bonding because it dissolves its own oxide at temperatures above 700°C. Optimum conditions for composite production have been established; typically it is carried out for a few hours at about 900°C. There are some advantages and disadvantages by the addition of alloying elements to titanium; for example, the addition of Al, Mo, or V slows down the kinetics of interfacial reaction but also make the rolling of thin foils more difficult. While the addition of Ni reduces the diffusion bonding temperature, at the same time triggers the formation of coarse intermetallics such as Ti_2Ni, which impair the mechanical properties. Overall, the process is slow and there can be difficulties in obtaining very high fiber volume fractions. Fiber distribution is one of the important parameters in controlling the mechanical properties of composites. Fiber-to-fiber contact or very close spacing between the fibers can result in very high, localized stress concentrations. These stress concentrations are responsible for the premature damage, cracking, and failure of the composites during service. Furthermore, a significantly thick (~1 μm) interfacial reaction layer usually forms during the process. There is considerable interest to develop fiber coatings, which reduce the problem of interfacial reaction. Composites based on superplastic aluminum alloys have also been made by this diffusion bonding process.

5.2.2.3 Explosive Shock Consolidation

A fairly novel, high-strain-rate technique, mainly used for the consolidation of fine powders, is the explosive shock consolidation technique. In this technique, dynamic compaction is achieved by means of high-velocity impact from projectiles driven by explosives (Figure 5.15). The detonator is placed at the top and the datasheet booster is used to create a more uniform detonation front. The explosive is placed in a cylinder surrounding the powder mixture. The whole setup is surrounded by sandbags to prevent damage due

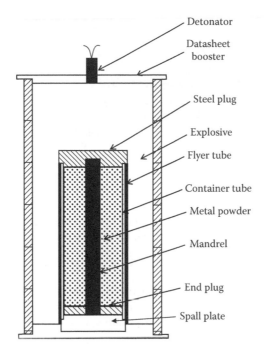

FIGURE 5.15
Explosive shock consolidation of MMCs. (Reprinted from Meyers, M.A. and Wang, S.L., *Acta Metall.*, 36, 1988, 925. With permission.)

to uncontrolled explosion. Once the explosive is detonated, the high pressure developed in the cylinder consolidates the powder mixture. This process is particularly useful for consolidating hard materials such as ceramics or ceramic fiber-/particulate-reinforced MMCs. This process is also very useful for the fabrication of intermetallic-based MMCs, such as TiB_2/Ti_3Al composites.

5.2.2.4 Roll Bonding and Coextrusion

Roll bonding is originally used to produce laminated composites from the sheets of reinforcement and matrix. This process with hot pressing can be used to make discontinuously reinforced MMCs. Figure 5.16 shows the schematic of roll bonding process to produce laminated MMCs. Co-extrusion process is used to produce niobium-based filamentary superconductors in copper matrix and high-T_c superconductors in silver matrix. There are two types in Nb-based superconductors: Nb–Ti and Nb_3Sn. Nb–Ti (~50–50) is a ductile material, whereas Nb_3Sn is an extremely brittle material. Hence, different processing approaches are used for making composites based on these materials. Rods of Nb–Ti are inserted in holes drilled in a block of copper, evacuated, sealed, and extruded in a series of steps interspersed with appropriate annealing treatments to obtain the composite superconductor of desired diameter. Nb_3Sn-based superconductors cannot be processed by the

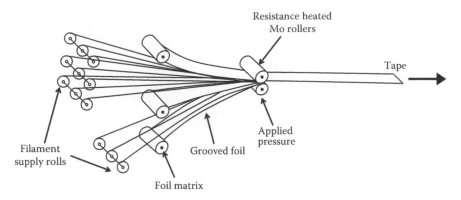

FIGURE 5.16
Roll bonding process to prepare a laminated MMC. (Reprinted from Schoutens, J.E., *Introduction to Metal Matrix Composite Materials*, Kaman Tempo, Santa Barbara, CA, 1982, p. 5. With permission.)

previously mentioned procedure, since Nb_3Sn is very brittle and these filaments will break during extrusion. Hence, a processing route called bronze route is used to make Nb_3Sn/Cu composite wires. In this process, pure niobium rods are inserted in holes drilled in bronze (Cu–13% Sn), evacuated, sealed, and subjected to drawing operations as in the case of Nb–Ti/Cu. The critical step in this route is the final heat treatment, which is normally carried out at ~700°C to drive out the tin from the bronze matrix to niobium for the formation of stoichiometric, superconducting Nb_3Sn.

The high-T_c ceramic superconductors are also brittle in nature, so it is desirable to form the ceramics only at the end of the process. The required metals (Y, Ca, Ba, Cu, and Ag) are melted together, spun into ribbon form, and pulverized to obtain a homogeneous alloy powder. This alloy powder is packed in a silver can, sealed, and extruded into a hexagonal rod. On controlled oxidation, the alloy is converted into a superconducting oxide. Another method of forming ceramic superconductor MMCs is the oxide powder in tube method. In this process, the oxide powders of appropriate composition (stoichiometry, purity, etc.) are packed inside a silver tube, degassed, and sealed. The sealed tube is subjected to swaging and drawing to form wires or subjected to rolling to form tapes. Intermediate and/or final heat treatments after deformation are given to form the desired phase and promote grain interconnectivity and proper crystallographic alignment of the oxide.

In a high-temperature superconductor with randomly oriented grains, the misoriented grains are the obstacles for the free flow of current. If the grains are aligned, then the flow of current will improve. YBCO superconducting material with aligned grains can be formed by epitaxial growth on a textured template of the substrate. A technique developed at Oak Ridge National Laboratory called RABiTS™ (rolling-assisted biaxial textured substrates) enables the formation of a substrate with a highly textured surface.

The process is quite complex. As a first step, a tape out of a metal, such as nickel, is prepared by special rolling and heat treatment. Next, a buffer layer is used to provide a chemical barrier between nickel and superconductor material while maintaining the texture. The buffer layer is usually a thin layer of palladium deposited using electron beam evaporation or sputtering. Metal oxide buffer layer, such as ceria- or yttrium-stabilized zirconia, is formed by pulsed-laser deposition. The high-T_c superconducting ceramic material is then deposited over the buffer layer by pulsed-laser deposition. This coating is highly textured and thus achieves the desired high super-conducting current density. This is the basic principle for making a second-generation high-temperature superconducting wire/tape.

5.2.3 Gaseous-State Processes

Several gaseous-state processes have been used in the fabrication of MMCs, and all these processes are relatively very slow. Among the vapor deposition processes, PVD processes are mainly used to produce MMCs. The three basic categories of PVD processes are based on evaporation, ion plating, and sputtering. Evaporation is generally carried out under high vacuum by means of resistive heating, electron beam, arc, high-energy radiation, and laser. Ion plating involves passing the vaporized metal through an argon flow gas discharge, and the ionized gas gets deposited into the substrate. In the sputtering techniques, a target made out of the coating material is bombarded by ions of a gas (usually Ar), and atoms are ejected from the target toward the substrate (fiber or preform). The main advantages of PVD techniques are versatility in the composition and microstructure of the coating, no chemical reaction/by-products during the process, superior bonding of the coating to the substrate, and excellent surface finish after coating. The disadvantages are that the processing equipment is complex and expensive, and it is a very slow process and hence may not be suitable for the production of large MMC parts.

The fastest PVD process involves thermal vaporization of the target metal in a relatively high vacuum. Continuous fiber-reinforced titanium can be fabricated using this process by passing the fiber through a chamber having a high partial vapor pressure of titanium, where condensation takes place so as to produce a relatively thick coating on the fiber. The metal vapor is produced by directing a high-power ($\sim 10\ \mu W$) electron beam onto the end of a solid feedstock. The main advantage of this technique is that a wide range of alloy compositions can be deposited. The differences in evaporation rate between different elements are compensated by changing the composition of the molten pool formed on the end of the bar. Titanium and aluminum alloys and Ti–Al intermetallic compounds have been deposited by this process. An interesting feature of this process is that there is little or no mechanical disturbance at the interfacial region. This is very important when the fibers have a diffusion barrier layer or a tailored surface chemistry.

This layer is often degraded by the droplet impacts in spray deposition or the frictional motion in diffusion bonding. The deposition rates in this PVD process are typically 5–10 μm min⁻¹. Composite fabrication is completed by assembling the coated fibers into the required shape and consolidating in a hot pressing or HIPing equipment. A very uniform fiber distribution can be obtained in this process with fiber contents up to about 80 vol.%. The fiber volume fraction can be accurately controlled by controlling the thickness of meal deposit.

Ion plating forms a denser deposit than evaporation, but at a slower deposition rate. This process has been used to deposit aluminum on C fibers. Sputter deposition is still slower, but has the advantage of being applicable to virtually all materials, even those with very low vapor pressure. It may not be a promising technique for matrix deposition, but may have the potential for thin coatings on fibers or depositing small quantities of elements with low vapor pressures.

PVD processes offer an extremely wide range of possibilities for fabricating nanocomposites with tailored chemistry, structure, and interface. Metal/ceramic nano-laminate systems exhibiting superior properties can be prepared by this process. Nanolayered Al/SiC composites have been prepared by RF magnetron sputter deposition (Deng et al. 2005). However, the removal of SiC from the substrate is more difficult because of strong chemical bonding, resulting in a lower deposition rate than aluminum deposition.

5.2.4 Deposition Techniques

Deposition techniques for MMC fabrication involve the coating of fibers with the required matrix material. The coated fibers are stacked and consolidated by applying pressure and temperature. Since the composite is made from identical units, the microstructure is more homogeneous than that of cast composites. However, it is a very time-consuming process. The other advantages of this process are the interfacial bonding is easily controllable, and diffusion barrier coatings and compliant coatings can be formed prior to matrix deposition.

Thin, monolayer tapes can be produced by combining filament winding with deposition process. These products are easier to handle and arrange into structural shapes than other precursor forms. Unidirectional or angle-plied fiber composites can be easily fabricated using these tapes.

There are many deposition techniques available and some of the techniques applicable to MMCs are immersion plating, electroplating, CVD, and PVD. In the immersion plating process, fiber tows are continuously passed through a bath containing molten metal or powder slurry. Electroplating involves coating from a solution containing the ion of desired matrix material under the influence of electric current. Fiber is wound on a conducting mandrel, which serves as the cathode, and immersed in the plating bath with an anode of the desired matrix material. The anode dissolves while passing current and the metallic cations deposit over the cathode (i.e., in between fibers). The main advantage of this process is that the deposition is carried out at moderate temperatures, which will not damage the fibers. However, there are some problems with this technique, which include the formation of voids between fibers and fiber layers, poor adhesion of deposit to the fibers, and the possibility to deposit only a limited number of alloy matrices. In CVD, a vaporized component decomposes or reacts with another gas and deposit into the substrate. The reaction/decomposition to form the matrix usually happens at elevated temperatures. Depending on the deposition conditions, the deposit can grow into amorphous, single-crystal, or poly-crystalline form. When the matrix material is deposited into a reinforcement preform, such as woven mat/3-D fiber preform, the process is called chemi-cal vapor infiltration. The details of PVD processes are already explained in the previous section.

5.3 Secondary Processing

A number of secondary processes can be used to MMC material for eliminating porosity, generating fiber alignment, and/or forming into a required shape. These processes involve high temperatures and large-strain deformations.

5.3.1 Extrusion and Drawing

Extrusion process has been used extensively as a means of secondary pro-cessing to reduce porosity in MMCs. It is particularly advantageous because some shear forces are created during the process, which results in the break-age of reinforcement clusters as well as the fracture of oxide skin on the unbonded matrix particles leading to enhanced bonding between reinforce-ment and matrix. This process is primarily used to consolidate composites with discontinuous reinforcement in order to minimize reinforcement frac-ture due to the large strains involved. The fracture of short fibers or par-ticles often takes place even in discontinuously reinforced MMCs, which can adversely affect the properties of the composites. Extrusion may be carried out for discontinuously reinforced MMCs produced by squeeze infiltration or powder blending. Extrusion pressures required are generally higher than that required for unreinforced material. Fiber alignment may take place dur-ing extrusion, but at the expense of progressive fiber fragmentation. The degree of fiber fracture decreases with increasing temperature and decreas-ing total strain rate. The most effective way to minimize the peak strain rate is to use the dies with tapered or streamlined shapes, and these shapes are now widely employed for MMC extrusion.

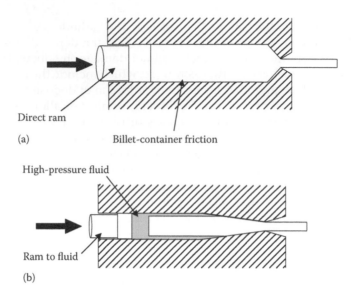

FIGURE 5.17
Extrusion processes for MMCs: (a) conventional extrusion and (b) hydrostatic extrusion. (Reprinted from Chawla, N. and Chawla, K.K., *Metal Matrix Composites*, Springer-Verlag, Inc., New York, 2006, p. 91. With permission.)

In conventional extrusion process, die friction leads to higher deformation stresses at the metal/die interface, and this effect is more pronounced due to the presence of particles. The large shear stresses present at the metal/die interface result in material discontinuities. This can result in ragged edges, known as "Christmas tree" structure. This effect is more pronounced with an increase in reinforcement content because of the increase in die friction. The problem of die friction can be minimized by carrying out the extrusion process inside a high-pressure fluid. This results in a close to hydrostatic stress state and minimal friction in the billet because the contact area between material and die is minimized. The schematic of these two types of extrusion is shown in Figure 5.17.

Several microstructural changes take place in the composites during extrusion, which include the alignment of reinforcement along the extrusion axis, reinforcement fracture, and refinement and recrystallization of matrix grains. The degree of alignment decreases with increasing volume fraction of reinforcement. The larger the volume fraction of reinforcement, the lower will be the mean free path. This reduced distance between the reinforcements restricts the rotation and alignment. The alignment of reinforcement has a profound influence on Young's modulus and tensile strength of composites, and it leads to anisotropic behavior. Extrusion-induced reinforcement fracture can reduce the strength of the composite significantly; in some cases the strength will be lower than that of unreinforced alloy.

The incorporation of reinforcement in a metal matrix also has an influence on the matrix grain size after extrusion. A larger degree of plastic flow is required to deform the matrix around the hard reinforcement particles, and this result in an overall refinement of matrix grain size. Hence, the grain size in the composite is inhomogeneous, being smaller near the particle/matrix interface and larger far away from the particles. The larger degree of shear produced in between particles and matrix is conducive for the formation of strong mechanical bond between the particle and matrix. A strong bond is highly desirable for the strengthening of MMCs, because it maximizes the degree of load transfer from the matrix to the reinforcement.

During hot extrusion, the reinforcement fibers/particles also act as nucleating sites for recrystallization of new matrix grains. This is another reason for the presence of finer-grain size along the reinforcement/matrix interface. However, the grains at the interface are randomly oriented, due to this dynamic recrystallization.

The two-stage, conventional P/M process followed by extrusion is an expensive process. While this process has been used to fabricate composites for very demanding applications, such as aerospace applications, it is not suitable for applications, where low cost and high volume are also important as performance.

The extrusion process can be carried out while the matrix is in a partially liquid state, and this reduces fiber fracture substantially. The problem of defects induced by the presence of liquid can be largely eliminated by the use of cladding material, which will remain solid during extrusion. The main difficulties in this process are the selection of suitable die materials, which should have high wear resistance at the extrusion temperature, and the control of heat flow during extrusion so as to maintain the required temperature field. Fiber strength also determines the final aspect ratio of fibers, and, for example, whiskers normally have higher values than staple fibers extruded under the same conditions.

Drawing involves a rather similar strain field as extrusion, but the stress state in the process zone has a smaller compressive component. It is normally carried out at a lower temperature than extrusion. The lower temperature may lead to the formation of internal cavities during drawing, especially when the interfacial bond strength is low. On the other hand, superior surface finish is often obtained.

5.3.2 Rolling, Forging, and Hot Isostatic Pressing

In general, the microstructural changes observed in MMCs after rolling and forging are on similar lines to those outlined for extrusion. However, high strains are generally imposed relatively quickly in these processes and hence can cause damages, such as cavitation, fiber fracture, and macroscopic cracking, particularly at low temperature. Rolling in particular involves high localized strain rates with significant reduction in temperature caused by contact with cold rollers. Hence, this process is generally unsuitable for consolidation operations. Forging is another common secondary processing technique used to remove defects in MMCs and/or to shape the MMC produced by some other processing methods. This technique is restricted to MMCs with discontinuous reinforcement. Forging operation can be more easily carried out at relatively low strain rates and elevated temperatures. At very high temperatures, the possibility of matrix liquefaction can quickly lead to the formation of macroscopic defects such as hot tearing or hot shortness.

In the sinter-forging process, a mixture of reinforcement and matrix powder is cold compacted, sintered, and forged to near-full density. Since the forging process can produce a near-net-shaped material, machining operations and material wastage are minimized.

The microstructure of forged composites exhibits some preferential alignment of reinforcement perpendicular to the forging direction. However, the alignment is not significant as that produced by extrusion, since a much larger amount of plastic deformation is induced during extrusion. The composites subjected to forging have comparable tensile and fatigue properties to those of composites produced by extrusion. The microstructure of matrix grains is also affected by forging. The grains are flattened like pancake, with the major axis of the grains being perpendicular to the forging axis.

HIPing generates symmetrical stresses from all directions, and so is unlikely to give rise to either microstructural or macroscopic defect. It is a more suitable method for removing residual pores, including surface-connected pores when the part is encapsulated in a can. However, it is difficult to remove residual porosity in the regions of reinforcement particle clusters. Such clusters cannot be dispersed during HIPing because of the absence of any shear.

5.3.3 Superplastic Processing and Sheet Forming

There has been considerable interest in the superplastic processing of MMCs, as this offers promise to carry out forming operations without any microstructural damage. A superplastic aluminum alloy can be used for making long-fiber-reinforced MMCs with low porosity by diffusion bonding. Even with the good flow characteristics of matrix, the composite is not expected to show superplasticity because it is presumed that the fibers inhibit flow. However, it is now well established that the presence of reinforcement can promote superplastic behavior by the generation of internal stresses during thermal cycling. This type of superplastic behavior is different from the conventional superplasticity that depends on fine-grain size and high-strain-rate sensitivity exponent. The superplasticity in composites is termed as internal stress superplasticity or mismatch-induced superplasticity. It was found that a thin sheet of MMC (SiC_p (10 vol.%)/6061 Al) subjected to thermal cycling between 125°C and 450°C could be deformed to a strain of 120%

while applying a pressure of about 0.8 MPa (~120 psi). Such superplasticity offers considerable promise for the forming of particulate and short-fiber MMCs into complex and convoluted shapes.

One of the most interesting phenomena observed in discontinuously reinforced MMCs after deformation processing is their ability to undergo extensive straining in excess of 300% during subsequent high-temperature deformation. This superplastic behavior is normally observed at temperatures near and in some cases slightly above matrix solidus and at strain rates, two to three orders of magnitude higher than used in aluminum alloys. Based on the detailed microstructural analysis, it has been suggested that this behavior is due to the combined action of grain boundary and interface sliding and enhanced relaxation of localized stresses in the presence of liquid films at reinforcement–matrix interfaces (Clyne 2000).

5.4 Machining and Joining of MMCs

The development of tertiary processing operations, such as machining and joining, for MMCs is still in its infancy stage. However, it is attracting increasing interest as more applications are being explored. In this section a brief summary of factors to be considered while using such processes to MMCs is given.

5.4.1 Mechanical Operations

Conventional cutting, turning, milling, and grinding operations can also be used on MMCs, but there is a problem of excessive tool wear because of the presence of hard ceramic reinforcements. In general, such a problem becomes more severe with increasing reinforcement volume fraction and size. At higher volume fractions the tool encounters more hard material for the same distance covered, while larger particles or fibers more strongly resist pullout and hence stress the tooling more. Diamond-tipped or diamond-impregnated tools are usually needed for monofilament-reinforced MMCs, while cemented carbide or even high-speed steel tooling may be adequate for short-fiber and particulate MMCs. The strength of the reinforcement also has an influence on machining: SiC whisker-reinforced aluminum composite has been found to be more difficult to machine than other aluminum composites containing weaker fibers (Matsubara et al. 1987). For most MMCs, the best machining is obtained with a sharp tooling, an appropriate cutting speed, copious cooling, and a high material feed rate. Diamond tools have been found to give better performance than cemented carbide and ceramic tools and allow relatively high machining speeds.

5.4.1.1 Electrical Cutting

Several cutting processes involving an electrical field between the tool and workpiece can be used for MMCs, since they are electrically conducting materials. Electrochemical machining involves the removal of material by anodic dissolution using a shaped cathode. An electrolyte is flushed through the cathode/workpiece gap, and this carries away the undissolved debris such as fibers. Electrochemical machining involves no mechanical contact between electrode and workpiece, and hence very little damage is induced in the remaining material. There would be difficulties in cutting and removing long fibers, but this could be done for certain geometries by combining electrochemical action with mechanical grinding, using a moving cathode with abrasives.

Electrical discharge machining (also known as spark erosion) involves cutting with a moving wire electrode immersed in a stream of dielectric fluid. Material removal results from the local temperatures generated by spark and liquid pressure pulses generated on the workpiece surface. In this process also the workpiece does not come into mechanical contact with the wire. As cutting is by thermal means, ceramic fibers can also be cut quite readily. The process, however, is rather slow and the damage caused in this process is relatively severe.

5.4.1.2 High-Energy Beam Cutting

MMCs can be cut using various high-energy beams, such as lasers and electron beams. For cutting along the fiber axis or for discontinuous fiber composites, the fibers need not be cut; melting or volatilization of the matrix is all that required and that is achieved by directing high-energy beams. When the power of the beam is sufficiently high, ceramic fibers can also be melted/volatilized or more probably fractured under the various mechanical and thermal stresses induced by the beam. A high level of microstructural damage may be induced during this cutting process, in the form of cracking, interfacial debonding, and heat-affected microstructures. High-energy laser cutting is also being used to cut engineering ceramics and ceramic matrix composites.

5.4.1.3 Fluid Jet Cutting

Low damage levels with high cutting rates can be realized by using a concentrated jet of high-velocity fluid, usually water, containing abrasive particles. This technique has been found to be useful to both continuous and discontinuous fiber-reinforced MMCs. As there is no significant temperature rise, damage is only due to the mechanical action and it is localized close to the machined area. The fluid jet cutting technique offers the best combination of cleanliness and maximum cutting speed. A related process is abrasive flow machining, which is mainly used to produce a good surface finish. In this process, a gel containing abrasive particles flows under pressure over workpiece surface. It is a very useful technique to produce polished surfaces on many composites, particularly on components of complex shape.

5.4.2 Joining Processes

Conventional welding processes may not be very useful for joining MMCs, particularly for fiber-reinforced MMCs, because the distribution of reinforcement will tend to be strongly disturbed in the fusion zone. Even in particulate MMCs, pushing effect in the welding zone will tend to generate marked inhomogeneities. Moreover, these problems of reinforcement distribution cannot be reverted by post-welding heat treatments in the manner that is often employed to modify the fusion and heat-affected zone microstructures in unreinforced materials. This inhomogeneity in reinforcement distribution may make the joint area very weak, and that leads to strain localization and failure. Therefore, it is better to consider processes, which produce relatively narrow joints, such as brazing, diffusion bonding, adhesive bonding, friction welding, laser or electron beam welding, and mechanical fastening.

Diffusion bonding is commonly used to join Ti alloys, and the process is suitable for the joining of Ti-based MMCs also, provided the temperature is controlled so as to limit the matrix/reinforcement interfacial reaction. There are some problems in the diffusion bonding of Al alloys, but Al-based MMCs can be diffusion bonded with the use of suitable interlayers. Friction welding is a more suitable process for Al–SiC_p composites, with a few problems in terms of particulate distribution and of post-welding control of the matrix microstructure. There is a good scope for controlling constitutional effects during joining. Tailoring the thermal expansion coefficient of MMCs by controlling the fiber content and orientation is important in making joints, especially between dissimilar materials. Graded composition MMCs may avoid the problem of thermal expansion mismatch.

5.5 Properties of Metal Matrix Composites

As in polymer matrix composites, the properties of MMCs are also affected by the type and volume fraction of reinforcement, properties of reinforcement and matrix materials, interface, and matrix microstructure. In the case of fiber reinforcement, the orientation of fiber will also affect the properties. The important physical and mechanical properties of MMCs are discussed in this section.

Improper SiC distribution is the main reason for the low strength of the SiC-reinforced composites. The voids present in between the SiC particle clusters act as stress raisers and are potential sites for the crack initiation. Liquid metal is not able to penetrate the voids. Hence, these voids are responsible for inferior mechanical properties of SiC-reinforced MMCs. The scanning electron microscope fractograph of a fractured SiC–aluminum alloy composite is shown in Figure 5.18. The presence of voids near the particle clusters can be clearly seen in the micrograph. Another MMC made with fly ash also

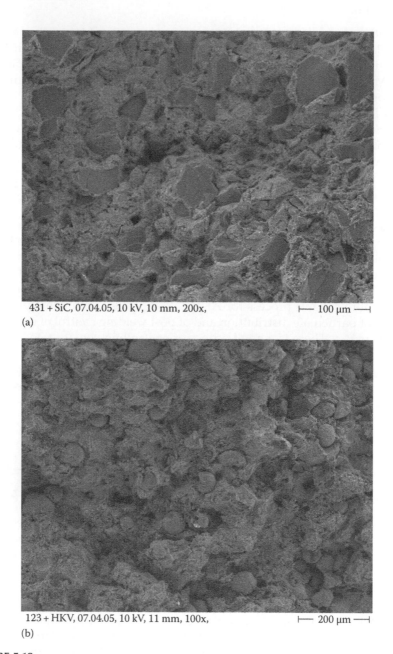

431 + SiC, 07.04.05, 10 kV, 10 mm, 200x,   ⊢— 100 μm —⊣
(a)

123 + HKV, 07.04.05, 10 kV, 11 mm, 100x,   ⊢— 200 μm —⊣
(b)

FIGURE 5.18
Microstructure of MMCs prepared by powder metallurgy technique: (a) SiC/Al alloy and
(b) fly ash/Al alloy. (Courtesy of Dr. B. Kieback and Dr. T. Schubert, TU-Dresden, Germany.)

showed the presence of unbonded particles (Abhishek 2005). Thus, proper mixing procedures must be adopted to ensure a thorough mixing of particles, and better bonding between particles and matrix should be ensured to realize better mechanical properties in MMCs.

5.5.1 Modulus

Unidirectional continuous fiber-reinforced MMCs show a linear increase in Young's modulus in the fiber direction on increasing the fiber volume fraction. The increase in Young's modulus in the longitudinal direction is agreement with the rule of mixtures. However, the modulus increase in the transverse direction is very low, compared to that in the longitudinal direction (Figure 5.19a). Particulate MMCs, such as SiC particle-reinforced aluminum, can offer a 50%–100% increase in modulus value over to that of unreinforced aluminum. This aluminum MMC has a modulus value equivalent to that of titanium but with 33% less density. Unlike the fiber-reinforced composites, the modulus enhancement in particulate composites is reasonably isotropic. Typical properties of some common fiber-reinforced MMCs are given in Table 5.3.

5.5.2 Strength

Prediction of strength of a MMC is more complicated than the prediction of modulus, since the strength of the matrix will vary due to the change in microstructure during processing. According to the rule of mixtures, the strength of the composite is a volume-weighted average of the strength values of the fiber and the matrix. In most MMCs, the fiber remains essentially elastic up to the point of fracture; hence the fiber strength in the composite is the same as the original strength of the fiber. However, the strength of the matrix in the composite will not be the same as that determined from an unreinforced matrix. As mentioned earlier, the metal matrix may undergo several microstructural alterations during processing, and, consequently, there will be a change in mechanical properties of the matrix. The extent of change is very difficult to determine.

In general, the coefficients of thermal expansion of ceramic reinforcements are lower than that of most metallic matrices, and hence thermal stresses will be generated in the two constituents on exposure to high temperatures during processing or service. A series of events can take place in the composite due to these thermal stresses, such as

1. Plastic deformation of the ductile metal matrix by slip or twining
2. Crack formation and failure of the brittle fiber
3. Failure of the fiber/matrix interface

It is very important to note that the matrix in fiber-reinforced composites is not merely a kind of binder to hold the fibers together. The characteristics of the

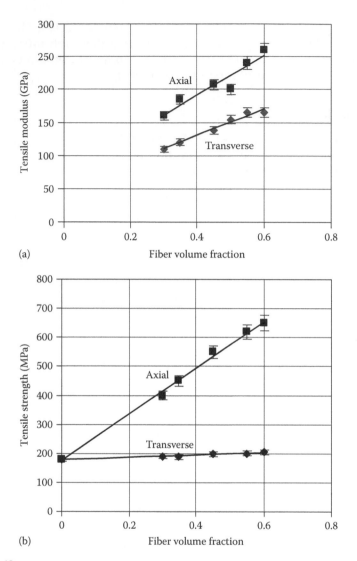

(a)

(b)

FIGURE 5.19
Properties of Al$_2$O$_3$–Al–Li composites as a function of fiber volume fraction: (a) axial and transverse Young's modulus and (b) axial and transverse ultimate tensile strength. (Reprinted from Champion, A.R. et al., in Norton, B.R., Signorelli, R.A, and Street, K.N. (eds.), *Proc. Int. Conf. Compos. Mater. (ICCM-2)*, TMS-AIME, New York, 1978, p. 883. With permission.)

matrix, as modified by the introduction of fibers, must be evaluated to obtain an optimum set of properties for the composite. The nature of fibers, fiber diameter and distribution, processing temperature, and solidification parameters influence the final matrix microstructure. Porosity is a common defect in cast MMCs, which is mainly formed due to the shrinkage of the metallic matrix during solidification. At higher fiber volume fractions, the large-scale

TABLE 5.3

Typical Properties of Particulate and Fiber-Reinforced Metal Matrix Composites

Composite	V_f (%)	Density (g cm⁻³)	σ_{max} (MPa)	E (GPa)
B$_f$/Al alloy (6061)	60	2.65	1490 (L)	214 (L)
			138 (T)	138 (T)
SiC$_p$/Al alloy (6061)	20	2.8	550	120
C$_f$ (T-300)/Al alloy (6061)	35–40	2.32–2.37	1030–1275 (L)	110–140 (L)
C$_f$ (GY-70)/Al alloy (201)	37.5	2.36	790 (L)	207 (L)
Al$_2$O$_3$(f)/Al–Li alloy	60	3.36	690 (L)	260 (L)
			170–210 (T)	150 (T)
SiC$_f$/Ti alloy	35	3.86	820 (L)	225 (L)
			380 (T)	—
SiC$_f$ (SCS-6)/Ti alloy	35	3.86	1455 (L)	240 (L)
			340 (T)	—
SiC/Mg alloy	38	1.8	510	—

Source: Adapted from Mallick, P.K., *Fiber-Reinforced Composites*, 3rd edn., CRC Press, Boca Raton, FL, 2008, p. 594.

movement of semisolid metal may not be possible. That can also lead to the formation of porosity. More importantly, the microstructure of metallic matrix in a fiber-reinforced composite will differ significantly from that of the unreinforced metal due to the modification of solidification parameters by the presence of fibers. In any case the improvement of strength by the incorporation of fibers is insignificant in the transverse direction (Figure 5.19b).

The influence of fibers on the solidification of matrix alloy has been studied extensively. It was found that the dendritic morphology of matrix was controlled by the fiber distribution in a SiC/Al–4.5% Cu composite, and the second phase, θ, appeared preferentially at the fiber/matrix interface or in the narrow interfiber spaces (Mortensen et al. 1986). In another study on SiC/Al composites, it was found that the shrinkage cavities present in the matrix were the predominant crack initiation sites and also found a wavy interface structure indicating some mechanical bonding (Kohyama et al. 1985). In the SiC/Mg (AZ 19C) system, it was found that a magnesium-rich interfacial layer acted as the fracture-initiating site.

Superior high-temperature mechanical properties have been observed in a number of MMC systems. For example, SiC whiskers improve the high-temperature mechanical properties of aluminum alloys considerably (Phillips 1978). The elastic modulus, yield stress, and ultimate tensile strength of SiC$_w$ (21 vol.%)/2024 Al alloy are better than those of unreinforced aluminum alloy up to about 250°C (Figure 5.20). In another study on Nicalon SiC/aluminum composites, it was found that the fracture behavior changed at high temperature (Yajima et al. 1981). The sample fractured at room temperature showed a more or less planar fracture, whereas there was a loss of adhesion between the fiber and the matrix and that led to fiber pullout.

High-modulus carbon fibers impart a very high stiffness with a very low thermal expansion to the aluminum matrix. However, carbon/aluminum composites are susceptible to galvanic corrosion between carbon (cathodic) and aluminum (anodic). Hence, galvanic corrosion will be a serious problem in aluminum and carbon/aluminum composite joints. A common solution is

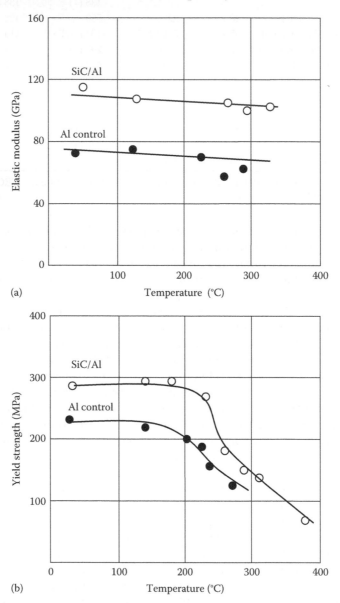

(a)

(b)

FIGURE 5.20

Comparison of high-temperature properties of SiC$_w$/Al composites and aluminum: (a) elastic modulus, (b) yield strength.

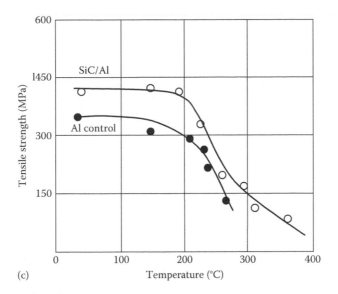

(c)

FIGURE 5.20 (continued)
(c) ultimate tensile strength. (Reprinted from Phillips, W.L., in Norton, B.R., Signorelli, R.A, and Street, K.N. (eds.), *Proc. Int. Conf. Compos. Mater. (ICCM-2)*, TMS-AIME, New York, 1978, p. 567. With permission.)

to introduce an insulating layer of glass between the aluminum and carbon/aluminum composite. Any welding or joining involving localized heating could also be a problem, since aluminum carbide may form during this process, which will be detrimental to the mechanical and corrosion properties.

Among the factors controlling the strength of MMCs, the type of matrix is the most important factor. The stress–strain behavior of SiC_w/Al composites made with different aluminum alloys is shown in Figure 5.21. The MMCs made with higher-strength aluminum alloys show higher strength but lower ductility. The MMC made with 6061 Al alloy has good strength and higher fracture strain (Figure 5.21).

As mentioned earlier, in situ composites can be made by directional solidification of a eutectic alloy. The strength of such an in situ MMC is given by a relationship similar to the Hall–Petch relationship:

$$\sigma_c = \sigma_o + k\lambda^{-1/2} \tag{5.18}$$

where
σ_o is a friction stress term
k is a material constant
λ is the spacing between reinforcements (rods or lamellae)

The inter-reinforcement spacing can be easily controlled by controlling the solidification rate and hence the strength.

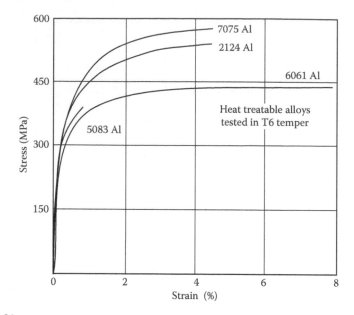

FIGURE 5.21
Stress–strain behavior of different aluminum alloy composites prepared with 20 v/o SiC$_w$. (Reprinted from McDanels, D.L., *Metall. Trans. A* 16, 1105, 1985. With permission.)

Compressive strength of MMCs is generally higher than that of PMCs, because of the higher modulus of metal matrices. Deve (1997) observed a compressive strength of ≥4 GPa for Nextel 610 alumina fiber- (55–65 vol.%) reinforced aluminum alloy composite.

There is some controversy on the strengthening mechanisms operating in particulate MMCs. In the absence of shear lag strengthening mechanism (i.e., load transfer mechanism), which would be operational in the long-fiber-reinforced composites, the following strengthening mechanisms have been proposed for particulate MMCs (Arsenault and Shi 1986, Shi and Arsenault 1991, 1993):

Orowan strengthening: It is given as

$$\sigma_o = \frac{Gb}{l} \tag{5.19}$$

where

G and b are the shear modulus and Burgers vector of the matrix, respectively
l is the interparticle spacing

Strengthening due to grain refinement: A Hall–Petch-type relationship gives the strengthening due to this mechanism:

$$\sigma_{gr} = kd^{-1/2} \tag{5.20}$$

where d is the size of the matrix grains or subgrains. Thus, a grain diameter of less than 10 μm will give significant strengthening, while $d < 1$ μm can give very good strengthening effect. In most MMCs, the presence of particulates restricts the matrix grain growth, resulting in strengthening.

Quench strengthening: Thermal strain generated in the matrix is given as $e_m = \Delta\alpha\Delta T$, where $\Delta\alpha$ is the difference in CTE between reinforcement and matrix and ΔT is the temperature difference. The dislocation density resulting from the CTE mismatch can be theoretically calculated as

$$\rho_{CTE} = \frac{Ae_m V_p}{b(1-V_p)d} \tag{5.21}$$

where
  A is a geometric constant related to particle
  V_p is the particle volume fraction

The corresponding strengthening is given by

$$\sigma_{qs} = \alpha Gb(\rho_{CTE})^{1/2} \tag{5.22}$$

where α is a constant.

Work hardening of the matrix (σ_{WH}): The presence of particles affects the matrix work-hardening rate. Thus, the total strength of composite from all these contributions can be written as

$$\sigma_c = \sigma_o + \sigma_{gr} + \sigma_{qs} + \sigma_{WH} + \sigma_m \tag{5.23}$$

where σ_m is the strength of the matrix without the particles.

5.5.3 Toughness

Toughness is a measure of energy absorbed in the process of fracture. The toughness of MMCs depends on the following factors:

- The type of matrix and microstructure
- The reinforcement type, content, size, and orientation
- Processing parameters, as they affect the microstructure (e.g., distribution of reinforcement, porosity, matrix grain size, and segregation of alloying elements in the matrix)

For a given volume fraction of fiber, the larger the diameter of the fiber, the tougher will be the composite. This is because of the presence of a relatively thicker metallic matrix in the interfiber region of the large-diameter fibers. The thick matrix can undergo more plastic deformation and thus contribute to the toughness.

The crack initiation and propagation is easier in unidirectional fiber-reinforced composites than in the unreinforced alloy. However, the crack propagation is very difficult in the braided fiber composites due to extensive matrix deformation, fiber bundle debonding, and fiber pullout. The typical K_{IC} values for particulate MMCs are in the range of 15–30 MPa m$^{1/2}$, whereas for the short-fiber- or whisker-reinforced composites, the values are in the range of 5–10 MPa m$^{1/2}$.

The reasons for the low toughness values of MMCs are the formation of intermetallics and inhomogeneous internal stress and particle or whisker distribution. Improvements in fracture toughness of SiC$_w$/Al composites equivalent to that of 7075 aluminum alloy have been obtained by using a cleaner matrix powder, better mixing, and increased mechanical working during fabrication (McDanels 1985).

In a MMC, the work of fracture is mainly due to the work done during plastic deformation of the matrix. The work of fracture of MMC is proportional to $d(V_m/V_f)^2$ (Cooper and Kelly 1967), where d is the fiber diameter. In the case of large-diameter fibers, the advancing crack will have to pass through a greater plastic zone of the matrix, and that will result in a larger work of fracture.

Fiber pullout also increases the work of fracture by causing a large deformation of composite before fracture. The work of fracture due to fiber pullout is proportional to the ratio d/τ_i, where d is the fiber diameter and τ_i is the interface shear strength. The work of fracture can be increased by increasing the fiber diameter, because of the increase in spacing between the defects (Cooper 1970, Kelly 1971). In the case of short-fiber-reinforced composites, the work of fracture due to fiber pullout increases with the fiber length and reaches a maximum at l_c.

5.5.4 Thermal Characteristics

In general, ceramic reinforcements (fibers, whiskers, or particles) have lower CTE expansion than most metallic matrices. Hence, thermal stresses will be generated in both the components, when the composite is exposed to a temperature change. It was found that the dislocation density in a single-crystal copper matrix was much higher near the tungsten fibers than that far away from the fiber. The enhanced dislocation density near the fiber arises because of the plastic deformation in response to the thermal stresses. The dislocation density gradient depends on the interfiber spacing, and it will decrease with a decrease in the interfiber spacing. The existence of dislocation density gradient has been found by a number of researchers, both in fibrous and particulate MMCs.

Thermal expansion mismatch is difficult to avoid in any composite. However, one can control the overall thermal expansion characteristics of a composite by controlling the relative amount of reinforcement and matrix and the distribution of the reinforcement in the matrix. Prediction of CTE is much complicated because of the complex structure of the composites containing different types of reinforcement (particles, whiskers, or fibers), interface, and the matrix plastic deformation due to internal thermal stresses.

The thermal expansion characteristics of isotropic composites are a function of volume fraction, size, and morphology of the reinforcement. Phenomena such as hysteretic deformation during thermal cycling, microcracking, and fiber intrusion on the surface lead to increase in surface roughness.

The CTE of a material is not an absolute physical constant, but it varies with temperature. The change is significant in MMCs, especially when the matrix undergoes plastic deformation as a result of thermal stresses. In addition, there are many microstructural factors that come into play for the modification of CTE. For example, the reinforcement continuity can play an important role on the variation of CTE. From a study on the thermal expansion of SiC/Al composite, it was found that a high matrix yield strength, a high interfacial strength, and a convex reinforcement shape would minimize variations in CTE during any thermal excursion. In a composite foam, the presence of voids minimizes the average CTE of the composite, but hysteretic thermal expansion would be present.

5.5.5 Aging Characteristics

Several matrix alloys used in MMCs have precipitation hardening characteristics, that is, such alloys can be hardened by a suitable heat treatment. It has been shown by many researchers that the microstructure of metallic matrix is modified by the presence of reinforcement, and consequently the standard aging treatment given for an unreinforced alloy may not be valid for the MMC. The particle- or whisker-type ceramic reinforcements are unaffected by the aging treatment; however, they can significantly affect the precipitation hardening behavior of a matrix alloy. As mentioned earlier, a higher dislocation density is produced in the matrix metal or alloy, in the presence of reinforcements, due to thermal mismatch. It is expected that the high dislocation density will affect the precipitation kinetics of a precipitation-hardenable matrix, such as 2xxx series aluminum alloy. Indeed, faster precipitation kinetics has been observed in the presence of reinforcement (Suresh and Chawla 1993). Hence, one should not use the standard heat treatments applicable for the unreinforced alloy to the MMC made with the same alloy.

5.5.6 Fatigue

Fatigue is the failure of a component under cyclic loading. Many MMC components are subjected to cyclic loading, for example, automobile components. It was found that the unbonded SiC_w and non-SiC intermetallics were the fatigue crack initiation sites in a SiC_w/2124 Al alloy composite subjected to cyclic loading (Williams and Fine 1985a). Thus, reducing the clustering of SiC_w and the number and size of the intermetallic particles can result in increased fatigue life. An interesting phenomenon has been observed in a MMC subjected to cyclic loading, which is the uniform distribution of dislocations in the matrix after fatigue testing (Williams and Fine 1985b).

Fatigue behavior of B/6061Al, alumina/Al, and alumina/Mg has been investigated by Champion et al. (1978). It was found that these unidirectional fiber-reinforced composites showed better fatigue properties than the matrix when loaded parallel to the fibers. The fatigue-to-tensile strength ratio of the composite was about 0.77 at 10^7 cycles, which is almost double to that of the matrix. Gouda et al. (1981) observed early crack initiation during the fatigue testing of unidirectional boron fiber-reinforced aluminum composites. These cracks grew along the fiber/matrix interface, which accounted for the major portion of the fatigue life, as would be the case in a composite with a high fiber-to-matrix strength ratio. In composites with a low fiber-to-matrix strength ratio also, the crack propagation may take up a major portion of fatigue life, but the crack would be expected to grow through the fibers resulting in poor fatigue resistance. McGuire and Harris (1974) investigated the fatigue behavior of W/Al–4 Cu under tension–compression cyclic loading. It was found that on increasing the fiber volume fraction from 0 to 0.24 resulted in increase in the fatigue resistance. This is directly related to the strength increase by increasing the fiber volume fraction. Due to the highly anisotropic nature of the fiber-reinforced composites in general, the fatigue strength of the composite will be expected to decrease with increasing angle between the fiber axis and the stress axis. This has been confirmed by the fatigue studies on alumina fiber-reinforced magnesium composites (Hack et al. 1987, Page et al., 1987). It was found that the fatigue behavior followed the static tensile behavior. Increasing the fiber volume fraction resulted in enhanced fatigue life in the axial direction, but there was little or no improvement in the transverse direction. In the transverse direction, fatigue crack initiation and propagation occurred primarily through the magnesium matrix. Hence, alloy additions to increase the strength of the matrix and interface were tried. The alloy addition improved the off-axis properties, but decreased the axial properties because of decreased fiber strength by the alloy additions.

5.5.6.1 *Fatigue of Composites under Cyclic Compressive Loading*

Under monotonic compression of unidirectional fiber-reinforced composites in the fiber direction, a fiber-kinking mechanism leads to failure of the composites. The failure initiates at a weak spot, usually at a point where the fiber/matrix bonding is weak. This initial failure will, in turn, destabilize the neighboring fiber bonding and cause more kink failure. Eventually, various kink failure sites can coalesce and the situation is similar to transverse tensile loading, and that can cause longitudinal splitting. Similarly, the fatigue resistance of fiber-reinforced composites is also poor in compression than in tension. Pruitt and Suresh (1992) found a single mode I crack perpendicular to the fiber axis (in addition to kink band). The origin of a mode I crack was the presence of residual tensile stresses formed as a result of a variety of permanent damages involving fiber, matrix, and interface. Defects to simulate a delamination are introduced in composites by the following means:

1. Introducing single circular inserts of different diameters at different layer interfaces
2. Introducing a hole in the laminate
3. Introducing defects by controlled, low-velocity impacts

5.5.6.2 Fatigue of Particle- and Whisker-Reinforced Composites

Silicon carbide- or alumina particle-reinforced aluminum alloys can have improved fatigue properties compared to unreinforced aluminum alloys, which can make these composites useful in applications where aluminum alloys may not be suitable. Such an improvement in stress-controlled fatigue or high-cycle fatigue is attributed to the higher stiffness of the composite. However, the low-cycle fatigue behavior of the composite is inferior to that of the unreinforced alloy, and this is attributed to the lower ductility of the composite.

Particle or short fibers can provide easy crack initiation sites during fatigue loading. The fatigue property of a composite varies depending on the volume fraction of the reinforcement, shape, size, and most importantly the reinforcement/matrix bond strength. For example, fatigue crack initiation was observed at the poles of SiC_w in 2124 aluminum alloy (Williams and Fine 1987). The arrest of short cracks at the whisker/matrix interfaces was also observed. Aluminum alloy matrix composites, especially those made by melt processing, often have particles other than SiC, such as $CuAl_2$, $(Fe, Mn)_3 SiAl_{12}$, and $Cu_2Mg_5Si_6Al_5$ (Kumai and Knott 1991). The phenomenon of particle pushing ahead of the solidification front results in the decoration of cell boundaries of aluminum alloy matrix by SiC and other particles. Levin et al. (1989) observed superior fatigue crack growth resistance in SiC (15 vol.%)/6061 Al composite compared to 6061 Al alloy, which was attributed to a slower crack growth rate in the composite because of crack deflection by the SiC particles. It is necessary to choose the optimum particle size and volume fraction, together with a clean matrix alloy to get improved fatigue characteristics. Shang et al. (1988) investigated the effect of particle size on fatigue crack propagation as a function of cyclic stress intensity in a SiC_p/Al composite. It was observed that the initial fatigue crack growth resistance of the composite was less than that of the unreinforced alloy in the composite with fine particles. In the composite with coarse particles, the threshold intensity value was comparable to the unreinforced alloy, while at very high values of the cyclic stress intensity, the fatigue crack growth in the composite was less than that of the unreinforced alloy.

5.5.6.3 Thermal Fatigue

There is a very fundamental physical incompatibility exists between the reinforcement and the matrix, that is, the difference in their thermal expansion coefficients. Thermal stresses are produced in composite materials due to this difference in CTEs. It should be noted that thermal stresses in composites will arise even if the temperature change is uniform throughout the volume of the composite. Such thermal stresses in composites can arise during cooling from high fabrication, annealing, or curing temperatures or during any temperature excursions during service. The magnitude of thermal stresses produced in composites is proportional to $\Delta\alpha\,\Delta T$, where $\Delta\alpha$ is the difference between the CTEs of fiber and matrix, and ΔT is the amplitude of thermal cycle.

In general, the matrix has a much higher CTE than the fiber in a composite. Hence, rather large internal stresses can be generated when fiber-reinforced composites are exposed to large temperature changes. When the temperature changes are happening in a repeated manner, the phenomenon is called thermal fatigue, because the cyclic stress is thermal in origin. Thermal fatigue can cause cracking in a brittle matrix or plastic deformation in a ductile matrix. Cavitation in the matrix and fiber/matrix debonding are the other damages in the composites due to thermal fatigue. Xu et al. (1995) studied the damage evolution as a function of thermal cycles in three MMCs: Al_2O_3 (fiber)/Mg alloy, B_4C_p/6061 Al alloy, and SiC_p/8090 Al alloy. These samples were thermally cycled between room temperature and 300°C, and it was observed that the reduction in elastic modulus caused by thermal cycling was more severe in the fiber-reinforced composites than in the particle-reinforced composites.

5.5.7 Creep

Creep refers to time-dependent deformation of a material; usually it becomes significant at high temperatures. The creep rate becomes high for most metals and ceramics above certain temperature, and that sets a limit on the maximum service temperature of that particular material. In general, this limit increases with the melting point of the material.

The creep curve of SiC_w/2124 Al alloy composite showed the primary, secondary or steady state, and tertiary stages (Lilholt and Taya 1987). In another study, it was found that the creep rate of Saffil alumina fiber-reinforced aluminum alloy was lower than that of the matrix alloy, but the stress exponent or slope for the composite was much higher than that of the unreinforced alloy.

5.5.8 Wear

The incorporation of hard particulates generally improves the wear resistance of metal matrices. The wear resistance of aluminum alloy with fly ash, SiC and alumina particles was studied by ball on disc wear testing. The discs were made

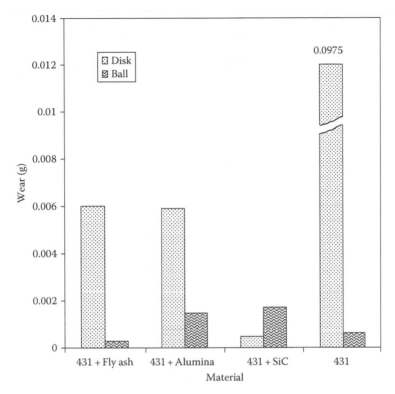

FIGURE 5.22
Wear behavior of SiC-, Al$_2$O$_3$-, and fly ash-reinforced aluminum alloy composites and aluminum alloy. (From Abhishek, A.D., Sintering behavior and mechanical properties of aluminum alloy powders and composites, MS dissertation, IIT Madras, Chennai, India, 2005.)

from aluminum alloy and its composites, while the balls were made from steel. A comparison of the wear loss of discs and balls under identical wear testing conditions is shown in Figure 5.22. The wear resistance of aluminum alloy (431) is significantly increased by the incorporation of hard particles. Although the wear resistance of SiC particulate composite is high, the wear loss of steel ball slid against this material is also high. The wear resistance of alumina and fly ash MMCs are comparable, but the counter-face wear is very less with fly ash MMC.

5.6 Applications of Metal Matrix Composites

The applications of MMCs can be divided into aerospace and non-aerospace applications. In the aerospace applications, low density, tailored thermal expansion and conductivity, and high stiffness and strength are the main drivers. Performance rather than cost is very important for these applications. Continuous fiber-reinforced MMCs usually deliver superior performance than short-fiber-reinforced and particulate MMCs, and hence the former are frequently used in aerospace applications. In the non-aerospace applications, both the cost and performance are equally important; hence an optimum combination of them is selected. Thus, short-fiber-reinforced or particulate MMCs are increasingly finding applications in non-aerospace sector.

Reduction in the weight of a component is a major driving force for any application in the aerospace industry. For example, in the Hubble telescope, C_f/Al composite is used for waveguide booms because this MMC is very light and has a high elastic modulus and a low CTE. Another aerospace application of MMCs involves replacement of light but toxic beryllium. For example, in the U.S. Trident missile, beryllium has been replaced by SiC_p/Al composite.

One of the important applications of MMCs in the automobile industry is in diesel engine piston crowns. It involves the incorporation of short alumina or alumina–silica fibers in the crown of the piston. The conventional diesel engine piston has been made with Al–Si casting alloy containing nickel cast-iron crown. The replacement of nickel cast iron by aluminum matrix composite has resulted in a lighter and more wear-resistant product. Another application in the automotive sector involves the combined use of carbon and alumina fibers in aluminum alloy matrix for use as cylinder liners in the Prelude model of Honda motor company.

An important potential commercial application of particle-reinforced aluminum MMC is automotive driveshafts. The driveshaft becomes dynamically unstable above certain speed. This critical speed depends on the geometric parameters of the shaft, such as length and diameter. A shorter shaft length and larger diameter (i.e., a higher stiffness) will give a higher critical speed. The diameter of shaft cannot be increased beyond a limit because of under-chasis space limitations. The best solution would be to increase the stiffness by using a high-modulus material, that is, instead of aluminum alloy, a particulate aluminum alloy composite.

Particulate MMCs, particularly light MMCs made with aluminum or magnesium alloys, also find applications in sporting goods. An excellent example is the use of Duralcan particulate MMCs to make mountain bicycles. The bicycle frames are being made from extruded tubes of 6061 aluminum alloy containing about 10 vol.% alumina particles. The main advantage is high stiffness.

MMCs can be tailored to have optimal thermal properties for electronic packaging systems such as cores, substrates, carriers, and housings. Continuous boron fiber-reinforced aluminum composites have been used in chip carrier multilayer boards. Unidirectionally aligned, pitch-based carbon fiber-reinforced aluminum composites also have high thermal conductivity along the fiber direction. Such a C_f/Al composite can find applications in electronic equipment where weight reduction is an important consideration. For example, this composite is suitable for high-density, high-speed integrated-circuit packages for computers and base plates in electronic equipment.

5.7 Recycling of Metal Matrix Composites

The wide usage of particulate composites can be attributed to the fact that particulate MMCs are more cost-effective than continuous or short-fiber-reinforced MMCs. Continuous fiber-reinforced MMCs are expensive mainly because of the high cost of fibers and processing. Moreover, secondary processing is very difficult with continuous fiber-reinforced composites. On the recycling point of view also, particulate MMCs are better than fiber-reinforced MMCs, because recycling of particulate MMCs is relatively easy. There are two aspects on recycling of MMCs:

1. Recycling for reuse as a composite
2. Reclamation to obtain the individual components from the composite, that is, matrix metal and reinforcement particles, separately

Unlike the homogeneous metal/alloy, the recycling of composite is not that simple or straightforward. One of the problems with aluminum-based composites is the difficulty in segregation since the density of unreinforced metal is indistinguishable from the density of reinforced material. Crushing followed by gravity separation may be useful to separate the metal and ceramic particles. A composite material with a desired volume fraction of particles can be obtained by adding virgin metal/alloy particles to the crushed composite powder.

It is also possible to reuse the cast MMCs by remelting. Although some of the issues, such as particle/molten metal reaction, are the same in remelting, the scrap sorting and melt cleanliness are additional problems. According to Lloyd (1994), by using combined argon and salt fluxing, SiC particles and 85%–90% of aluminum can be recovered from the SiC/Al composites.

5.8 Concluding Remarks

Costs are sometimes inherently high for the MMCs, and this must be balanced against performance benefits. Shortcomings of the currently available MMC database on predictive capacity and reliability are responsible for much of the caution expressed by engineers concerning wider MMC usage. For the composites, upper and lower limits of many properties can be found from the corresponding properties of the constituents. Since the limits are widely separated in many cases, always there is constant effort on establishing composite properties to a greater precision. Very often, the properties of metallic matrix materials are affected by the presence of reinforcements, which makes the prediction further difficult. Nevertheless, reliable approaches have been developed for the prediction of many basic properties of MMCs, such as modulus, thermal expansion, the onset of plasticity, and work hardening.

Questions

State whether the following statements are true or false, and give reasons:

5.1 By modifying the properties of matrix, it is possible to improve the transverse strength.

5.2 Toughness of metals can be improved by incorporating ceramic fibers.

5.3 Very fine reinforcements can be obtained in in situ composites at faster solidification rates.

5.4 The toughness of metals is improved by forming composites.

5.5 Problems with thermal expansion coefficient are minimized by forming MMCs by deposition techniques.

Answer the following questions:

5.6 Compare liquid infiltration and squeeze casting techniques.

5.7 What are the main advantages of squeeze casting technique over stir casting in making MMCs?

5.8 Compare liquid infiltration and diffusion bonding processes for making MMCs.

5.9 Explain the factors controlling the toughness of MMCs.

References

Abhishek, A.D. 2005. Sintering behavior and mechanical properties of aluminum alloy powders and composites. MS dissertation, IIT Madras, Chennai, India.

Agarwal, A. et al. 2010. *Carbon Nanotubes: Reinforced Metal Matrix Composites*. Boca Raton, FL: CRC Press, p. 33.

Arsenault, R.J. and N. Shi. 1986. *Mater. Sci. Eng.* 81: 175.

Bryant, J.D., L. Christodoulou, and J.R. Maisano. 1990. *Scripta Metall. Mater.* 24: 33.

Bystricky, P., H. Bjerregard, and A. Mortensen. 1999. *Metall. Trans.* A30: 1843.

Champion, A.R., W.H. Krueger, H.S. Hartman, and A.K. Dhingra. 1978. In *Proc. 2nd Int. Conf. Compos. Mater. (ICCM/II)*. New York: TMS-AIME, p. 883.

Chen, Z.Y., Y.Y. Chen, Q. Shu, and G.Y. An. 1999. *Acta Metall. Sinica* 35: 874.

Chen, Y. and D.D.L. Chung. 1996. *J. Mater. Sci.* 31: 311.

Chrysanthou, A. and G. Erbaccio. 1995. *J. Mater. Sci.* 30: 6339.

Cisse, J. and G.F. Bolling. 1971. *J. Cryst. Growth* 11: 25.

Clyne, T.W. 2000. *Metal Matrix Composites.* Vol. 3. *Comprehensive Composite Materials,* Kelly, A. and Zweben, C. (Eds.). Oxford, U.K.: Elsevier Science Ltd., p. 696.

Cooper, G.A. 1970. *J. Mater. Sci.* 5: 645.

Cooper, G.A. and A. Kelly. 1967. *J. Mech. Phys. Solids* 15: 279.

Deevi, S.C. and V.K. Sikka. 1997. *Intermetallics* 5: 19.

Deng, X., C. Cleveland, T. Karcher, M. Koopman, N. Chawla, and K.K. Chawla. 2005. *J. Mater. Eng. Perf.* 14: 1.

Deve, H.E. 1997. *Acta Mater.* 45: 5041.

German, R.M. 1984. *Powder Metallurgy Science.* Princeton, NJ: Metal Powder Industries Federation, p. 250.

Gonzalez, E.J. and K.P. Trumble. 1996. *J. Am. Ceram. Soc.* 79: 114.

Gouda, M., K.M. Prewo, and A.J. McEvily. 1981. *Fatigue of Fibrous Composite Materials, ASTM STP 723.* West Conshohocken, PA: American Society of Testing and Materials, p. 101.

Hack, J.E., R.A. Page, and G.R. Leverant. 1987. *Metall. Trans.* A15: 1389.

Kellie, J. and J.V. Wood. 1995. *Mater. World* 3: 10.

Kelly, A. 1971. In *The Properties of Fiber Composites,* p. 5. Guildford, Surrey, U.K.: IPC Science & Technology Press.

Kohyama, A., N. Igata, Y. Imai, H. Teranishi, and T. Ishikawa. 1985. In *Proc. 5th Int. Conf. Compos. Mater. (ICCM/V).* Warrendale, PA: TMS-AIME, p. 609.

Kumai, S. and J.F. Knott. 1991. *Mater. Sci. Eng.* A146: 317.

Levin, M., B. Karlsson, and J. Wasen. 1989. In *Fundamental Relationships between Microstructure and Mechanical Properties of Metal Matrix Composites,* ed. P.K. Liaw and M.N. Gungor, Warrendale, PA: TMS, p. 421.

Lilholt, H. and M. Taya. 1987. In *Proc. 6th Int. Conf. Compos. Mater. (ICCM/VI),* Vol. 2. London, U.K.: Elsevier Applied Science, p. 234.

Lin, Y., R.H. Zee, and B.A. Chin. 1991. *Metall. Trans.* A22: 859.

Lloyd, D.J. 1994. *Int. Mater. Rev.* 39: 1.

Lu, L., M.O. Lai, and F.L. Chen. 1997. *Acta Mater.* 45: 4297.

Luo, A. 1994. *Scripta Metall. Mater.* 31: 1253.

Lynch, C.T. and J.P. Kershaw. 1972. *Metal Matrix Composites.* Boca Raton, FL: CRC Press.

Maity, P.C., S.C. Panigrahi, and P.N. Chakraborty. 1993. *Scripta Metall. Mater.* 28: 549.

Matsubara, H., Y. Nishida, M. Yamada, I. Shirayanagi, and T. Imai. 1987. *J. Mater. Sci. Lett.* 6: 1313.

McDanels, D.L. 1985. *Metall. Trans.* A16: 1105.

McGuire, M.A. and B. Harris. 1974. *J. Phys. D: Appl. Phys.* 7: 1788.

Meyers, M.A. and S.L. Wang. 1988. *Acta Metallurgica* 36: 925.

Mortensen, A. 1993. *Mater. Sci. Eng.* A173: 205.

Mortensen, A. and M.C. Flemings. 1996. *Metall. Trans.* A27: 595.

Mortensen, A., M.N. Gungor, J.A. Cornie, and M.C. Flemings. 1986. *J. Metals* 38(3): 30.

Mortensen, A. and I. Jin. 1992. *Int. Mater. Rev.* 37: 101.

Nakata, H., T. Choh, and N. Kanetake. 1995. *J. Mater. Sci.* 30: 1719.

Page, R.A., J.E. Hack, R. Sherman, and G.R. Leverant. 1987. *Metall. Trans.* A15: 1397.

Phillips, W.L. 1978. In *Proc. 2nd Int. Conf. Compos. Mater. (ICCM/II),* ed. B.R. Noton, R.A. Signorelli and K.N Street, New York: TMS-AIME, p. 567.

Pruitt, L. and S. Suresh. 1992. *J. Mater. Sci. Lett.* 11: 1356.
Rajan, T.P.D. et al. 2010. *Mater. Char.* 61, 923.
Ranganath, S., M. Vijayakumar, and J. Subrahmanyam. 1992. *Mater. Sci. Eng.* A149: 253.
Schoutens, J.E. 1982. *Introduction to Metal Matrix Composite Materials.* Santa Barbara, CA: Kaman Tempo, p. 5.
Shang, J.K., W. Yu, and R.O. Ritchie. 1988. *Mater. Sci. Eng.* A102: 181.
Shi, N. and R.J. Arsenault. 1991. *J. Compos. Tech. Res.* 13: 211.
Shi, N. and R.J. Arsenault. 1993. *Metall. Trans.* A24: 1879.
Suresh, S. and K.K. Chawla. 1993. In *Metal-Matrix Composites*, ed. S. Suresh, A. Mortensen, and A. Needleman, Boston, MA: Butterworth-Heinemann, p. 119.
Thomas, D.G. 1965. *J. Colloid Sci.* 20: 267.
Tjong, S.C. and Z.Y. Ma. 2000. *Mater. Sci. Eng.* 29: 49.
Walter, J.L. 1982. In Situ *Composites IV*. Amsterdam, the Netherlands: Elsevier, p. 85.
Wang, W., F. Ajersch, and J.P.A. Löfvander. 1994. *Mater. Sci. Eng.* A187: 65.
Warren, R. 2000. *Comprehensive Composite Materials*, Vol. 4, Kelly, A. and Zweben, C. (Eds.). Amsterdam, the Netherlands: Elsevier, p. 658.
Williams, D.R. and M.E. Fine. 1985a. In *Proc. 5th Int. Conf. Compos. Mater. (ICCM/V)*. Warrendale, PA: TMS-AIME, p. 275.
Williams, D.R. and M.E. Fine. 1985b. In *Proc. 5th Int. Conf. Compos. Mater. (ICCM/V)*. Warrendale, PA: TMS-AIME, p. 369.
Williams, D.R. and M.E. Fine. 1987. In *Proc. 6th Int. Conf. Compos. Mater. (ICCM/VI)*, Vol. 2. London, U.K.: Elsevier Applied Science, p. 113.
Willis, T.C. 1988. *Metals Mater.* 4: 485.
Wood, J.V., P. Davies, and J.L.F. Kellie. 1993. *Mater. Sci. Technol.* 9: 833.
Xu, Z.R., K.K. Chawla, A. Wolfenden, A. Neuman, G.M. Liggett, and N. Chawla. 1995. *Mater. Sci. Eng.* A203: 75.
Yajima, S., J. Tanaka, K. Okamura, H. Ichikawa, and T. Hayase. 1981. *Revue de Chimie Minerale* 18: 412.
Yang, N., J. Boselli, and I. Sinclair. 2001. *J. Microsc.* 201: 189.
Zhang, E., S. Zeng, B. Yang, Q. Li, and M. Ma. 1999. *Metall. Mater. Trans.* 30A: 1147.

Bibliography

ASM Handbook. 2001. *Composites*, Vol. 21. Materials Park, OH: ASM International.
Chawla, K.K. 1998. *Composite Materials: Science and Engineering*, 2nd edn. New York: Springer-Verlag Inc.
Chawla, N. and K.K. Chawla. 2006. *Metal Matrix Composites*. New York: Springer.
Clyne, T.W. 2000. *Metal Matrix Composites*. Vol. 3. *Comprehensive Composite Materials*, Kelly, A. and Zweben, C. (Eds.). Oxford, U.K.: Elsevier Science Ltd.
Clyne, T.W. and P.J. Withers. 1993. *An Introduction to Metal Matrix Composites*. Cambridge, U.K.: Cambridge University Press.
Mallick, P.K. 2008. *Fiber-Reinforced Composites*, 3rd edn. Boca Raton, FL: CRC Press.
Schwartz, M.M. 1997. *Composite Materials*, Vol. II. Upper Saddle River, NJ: Prentice Hall.

6

Ceramic Matrix Composites

Ceramic materials in general have very attractive properties such as high hardness and modulus, strength retention at high temperatures, chemical inertness, and low density. Even then these materials are not very popular as metallic materials for engineering applications, mainly because of their low toughness. This type of brittle materials is prone to catastrophic failures. They are also extremely susceptible to thermal shock though they can withstand high temperatures. Most of the ceramic components are exposed to thermal cycles during fabrication/service, and this may introduce defects in them because of the poor thermal shock resistance. Unlike polymer and metal matrix composites (PMCs and MMCs), the main consideration in developing a ceramic matrix composite (CMC) is to increase the toughness. Hence, for most CMCs, the driving force to form composites is to enhance the mechanical reliability of inherently brittle ceramic materials, while retaining their attractive properties.

This can be achieved by introducing a second phase in the form of particles or fibers/whiskers in the ceramic matrix. Once their toughness is improved, then it is possible to exploit the attractive high-temperature strength and environmental resistance of these materials without the risk of catastrophic failure. The damage-tolerant behavior of a CMC is shown in Figure 6.1.

It should be noted that there are some basic differences between CMCs and other composites. In general, the fiber reinforcements bear a greater proportion of the applied load in non-CMCs. As mentioned in the first chapter, the ratio of loads in the fiber and matrix is directly proportional to the ratio of fiber to matrix elastic moduli (E_f/E_m). In non-CMCs, the ratio is usually very high, and hence, the fibers bear the maximum applied load. However, in CMCs, the ratio is rather low and can be as low as unity in many CMCs. That means the applied load is almost equally shared by the fibers and matrix. Another distinction is that the thermal mismatch between the constituents in a CMC has a significant influence on the performance of CMC because of limited matrix ductility and generally high fabrication temperature. Another important difference is related to the bonding between reinforcement and matrix. Generally, strong interfacial bonding is preferred in non-CMCs for efficient load transfer to the fiber reinforcements, while in CMCs, a tailored interfacial bonding (neither strong nor weak) is warranted for many toughening mechanisms to operate.

FIGURE 6.1
Picture showing the improved toughness of ceramic composites. (SiC$_f$/SiC composite produced by the CVI process. Photo courtesy of A. Udayakumar, NAL, Bangalore, India.)

Although the market share of CMCs is lower compared to PMCs and MMCs, the growth is higher. Quality reinforcements are still not available at competitive prices, and composite components are more expensive to produce. However, reductions in cost are being pursued by developing and improving fiber and composite manufacturing processes.

Ceramic composites attract increasing attention mainly because of the high-temperature properties they can provide. However, the properties depend upon achieving a specific range of microstructures. Most modern ceramic composites of course go well beyond monolithic ceramics in microstructure control and hence properties.

The motivation for the incorporation of the reinforcement phase is to modify or tailor the properties, especially the mechanical properties. The reinforcement phase can be a continuous fiber or a discontinuous fiber, whisker, platelet, or particle. The continuous fiber-reinforced CMCs are expensive to produce due to high costs, limited availability, and forming difficulty of continuous fibers. Moreover, the properties of these composites tend to be highly anisotropic. Hence, continuous fiber-reinforced CMCs have found uses only for certain highly specialized applications. On the other hand, discontinuously reinforced composites are less expensive to fabricate and their properties are nearly isotropic. A major advantage of this type of composites is that the conventional processing techniques utilized for monolithic ceramics can be used to make these composites. Hence, the overall cost associated with the production of these composites is only slightly higher than those associated with the production of monolithic ceramics.

The carbon–carbon (C/C) and CMCs are the two families of high-temperature composite materials. They are being developed for use at high or very high temperatures (400°C–3000°C). The constituents, mainly continuous fibers and matrix, are made of carbon or ceramic compounds with outstanding characteristics, such as low density, high melting or decomposition temperatures, and high hardness and stiffness. A few properties such as toughness and chemical resistance can be tailored in these composite materials by selecting the appropriate fiber and matrix but also by defining a specific design for the composite, including interphase and internal or external oxidation protection.

Initially these materials were developed for aerospace applications. Now they are being introduced into more and more new fields, and their range of applications will expand significantly if their cost is lowered substantially. C/C composites are extensively used in aircraft brake disks. Other specialized applications are rocket nozzles, thermal shields of reentry vehicles, and space shuttles. Flaps of military aircraft engines are produced industrially using C/SiC composites. Several other parts of aircraft engines such as combustion chambers and turbines and energy generation components such as hot filtration systems, burners, and heat recuperators are being developed. This development will increase further for many applications if the cost of processing of these composites is lowered. Even though C/C composites are a class of CMCs, because of the growing importance of these materials, a separate chapter has been devoted.

6.1 Failure Behavior of CMCs

In non-CMCs (MMCs and PMCs), the matrix failure strain is considerably greater than that of fibers. Hence, fiber failure strain controls the composite failure strain in these composites. In contrast, the strain-to-failure value of a ceramic matrix is lower than that of fibers. The failure strain values for most ceramic matrix materials are less than 0.05%, whereas the failure strain values of fibers such as boron, carbon, and silicon carbide are about 1%. In the case of non-CMCs, fibers fail first at various weak points and the composite will fail along a section that has a large number of fiber fractures. However, the failure behavior is entirely different in CMCs. In the CMCs, the failure starts first in the matrix. In a strongly bonded CMC, the crack in the matrix propagates through the fibers and the composite will fail catastrophically, whereas in a weakly bonded CMC, the matrix crack cannot grow through the fibers and the fibers will be bridging the fractured matrix blocks. Thus, a strong bonding in a CMC is not preferable from a toughness point of view.

Consider a CMC, in which continuous fibers are aligned in the loading direction and interface bonding is weak. On applying tensile load to the composite, a crack initiates in the matrix at a weak region and starts propagating normal to the loading direction. As it approaches the fiber, the crack is momentarily stopped by the fiber. Since the fibers have enough strength to support the load, the composite does not fail at this load. If the fiber/matrix interface is weak, then interfacial shear and lateral contraction of the fiber leads to debonding and the crack gets deflected away from its principal direction. On increasing the load further, the fibers start failing at weak points anywhere along their length. Most of the broken fibers are embedded in the matrix. On further straining the composite, the broken fibers are pulled out from the matrix against the frictional resistance of the interface. Only after this the complete separation (failure) of composite occurs. The tensile stress–strain behavior of this type of composite and failure progression are shown schematically in Figure 6.2.

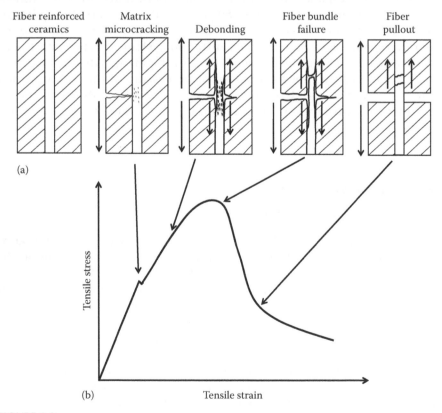

FIGURE 6.2
The progression of a crack in a fiber-reinforced CMC (a) and the tensile stress–strain curve (b) of a continuous fiber-reinforced composite in the longitudinal direction. (Reprinted from Evans, A.G., *Mater. Sci. Eng.*, 71, 3, 1985. With permission; From Harris, B., *Metal Sci.*, 14, 351, 1980. With permission.)

Assuming elastic behavior for both the matrix and fiber in a CMC, the ratio of stress carried by the fiber to that of the matrix can be written as

$$\frac{\text{Stress carried by fiber}}{\text{Stress carried by matrix}} = \frac{\sigma_f V_f}{\sigma_m V_m} = \frac{E_f V_f}{E_m V_m} \tag{6.1}$$

Since the matrix failure strain is usually lower than that of the fiber, cracks start appearing first in the matrix at a composite stress σ_c. The rule-of-mixtures-type relationship for this condition is

$$\sigma_c = \sigma_f V_f + \sigma_m^u (1 - V_f) \tag{6.2}$$

where σ_m^u is the matrix stress at its failure strain. On rearranging,

$$\sigma_c = \sigma_m^u \left[\frac{\sigma_f V_f}{\sigma_m^u} + (1 - V_f) \right]$$

$$\sigma_c = \sigma_m^u \left[\frac{\sigma_f V_f}{\sigma_m^u (1 - V_f)} + 1 \right] (1 - V_f), \text{ On comparing Equation (6.1)}$$

$$\sigma_c = \sigma_m^u \left[\frac{E_f V_f}{E_m (1 - V_f)} + 1 \right] (1 - V_f)$$

$$\sigma_c = \sigma_m^u \left[\frac{E_f V_f}{E_m} + (1 - V_f) \right]$$

$$\sigma_c = \sigma_m^u \left[1 + V_f \left(\frac{E_f}{E_m} - 1 \right) \right] \tag{6.3}$$

From this equation, it can be noted that in composites with high-modulus matrix, matrix cracking will occur at much lower stresses. That means low-modulus glasses and ceramics are more preferred over high-modulus ceramic matrices, with respect to the formation of microcracks. A CMC, even with cracks in the matrix, can retain some reasonable strength $\left(\sigma_c^u \cong \sigma_f^u V_f \right)$. There are some applications where such a damage-tolerant behavior would be very valuable to avoid catastrophic failure.

Fiber aspect ratio, fiber orientation, thermal expansion mismatch, matrix porosity, and fiber defects are the other important variables that control the performance of CMCs. In a CMC containing short, randomly oriented fibers, a weakening effect may occur rather than a strengthening effect. This could be attributed to the stress concentration effect at the ends of short fibers and thermal expansion mismatch. With aligned, continuous fibers, the stress concentration effect at the fiber ends is minimized and higher fiber volume fractions can be obtained. However, it is difficult to avoid matrix porosity at very high fiber volume fractions.

6.2 Toughening Mechanisms in CMCs

Many mechanisms have been proposed for the improvement in toughness of ceramic materials in the presence of a second phase, which include crack deflection, crack bridging, reinforcement pullout, microcracking, crack pinning, and phase transformation. At any instance, two or more toughening mechanisms operate in CMCs, and the resultant toughening is due to these combined effects. The toughening mechanisms operating in a CMC depend on a number of microstructural parameters, which are affected by the fiber, matrix, and interface properties. Some of the common toughening mechanisms are listed in Table 6.1.

Microcrack formation in matrix, fiber/matrix debonding leading to crack deflection and fiber pullout, and phase transformation are all basically energy-dissipating processes that can result in an increase in toughness or work of fracture of the ceramic matrix. Some of these toughening

TABLE 6.1

CMC Toughening Mechanisms

Sl. No.	Mechanism	Requirement
1.	Compressive prestressing of the matrix	CTE of fiber > CTE of matrix
2.	Crack impeding	Fracture toughness of the second phase (fibers/particles) is greater than that of the matrix locally. Crack either is arrested or bows out.
3.	Fiber (whisker) pullout	Fibers (whiskers) having high transverse fracture toughness will cause failure along the fiber/matrix interface leading to fiber pullout on further straining.
4.	Crack deflection	Weak fiber/matrix interfaces deflect the propagating crack away from the principal direction.
5.	Phase transformation toughening	The crack tip stress field in the matrix can cause the second-phase particles at the crack tip to undergo a phase transformation causing expansion ($\Delta V > 0$). The volume expansion can squeeze the crack shut.

Source: Data from Chawla, K.K., *Composite Materials: Science and Engineering*, 2nd edn., Springer-Verlag, Inc., New York, 1998, p. 237. With permission.

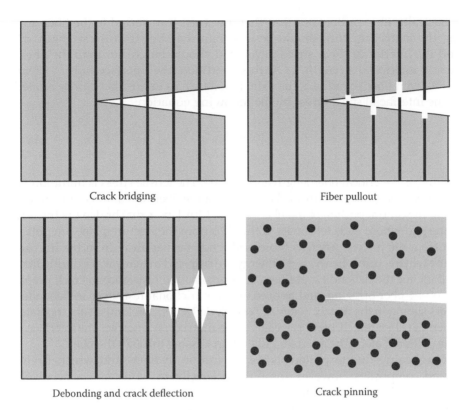

Crack bridging   Fiber pullout

Debonding and crack deflection   Crack pinning

FIGURE 6.3
Schematic of some toughening mechanisms operating in CMCs.

mechanisms operating in CMCs to improve toughness are shown schematically in Figure 6.3. The crack deflection mechanism improves the toughness by deflecting the path of a propagating crack and causing the crack to travel a longer distance. Deflection is caused by the strong nature of reinforcement and/or by the relatively weak reinforcement–matrix interface. This mechanism is common in almost all CMCs, although it may not be the dominant mechanism. In general, the toughness increase by crack deflection depends on the reinforcement shape and content, but not on the reinforcement size (Faber and Evans 1983a,b); for example, rod shape is the most effective morphology for increasing toughness. In the case of discontinuously reinforced CMCs, the toughness increases up to the reinforcement content of 20 vol.%, beyond that it increases very little. In crack bridging, unbroken reinforcements bridge the material behind the crack tip. The presence of these bridges makes further crack opening difficult and thus inhibits the crack propagation, thereby increasing fracture toughness. Whiskers, fibers, and ductile metallic particles can act as bridging elements. Generally, toughening by crack bridging is enhanced by an increase in the strength, diameter, and/or amount of reinforcement.

Basically, fiber pullout requires that the stress transferred to the fiber during the matrix fracture should be less than the fiber fracture strength, σ_f^u, and the interfacial shear stress developed should be greater than the fiber/matrix interfacial strength, τ_i. At this condition, the interface fails in shear. For a given fiber of radius r, the axial tensile stress in the fiber can be related to the interfacial shear stress by the following equation:

$$\sigma_f = 2\tau_i \left(\frac{l_c}{r} \right) \tag{6.4}$$

where l_c is the critical fiber length and $\sigma_f < \sigma_f^u$. The tensile stress is minimum at both fiber ends, increases toward the center of the fiber, and attains a maximum along the central portion of the fiber (Figure 1.10). Fibers that bridge the fracture plane and whose ends terminate within $l_c/2$ from the fracture plane are pulled out from the matrix, whereas those with ends farther away from the fracture plane will fracture when $\sigma_f = \sigma_f^u$. Energy is dissipated in many ways during fiber pullout: for the debonding of fibers, to overcome the resistance to crack propagation by fiber bridging, and to overcome the frictional resistance to debonded fibers sliding in the matrix material. A crack deflection mechanism also operates when the fiber/matrix bond is weak. As the matrix crack reaches the interface, it gets deflected along the interface rather than passing through the fiber.

Toughening due to pullout is very common in fiber- and whisker-reinforced composites; however, the pullout lengths of whiskers are relatively small, and hence, their toughening effects are correspondingly less. When pullout is the dominant toughening mechanism, a balance between composite strength and fracture toughness must be achieved by carefully engineering the reinforcement–matrix interface such that it is sufficiently weak to maximize pullout and at the same time not so weak as to reduce the strength of the composite. Interfacial strength is controlled by the chemical reactions taking place at the interface and the degree of mechanical bonding.

A mismatch in the coefficient of thermal expansion (CTE) between the matrix and reinforcement can produce stresses while the composite is cooled from the processing temperature. If the stresses are sufficiently high, they will lead to the formation of spontaneous microcracks. Toughening can occur if these microcracks are not formed until an external stress is applied to the composite. When the microcracks are forming after the external stress is applied, they form only at the tip of the crack, reducing the stress intensity at the crack tip and therefore increasing fracture toughness. However, the increase in fracture toughness by this mechanism is not more than 10%, and if the microcrack formation is not carefully controlled, excessive microcracking can lead to a decrease in fracture toughness as well as the strength of the composite.

In the crack pinning mechanism, the propagating crack is pinned by the reinforcement phase, causing an increase in the fracture energy as the crack front has to bow out between the pinning points. For effective crack pinning, the reinforcement–matrix interface and the reinforcements should be strong. In addition, the spacing between the reinforcements must be small relative to the crack size.

Transformation toughening occurs mainly by the addition of tetragonal zirconia particles. When the tetragonal zirconia particles undergo martensitic transformation to the monoclinic phase in the presence of external stress, the resulting volume expansion exerts closure stress on the crack tip. Transformation toughening is maximized when the zirconia particles are just below the critical size for spontaneous transformation. Microcracking, which may form during the phase transformation, can also improve the composite fracture toughness by deflecting the main crack path. These mechanisms are illustrated in Figure 6.4.

It is important to note that while these toughening mechanisms are active at low to medium temperature, almost all of them may not be effective on prolonged exposure to high temperatures above 1000°C. At least the toughening mechanisms that rely on the mismatch in CTE and phase transformation will not have the same influence at elevated temperatures. Long-term exposure to high temperatures is also more likely to change reinforcement morphology and modify the chemical nature of the reinforcement–matrix interface, potentially reducing the contributions from the mechanisms such as pullout, bridging, and crack deflection to the overall toughness of the composite.

Avoiding processing-related flaws in the matrix and fiber is necessary to attain high strength in the composite. If large flaws are formed in the matrix during processing, then the composite fracture strain will be low. However, the bridging of cracks by fibers restricts the flaw size in the matrix. This, in turn, will help to achieve higher failure strains in the matrix than in an unreinforced, monolithic ceramic material. The use of a high volume fraction of continuous

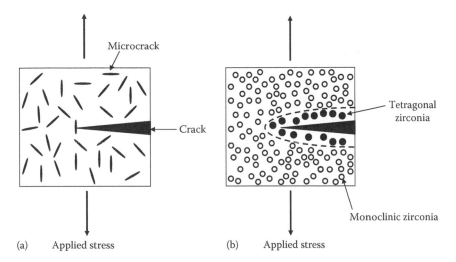

FIGURE 6.4
Toughening mechanisms operating in zirconia-toughened ceramics: (a) microcrack toughening and (b) transformation toughening. (Reprinted from Bhaduri, S.B. and Froes, F.H., *J. Metals*, 43(5), 16, 1991. With permission.)

fibers stiffer than the matrix will increase the stiffness of the composite. As discussed earlier, this results in a higher stress level being needed to produce matrix microcracking and hence increases composite ultimate strength, as well as creep resistance. A high volume fraction of small-diameter fibers provides a large number of fibers for crack bridging and hence delay crack propagation to higher strain levels. The critical fiber length needed for effective load transfer from matrix to fiber is also getting reduced with small-diameter fiber. Although a weak interface is desirable from a toughness point of view, it provides an easy path for environmental attack; the corrosive chemicals can ingress through the debonded region. The ability to maintain a high strength level and high inertness at high temperatures under an aggressive environment is also very important for the CMCs. Hence, the appropriate reinforcement, matrix, and interface should be selected for any particular application.

6.3 Processing of Ceramic Matrix Composites

The selection of appropriate reinforcement and matrix for the intended application is very important, before selecting the processing technique. Among the properties of reinforcement and matrix, one should consider the following properties during selection: melting point, density, elastic modulus, strength, fracture toughness, and compatibility. Compatibility means the chemical compatibility between the reinforcement and matrix, the thermal compatibility (mainly determined by the thermal expansion coefficients), and the compatibility with the environment during processing and service. In addition to these, the availability of reinforcement in different forms should also be considered. Above all these items, the cost and availability are also more important. Most of these characteristics will have an important bearing on the processing route as well.

CMCs can be made either by conventional processing techniques used for making monolithic ceramics or by some new, non-conventional techniques. Most of the techniques are simply the variants of the processing techniques used for monolithic ceramics. The processing of a CMC should be considered as an integral part of the whole process of designing a CMC component. For example, any damage to the reinforcement during processing will result in inferior performance to the final product. The orientation of fiber or whisker in the matrix is an integral part of the fabrication process, and it will have a significant influence on the mechanical response of a CMC to an applied load. The selection of a processing method to make a CMC is based upon the characteristics of the composite, its size and shape, and its cost. Design or tailoring of composite microstructures to achieve the desired properties presents processing challenges.

A combination of uniformly dispersed phases and high volume fractions is often desired, which makes achieving a high density and degree of homogeneity a challenge. These challenges of CMC processing have resulted in a significant shift in the emphasis of processing technologies in comparison to the processes used for monolithic ceramic materials. The ensuing sections will cover the processing of both particulate and fiber composites.

Processing of CMCs with particulates has overall similarity to most conventional ceramic processing methods. The most generic difference is obtaining a uniform distribution of dispersed phase. The densification of the ceramic matrix in the presence of a dispersed phase can also be a problem.

There are more significant differences between the processing of fiber-reinforced CMCs and that of monolithic ceramics; that is, fiber orientation and architecture are major concerns while processing CMCs. Although unidirectional fiber-reinforced CMCs have limited commercial use, most of the processing studies are related to these composites only. Laminating with tapes of uniaxially aligned fibers or various woven fabrics in controlled orientation between the layers is important to achieve the desired properties. The type of weave or the type of fiber in each layer may be varied to balance different processing opportunities. Mats of chopped or short fibers can also be used in CMCs. There are also some specialized processing methods with multidimensional weaves, braids, and so on.

An important way in achieving any fiber architecture with fine fibers is the use of fiber tows. Hence, the handling of fiber tows becomes a major factor in fiber-reinforced CMC processing. The most common approach in introducing the matrix is to pull fiber tows through a bath containing the matrix material or its precursor and then lay this up directly by filament winding or another related technique. Most commonly, the bath consists of a slurry made out of matrix powder, but it may also be a sol or a preceramic polymer. A challenge is to achieve the distribution of matrix within the tows (between individual fibers) similar to that between tows. Woven cloth may also be impregnated and then laminated but may often present greater challenges in controlling the quantity and uniformity of the matrix.

The key to the successful performance of CMCs lies in the control of the fiber/matrix interface. It is essential to produce a desirable level of interfacial strength to permit fiber debonding during the fracture process, which in turn improves the fracture toughness of the CMC. This is in contrast to the PMCs and MMCs, which generally require strong fiber–matrix bonding. In many high-toughness CMC systems, the value of interfacial shear strength has been found to be on the order of 2–3 MPa. Apart from having a controlled strength, the interface must prevent fiber–matrix reactions not only during processing but also during service at high temperatures and in aggressive environments. A number of materials have been used with success for interface control, which include C and NbC coatings on Nicalon SiC_f in glass matrix; BN coatings on Nicalon SiC_f in SiO_2, ZrO_2, $ZrSiO_4$, and $ZrTiO_4$ matrices; amorphous C coatings on Si_3N_4 fiber and Nicalon SiC_f in aluminosilicate matrices; and BN/SiC bilayer coatings on Nicalon SiC_f in $ZrSiO_4$ and $ZrTiO_4$ matrices.

TABLE 6.2

CMC Fabrication Processes and Examples of Typical Composite Systems Fabricated by Each Process

Process	Examples
Slurry/sol impregnation, ply stacking, and hot pressing	SiC_f/glass–ceramic, C_f/glass–ceramic, C_f/glass, mullite/glass
Powder ceramics: mixing and hot pressing	SiC_w/Al_2O_3, SiC_p/Al_2O_3, $Al_2O_{3(w)}$/Al_2O_3, SiC_w/Si_3N_4, SiC_p/Si_3N_4
Gas–liquid metal reaction (DIMOX™)	SiC_f/Al_2O_3, SiC_f/SiC, SiC_w/Al_2O_3, SiC_p/Al_2O_3, $Al_2O_{3(f)}$/Al_2O_3, $Al_2TiO_{5(p)}$/Al_2O_3, $ZrB_{2(p)}$/ZrC, TiC-coated graphite/Al_2O_3
Sol/slurry infiltration: sintering/hot pressing	C_f/glass, SiC_f/glass, $mullite_f$/mullite
CVI	SiC_f/SiC, C_f/SiC, SiC_f/Si_3N_4, C_f/C
Polymer infiltration and pyrolysis	C_f/C, C_f/SiC, SiC_f/Si–C–N
Liquid metal infiltration and reaction	SiC_f/SiC, C_f/SiC
In situ	ZrO_2/mullite, Al_2TiO_5/Al_2O_3, TiB_2/SiC
Reaction bonding	SiC_f/Si_3N_4, SiC_w/Si_3N_4

Source: Adapted from Chawla, K.K., *Ceramic Matrix Composites*, Kluwer Academic Publishers, Boston, MA, 2003, p. 135.

Among the processing techniques used for the fabrication of CMCs, the most important and promising processes are summarized in Table 6.2. Some of these techniques are mere extensions of the conventional processing techniques used for monolithic ceramics. Others are novel techniques developed specifically for the processing of CMCs/MMCs.

6.3.1 Ceramic Particle–Based Processes

Powder-based processing (the abbreviation P/C is used instead of P/M, for ceramics) is ideal for fabricating CMCs because of the following reasons: (1) processing is carried out in the solid state, (2) easy incorporation of reinforcement into the matrix, and (3) possibility of near-net-shape (NNS) processing. P/C processing is performed in the solid state, and hence, the difficulties associated with melting of the high melting point ceramics are avoided. Since the P/C technique offers the possibility for NNS component fabrication, expensive and difficult finishing operations, such as machining the hard and brittle ceramic composites, can be avoided during component manufacturing.

P/C technique of discontinuously reinforced CMCs involves four basic steps: (1) the selection of appropriate raw materials, (2) mixing, (3) green body formation (shaping), and (4) consolidation (densification). Mixing involves the uniform distribution of reinforcement phase within the matrix powder. Green body formation involves compacting the mixture into a shape by mechanical pressing in a die or the use of plastic-forming techniques that utilize a fugitive binder, such as tape casting, slip casting, extrusion, or injection molding. Consolidation entails the sintering of the shaped body into a dense component.

The simplest method to manipulate the microstructure of the composite formed in this technique is by altering the particle size of the starting matrix powder and the size of reinforcement elements. This has been demonstrated in Si_3N_4 matrix composites containing fine (3 μm) and coarse (10 μm) $MoSi_2$ powders as reinforcement (Petrovic et al. 1997). It has been found that there is interconnectivity between $MoSi_2$ particles when the fine powders are used, whereas the $MoSi_2$ particles are more discrete when the coarse powders are used. As a consequence, the properties of the composites are different; the composite with coarser $MoSi_2$ particles shows higher fracture toughness due to the reinforcement size effects on the internal stresses that develop in the composite due to thermal expansion mismatch between the phases. It clearly indicates that it is feasible to manipulate the microstructure, and hence properties, of the composites through judicial selection of the size of starting materials.

Models have been proposed to predict the minimum volume fraction of reinforcement phase required to produce a connected microstructure for different sizes of the starting powders. A large volume fraction of the reinforcement phase is needed for connectivity when the reinforcement and matrix particles are of the same size and spherical in shape. The particles can arrange in an orderly fashion or random fashion. For orderly packing, depending on the type of packing (face-centered cubic, body-centered cubic, or simple cubic), the critical volume fraction for percolation (connectivity) varies between 0.147 and 0.167. These values are derived by multiplying the packing factor and the critical probability for site occupancy. Similarly, for random packing arrangement of uniform-sized spheres, the critical volume fraction for connectivity is 0.162. Packing factor can be increased by using particles of different sizes; as a result, lower concentrations of reinforcement phase will be sufficient for connectivity. Smaller particles can effectively coat the surface of large particles and form networks along their surfaces, resulting in connectivity at lower volume fraction than equisized particles. The critical volume fraction for connectivity (V_{con}) of spherical particles can be expressed as a function of the particle size ratio of the matrix and reinforcement phase as follows:

$$V_{con} = \left[1 + \frac{P_f D_m}{4 V_{smf} D_r}\right]^{-1} \tag{6.5}$$

where
    D_m and D_r are the particle sizes for the matrix phase and reinforcement phase, respectively
    P_f is the effective packing factor
    V_{smf} is the critical volume fraction for surface coverage needed for percolation

It shows that the reinforcement phase connectivity occurs at relatively low volume fractions when the particle size ratio (D_m/D_r) is large.

Particle shape will also have an influence on connectivity. As the aspect ratio of reinforcement particle increases, the critical volume fraction required for connectivity decreases. For example, a fiber reinforcement needs only two contacts per fiber to establish a continuous network throughout the composite; hence, the critical volume fraction required for connectivity is lower when compared to spherical particles.

As discussed earlier, the packing efficiency of a mixed powder (reinforcement + matrix) can significantly affect the connectivity. The critical volume fraction for connectivity is lower for a powder mixture containing small and large particles compared to a mixture of equisized particles. In fact, the packing efficiency will have a critical influence not only on the microstructure of the composite but also on the green body formation and densification during sintering. In general, the powder mixtures with higher packing efficiency will require lower binder content for green body formation. The higher packing efficiency will maximize green density and promote densification during sintering. Powder mixtures having broad particle size distribution will increase packing efficiency. This is due to the filling of interstices formed between large particles by smaller particles.

The volume fraction, the diameter, and the aspect ratio of the fibrous material will affect the packing efficiency of a mixture containing powder and fibrous materials (fiber/whisker). Generally, maximum packing efficiency occurs at low fiber contents (<25 vol.%) and low fiber aspect ratio. The packing efficiency with high aspect ratio fibers can be increased by using small-diameter matrix powders. However, for low aspect ratio fibers, packing efficiency is minimized when the particles and fibers have the same diameters. The packing efficiency of a mixture containing low volume fractions of low aspect ratio fibers is maximized when the particle size of a matrix powder is much greater than the diameter of fibers. However, for low aspect ratio fibers at higher volume fractions, the packing efficiency of a mixture is maximized when the particle size of a matrix powder is much smaller than the diameter of fibers.

Mechanical mixing devices such as high-shear mixer, ball mill and turbo mill, or ultrasonic homogenization can be used to produce blends with uniform distribution of matrix and reinforcement phases. For slurry mixing, the powders of the matrix and reinforcement (powder/whisker) are added to the liquid medium to produce a suspension (typical solid loading is about 15 wt.%). It is often necessary to control the surface chemistry of the constituent phases to get a well-dispersed suspension. Well-dispersed suspensions are usually obtained when the electrostatic repulsive forces between the particles are large. It is necessary to determine the conditions at which both phases acquire similar charges. Modification of the surface charge can be accomplished by controlling the pH of the slurry or by the addition of chemical agents.

The typical particle size of powders used to produce composites by slurry processing is of the order of one micrometer. For such small particles, the surface to volume ratio is high and it is the surface forces that determine the colloidal behavior. When polar particles are suspended in a polar medium containing ions, they tend to accumulate ions on their surfaces and become

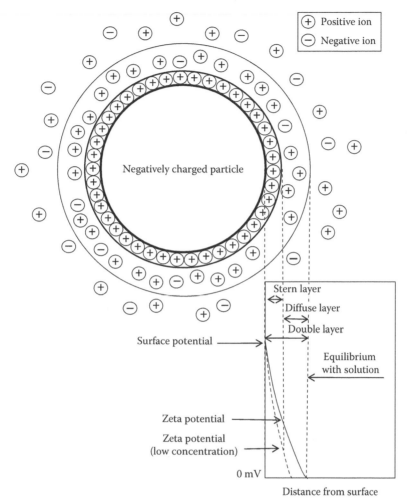

FIGURE 6.5
Particle in a fluid, surrounding double layer, surface potential, and zeta potential with respect to distance from the particle surface. (Reprinted from Gupta, R.K. et al., Eds., *Polymer Nanocomposites Handbook*, CRC Press, Boca Raton, FL, 2010, p. 47. With permission.)

electrically charged (Figure 6.5). Around the charged colloidal particle surface, counterions accumulate and form an electrical double layer, called the stern layer and the diffuse layer. The stern layer is rigidly attached to the particle, whereas the diffuse layer is not. Consider isolated particles having identical charges that are suspended in a liquid. The electrostatic potential linearly decreases from stern layer to diffuse layer. The electrostatic potential decays according to the following equation.

$$V_R < e^{-\left[\kappa(h-H)\right]}$$

(6.6)

where κ is the Debye–Hückel screening length given by

$$\kappa = \sqrt{\frac{F^2 \Sigma_i c_i z_i^2}{\varepsilon \varepsilon_o k_B T}} \qquad (6.7)$$

where
F is Faraday's constant
c_i is concentration
z_i is the valence of ion i
ε is the permittivity of the solvent
ε_o is the permittivity of free space
k_B is Boltzmann's constant
T is temperature

When the κ value is low, it indicates a wide, uncompressed electrical double layer. The slip plane separates the region of fluid that moves with the particle from the bulk, and the magnitude of potential at this plane is called zeta potential (φ_ε), which is correlated to the suspension stability. Although the zeta potential is an intermediate potential value, it is more important than the surface potential as far as electrostatic repulsion is concerned. The zeta potential can be measured by tracking the movement of particles with a microscope as they migrate under a known applied voltage.

The zeta potential is the electrostatic potential generated by the accumulation of ions at the surface of a charged colloidal particle. The concentration of counterions within the double layer depends on the type of charged particle. The thickness of the double layer also increases with the increase in concentration of counterions in the solution. These counterions are temporarily bound, and they shield the surface charge of the particle. Charge density is greatest near the particle and gradually decreases toward zero as the concentration of positive and negative ions becomes equal.

The formation of double layer neutralizes the charge on a particle. This double layer leads to an electrokinetic potential difference between the surface of the particle and any point in the suspending fluid. This voltage difference, which is known as the surface potential, is of the order of millivolts. The magnitude of the surface potential depends on the surface charge and the thickness of the double layer. As the distance from the particle surface increases, the surface potential decreases almost linearly in the stern layer and then exponentially through the diffuse layer, finally approaching zero at the imaginary boundary of the double layer. The electrokinetic potential curve can be used to find the strength of the electrical force between particles and the distance at which this force arises.

Charged particles with counterions repel each other, while neutral particles attract each other. Nanoparticles are increasingly being used for many applications in electronics, pharmaceutical, biomedical, chemical, and energy generation

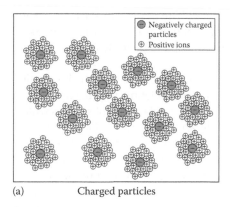 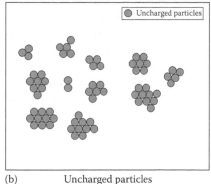

(a)   Charged particles   (b)   Uncharged particles

FIGURE 6.6
Collision behavior of (a) charged particles and (b) uncharged particles. (Reprinted from Gupta, R.K. et al., Eds., *Polymer Nanocomposites Handbook*, CRC Press, Boca Raton, FL, 2010, p. 46. With permission.)

industries. When these particles are charged, they repel each other, whereas uncharged particles are free to collide and agglomerate as shown in Figure 6.6.

To keep the particles separate and prevent them from gathering into agglomerates, the attractive forces between particles must be reduced. When the particle size is very small, the surface force is very large. One of the major surface effects is electrokinetic force. When the particles have the same electrical charge (either positive or negative), they produce a force of mutual electrostatic repulsion between adjacent particles. When the degree of charge is sufficiently high, the particles will remain separate, diffuse, and suspended. When the degree of charge is low, it will create the opposite effect and result in the formation of agglomerates. The charge on the particles can be modified and controlled by changing the pH of the liquid medium or changing the ionic species in the solution. The most widely used technique to prevent agglomeration is the use of surfactants. The surfactants adsorb or chemically bond on the particle surface and change its charge characteristics.

The Derajwin, Landan, Verwey, and Overbeek (DLVO) theory can be used to predict the stability of colloid made with a polar solvent. Based on this theory, the stability of a colloid is determined by the balance of double-layer repulsions and van der Waals attractions:

$$V_B = V_A + V_R \tag{6.8}$$

where
V_A is the attractive potential due to van der Waals forces
V_R is the sum of all repulsive potentials
V_B gives the resultant magnitude of attractive and repulsive potentials

When van der Waals attraction is dominated in a suspension, the colloidal particles will stick together or flocculate, forming aggregates or agglomerates, which settle at the bottom of the suspension. The aggregates are the groups of particles joined through their faces, and hence, it is difficult to break the aggregates, whereas the agglomerates are groups of particles joined through corners or edges and they are easy to break.

Electrostatic forces occur even in a nonpolar solvent, in which the particles can become charged by the desorption of ions or electrons. Repulsive forces are developed between like charged particles. When a polar solvent such as water is used for the suspension, both the desorption and adsorption of ions may cause surface charging of particles. The mechanisms acting on the surface of particles include hydrolysis, dissociation, and complexation. Regardless of the mechanism, a surrounding electrical double layer forms as illustrated in Figure 6.5.

If the ionic strength is increased by adding an electrolyte (acid or base) to the suspension, κ becomes larger and V_R falls off more quickly with distance. This lowers the barrier to the close approach of particles. Eventually V_A becomes dominant causing attraction between particles when the ionic strength is very high.

Varying the pH of suspension alters the charge density on the particle surface and so also alters the zeta potential. The pH value corresponding to zero surface charge and where the zeta potential is zero (isoelectric point, IEP) can be identified. This pH value will vary with ionic strength, and the accurate determination requires the control of the concentration of electrolytes and the prevention of active participation of ions in the surface dissolution/adsorption.

The electrostatic stability of a suspension can thus be manipulated by altering the ionic strength of the suspension, by the addition of a salt to alter the electrical double layer, or by changing the degree and even the sign of the surface charge of the particle by the addition of an electrostatic surface-active agent (surfactant). Large nonionic polymer molecules that associate with the particle surface and sterically hindering the close approach of particles may also be used as surfactants. The stabilization effect of a polymer additive is insensitive to the ionic strength of the suspension, but relatively large amounts of polymer are required, which may cause problems during sintering. Polyelectrolyte additives combine the features of both the systems in that they are adsorbed by ionic means but stabilize the suspension by steric hindrance.

6.3.1.1 Cold Compaction

The composite powder mixture is formed into an NNS part suitable for sintering by using shaping techniques. The simplest shaping method is cold compaction. The appropriate quantity of powder mixture is filled into a die cavity. Pressure is applied through the punches of the die and the powder mixture is compacted to form an NNS of the product. The pressure is released and the green compact is ejected from the die using the lower punch. Friction of the powder with die wall is a main problem in uniaxial powder compaction. The friction causes a decrease in applied pressure with depth in the powder bed. Double-acting compaction (pressure applied simultaneously through top and bottom punches) will result in a more homogeneous pressure distribution in the compact. In either case, the pressure decrease depends on the compact height to diameter ratio. For a large ratio, the pressure decreases more rapidly with depth. Hence, for a homogeneous compaction, small height to diameter ratio is desirable. The problem of pressure variation, and hence density variation, at different depths can be avoided by cold isostatic pressing (CIP), in which pressure is applied from all sides. In this method, the powder mixture is placed in a flexible mold and the mold is placed in a pressure vessel. A liquid pumped into the vessel at high pressure is isostatically compressing the mold and thus compact the powders. A binder or lubricant may be added to the powder to facilitate contact between particles and reduce the friction.

The compaction of the powder occurs by particle rearrangement, particle deformation, and particle fragmentation. Initially, the particles are loosely packed with a low coordination number (the number of particles in contact). There is an excess of void space between the particles and the highest density achievable after filling the die is the tap density. As the pressure is applied to the powder, the first response of the particles is to fill the voids by rearrangement, which increases the packing density of the powder. As the pressure increases further, the particles will undergo elastic and plastic deformation. For brittle ceramic powders and reinforcements, extensive plastic deformation cannot occur; instead the particles will undergo fragmentation. These fragments will fill the voids between the particles and enhance the green density of the component. Only low green densities (typically 40%) are achievable for whisker-, platelet-, or fiber-reinforced ceramic composites by die compaction.

6.3.1.2 Slurry Impregnation

The slurry impregnation process is perhaps the most important technique used to produce continuous fiber-reinforced glass, glass–ceramic, and other CMCs. This process is analogous to the PMC processing and the main difference being the requirement of consolidation step. This process involves two main stages: (1) impregnation of fiber by passing through a slurry made out of matrix powder and (2) consolidation by hot pressing. Figure 6.7 shows a schematic of the slurry impregnation process. Slurry is generally prepared by ball milling the matrix powder, binder, and dispersant in a suitable solvent (e.g., ethanol or distilled water) using appropriate milling media that matches with the matrix material in terms of composition and hardness. Wetting agents may be added for better impregnation of fiber tow or preform. The fiber tow or preform is impregnated with the slurry by passing it through the slurry tank. The slurry is required to be agitated during impregnation in order to avoid settling of the matrix powder. The impregnated tow or preform is wound on a drum and dried and this impregnated sheet is similar to the prepreg used in the fabrication of PMCs. This is followed by cutting and stacking of sheets and consolidation by hot pressing. This process has the advantage of stacking the impregnated sheets in a variety of sequences, for example, unidirectional, cross-plied ($0°/90°/0°/90°$, ...), or angle plied ($+\theta/-\theta/+\theta/-\theta$, ...).

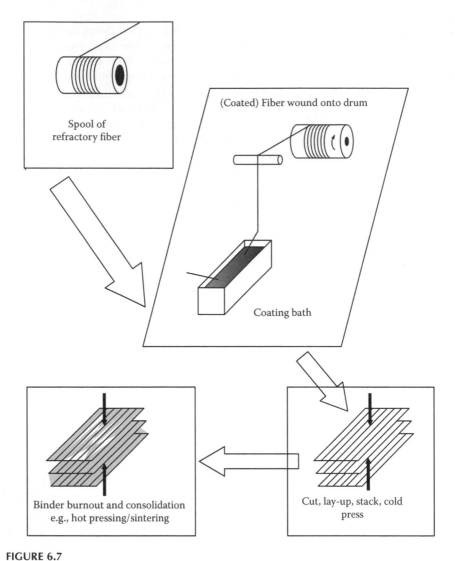

FIGURE 6.7
Schematic of processes involved in filament winding of ceramic composites. (Reprinted from Warren, R. (volume ed.), in *Comprehensive Composite Materials*, Vol. 4, eds. A. Kelly and C. Zweben, Elsevier Science Ltd., Oxford, U.K., 2000, p. 647. With permission.)

The binder in the green body is removed by controlled heating to a temperature at which it volatilizes (usually at 500°C). The temperature is maintained until full binder burnout is achieved. Faster binder removal should be avoided since it can lead to the formation of cracks. The binder removal rate should be optimized depending on the size of the part. It is also necessary to optimize the binder content, since excessive binder can cause problems during binder burnout such as slumping of the part. The slumping may cause fiber misalignment in the green compact. It can be avoided by adding a plasticizer to the slurry composition, and the plasticizer can also impart some flexibility to the sheet when dry. Binders, being generally organic materials, can leave a carbon residue on incomplete burnout. The carbon, though usually inert, can react with some materials at high consolidation temperatures, which is detrimental to the properties of the final component. Alternatively, the binder can also be removed during hot pressing, which is the primary method used for densification. Here also a slow heating rate must be used to remove all the binder from the part without causing any cracks.

The slurry impregnation followed by hot pressing is well suited for glass and glass-ceramic matrix composites, mainly because the hot pressing temperatures for these materials are much lower than those used for other ceramic materials. The hot pressing process has the limitation of not being able to produce complex shapes, and hot isostatic pressing can be used in that case. Very-high-pressure application can easily damage the fibers and decrease the strength of the composite. The fibers may also be damaged by the mechanical contact of hard matrix particles or by the reaction with the matrix at very high consolidation temperatures. The composite should have minimum porosity as possible, since porosity in a structural composite is highly undesirable. To effect this, it is important to completely remove the fugitive binder and use the matrix powder having particles smaller than the fiber diameter. Controlling the hot pressing operational parameters is also very important to produce a quality composite. The pressure and temperature cycle used during hot pressing of these composites is shown in Figure 6.8. Precise control of temperature within a narrow range, minimization of processing time, and utilization of a pressure low enough to avoid fiber damage are the important factors of hot pressing process. The slurry impregnation process with hot pressing generally results in a composite with fairly uniform fiber distribution, low porosity, and relatively high strength.

Slurry impregnation processes have been used successfully to produce dense composites of carbon and silicon carbide fibers in borosilicate glass, lithium aluminosilicate, magnesium aluminosilicate and barium magnesium aluminosilicate glass–ceramics, and silicon nitride, SiALON, silicon carbide, zircon, and mullite ceramic matrices.

The main advantage of filament winding is its relatively short fabrication time. It allows the fabrication of cylindrical shapes with different orientation of fibers in each layer. The main limitation of this process is that it does not allow the fabrication of complex-shaped parts. Fibers should not be damaged while handling, otherwise the damages such as surface notches will serve as stress raisers and initiate cracks under load. The presence of inhomogeneities in terms of matrix-rich and fiber-rich regions in hot pressed composites is very common, which needs to be addressed, otherwise it can weaken the composites.

A more recent variation of the slurry impregnation process is the infiltration of 2-D or 3-D fiber preforms. Textile production techniques such as weaving, stitching, knitting, and braiding have been employed to prepare preforms with various fiber architectures. Wetting characteristics of the

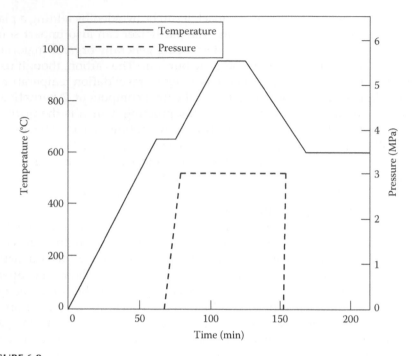

FIGURE 6.8
Pressure and temperature cycles used during hot pressing of alumina fiber-reinforced glass composite. (Reprinted from Chawla, K.K., *Ceramic Matrix Composites*, Kluwer Academic Publishers, Boston, MA, 2003, p. 111. With permission.)

slurry and particle size distribution of the ceramic powder are critical issues during the infiltration of slurry into these preforms. If these issues are not properly addressed, then the composites exhibit large local variations in fiber volume fraction and relatively large regions of matrix-free fibers. A configuration, which retains the fiber separation of the fiber network, is necessary for efficient infiltration to achieve dense composites with high fracture toughness. Inhomogeneous distribution of the powder may be a problem with infiltration, since the preform may act as a filter, especially when the thickness of the cross section is high.

6.3.1.3 Sol–Gel Processing

Sol–gel processing has been used for many years for the production of ceramics, but only recently it has been used to make glass, glass–ceramics, and CMCs. The sol–gel processing technique has already been described in detail in Chapter 2 and a brief description is given here. The sol is a colloidal system in which fine solid particles are dispersed in a liquid medium. In the sol, the particles are usually 1–100 nm in size. At sufficient concentrations, these very small particles link together and form chains and then a 3-D network, which is called gel. The sol is converted to gel by the controlled evaporation of the solvent or by changing the pH, which in turn is subjected to controlled heating to produce the desired ceramic product. It can be a glass, glass–ceramics, or a crystalline ceramics depending on the starting material and heat treatment. The gel to ceramic conversion temperature is usually much lower than that required in a melt or sintering process.

Sol–gel processing is an attractive method to produce some special ceramic materials. Some of the advantages of this process are lower processing temperatures, greater compositional homogeneity in the matrix, and potential for producing unique multiphase matrix. These attributes will be useful for CMCs as well; in particular, the reduction in matrix densification temperature is more important for ceramic composites, since this will reduce the possibility of deleterious reactions between the fibers or fiber coatings and the matrix. However, sol–gel processing is not widely used for the fabrication of ceramic composites because of its inherent problems. The main disadvantages are high shrinkage and low yield compared to powder slurry techniques. A large density of cracks is formed in the matrix because of this high shrinkage during drying. This warrants repeated infiltrations to produce a substantially dense matrix. Linear shrinkages of the order of 30% are more common in this process. Although such a large shrinkage may not give any problem to the monolithic ceramics if processing conditions are carefully controlled, the fibers in a sol-impregnated compact restrict the possible change in external dimensions, and hence, cracking is inevitable.

Sol–gel technique can be used to prepare prepregs similar to slurry impregnation. Many polymer composite processing techniques can be used with sol to prepare CMCs. Similar to filament winding of PMCs, fiber tows or rovings are passed through a tank containing sol and the impregnated fiber is wound on a mandrel to the desired shape and thickness. After the gelation, the part is removed from the mandrel. A final heat treatment at high temperature converts the gel into a glass or crystalline ceramic matrix. The sol–gel technique allows the infiltration of low-viscosity liquids, such as the sols derived from alkoxides. Further reduction of the viscosity of sol is also possible, by adding more solvent. Although this process is more suitable for producing oxide ceramic matrices, non-oxide ceramic matrices can also be produced by using the polymeric precursor itself instead of sol. The yield of non-oxide ceramics is higher than those in chemical vapor deposition (CVD) processes.

Sol–gel process, which has been widely used to prepare various forms of ceramic materials, can also be used to form a ceramic matrix in the interstices of a porous preform. More information for preparation of preform can be found in the Chapter 5. Composites are made by infiltrating the sol into the fiber preform and gelled by drying at low temperature, usually less than 150°C. The gel can then be converted to a glass, glass–ceramics, or ceramics by subjecting it to a controlled high-temperature treatment. The heat treatment is carried out at temperatures hundreds of degrees centigrade lower than that required for sintering of powder compacts. This is attributed to high reactivity of the gel because of high surface area. The use of lower temperatures suppresses any adverse interfacial reactions between the matrix and the fiber. In addition, the materials have a uniform composition (if more than one oxide is used) due to intimate mixing of fine ceramic particles in the colloidal state. Other advantages include the requirement of simple, low-cost tooling due to the low temperatures involved. The tooling is removed after the initial drying step, and the component is heat-treated in a free-standing condition. It is a net-shape processing technique and can lead to substantial reduction in machining costs. The process is also easy to scale up, which requires only larger infiltration tanks, drying ovens, and furnaces. Often no changes to the sol chemistry and physical processes are needed.

The main disadvantage of this process is the tendency of the gel to shrink substantially during the drying stage, which can lead to matrix cracking. Many approaches have been adopted to overcome this problem, which include the use of drying control chemical additives (DCCAs) and the addition of filler particles to the sol. Some of the effective DCCAs are formamide, glycerol, and oxalic acid. It was found that these additives produced a narrower pore size distribution, thus reducing differential drying stresses. High-density SiC fiber-reinforced alumina and mullite composites have been produced using filler in the sol (Chen et al. 1989, 1990). Boehmite powder and colloidal silica were used to prepare alumina and silica sol, respectively, and α-alumina powder was used as filler. The filler was deflocculated ultrasonically in dilute nitric acid at a pH of 3.5. To prepare the mullite sol, an appropriate amount of silica sol was added to the alumina sol. The fibers were incorporated by arranging them in a mold and then infiltrating with alumina sol or mullite sol mixed with filler to produce SiC/alumina or SiC/mullite. The composite was dried in the mold, removed from the mold, and then sintered at ~1300°C for 1 h. The mullite matrix was found to sinter well and yielding a high-density composite.

The type, amount, and particle size of the filler are the important parameters to be considered before adding into a sol. Increasing the filler content will increase the viscosity of the sol and the strength of gel and reduce the porosity. For better infiltration, smaller particles are preferred. Another advantage of fine particles is that they can remain dispersed in the suspension better than coarser ones, which need to be agitated to prevent sedimentation. There is no doubt that the fillers can improve the yield, but often they also increase the final heat treatment temperature. Some fillers can degrade the fibers and affect the bonding between the fiber and matrix. For example, the addition of glass–ceramic fillers containing lithium degrades mullite fiber and causes fiber–matrix bonding.

Another method to minimize the problem of microcracks due to shrinkage is hot pressing of the sol-impregnated compact. However, this entails the inherent limitations of hot pressing such as specimen size, shape, and complexity, as well as cost.

Sol–gel processing is one of a few methods currently available to infiltrate 3-D woven preforms without damaging or altering fiber architecture. Chemical vapor infiltration (CVI) technique can also be used, but this technique is restricted mainly to form non-oxide matrices. A major disadvantage of the sol–gel processing is that it needs multiple infiltrations to obtain a substantially dense matrix. Therefore, the most successful use of sol–gel technology is the modified slurry infiltration process, in which the sol is used as the carrier liquid. However, there are considerable advantages in using sol rather than powder slurry for infiltration, especially for hard ceramic matrices. Mechanical damage of fibers by particles is minimized and the infiltration is more homogeneous when sol alone is used for infiltration.

The problem of restricted shrinkage in the presence of fibrous materials has been taken as an advantage in forming lightweight 20 vol% SiC_w/SiO_2 composites with a reinforced cellular structure. This type of composite can be fabricated by adding a foaming additive to the sol and sintering to obtain relative densities as low as 10%. The use of fiber reinforcements permits the fabrication of significantly larger sol–gel-derived foams because of the large reduction in drying shrinkage by the presence fibrous materials. The reinforcement also significantly increases the strength of the sintered SiC_w/SiO_2 composite foams relative to complete silica foams. It is expected that whisker and fiber reinforcements will lead to similar enhancements in the mechanical properties of other lightweight materials. Thus, the structural applications and performance of lightweight cellular materials can be considerably expanded by the use of whisker or fiber reinforcement.

When the V_f of continuous reinforcing fiber is above 0.05–0.10, the mechanical performance of the CMC is dominated by the fiber properties and the matrix has little influence. For applications where the composite product must be gas-tight, glazes can be applied to seal the surface, but they may not be useful at high-temperature applications. With short, random fiber reinforcement, however, the mechanical properties of the composite are dominated by the matrix characteristics and the residual porosity limits the maximum strength below 50 MPa. The properties of some sol–gel-processed CMCs are given in Table 6.3.

If the production of low-cost CMCs by freeze-drying or other methods is developed, then the prospects for their wide applications are likely to be enhanced to greater levels. The advantages they offer over conventional engineering materials are similar to monolithic ceramics but with enhanced mechanical performance. It should be noted that in many applications the actual mechanical loading to components is much lower than the presently used materials can safely accommodate, but the selection of which has been based on the other properties. For example, in a biochemical engineering application, the part may be exposed to pressures of a few atmospheres and temperatures below 300°C and the prime requirements of material are chemical inertness and corrosion resistance. Stainless steels are generally used for these applications, and CMCs show potential substitutes for these materials with the added advantage of lower density that allows reduction in the size of support structures. Hence, the ability to control the size and amount of porosity, and obtain adequate mechanical properties and good corrosion resistance, suggests the potential for using these sol–gel CMCs

TABLE 6.3

Summary of Mechanical Properties of Typical CMCs Prepared by Sol–Gel Processing through Freeze Gelation

Fabrication Route	Fiber	Main Filler Powder	Density (g cm^{-3})	Dynamic Modulus (GPa)	Flexural Strength (MPa)	ILSS[a] (MPa)	Work of Fracture (kJ m^{-2})	Max. Processing Temperature (°C)
1. Filament winding	FP alumina	Mullite	2.19	—	220	—	2.90	1000
2. Filament winding	Nextel	Silica	1.91	48.1	202	>25	4.88	700
3. Filament winding	Nextel	Silica–zirconia	1.95	40.7	104	>20	1.03	1000
4. Filament winding	Carbon	Low-expansion glass–ceramics	1.70	45.4	400	31.4	10.30	750 (in Ar)
5. Casting	Saffil	Mullite	1.88	30.2	46	—	0.23	1050
6. Casting	Saffil	Silica–zirconia	2.38	29.0	32	—	0.15	1150

Fiber volume fractions in continuous filament-wound material are typically 0.15–0.35 (0.55 max.), and in short Saffil fiber-reinforced materials are typically 0.08–0.15 (0.30 max.).

The residual porosity in continuous filament-wound material is typically 20%–25% and 25%–32% in short Saffil fiber CMCs. This can be increased to about 80%, though with a consequential reduction in mechanical performance.

Material 3 shows no reduction in mechanical properties after repeated water quenching from 550°C. Material 5 exhibits a gradual (near-linear) reduction in performance. As the quenching temperature is increased to 600°C, the flexural strength had declined by about 30%.

Source: Data from Russell-Floyd, R.S. et al., *Ceramic International Technology 1994*, Birkby, I., Ed., Sterling Publications Ltd., London, U.K., 1993, p. 62.

[a] ILSS, interlaminar shear strength.

in a variety of applications requiring a porous, inert material. Some of the potential applications include hydraulic oil filtration devices, catalyst supports, precision refractories including metal casting molds, and low-density, fire-retardant boards.

The sol can also be used as a medium to disperse the ceramic powder to prepare slurry. The advantage of this sol-based slurry is that the sol can act as a binder and the binder burnout step can be eliminated because the material formed from the sol can become part of the composition of the final ceramic matrix.

An alternative to the conventional gelation is freeze gelation, which allows the formation of crack-free, essentially zero-shrinkage composites with a variety of shapes. The flow diagram for the production of CMCs by this sol–gel processing is outlined in Figure 6.9. A sol is formed from a metal alkoxide, acetate, or halide by a suitable chemical reaction. Generally, the sol prepared from inorganic salts is cheaper than that prepared from metallo-organic compounds. However,

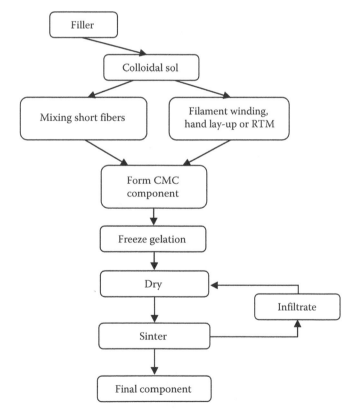

FIGURE 6.9
Application of sol–gel processing techniques for the manufacture of fiber-reinforced ceramics. (Reprinted from Russell-Floyd, R.S. et al., *Ceramic International Technology 1994*, Birkby, I., Ed., Sterling Publications Ltd., London, U.K., 1993, p. 62. With permission.)

the preparation is easy and better control of microstructure is possible with metallo-organic compound-derived sols. Numerous reports are available for the preparation of sol from different materials. Somewhat larger filler particles (1–10 μm) can be dispersed into the sol to increase the solid yield. This mixed sol is then combined with the reinforcing fibers to form the shape by using the well-known processing techniques of fiber-reinforced plastics, in which the polymeric resin is replaced by the sol. Either liquid nitrogen is poured over the mandrel or the mandrel is soaked in a liquid nitrogen bath to cause freeze gelation of the matrix. Fiber/gelled matrix sheets are removed from the mandrel, thawed to room temperature, and dried at 50°C. This results in a product almost without shrinkage and certainly none of the cracking inherent in the usual sol–gel processing. The dried sheets are cut to the required dimensions. Although the green samples are porous, it is easy to handle them. The sheets are again infiltrated with sol and then dried. This is usually repeated several times till appreciable density is achieved. Because of the high reactivity of sol–gel-derived matrix, much lower sintering temperatures are required, typically less than 850°C with continuous fiber reinforcement. In order to enhance the mechanical properties, the inherent porosity within the matrix would be reduced below 20% by volume through repeated sol infiltration assisted by a vacuum and small overpressure. The main requirement for a sol to be freeze-gelled is that it should be free from aggregates as these can inhibit the formation of dense composite.

The freeze-gelation technique has been used extensively on systems based on colloidal silica. However, it is also applicable to other oxide systems, such as alumina and mullite. Various colloidal oxides are available commercially, but multi-oxide colloids are not available in large quantities; hence, they are expensive.

A number of factors should be considered while using sol–gel processing involving freeze gelation. They are

- The sols must have very fine particles (<100 nm in diameter) to successfully infiltrate fiber tows without closing surface pores (once the surface pores are sealed, no further infiltration can occur).
- The sol must be free from aggregates, since the aggregates are known to retard densification.
- The influence of freezing rate, freezing temperature, and thawing rate is very complex.
- Repeated infiltration can be carried out in either the dried or post-sintered condition. The best infiltrations are achieved by applying vacuum.
- To improve the density of composites, the sintering temperatures can be increased (however, the possibility of fiber–matrix interfacial reaction with consequent reduction in composite toughness also increases).

The main advantages of this process are the ability to produce large or small complex-shaped components to NNS with multidirectional fiber reinforcement. To obtain different shapes, simple casting can be used with short-fiber reinforcements, while filament winding and even hand lay-up can be used with long-fiber reinforcement. Minimal machining is required for the product, which normally can be completed at the dried stage (before sintering). The volume fraction of porosity in the product can be varied between 0.2 and 0.8, and the pore size can range from 1 to 20 μm in diameter, which gives potential for use in applications requiring a microporous material. Moreover, the precursor materials of sols and filler powders are inexpensive and widely available compared to the polymer precursors used in polymer impregnation/infiltration and pyrolysis (PIP) process. The two main disadvantages are that the solid content in the sol is very less and some residual porosity always remains. The development of other freeze-dryable sols and improvement in the infiltration process may reduce these problems.

6.3.1.4 Electrophoretic Deposition

The phenomenon of electrophoresis has been known since the beginning of the nineteenth century, but its applications to the processing of ceramics and CMCs are relatively recent. Electrophoretic deposition (EPD) is used to form shaped ceramics directly from a colloidal suspension. EPD is applied commercially to produce a wide variety of structural and functional monolithic ceramics and functionally graded composites. Composites are produced by this method by substituting the depositing electrode with a conducting fiber preform. The controlled deposition technique produces homogeneous green bodies with better control of flaw population. A dense homogeneous green body produced by this process allows densification to take place at lower temperatures without the application of pressure. In a polar liquid, the suspended particles acquire a charge whose sign and magnitude depend on the interaction between the particle surface and the liquid surrounding it. The pH of the suspension can have a large effect if hydrogen ions in the liquid play a role. When an electric potential is applied into the suspension, the charged particles move and this phenomenon is called electrophoresis. The movement of particles in an aqueous suspension under an electric field is governed by the field strength, pH, ionic strength, and viscosity of the suspension (Illston et al. 1993). The electrophoretic mobility of charged particles in a suspension is given by the Smoluchowski equation (Brown and Salt 1965):

$$\text{Electrophoretic mobility} = \frac{u}{E} = \frac{\varepsilon\zeta}{4\pi\eta} \tag{6.9}$$

where
 u is the velocity
 E is field strength
 ε is the dielectric constant of the medium
 ζ is zeta potential
 η is viscosity

A suitable suspension for the EPD process should have high electrophoretic mobility. For better electrophoretic mobility, the suspension should have

1. High particle surface charge
2. High dielectric constant for the liquid phase
3. Low viscosity
4. Low conductivity to minimize solvent ion transport

The stability of suspension is generally very high when the surface charge on the particles is high, which leads to greater repulsion between the particles. The surface charge of the particles can be engineered by the addition of surfactants to achieve a high charge of required sign. Since the surfactant is retained in the green body, it should be a clean burning material, that is, it should not produce any residue during high-temperature densification. Deposition of particles occurs at either the anode or cathode depending on the polarity of the charge on the particles. Electrolysis of the suspending medium may also occur at the same time during EPD. The liquid medium should be selected in such way that it is not producing any gases at the depositing electrode; otherwise it may disrupt the formation of the green body.

EPD makes use of nanoscale ceramic particles in a stable non-agglomerated form, such as sol, and exploits their net surface electrostatic charge characteristics. On application of an electric field, the solid particles will migrate toward an electrode and deposit there. If that electrode is replaced by a conducting preform, the suspended particles will be attracted by the preform and deposited within it. This provides an appropriate means of effectively infiltrating densely packed fibrous bundles. A schematic of a simple EPD process is shown in Figure 6.10. EPD is different from electroplating in many aspects. In electroplating, metallic ions move to the cathode, get reduced, and deposit over the cathode. On the other hand, in the EPD process, solid particles with surface charges migrate toward the corresponding electrode and get deposited. Moreover, the deposition rate in EPD is of the order of 1 mm min^{-1}, while it is of the order of 0.1 mm min^{-1} in electroplating. EPD is a relatively simple and inexpensive technique, which can also be used for the infiltration of tightly woven fiber preforms. A mixed sol containing two different solid particles to form a compound can also be deposited by this process. For example, to form a mullite composition, a mixed sol containing silica and alumina is used. The primary requirement for this process is the availability of reasonably conducting preform. If the material of the preform is not having good electrical conductivity, then a coating of electrically conducting material can be given, to make it conductive. For example, nickel coating is given to Nicalon SiC fiber before EPD.

A nonconducting fiber preform can also be infiltrated by placing it in front of the depositing electrode. EPD for producing continuous fiber-reinforced CMCs was originally applied to electrically conducting fibers and oxide matrix

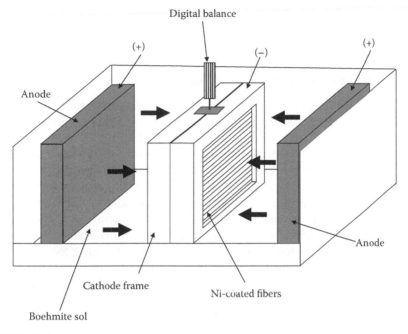

FIGURE 6.10
Schematic of a basic EPD cell. (Reprinted from Kaya, C. et al., *J. Am. Ceram. Soc.*, 83, 1885, 2000. With permission.)

materials. Nicalon SiC fibers with 100 nm thick carbon coating have been used to make a preform for the EPD of silica and alumina using their respective sols. A novel EPD technique in which a filter was introduced between nonconducting Nextel-mullite fiber preform and depositing electrode. This allowed the application of potential up to 60 V to infiltrate particles within the nonconductive weave, while isolating the deposit from the disruptive influence of hydrogen gas evolved from the aqueous medium at this voltage.

Electrophoretic infiltration from essentially a single-phase suspension into carbon or carbon-coated SiC fiber preform could be easily achieved. The technique has more recently been applied to infiltrate mixed oxides into the preform. A mullite matrix can be formed by infiltrating a mixed sol of submicron-sized alumina and silica particles into fiber-weave (Kaya et al. 1998). Careful manipulation of the surface properties of particles is required to achieve consistent deposits of the required composition. The mobility of different powders should be matched by the use of appropriate surfactants and then combined in the compositional ratio required to form the desired matrix material. Using similar principles, three-component deposition has also been achieved. Mullite is added as a third additive to act as a seed to promote transformation during the infiltration of alumina and silica.

Densification of these oxide-based composites by pressureless sintering has been demonstrated, but the temperatures required are between 1350°C and

1400°C, which are still at a level to induce damage into the fibers. Full conversion to mullite is not achieved at these temperatures, when the required amount of alumina and silica is infiltrated.

6.3.1.5 Other Ceramic Particle–Based Methods

Slip casting and tape casting are slurry-based techniques and extrusion and injection molding are plastic-forming techniques. All these methods consist of four basic steps: (1) feedstock preparation, (2) shaping, (3) binder removal, and (4) consolidation. The feedstock, typically slurry or a paste, is a mixture of the matrix powder, reinforcement, and an appropriate binder. The role of the binder is to impart the proper rheological properties required for shaping and give sufficient strength for the green body. Binders soluble in water or nonaqueous solvents are used for the slurry-based techniques, and organic polymers are used in plastic-forming techniques. Slurry-based techniques are performed at room temperature, whereas plastic-forming techniques are carried out above the melting temperature of the polymer.

The viscosity is an important property of the feedstock for shape forming. Binder content is one of the most important parameters to modify the viscosity of the feedstock, the binder physically separates the individual particles, and also it acts as a lubricant during injection molding or extrusion. If too much binder is added, the feedstock will be too soft and pliable and most likely the part will slump. Conversely, if insufficient binder is added, the feedstock will not flow easily and high pressures are required for injection molding and extrusion. Hence, the minimum amount of binder required for flow should be used, as this will produce a green specimen with the highest possible density.

Shaping involves pouring the feedstock into a mold or forcing it through a shaped die. In the slip casting process, the slurry is poured into a porous plaster of Paris mold. Capillary action associated with the porous mold absorbs liquid from the slurry, resulting in the formation of a rim of solid green compact along the mold wall. Particle size, size distribution, and shape influence the green density, and those factors that promote packing efficiency of powders also promote high green densities. The solid loading and rheology of the slurry will also influence the green density. In general, deflocculated slurry will form a dense component with low permeability at a slower rate, whereas flocculated slurry will form an open and permeable compact with low green density but at a faster rate. On commercial point of view, the processes need to be fast; however, green structures with agglomerated particles and low densities tend to have poor sintered strength. The permeability of the green structure also influences the drying rate. Hence, the properties of the slurry must be carefully controlled during slip casting. Relatively large and complex shapes can be made by slip casting. Tape casting is similar to slip casting, except that the slurry is allowed to flow through a "doctor blade," which produces thin green sheets.

Whisker-reinforced CMCs are generally made by dispersing the whiskers in ceramic powder slurry, followed by casting, drying, and hot pressing. Sometimes hot isostatic pressing rather than hot pressing is used, especially when complex shapes are made. Whisker agglomeration is a major problem in this process, but mechanical stirring and adjustment of pH level of the slurry can help to minimize the problem. The addition of whiskers into the slurry increases the viscosity and also whiskers with high aspect ratios (>50) tend to form bundles and clumps. Obtaining well-distributed and deagglomerated whiskers is of great importance for achieving reasonably high densities. The use of organic dispersants, deflocculation by pH control, and agitation by an ultrasonic probe are of great help in achieving uniform distribution. SiC whisker-reinforced composites are generally hot pressed in the temperature range of 1500°C–1900°C and pressures in the 20–40 MPa range.

In extrusion and injection molding processes, the feedstock is forced through a die to get shaped components. In extrusion, an open die is used to produce shapes with regular cross sections, such as tubes and rods. In injection molding, the feedstock is forced into a closed die cavity to form complex shapes.

Binder removal is one of the critical steps in these processes. Binder removal can be performed by chemical or thermal means. For aqueous and nonaqueous solutions, the binder removal is easy; a simple drying can do. Here the permeability of the green structure is a controlling factor, and drying should be carried out at a slower rate, depending on the thickness of the component. Polymer-based binders may be removed by thermal treatment in reactive atmosphere or leaching in chemical solutions. The binder removal by thermal treatment involves the thermal degradation of the organic compounds into volatile species and the diffusion of these species to the surface of the component. The time required for binder removal is proportional to the square of the cross section of the component; hence, the process can be quite time consuming for thick parts. Partial binder burnout or residues remaining after pyrolysis can be detrimental during the subsequent sintering or become sources of property-limiting defects in the consolidated component. The binder removal by chemical leaching involves the dissolution of binder selectively by an appropriate solvent, without affecting the integrity of the component.

Consolidation of the shaped components occurs by sintering at high temperatures. Pressureless sintering may not be suitable for many discontinuously reinforced CMCs. Unfortunately, these shaped components are not readily amenable to consolidation by pressure-assisted sintering methods.

A preferential orientation of the reinforcement phase may develop during the shaping process. Whiskers or fibers will tend to align perpendicular to the direction of die punches and doctor blade during die compaction and tape casting, respectively, and parallel to the mold/die surface during slip casting, extrusion, and injection molding. The preferred orientation can result in anisotropic shrinkage during sintering.

It is possible to control the degree of orientation during extrusion and injection molding by careful die design and control of the feedstock flow patterns. Particles with aspect ratio >1 will tend to align under certain flow conditions. If the flow converges, then the fibrous materials will align parallel to the flow direction, whereas they align perpendicular to the flow direction when the flow diverges. However, for the alignment of fibers, the maximum particle size of matrix powder should not be greater than the edge-to-edge distance between the fibers. In addition to that, the fiber diameter and volume fraction also control the maximum powder particle size required for alignment. A simple geometric model has been derived, and it reveals that the matrix powder size required for alignment is inversely proportional to the fiber volume fraction and directly proportional to the fiber diameter.

6.3.1.6 Layered Ceramic Structures

A simple way to prevent the growth of larger flaws at a reasonable cost is to make layered ceramics with crack-deflecting interfaces. Any crack that starts to grow in the material is deflected at the first interface that it encounters. The presence of different materials in the adjacent layers may give rise to residual stresses, which are developed while cooling from the sintering temperature due either to differences in the thermal expansion coefficients or to the occurrence of phase transformations. The result is the presence of compressive stress in a layer is balanced by tensile stresses in the adjacent layers. It is a well-known fact that the growth of flaws in a compressed layer is more difficult. The real benefit can be obtained by ensuring that the layers with tensile stress are much thicker than those with compressive stress, so that the magnitude of tensile stress is small compared to that of compressive stress. An example of layered structure is SiC/graphite laminate, consisting of 150 μm thick SiC layers and 3 μm thick graphite layers. While cooling from the sintering temperature, SiC contracts more, thus introducing compressive residual stress in the graphite layers.

6.3.1.6.1 Fabrication

Sheets for producing a layered ceramic material can be easily made from the corresponding powders. Ceramic powders can be mixed with viscous polymer solutions and then rolled to produce sheets or tapes. The individual sheets are then usually coated in some way, so as to ensure a sufficiently weak interlayer in the final composite. It is advisable to use the coating with some polymer to enable the coated layers to stick together on pressing. These stacked layers can then be sintered without any applied pressure.

Rather than stacking different layers, layered structures can also be built by alternatively depositing layers of different materials. One of the techniques used for this type of deposition is EPD. Layered structure can be deposited by placing the electrode in two different suspensions.

Another method to deposit individual layers is plasma spraying, which offers the advantage of rapid deposition, but the interfaces are relatively rough and the minimum thickness of deposit is around 20 μm. However, this process is well suited to the production of compositionally graded structures. Molecular deposition techniques can also be used, which have the potential for very accurate control of layer thickness, although they are very slow and expensive.

The layered compact must be sintered after the assembly. While sintering, the compact must withstand any stresses that might develop due to differences in sintering rates of materials in different layers. It should also withstand the stresses that will develop on cooling from the sintering temperature due to differences in the thermal expansion coefficients between the different materials or to any phase transformations that might occur.

Despite their simple geometry, layered ceramics can be produced with a range of properties tailored for specific applications. It has the potential to dramatically reduce the manufacturing costs of toughened ceramics, which is the major factor limiting the widespread use of fibrous composites.

6.3.1.7 Sintering

Composite forms produced by slurry method, EPD, or sol–gel processing are in a green state, in which the matrix phase is an assembly of powder particles rather than a continuous solid. A sintering process is required to effect the conversion to a continuous solid. The driving force for densification during sintering is the reduction in overall surface area. The densification can be achieved through solid-state diffusion in ceramics or viscous flow in glasses. Ceramics can be densified by liquid-phase sintering, where there is a combination of diffusion and flow acting to densify the matrix, but this method is seldom used for composite densification. Densification by diffusion and viscous flow at elevated temperatures is enhanced by the application of pressure; hence, hot pressing and HIPing are often used for CMC densification. The high temperatures required to densify ceramics are of a level that can cause degradation of fibers. The simultaneous use of pressure and temperature restricts the components to simple shapes. Hence, the requirement of densification processes is thus to achieve the highest density at the lowest feasible temperature without the application of pressure, if possible.

6.3.1.7.1 Liquid-Phase Sintering

Liquid-phase sintering is used especially when solid-state sintering is difficult or requires very high temperatures. A low-melting liquid phase is formed in this process, which facilitates density increase by accommodating particle rearrangement and providing a fast diffusion route for atoms dissolved from the solid phase. It is also possible to sinter the usually difficult-to-sinter covalent ceramics by this process. A disadvantage of this method is the persistence of a glassy intergranular phase after densification that may impair high-temperature mechanical properties. The allowable volume of liquid phase is generally restricted to below 5 vol.% for advanced ceramics, which may not be sufficient to fill the interparticle pores, a fact that has implications for the mechanism of densification.

6.3.1.7.2 Retardation Effects of Reinforcement during Sintering

Incorporation of a rigid, non-sintering, reinforcing phase into a matrix significantly affects the sintering behavior of matrix material. Initially it was thought that a densifying matrix shrinks around a non-sintering reinforcement, experiences a tensile stress, and retards the densification of matrix. Later it was found that only the interactions between the reinforcements rather than the tensile stresses developed in the densifying matrix control sintering. During sintering the matrix phase will tend to shrink, but the rigid reinforcements resist the shrinkage. The net result is that either the matrix phase separates from the reinforcement and forms pores adjacent to the reinforcement phase or it sticks to the reinforcement phase and the constrained densification results in the retention of pores in the matrix. It has been suggested that the sintering of the matrix by neck growth without shrinkage will give acceptable composite properties, though it may leave some amount of porosity.

At high volume fractions, the reinforcement phase can form a connective network throughout the composite. This network will resist densification and lead to a porous microstructure. For fibrous reinforcements, the critical volume fraction for connectivity decreases as the aspect ratio increases; hence, composites containing smaller aspect ratio reinforcements are easier to densify by pressureless sintering. In any case, hot pressing is needed for most composites to achieve good densification.

Sintering is an elevated temperature process, and hence, large residual stresses can develop in composites due to thermal expansion mismatch between the constituent phases. These stresses can sometimes be large enough to cause cracking in the composite. The stress developed in the matrix (σ_T) due to thermal expansion mismatch can be estimated as

$$\sigma_T = E_m \varepsilon_T \approx E_m \Delta_\alpha \Delta T \tag{6.10}$$

where
E_m is the modulus of the matrix
ε_T is strain in the matrix corresponding to σ_T
Δ_α is the difference between the thermal expansion coefficients of the constituents
ΔT is the temperature difference between processing and ambient temperatures

However, the severity of matrix cracking due to thermal expansion mismatch is dependent on the size of the reinforcement phase. Experimental observation and theoretical analysis revealed that the smaller reinforcements did not cause cracking of matrices during composite processing. A non-dimensional quantity S has been defined (Lu et al. 1991), which is related to thermal expansion mismatch, the fracture toughness of the matrix, K_m, and the size of the reinforcement phase, r, as

$$S = r\left(\frac{E_m \varepsilon_T}{K_m}\right) \tag{6.11}$$

It has been suggested that all forms of matrix cracking will occur profusely when S is greater than 10. Conversely, when S has a value equal to or less than unity, matrix cracking due to thermal expansion mismatch will be absent. This criterion establishes, to a first approximation, the maximum size of the reinforcement that can be incorporated into a matrix without causing thermal expansion mismatch cracks.

Another problem that can arise during consolidation is reaction between the constituents. In some instances, these reactions are utilized for in situ formation of reinforcement or matrix phase during processing. However, if these reactions are unintentional, the interphases formed at the phase boundaries can have a deleterious effect on the mechanical properties of the composite. During pressureless sintering, extensive reactions can also generate porosity. Porosity will be generated when the volume shrinkage is not equal to the concomitant decrease in volume associated with the new phase formation (usually the new phases have different densities compared to those of constituents). Hence, this type of in situ composite processing needs to perform under applied pressure to produce a dense composite.

Sintering is a lower-cost process with greater shape and size versatility than HIPing or hot pressing. However, several important qualifications and uncertainties must be considered before selecting this method. The cost of a green composite is generally much higher than that of a conventional green ceramic compact because of higher fiber and composite body formation costs. Hence, the cost difference between sintered and hot pressed fiber-reinforced ceramic composites may be substantially less than that of conventional ceramics. Sintering needs usually higher temperatures than hot pressing, which impose certain limitations on the composite system, such as fiber–matrix reactions that should be prevented and a fiber, which is stable at these temperatures, that should be selected. Such densification limitations are likely to be least for short-fiber- and whisker-reinforced composites and become progressively greater for continuous unidirectional, bidirectional, and multidirectional fiber architectures. Encouraging results have been obtained in the sintering of some discontinuous fiber- as well as whisker-reinforced ceramic composites. However, it is expected that intrinsic densification problems are likely to pose serious challenges in producing reproducible quality composites and that these problems are likely to become more serious with increasing composite size and shape complexity.

6.3.1.8 Hot Pressing

Hot pressing is a versatile technique for CMCs, because the simultaneous application of pressure and temperature can significantly accelerate the rate of densification, resulting in a pore-free and fine-grained product. An example of a common hot pressed CMC is SiC whisker-reinforced Al_2O_3, which is used in cutting tool applications. Hot pressing is used in a combination of steps or in a single step for the consolidation of CMCs.

Consolidation of matrix powder with reinforcement, primarily by hot pressing, has been most extensively used for both glass-based and crystalline matrices. Hot pressing is a suitable choice for the consolidation of powder-impregnated preforms, which is carried out after drying and/or calcining.

Glass and glass-ceramic matrix composites are typically hot pressed at temperatures close to or above the softening point of the glass under a pressure of 25–40 MPa. At these temperatures, the viscosity is low enough to permit the glass to flow into the interstices between individual fibers within the fiber tows, and the pressure aids this. The resulting composites generally have densities more than 98% of theoretical density. Glass-ceramic matrix composites can be produced using either a glass–ceramic powder or glass powder. In the latter case, the resulting composite is heat-treated to bring about devitrification. This process offers the prospect of lower processing temperature, but the fabrication is more difficult to optimize due to the dimensional changes occurring during devitrification.

Hot pressing has also been used to consolidate CMCs made with SiC, Si_3N_4, Al_2O_3, and mullite matrices. Composites with the highest densities, that is, with the lowest porosity, can be produced by hot pressing. However, substantial densification is obtained only at higher temperatures, typically at 1400°C–1600°C, due to the absence of viscous flow, which seriously limits the type of fiber to be used as well as the various fiber–matrix combinations. The use of higher processing temperatures can lead to fiber degradation through grain growth, oxidation, or fiber–matrix chemical reaction. Nicalon SiC fibers, which are known to degrade beyond 1200°C, can no longer be used at such high processing temperatures, but Hi–Nicalon SiC fibers may be a suitable choice. This problem can be alleviated by using liquid-phase sintering additives and applying barrier coatings on the fibers. Sintering additives, such as TiB_2 or TiC and Y_2O_3, reduce hot pressing temperatures of SiC and Si_3N_4, respectively. Barrier coating on fiber helps to suppress high-temperature chemical reactions between the fiber and matrix. The coating can also provide a weak fiber–matrix interface, which helps to improve toughness by activating fiber debonding and fiber pullout. This has been achieved with C, BN, or C/BN dual-layer coating on fiber. The primary coating method to apply these coating materials is CVD. Out of these coating materials, BN being the first and one of the most extensively used coating materials.

There are some limitations with hot pressing of ceramic composites. Certain limitations are due to the basic characteristics of hot pressing, namely, it is mainly applicable to rather simple shapes, such as plates, cubical blocks, and cylinders, and is not a particularly low-cost process. The other limitations stem from the temperatures commonly used to achieve low levels of porosity. Such temperatures, which are commonly 100°C–200°C higher than that used for hot pressing of the matrix alone, lead to reaction between the fibers and matrix and degradation of the fibers. A potential way to reduce processing temperatures, as well as to provide much greater versatility in shape, is to use HIPing with such composites. However, this process requires a simple, effective canning method to be practical.

While hot pressing has been applied predominantly to unidirectional fiber composites, it should be applicable to composites with fibers in two directions in a layer, for example, woven cloth laminates, but adequate infiltration of the matrix between the fibers within the cloth should be ensured. Its applicability to three- and multidirectional fiber composites will be more limited because of possible fiber buckling and interference with densification caused by the fibers in the axial direction. Hot pressing temperatures can be substantially reduced by obtaining higher initial matrix densities, using finer matrix particulates, and the use of densification aids.

Most zirconia-toughened alumina (ZTA) composites are hot pressed; however, the stresses probably result from gradients in reduction of ZrO_2 at different regions, which may have been the cause of delamination and cracking and thus may limit scaling up this process to large bodies.

Fabrication with a non-oxide matrix, such as Si_3N_4 or SiC, requires high temperatures during hot pressing, for example, SiC requires about 2000°C. The high temperatures and reducing conditions of such pressing can cause reactions with the ZrO_2, resulting in the formation of oxynitride with Si_3N_4 matrix, which leads to oxidation problems. However, some composites, such as ZrO_2/TiB_2, show good promise for hot pressing despite the requirements of high temperatures and the long durations of hot pressing (2000°C for 3 h).

Composites containing non-oxide dispersants have also been more extensively consolidated by hot pressing. Although the sintering of TiC/Al_2O_3 composites is reasonably successful, most production has been by hot pressing around 1600°C.

Composites of Al_2O_3 with varying amounts of SiC have been commercially hot pressed for use as controlled microwave absorbers. Similar composites have been prepared with BeO and MgO matrices. These materials show strengthening as well as toughening due to the presence of SiC, and the mechanical properties are maximum at ~50% SiC. Limited studies have been carried out for other carbide–oxide systems, such as WC/Al_2O_3 and diamond/Al_2O_3.

Most Al_2O_3 composites containing BN particulates have been made by hot pressing. Earlier composites are made with large BN flakes (~100 μm diameter); however, much finer BN flakes (2–5 µm diameter) are used later. These composites have better thermal shock resistance than Al_2O_3 ceramics. While hot pressing of such oxide matrix composites produces high densification, only limited densities are achieved at higher levels of non-oxide additions. A BN/SiO_2 composite has also been developed and sold commercially, and the hot pressed composite offers good thermal shock resistance along with a low dielectric constant and reasonable conductivity.

A wide variety of non-oxide matrix composites has also been made by hot pressing. While some of these composites have higher strength than that of the matrix, some have lower strength, which often reflects, at least in part, the nature of the reinforcement. In a few cases, strength decreased because of the larger size of reinforcement particulates, for example, 9 and 32 µm SiC_p decreased the strength of Si_3N_4. Moreover, the use of weaker second-phase particles, such as graphite or BN (hexagonal), generally decreases strength but improves thermal shock resistance. In addition to these factors, the processing conditions also have a major influence on the properties of composites.

Hot pressing is more suitable for simple, flat shapes such as cutting tools and certain wear parts. For composites that require processing in non-oxidizing atmospheres also, hot pressing is the best choice. However, pressureless sintering has many advantages over hot pressing. Heating facilities are lower in cost and can be operated in any atmospheric conditions, such as air, nitrogen, argon, or vacuum, and the throughput available for air firing is high, for example, using tunnel kilns.

6.3.1.9 Hot Isostatic Pressing

HIPing clearly offers the potential of achieving high densities at somewhat lower temperatures than are required for hot pressing because of the possibility of applying higher pressures during HIPing. A much broader range of shapes can be made by HIPing, though the ultimate size of the product is limited. The preferred way to process ceramics and ceramic composites by HIPing involves glass encapsulation. A novel process to form glass encapsulation has been developed, in which the glass is applied over the body as particulate coatings. The coated green body is placed in the HIPing equipment, vacuum is applied, and the temperature is raised until the glass softens and forms a continuous layer. This is followed by densification of the green body by the normal HIPing cycle. It should be noted that the development of encapsulation technology is based on long periods of trial and error and the specific details are not available in open literature.

Fully dense silicon nitride composites containing carbon fibers have been successfully produced by HIPing (Lundberg et al. 1986). Unidirectional carbon fibers are stacked in a plaster of Paris mold and infiltrated with silicon nitride powder slurry. The dried green bodies are vacuum treated, glass encapsulated, and HIPed at 1600°C under 200 MPa pressure. This technique is shown as a promising method to infiltrate large 3-D fiber preforms. Another composite made up of SiC monofilament and mullite matrix has also been produced using encapsulated HIPing at 1450°C under 200 MPa pressure (Holmquist et al. 1997). These works clearly indicate that the higher pressures available in HIPing can reduce the processing temperatures significantly and the composites produced at lower temperatures have better mechanical properties than those produced at higher temperatures, probably due to a reduction in degradation of fibers.

Several types of particulate CMCs have also been hot isostatically pressed, including partially stabilized zirconia (PSZ), ZTA, and TiC/Al$_2$O$_3$. Most of these have been HIPed after sintering, but without canning. The closed porosity level during initial sintering allows HIPing without any canning and therefore is much more practical. This type of processing is being practiced on an industrial scale to produce WC/Co cutting tools. Such sintering followed by HIPing process greatly reduces isolated pores and pore clusters, which are often responsible for lower-strength failures. This combined process can be applied to produce large numbers of small components in a single run, which is the most economical way of using HIPing. Typically, HIPing results in somewhat higher and more uniform strengths than both the pressureless sintering and hot pressing processes.

The usage of HIPing for composite processing is rather limited, because the practical method of canning needs further development. The processing of multidirectional fiber composites also has severe limitations, probably more problematic than hot pressing.

6.3.2 Reaction Bonding Process

This is a relatively low-temperature process compared to pressureless sintering or HIPing. The processing of CMCs by this method is similar to that used for preparing monolithic ceramics. The only difference is that the reinforcement phase is also incorporated in the precursor material. In this process, a suitable ceramic matrix is formed by the reaction of a precursor solid matrix with a suitable gas. For example, silicon nitride matrix is formed by the reaction of silicon with nitrogen gas. This process is mostly used to prepare SiC or Si$_3$N$_4$ matrix composites. The main advantages of this process are

1. Little or no matrix shrinkage occurs during the formation of CMC.
2. High volume fraction of reinforcement can be used.
3. Multidirectional, continuous fiber-reinforced composites can be made.
4. The processing temperatures for most systems are generally lower than the pressureless sintering/hot pressing temperatures; hence, fiber degradation can be avoided.

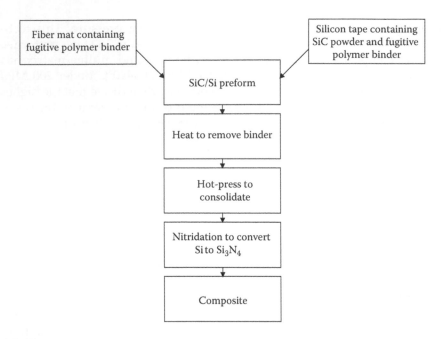

FIGURE 6.11
Flow diagram for the preparation of fiber-reinforced Si_3N_4 composites by reaction bonding process. (Reprinted from Bhatt, R.T. and Phillips, R.E., *J. Mater. Sci.*, 25, 3401, 1990. With permission.)

One major disadvantage of this process is high porosity, which is difficult to avoid. A hybrid process involving a combination of reaction bonding and hot pressing can be used to overcome the problem of porosity.

Figure 6.11 shows the flow diagram of the reaction bonding process. A mixture of silicon powder and a polymer binder is milled in an organic solvent to obtain dough of proper consistency. This dough is then rolled to make a silicon sheet of desired thickness. Fiber mats are made by filament winding of fiber tows with a fugitive binder. Either these fiber mats or fiber cloths and silicon sheets are stacked in an alternate sequence to prepare the preform. The preform is then heated slowly under nitrogen atmosphere up to about 1400°C. At this stage, the silicon matrix is converted to silicon nitride on reaction with nitrogen gas. The resulting silicon nitride matrix usually will have about 30% porosity.

Reaction bonding process can also be used to produce alumina and mullite matrix composites. In the case of an oxide matrix, reaction bonding uses direct oxidation of starting matrix material to create the ceramic matrix. This process requires much lower temperatures than the melt infiltration techniques. To prepare an alumina matrix composite, a compact made with Al and Al_2O_3 powders or a metal powder slurry infiltrated compact is heat-treated in an oxidizing atmosphere. The aluminum powder oxidizes to Al_2O_3, and in the process about 28% volume expansion occurs in the matrix, which can partially compensate the sintering shrinkage. However, oxidation is usually completed below 1100°C, where sintering does not take place. Mullite matrix is obtained indirectly by adding SiC to the starting powder. On oxidation, SiC forms SiO_2, which reacts with Al_2O_3 formed by the oxidation of Al, to form mullite. An alternative technique is to use an Al–Si alloy as the starting material. The reaction bonding process is an attractive and relatively fast method to produce fiber-reinforced ceramic composites, because the problems related to matrix shrinkage are avoided in this process.

6.3.3 Self-Propagating High-Temperature Synthesis

This technique can be used to produce a variety of high-temperature materials, such as intermetallics and carbides. It involves synthesis of compounds without an external source of energy. Once the reaction is initiated, the heat released from the reaction is sufficient for the propagation. This combustion synthesis is a potential new method to prepare unique CMCs in a simple, rapid, energy-saving way. The salient features of self-propagating high-temperature synthesis (SHS) process are

1. High combustion temperatures (up to 4000°C)
2. Simple and low-cost process
3. Better control of chemical composition
4. Amenable to produce different shapes and forms

The main disadvantage of this process is that the products are very porous, because of the fairly high porosity present in the original mix of reactants and the large volume change associated with the transformation of reactants to products. Any gases adsorbed during the reaction can also add to the porosity. Synthesis concomitant with densification can overcome the problem to some extent. This involves the application of high pressure either during the combustion reaction or immediately after the reaction when the product temperature is quite high. Hot pressing and shock wave compaction are some of the techniques used to apply the required pressure.

In this technique, the heat released from the exothermic reaction is exploited to synthesize unique ceramic compounds that are difficult to produce by conventional techniques. While such reactions are generally used to produce single-phase compounds, such as TiC and TiB_2, they can also be used to produce composites (Schwartz 1997a). For example, titanium powder is mixed with carbon black, cold pressed, and ignited at the top in a cold-walled vessel. The heat generated from the combustion reaction will pass through the compact, and the whole compact is converted to titanium carbide. A classical example for the composite formation is the thermite reaction to produce a metal ceramic composite product:

$$Fe_2O_3 + 2Al \rightarrow 2Fe + Al_2O_3 \qquad (6.12)$$

A variety of ceramic composites have also been produced by this type of reactions, for example,

$$10Al + 3TiO_2 + 3B_2O_3 \rightarrow 5Al_2O_3 + 3TiB_2 \tag{6.13}$$

$$4Al + 3TiO_2 + 3C \rightarrow 2Al_2O_3 + 3TiC \tag{6.14}$$

Some of the reactions appear to produce whiskers, which open an interesting possibility to produce in situ whisker-reinforced composites besides the possibility of producing a product directly; such processes can be used to prepare composite powders with intimate mixing. This in turn introduces another opportunity for the control of composite microstructure by controlling the size of powders and densification parameters.

Consolidating (pressureless sintering, hot pressing, or HIPing) reactant compacts and then initiating the reaction may provide opportunities for controlling composite microstructures by controlling nucleation and growth of the new phases, which is providing some, possibly all, of the opportunities for microstructural control offered by glasses.

Boride, carbide, nitride, and oxynitride ceramics as well as composites such as SiC_w/Al_2O_3 have been produced by means of SHS. As mentioned earlier, a weakly bonded, porous product is usually obtained in this process. Therefore, the process is generally followed by breaking the compact, milling, and consolidation using some techniques such as hot pressing or HIPing. Explosive or dynamic compaction can also produce a relatively dense product. Reinforcement in the form of particles, platelets, or whiskers can be mixed either with the reactants or with milled products to make a composite.

Fine SiC/Si_3N_4 composite powders with varying SiC contents from 8 to 46 vol.% have been synthesized by nitriding combustion of Si and C at room temperature under 10 MPa nitrogen gas pressure (Zeng et al. 1991). The powders composed of α-Si_3N_4, β-Si_3N_4, and β-SiC in various proportions, and the sintered samples consisted of uniformly distributed grains of β-Si_3N_4, β-SiC, and a few Si_2N_2O. The heat generated from silicon nitridation increases the dissociation rate of Si_3N_4 and assists the formation of SiC, which is more stable than Si_3N_4 at the reaction temperature.

Ceramic composites based on TiC/Al_2O_3 have been produced by means of SHS (Chrysanthou et al. 1994). The rate of reaction between Ti and carbon black, the ignition temperature, and the maximum temperature attained are dependent on the amount of Al_2O_3 present in the mixture. The reaction does not undergo in a self-propagating mode when the Al_2O_3 content is more than 41%, unless the reactants are not homogeneously mixed. However, it is possible to generate enough heat for self-propagation of reaction in the samples containing 50%–70% Al_2O_3 by introducing regions of high titanium and carbon concentration. The aluminothermic reduction of TiO_2 in the presence of carbon black also undergoes in a self-propagating mode. The reaction of

TiO_2 and Al starts at 1310°C and releases exothermic heat. The Ti formed due to the reduction reaction reacts with carbon black and the temperature increases to a maximum of 2100°C. The reaction proceeds to the entire compact because of this enormous heat release. The microstructure of the product formed by this route is entirely different from that of the product formed by the previous reaction (reaction of Ti and C in the presence of Al_2O_3).

Another approach is reaction synthesis followed by dynamic consolidation of the hot products using explosives. By this method, it is possible to achieve near-full density at a shock pressure as low as 1 GPa. However, care must be taken to allow the escape of gases evolving during reaction and to avoid rapid cooling of the consolidated composite to prevent cracking (Rabin et al. 1990).

Another composite successfully prepared by the SHS process is $MoSi_2$/SiC composite (Subrahmanyam and Rao 1995). It is formed from elemental powder of Mo, Si, and C by the thermal explosion mode of SHS. The morphology of $MoSi_2$ indicates that it is formed from the melt at the combustion temperature and SiC is formed as very fine particles. Hence, this SHS process can be used very effectively to produce composites based on $MoSi_2$ with complete conversion and desired morphology.

6.3.4 In Situ Ceramic Composite Processing

The processing of metal or polymer matrix composites is fairly easy because of lower melting point or plastic deformation of the matrix. However, the fabrication of CMCs is very difficult since the addition of foreign materials decreases the sinterability of ceramics. It is a well-known fact that the sinterability of ceramic powder compact decreases with increasing aspect ratio of the dispersed phase, that is, particle > whisker > fiber. The application of external pressure is generally necessary to achieve near-full density. Low sinterability is a more serious problem with more effective reinforcing materials like fibers, and many efforts have been made so far to overcome the problem. The formation in situ reinforcement could be one of the possible solutions.

In situ composites are the one in which the reinforcing particles, whiskers, or fibers are formed during the fabrication of composites. Hence, it is not necessary to produce reinforcing materials separately before the fabrication of composites. Typical microstructural designs for the fabrication of in situ formed particle, whisker, and continuous secondary-phase composites are illustrated in Figure 6.12. The particles are expected to be at the grain boundaries or in the grains, depending on the size of particles and the type of matrix. Whiskers are not likely to form during processing, but large elongated or platelet grains can form. The formation of fibers in situ process is more difficult than that of whiskers, whereas the formation of continuous secondary phases might be much easier in processing design. The fabrication of this type of composite is mainly focused on the interpenetrating

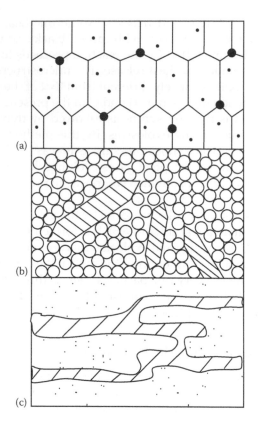

FIGURE 6.12
Microstructural design for in situ reinforcement by (a) particles, (b) elongated or platelet grains, and (c) interpenetrating grains. (Reprinted from Warren, R. (volume ed.), in *Comprehensive Composite Materials*, Vol. 4, eds. A. Kelly and C. Zweben, Elsevier Science Ltd., Oxford, U.K., 2000, p. 473. With permission.)

network, in which both the reinforcing material and matrix are continuous. This type of composites differs from the normal composites, in which the reinforcement phase is discontinuous.

Many processes have been developed for the fabrication of in situ composites. While processing, it is necessary to keep the microstructure fine and uniform at the initial stage to maintain enough sinterability and allow the heterogeneous reinforcing material to appear in the latter stage. The fabrication of in situ ceramic composites is basically the same as that of monolithic ceramics. It provides a better opportunity to reduce the cost of production of composites. Moreover, a better control of mechanical properties might be possible by controlling the amount, size, shape, and distribution of reinforcing material during processing. The investigations on in situ ceramic composites have mostly focused on the microstructures similar to whisker-reinforced ceramics, based on the balance between ease of fabrication and better mechanical properties.

6.3.4.1 Mullite Matrix Composites

Mullite with a composition of $3Al_2O_3 \cdot 2SiO_2$ has a low thermal expansion coefficient and a high melting temperature. However, the engineering applications of mullite are quite limited because of poor fracture toughness and strength. The incorporation of zirconia can improve the fracture toughness of mullite. In situ zirconia-toughened mullite has been produced by the reaction of zirconium silicate with alumina (Claussen and Jahn 1980, Rupo and Anseau 1980):

$$2ZrSiO_4 + 3Al_2O_3 \rightarrow 3Al_2O_3 \cdot 2SiO_2 + 2ZrO_2 \tag{6.15}$$

This chemical reaction and sintering took place simultaneously leading to the formation of sintered zirconia/mullite composite. The energy released during the reaction can contribute to the promotion of sintering. It is important to control the grain size of zirconia to retain the tetragonal structure. The bending strength of mullite (152 MPa) is increased to 400 MPa by the presence of zirconia.

6.3.4.2 Alumina Matrix Composites

The addition of a large amount of titania to alumina can lead to the formation of in situ aluminum titanate/alumina composites:

$$Al_2O_3 + xTiO_2 \rightarrow (1-x)Al_2O_3 + xAl_2TiO_5 \tag{6.16}$$

It was found that the Al_2O_3–20 vol.% Al_2TiO_5 composite had better flaw tolerance than monolithic alumina (Padture et al. 1991).

It has been reported that some aluminates show platelet or elongated grain morphology after sintering. $LaAl_{11}O_{18}/Al_2O_3$ composite has been prepared by the reaction of alumina with lanthanum oxide (Chen and Chen 1992). The fracture toughness of this composite is found to increase with increasing amount of $LaAl_{11}O_{18}$ platelets, which bridge the propagating cracks. Further increase in fracture toughness is achieved by the co-addition of silica, which resulted in a weaker interface and higher contribution to crack bridging (Yasuoka et al. 1997).

Alumina matrix composites containing continuous yttrium aluminum garnet (YAG) have been produced by unidirectional solidification of the molten eutectic (Waku et al. 1998). It has been found that the in situ composite retained the flexural strength up to 1725°C, whereas the sintered YAG/Al_2O_3 composite exhibits a reduction in flexural strength at high temperatures because of the presence of a glassy phase at YAG/Al_2O_3 phase boundaries. The high strength of in situ composites at high temperature is due to the interpenetrating microstructure and the absence of glassy phase.

6.3.4.3 SiC Matrix Composites

The addition of metal carbides and borides into a SiC matrix has been considered to improve high-temperature strength and corrosion resistance. However, the commercially available non-oxide powders are very coarse and decrease the sinterability of SiC. Hence, several attempts have been made to form fine reinforcing particles through chemical reaction in the SiC matrix. The starting materials for titanium boride are titanium oxide, boron carbide, and carbon. These materials are thoroughly mixed and then mixed with SiC powder. When a compact made out of these powders is sintered, the following reaction leads to the formation of TiB_2 in the SiC matrix:

$$2TiO_2 + B_4C + 3C \rightarrow 2TiB_2 + 4CO \qquad (6.17)$$

Similarly several borides can be formed in situ in SiC matrix.

Liquid-phase sintering of SiC with oxide additives leads to β- to α-phase transformation, which is accompanied by the anisotropic grain growth of α-grains. Based on this phenomenon, tough SiC ceramics can be prepared by high-temperature sintering at 1950°C–2100°C, at which the β- to α-phase transformation with accelerated abnormal grain growth resulting in the formation of in situ reinforcement. In situ formation of SiC composite was even more successful in liquid-phase-sintered SiC with non-oxide additives. The lowest liquidus temperature in this system is about 1800°C, so the elongated α-grain can form at relatively low temperature. Highly tough SiC ceramics with fracture toughness values more than 9 MPa m$^{1/2}$ have been prepared in this system (Cao et al. 1996).

6.3.4.4 Si₃N₄ Matrix Composites

Similar to SiC, the liquid-phase sintering of Si_3N_4 with oxide additives leads to the phase change from low-temperature α-form to high-temperature β-form, which facilitates abnormal grain growth as elongated grains. This in situ composite consists of both elongated and fine matrix grains of Si_3N_4 (Mitomo and Uenosono 1992). The rodlike elongated grains have a hexagonal cross section and the aspect ratio is typically around 4. The fracture toughness of sintered Si_3N_4 is increased by the formation of elongated grains.

6.3.4.5 Advantages of In Situ Composites

The primary advantage of in situ composites is the ease of fabricating composites or composite-like microstructures. The P/C techniques used for the fabrication of monolithic ceramics can be equally well applied to fabricating dense composite materials of this type without any increase in cost of production. Although the amount of reinforcing grains is less compared to that of the matrix, the improvement of mechanical properties is significant. Still there are many possibilities for optimizing microstructural and processing designs to achieve further improvement. The reliability of in situ compounds is also found to be improved. Another important advantage of in situ composites is good phase compatibility, which is very difficult to achieve when separately formed reinforcements are incorporated in the matrix.

6.3.5 Melt Processing

The melt processing of ceramic composites is a real challenge because of the high melting temperatures of most ceramics. More importantly, it is very difficult to avoid fiber–matrix reactions at those high temperatures. The very high solidification shrinkages of many ceramic materials (typically 10%–25%) can also be a problem. Despite these problems, some selected ceramic composites can be processed using melting techniques. A melt processing technique of composites involves the uniform mixing in the molten phase and subsequent solidification of eutectic structures or precipitation during solidification or after solidification. One of the important applications of melt processing is to produce composite powders such as ZrO_2 or PSZ in Al_2O_3. Mixing in the molten stage provides a significantly different and much more homogeneous mixing of the constituents than simple mixing of powders. But the challenge is to obtain sufficiently small particles for good sinterability but larger than the precipitate or lamellae. Since the precipitate size required is typically less than 1 μm for ZrO_2 toughening, this process should be feasible. A possible practical aid in obtaining the required particle size is atomization of a molten stream. The melt atomization procedure also increases cooling rates, giving finer structures than solidification of bulk ingots.

The second approach to melt processing of composites is to infiltrate compacts consisting of the reinforcement phase with a reacting melt. This may be the most limited melt approach since it is applicable to a very few composite system. The infiltrations of SiC + C and B_4C + C compacts with molten silicon to make SiC/SiC and B_4C/SiC composites, respectively, are some practical examples.

Another approach to melt processing of composites is the formation of a composite from the melt. The challenge in this process is to obtain sufficiently fine grain sizes and control the amount, the size, and especially the location of solidification pores. A more suitable method of controlling solidification porosity is directional solidification. Such directional solidification of eutectics results in the formation of unidirectional lamella- or rod-reinforced composites, wherein the lamella or rod diameter is inversely proportional to the solidification rate. Interestingly, the chemical compatibility between the two constituents is very good and they have particular crystallographic relations. This process has been explored mostly in oxide systems, with zirconia-containing compositions, especially ZrO_2/Al_2O_3, being very common. However, some non-oxide systems such as ZrB_2/ZrC have also been studied.

The fourth method of melt processing of composites is preparing a glass matrix and then crystallizing it. For example, ZrO_2-containing cordierite-based glass–ceramic composites have been prepared by this method (Rice 1990). These composites are made by melting at 1650°C and then annealing between 860°C and 1100°C. The fracture toughness of glass–ceramics is increased by ~100% on the addition of 12.5% ZrO_2. In another composite containing ~12 vol.% fine tetragonal ZrO_2 precipitates in $Li_2O/SiO_2/ZrO_2$, the fracture toughness is increased by ~50% compared to the monolithic matrix material. Another interesting composite system is the directionally solidified CaO/P_2O_5 glass ($CaO/P_2O_5 = 0.94$), which is solidified at 560°C with a solidification rate of 20 μm min^{-1}. This composite has strength of up to 600 MPa and showed non-catastrophic failure.

6.3.5.1 Melt Infiltration

The term melt infiltration is used to refer pressureless infiltration process, which is capillary-driven infiltration by a liquid above its freezing point into a porous body. When it is cooled, a composite results, in which the matrix is fully solidified. This technique is very similar to the infiltration of a liquid polymer or liquid metal. As in the other matrix systems, the proper control of the fluidity of the melt is the key to this technique.

The main advantage of melt infiltration process is that it has a potential for producing complex shapes with convenience and precision. The other advantage of this technique is the formation of dense matrix without any porosity. Almost any reinforcement geometry can be used to make a preform for infiltration. There are many disadvantages with this process.

The temperatures involved in ceramic melt infiltration are much higher than those encountered in polymer and metal infiltration. Since the temperatures involved are very high, it may not be possible to completely avoid the reactions between reinforcement and matrix. Thermal expansion mismatch between the constituents, the large temperature difference between processing and ambient temperatures, and the low strain-to-failure of ceramics can add up to a formidable set of problems in producing a defect-free CMC. Viscosities of ceramic melts are generally very high, which is another major problem since the infiltration becomes difficult with a high-viscosity melt. Hence, it is necessary to apply pressure or carry out the processing under vacuum for better infiltration. Wettability of the reinforcement by the ceramic melt is another concern. A summary of advantages and disadvantages of melt infiltration techniques are given in Table 6.4.

A preform made from any geometric form of reinforcement (fiber, whisker, or particle) having a network of pores can be infiltrated by a ceramic melt due to capillary pressure. Assuming that the preform consists of regularly

TABLE 6.4

Merits and Demerits of Melt Infiltration Techniques

Advantages	The matrix is formed in a single step.
	A homogeneous matrix without porosity can be obtained.
Disadvantages	Greater chance of reaction between reinforcement and matrix due to high melt temperature.
	Infiltration is relatively difficult because of higher melt viscosities of ceramics than that of metals.
	Matrix is likely to crack because of the differential shrinkage between the matrix and the reinforcement after solidification, especially when their respective thermal expansion coefficients differ widely.

spaced, parallel channels, Poiseuille's equation can be used to obtain the infiltration height, h (Washburn 1921):

$$h = \sqrt{\frac{\gamma r t \cos \theta}{2 \eta}} \tag{6.18}$$

where
  r is the radius of cylindrical channel, which depends on the size and volume fraction of reinforcement
  t is time
  γ is the surface energy of infiltrant
  θ is contact angle
  η is the viscosity of the melt

It should be noted that the infiltration height is directly proportional to the square root of time and inversely proportional to the square root of viscosity. Infiltration will be easier, if the contact angle and the viscosity of melt are low and the surface energy and pore radius are large. However, if the radius of the channel is too large, the capillary effect may not exist at all.

The melt infiltration processes can be subdivided into reactive and nonreactive processes. In reactive melt infiltration, a constituent of the preform reacts with the melt, often with considerable heat evolution. The most prominent example of reactive melt infiltration is the formation of SiC matrix by infiltrating Si or Si alloys into a porous preform that contains C_p and/or C_f.

Many CMCs have been successfully produced by nonreactive infiltration of various silicates, aluminosilicates, or refractory fluorides into porous SiC preforms. Particulates, whiskers, and discontinuous and continuous fibers can be used to prepare the preform. The incorporation of particulates in a fiber preform serves to adjust the spacing of the fibers. Thus, the matrix that envelops the fibers may be considered as a particulate composite, since the particulates incorporated in the preform will go into the matrix during infiltration. A ceramic melt is infiltrated into the dried preform, and the driving force for the infiltration is surface tension. The melt solidifies on cooling and a fully dense composite structure is produced. However, some closed porosity may result due to volumetric changes during freezing.

6.3.5.1.1 Reactive Infiltration

This is another variant of liquid infiltration process, in which the infiltrating liquid reacts with a component in the preform to form the ceramic matrix. Polymers are used to bind fiber reinforcements, such as carbon or SiC fibers. A fiber preform is made with an appropriate polymer binder, and on pyrolysis the polymer is converted to carbon mass, which is surrounding the reinforcements. Molten silicon is then infiltrated into the preform containing carbon. The molten silicon reacts with carbon and forms the SiC matrix. Thus, a CMC is produced by this reactive infiltration process. In many cases, some amount of residual silicon will be present in the matrix. It is important to control pore size and carbon content in the preform to avoid the presence of unreacted silicon. The presence of unreacted silicon will limit the service temperature to about 1400°C. A great advantage of this process is that the constituents are in chemical equilibrium and they have closely matching thermal expansion coefficients. Carbon fiber in the form of cloth, tow, felt, or mat can be used to prepare the preforms.

Reactive infiltration technique has also been used to prepare a type of Si/SiC composites, known as biomorphic SiC (Kovar et al. 1997, Naslain 1999). The fabrication of these composites involves rapid and controlled mineralization of the wood in two steps. The first step is the pyrolysis of wood under inert atmosphere to form a carbon preform and the second step is reactive infiltration of this carbon preform with silicon vapor or melt. An attractive feature of this process is that the microstructure of wood will be retained in the composite. Hence, a variety wood with different density and microstructure can be selected, which results in a variety of composites with differing density and microstructure. The resulting composites will have anisotropic properties, since the wood structures in the parallel (axial) and perpendicular (radial) directions to the growth of the tree are different. The fabrication of composites by this method is relatively fast and it is limited only by the heating and cooling cycles. It is an NNS fabrication technique, and the fabrication of a part usually takes less than two days. The size and number of parts produced are limited only by the size of the infiltration furnace; however, larger parts may require slower heating and cooling cycles.

The most important steps in the fabrication of biomorphic SiC are given in the flow diagram (Figure 6.13). Tree wood is cut to the required shape (complex shapes can be made by wood carving), and then the wood piece is pyrolyzed at 1000°C under inert atmosphere. The water and volatiles present in the wood evaporate, leaving a carbon skeleton (carbon preform) that keeps the basic microstructural features of the natural wood, and at the same time interconnected pores of different sizes and configurations ideal for fluid infiltration are formed. During this pyrolysis, the weight loss is approximately 75% and the volume reduction is about 60%, which depend on the type of wood.

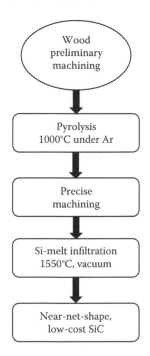

FIGURE 6.13

Flow diagram for the preparation of biomorphic SiC from wood. (Reprinted from Varela-Feria, F.M., *J. Eur. Ceram. Soc.*, 22, 2719, 2002. With permission.)

This carbon preform is machined to the final size and shape of the component. It is then infiltrated with molten silicon in a vacuum melting furnace. The furnace temperature should be more than the melting temperature of silicon (1410°C), but a higher temperature is preferable for high infiltration rate due to the low viscosity of melt. Generally, the infiltration is carried out at 1550°C. The molten silicon infiltrates the preform by capillary action and fills the wood channels of size 5–100 μm in diameter. At this temperature, there is a reaction between silicon and carbon, which leads to the formation of SiC. A holding time of about 30 min is sufficient for complete reaction (Smirnov et al. 2003).

Complete reaction of carbon with silicon during melt infiltration is achieved when the amount of silicon infiltrated exceeds the stoichiometric quantity by a factor of 1.2–2, but it depends on the porosity in the carbon preform. A denser preform needs only a little excess, because less silicon is needed to fill the less quantity of open pores available. The SiC formation reaction is spontaneous and exothermic and it progresses until either all of the Si or C is consumed. However, the rate of reaction is several orders of magnitude slower than the infiltration rate and that implies that the SiC formation reaction is the rate-limiting step in this process (Greil et al. 1998). Depending on the initial cellular structure of wood, SiC-based composites with different anisotropic structures can be obtained. Greil et al. (1998) have prepared SiC ceramics from ebony, beech, maple, oak-, pine-, and balsa wood. The density of wood, char, and ceramic composites were 0.47, 0.34, and 1.0 g cm^{-3}, respectively, when pinewood was used to form the SiC composites. High-temperature compressive strength of these composites was found to be very good. The compressive strength values parallel to the fiber direction (tree growth direction) were 750 and 300 MPa at 1100°C and 1350°C, respectively, whereas the strength values perpendicular to the fiber direction were 215 and 120 MPa at 1100°C and 1350°C, respectively, when African bubinga wood was used (Munoz et al. 2002).

When the large pores (>30 μm) in the carbon preform is infiltrated, some unreacted Si remains in the pore, whereas some unreacted C will be present when small pores near the thick carbon sections are infiltrated. In the natural wood, it may not be possible to control the pore structure, and hence, the presence of unreacted Si and C is unavoidable. One possibility is to select the appropriate wood microstructure from a particular tree, but that will also change with the growing conditions of the tree. Si/SiC composites without residual carbon could retain the mechanical properties up to 1350°C, but there was a drastic reduction in mechanical properties at 1350°C, in the composites containing residual carbon (Presas et al. 2005).

Although the mechanical properties of Si/SiC composites prepared using wood are good, the main drawback is anisotropic mechanical properties. In many applications, isotropic mechanical properties are required. Since the microstructure of these composites is decided by the internal structure of wood, there is no flexibility in changing the microstructure. The best alternative is to use wood-based composites instead of solid wood to form the carbon preform.

Wood-based composites are low-cost materials produced using wood chips and phenolic resin. The microstructure of these composites can be varied by changing the composition of the constituents, particle size of wood chips, and the orientation of particles. A typical microstructure of Si/SiC composites formed by reactive infiltration using wood particles is shown in Figure 6.14. It contains SiC, Si, and C phases. Amirthan et al. (2010) have reported the preparation of Si/SiC composites using wood particles and phenolic resin. The dried wood particles were mixed with phenolic resin solution in a ball mill for 5 h. After removing the solvent (isopropyl alcohol), the mixture was compressed in a mold at 160°C. This wood–phenolic resin composite was pyrolyzed under N_2 atmosphere using a heating rate of 2°C min^{-1} up to 600°C and a heating rate of 5°C min^{-1} up to 1100°C and then held at this temperature for 2 h. The resulting carbon preform was infiltrated with molten silicon at 1600°C for 2 h. The flow diagram for the preparation of Si/SiC composites from wood particles and phenolic resin is shown in Figure 6.15.

The microstructure of a typical wood particle–phenolic resin composite after pyrolysis is shown in Figure 6.16a. It clearly indicates the porous nature of the carbon preform with hollow channels of various diameters and cellular

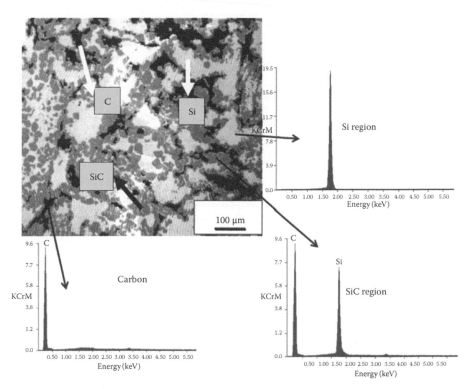

FIGURE 6.14
A typical microstructure of Si/SiC composites prepared from wood particle composite by reactive infiltration.

structure. The pores are circular in shape with size varying from 10 to 60 μm. The microstructure of the SiC-based composite after silicon infiltration is shown in Figure 6.16b. It consists of SiC (gray) with unreacted carbon (black) and silicon (white) phases. To maximize the SiC content, the microstructure of the carbon preform should be designed in such a way that the pores are large enough to allow infiltration but not too large to retain unreacted silicon. Similarly, the thickness of carbon between the pores should be as minimal as possible. This can be controlled by tailoring the microstructure of wood particle–phenolic resin composites, which is decided by the wood particle morphology and quantity. One possible option to improve the thermal stability of this type of composites is to give a SiC coating by the CVD process (Figure 6.16c). This coating can prevent the oxidation of residual carbon and silicon up to 1200°C.

6.3.5.1.2 Directed Oxidation

This is another version of liquid infiltration process, in which the infiltrating metallic melt undergoes reaction and forms the desired ceramic matrix. The process was developed by Lanxide Corporation and it is called DIMOX™ process. DIMOX™ process involves the growth of Al_2O_3 by the reaction of

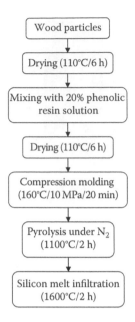

FIGURE 6.15
Flow diagram for the preparation of Si/SiC composites by reactive infiltration from pyrolyzed wood particle/phenolic resin composite.

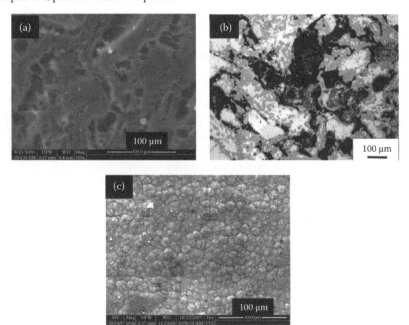

FIGURE 6.16
SEM micrographs of (a) pyrolyzed, (b) Si-infiltrated (BSE), and (c) CVI-treated samples. (Reprinted from Amirthan, G. et al., *Ceram. Int.*, 37, 423, 2011. With permission.)

molten aluminum and O_2 at the interstices in the preform. This process can also be applied to other matrix materials such as AlN, TiN, and ZrN by suitably selecting the reactive metal and gas. Since this is a growth process, the matrix material can grow around particulates as well as fibers to form composites. The amount of reinforcement and matrix in the composite can be controlled by controlling the porosity in the preform.

The first step in this process is to make a preform of suitable shape from the reinforcement. The same process can be used to produce either a CMC or an MMC, just by controlling the reaction between the infiltrating melt and surrounding gases. In any case, it may not be possible to produce 100% CMC, since some unreacted metal will always remain in the matrix. A barrier is placed around the preform surface to prevent the growth of matrix on the surfaces. The molten metal is then allowed to infiltrate from the bottom, and it is subjected to directional oxidation, that is, the ceramic matrix forms on the surface of the molten metal and grows outward. The molten metal flows continuously to the reaction front by a wicking action through channels in the oxidized product. The end product in this process is a 3-D, interconnected network of a ceramic matrix material with the reinforcement. Depending on the reaction conditions, about 5%–30% unreacted metal remains in the matrix. This must be removed, if the composite is intended for high-temperature applications above the melting point of metal. On the other hand, the presence of residual metal can improve the room-temperature fracture toughness of the composite.

Proper control of the reaction kinetics is very important in this process, to achieve complete infiltration as well as reaction. The process is potentially a low-cost process because NNS CMCs are produced at very low temperatures. However, there are some disadvantages with this process:

1. It is very difficult to control the chemical reaction to produce an all-ceramic matrix. Always some residual metal will remain in the matrix that is not easy to remove completely.
2. It is difficult to apply this technique for the production of large, complex parts.

It has been shown that Al_2O_3 matrix composites containing SiC_p and an inter-penetrating network of metal can be formed by directed metal oxidation (DMO) of a slightly modified A380 aluminum alloy at temperatures ranging from 950°C to 1100°C (Manor et al. 1993). The relative amount of residual metal in the product was found to decrease with increasing processing temperature. The oxidation rate of the modified A380 alloy was comparable to other Zn-bearing aluminum alloys. It was also found that the rate was significantly enhanced by the presence of SiC and the tendency to oxidation increased with decreasing particle size. The underlying mechanism involves an extended oxidation front arising due to wetting by the melt and secondary nucleation of alumina on the particle surfaces, perhaps by the reaction of the alloy with the SiO_2 layer present on the SiC. The total porosity in the composite was found to increase with increasing Mg content, processing temperature, and/or SiC_p size.

6.3.5.1.3 Fabrication of SiC_p/Al_2O_3 Composites by Directed Metal Oxidation

A schematic of DMO process is shown in Figure 6.17. A SiC_p preform is prepared by using one of the conventional ceramic processing routes such as uniaxial or isostatic pressing, injection molding, extrusion, or slip casting. It is then placed in contact with a molten aluminum alloy in a furnace at a temperature between 900°C and 1100°C. In the presence of oxygen (air) atmosphere, α-Al_2O_3 begins to grow starting from the alloy–preform interface into the preform after a short incubation period. The molten alloy, driven by surface tension, flows from the reservoir through an interconnected network of microscopic channels to the Al_2O_3 growth front. Growth proceeds as long as molten alloy, sufficient temperature, and oxygen are available and stops upon contact with a gas-permeable barrier layer. The barrier coating is applied prior to infiltration to the surfaces where matrix growth is to be eventually stopped, thereby enabling the fabrication to the desired shape.

The rate of growth and microstructure development are influenced by the molten alloy composition. The type of particulate material, particle size, and residual metal in the matrix influence composite properties. For example, the residual aluminum alloy in SiC_p/Al_2O_3 composite significantly contributes for the increase of room-temperature strength and toughness.

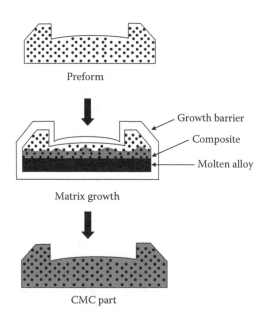

Preform

Growth barrier
Composite
Molten alloy

Matrix growth

CMC part

FIGURE 6.17
Schematic of the DMO process (DIMOX™). (Reprinted from Lanxide Corporation, Newark, DE.)

When the temperature is above the melting point of the residual metal, the softening of the metal leads to a reduction in the strength and toughness of the composite. SiC_p/Al_2O_3 composites produced by this process have excellent thermal shock resistance. Creep rates are also generally low for these composites containing less residual metal, which makes them attractive for high-temperature applications. Some of the composite systems produced by DMO are SiC_p/Al_2O_3, SiC_f/Al_2O_3, Al_2O_{3f}/Al_2O_3, Al_2TiO_{5p}/Al_2O_3, ZrB_{2p}/ZrC, and TiC-coated graphite/Al_2O_3.

6.3.6 Polymer Infiltration and Pyrolysis

Another important chemical method of producing ceramic matrices for fiber-reinforced CMCs is PIP. This process involves the impregnation of polymeric materials into the fiber networks, which on pyrolysis decompose to non-oxide ceramic matrix materials. However, a suitable polymeric material containing the elements to form the desired ceramics should be selected. The polymer should not undergo excessive polymer backbone loses (ideally H_2 losses only) during the conversion to a ceramic product. Many precursor polymers are commercially available, and new polymers are being developed for SiC- and Si_3N_4-based ceramics. Precursor polymers are also being developed to yield BN- or B_4C-based ceramics. The selection of a preceramic polymer suitable for this process is very important. The desirable characteristics of the preceramic polymer include

- High ceramic yield with low residual carbon
- Controllable molecular weight, which determines the solubility in a solvent and controls the viscosity for infiltration purposes
- Ability to cross-link at a low temperature, which retains the shape of the composite during pyrolysis
- Low cost and toxicity

Most preceramic polymers are prepared from chloro-organosilicon compounds. Some of the preceramic polymers are polycarbosilanes, polysilazanes, polyborosilanes, polysilsesquioxanes, and polycarbosiloxanes. The preparation of polycarbosilane from dichlorodimethyl silane has been described in Chapter 2, under SiC fibers. Chlorosilane monomer is inexpensive and readily available, since a large quantity of this monomer is formed as a by-product in the silicon industry.

All silicon-based polymer precursors lead to the formation of amorphous ceramic matrices, which on high-temperature treatments transform into crystalline ceramics with slight densification. Si–C matrices derived from polycarbosilanes begin to crystallize at 1100°C–1200°C, whereas Si–C–O and Si–N–C matrices derived from polysiloxanes and polysilazanes, respectively, remain amorphous even at 1300°C–1400°C.

A polymer with a suitable range of molecular weight is selected and the molten polymer is infiltrated into the preform. The polymer is then stabilized (to avoid remelting) and pyrolyzed in an inert atmosphere at temperatures between 800°C and 1300°C. During pyrolysis, there will be weight loss and pore formation because of the evolution of gaseous products. The porosity is of the order of 20%–30% after single impregnation and pyrolysis. Hence, multiple impregnations are needed to reduce porosity. Re-impregnation is generally carried out with a low-viscosity polymer so that the polymer effectively wets and infiltrates the micropores that exist in the preform. Re-impregnation with low-viscosity polymer is carried out by immersing the preform in the liquid polymer under vacuum, while higher-viscosity polymers require pressure impregnation. The porosity will reduce from 35% to less than 10% after about five cycles. Finally, it is heat-treated between 1300°C and 1800°C to form crystalline phases. At this stage, any residual amorphous phase present in the CMC is transformed to crystalline phase, and there is a reduction in the amount of oxygen due to evaporation of SiO and CO. Depending on the number of infiltration cycles and pyrolysis temperature, the porosity varies between 5% and 20% with pore sizes of the order of 1–50 nm. The average pore size and volume fraction of pores decreases with increasing pyrolysis temperature because of the densification of the matrix. The flow diagram for the preparation of CMCs by the PIP process is shown in Figure 6.18.

The mechanical properties of the composites prepared by this process are moderate, since in most cases the fibers and the matrix materials have related chemical compositions resulting in undesirable strong bonds.

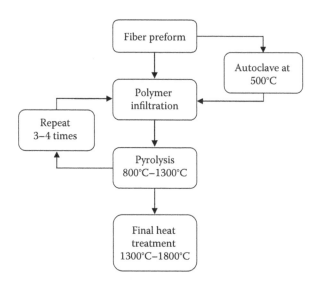

FIGURE 6.18
Flow diagram for the preparation of ceramic composites by polymer infiltration and pyrolysis.

Suitable fiber coatings will probably be required to prevent the formation of strong bonds between fibers and matrix.

This is an attractive processing route because of its relatively low cost compared to other processing routes to produce CMCs. The fiber degradation is minimum in this process, but a small amount of residual porosity is inevitable. Moreover, this approach allows the fabrication of NNS composites. Due to the relatively low yield of ceramics from the polymer, multiple infiltration and pyrolysis cycles are needed to obtain an acceptable density of the composite. There are two significant advantages with PIP process, which makes it as a competitor to the CVI process in generating ceramic matrices. The first advantage is that all but the final pyrolysis step is identical to the fabrication of a polymer composite, thus providing essentially the same flexibility in terms of handling, fiber architecture, product size, and shape. The other major advantage of the process is the relatively low temperatures used for pyrolysis.

Several factors strongly favor the use of the PIP process for fiber composites than particulate composites, which include cost, shrinkage, and low-temperature processing. Lower-temperature processing, as mentioned earlier, allows the use of a wide variety of fibers, which are not very expensive. The large density change, typically from about 1 g cm^{-3} for the polymer to over 3 g cm^{-3} for the theoretical density of resultant products, such as SiC and Si$_3$N$_4$, and the evolution of gases due to the decomposition of polymer result in large shrinkage, substantial porosity, and/or microcracking. With a high packing density of particulates and a small composite body of simple shape, it may be feasible to obtain some useful properties. However, in general, it may not be possible to produce particulate composites with reasonable strength and toughness by this process due to the formation of high density of defects. On the other hand, the strengthening and especially the toughening achievable in fiber composites can yield quite useful properties despite the formation of porosity and microcracking from polymer pyrolysis.

The PIP process is more restricted in terms of the matrices, since only a limited range of precursor polymers are available. An intrinsic limitation is that this method cannot yield a totally dense matrix, which in fact commonly results in considerable porosity, although this can be reduced by multiple impregnations. Another limitation is that the polymers available at present are relatively costly and the cost may reduce once the supply/use is increased. Like the CVI process, it also has some limitation on the cross-sectional thickness to which it can be infiltrated. However, this process is applicable to all fiber architectures in principle and is quite versatile in size and shape.

Significant gas evolution occurs during pyrolysis because the elements other than forming the ceramic matrix will leave the system as gases. Hence, it is advisable to allow these gases to diffuse out of the matrix slowly, especially from thicker parts. To avoid delamination due to gas evolution, the pyrolysis cycles ramp from 800°C to 1400°C over a period of 1–2 days. As mentioned earlier, the pyrolysis temperature should be below the crystallization temperature of the matrix as well as the degradation temperature of the reinforcing fibers. Most commonly argon or nitrogen atmosphere is maintained during pyrolysis, and in some cases ammonia atmosphere is used. The reaction for the formation of ceramic from the polymer can be written simply as (Greil 1995)

$$P_{(s)} \rightarrow C_{(s)} + G_{(g)} \tag{6.19}$$

where
   P is polymer
   C is ceramic
   G is the gaseous by-product

The ceramic yield, α, can be written as the ratio of ceramic formed and the initial amount of polymer:

$$\alpha = \frac{w_{(C)}}{w_{(P)}} = 1 - \frac{w_{(G)}}{w_{(P)}} \tag{6.20}$$

The density ratio of the (precursor polymer)/(ceramic product) is

$$\beta = \frac{\rho_{(P)}}{\rho_{(C)}} \tag{6.21}$$

Two extreme cases of polymer–ceramic conversion can be considered based on the final volume of ceramic matrix. If the volume change is not constrained, the diffusion of elements causes pore closure and a large amount of shrinkage will take place. The maximum volume change occurring during the conversion can be written as

$$\Delta V = \alpha\beta - 1 \tag{6.22}$$

Since the conversion usually leads to shrinkage, the ΔV value is negative. If the volume change is constrained (i.e., $\Delta V = 0$), then there is no shrinkage, but a large amount of residual pores will be present. The maximum volume fraction of pores can be written as

$$V_{(pores)} = 1 - \alpha\beta \tag{6.23}$$

Usually the product $\alpha\beta$ is less than 1; hence, it is almost impossible to obtain fully dense ceramic matrix from a polymer precursor without shrinkage. For example, during the pyrolysis of filler-free polysilazane, to form bulk Si_3N_4, either a large amount of porosity (>8%) results or a large amount of shrinkage (~20%) takes place.

Polymer-derived CMCs generally have a large number of cracks, voids, and pores in the matrix. The large amount of shrinkage during pyrolysis is mainly responsible for these defects. The shrinkage and hence the cracking in the matrix can be contained, to some extent, by the addition of particulate fillers to the infiltrating polymer (Greil 1995). The particulate fillers added in the polymer precursor can serve a variety of purposes:

- Reduce the formation of matrix cracks during the shrinkage of polymer
- Enhance ceramic yield by reacting with solid or gaseous decomposition products formed during pyrolysis
- Strengthen and toughen the amorphous matrix and increase the interlaminar shear strength of the CMC

The filler must be submicrometer in size in order to penetrate the fiber bundle and its thermal expansion coefficient must match to that of the precursor polymer. It should also be noted that the filler content must not be very high and the melt slurry should not be forced into the preform, since higher abrasion on fibers can take place. This is especially true with hard and angular fillers. Typically, the amount of filler recommended is 15%–25% by volume of the matrix. When an active filler is added to the polymer, it reacts with solid or gaseous decomposition products during pyrolysis and forms new ceramic phases.

Fiber architecture may also have an influence on the matrix microstructure formed during the PIP process. During pyrolysis, the precursor shrinks around the fibers and cracks are formed. The extent of cracking depends on the fiber architecture. For example, 2-D woven fabrics seem to have less tendency in developing interlaminar cracks than cross-ply or unidirectional fiber arrangement. Satin weave fabrics are preferred instead of plain weave fabrics because more uniform cracking is achieved and large cracks between fiber crossover points are avoided. The nature of weaving in satin weave fabrics facilitates better wetting and densification, although it makes the handling of fabrics difficult.

To address the problem of large mass loss and shrinkage during pyrolysis, much of the work has involved the use of either filler particles or multiple impregnation and pyrolysis cycles. It is expected that polymers with greater mass yields and lower shrinkages upon pyrolysis will be developed in the future, resulting in improved composite properties.

6.3.7 Chemical Vapor Infiltration

This is one of the infiltration techniques, in which gaseous reactants infiltrate the preform and form the solid matrix. The CVI process is used since the 1980s to produce C/C composites. This process is simply a variant of the CVD process, in which a solid thin film is deposited on a substrate by heterogeneous chemical reactions from one or more gaseous species. When the CVD technique is used to form bulk matrix inside a preform, then it is called CVI. A variety of ceramics such as oxides, and non-oxides, as well as glasses, can be formed by the CVI process. Common ceramic materials formed by this process are SiC, Si_3N_4, and HfC. The preforms for the CVI process can be made using yarns, or woven fabrics, or they can be filament-wound/braided 3-D shapes. This process has been used extensively for producing NNS CMCs. The main advantages of CVI techniques include

1. It is a pressureless process carried out at relatively low temperatures, and hence, the damage to the reinforcement is very minimal.
2. Large, complex parts to NNS can be produced.
3. Considerable flexibility in selecting fibers and matrices.
4. Very pure deposit can be obtained by carefully controlling the purity of gases.
5. CMCs with good high-temperature mechanical properties can be produced.
6. Specific microstructure in the deposit can be produced by controlling the deposition parameters.

The main disadvantages are that the process is very slow and the precursor materials are very expensive.

CVI is potentially one of the most important chemical methods of forming ceramic matrices for fiber-reinforced CMCs. It also has the potential for producing some of the largest fiber-reinforced CMCs achievable by any processing method. A wide variety of shapes can be made by this process and it can in principle be utilized with any fiber architecture. This process has already been used to produce large tubes consisting of one layer of braided oxide fibers infiltrated with SiC for flame tubes used in refining Al metal. Other examples of products made by this process are C/C aircraft brake disks, Al_2O_3/SiC heat exchanger tube sections and thermocouple well, and C/SiC rocket nozzle and exit cones. SiC_f/SiC composites prepared by the CVI process find many applications, and these composites have already been used for prospective bipropellant thrusters, space plane thermal protection systems, and gas turbine engine components. Although the CVI process has been utilized almost exclusively for SiC-based matrices, it has the potential for producing a widest range of matrices.

Originally this process was used to densify porous carbon composites. In fact, about half of the commercially available carbon–carbon composites are being made by the CVI process. In this process, the solid matrix forms over the reinforcement at a relatively lower temperature. For example, titanium diboride (TiB_2) with melting point of 3225°C can be deposited at a temperature, as low as 900°C by the CVI process. The chemical reaction involved for the deposition of TiB$_2$ is

$$TiCl_4 + 2BCl + 3H_2 \rightarrow TiB_2 + 6HCl \qquad (6.24)$$

The HCl gas is a common by-product in many CVI processes. The solid matrix materials are deposited into a hot preform from the gaseous reaction products. A typical CVI process reactor consists of the following components and the schematic of a simple reactor is shown in Figure 6.19:

1. A vapor feed system
2. A reaction chamber with a provision for heating the preform
3. An effluent system to handle exhaust gases

The CVI processes currently used for the fabrication of ceramic composites are isothermal diffusion-limited infiltration and forced-flow thermal gradient process. The isothermal method developed in France by the Societe Europeene de Propulsion produces high-quality NNS composites and places no size or shape limitations on the preform. However, preferential deposition in the outer regions can lead to premature pore closure, which requires process interruptions for machining operations to reopen the pores.

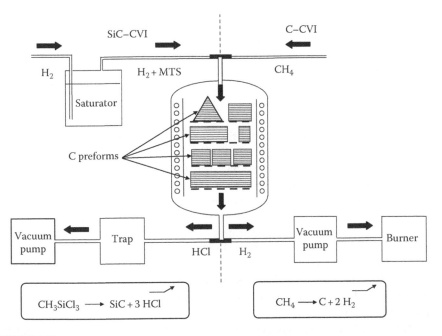

FIGURE 6.19
Schematic of the CVI process. (Reprinted from Naslain, R. and Langlais, F., *Proc. 21st Univ. Conf. Ceram. Sci.*, Plenum, New York, 1986, p. 145. With permission.)

Since this process is diffusion-limited, it may require weeks of processing time. The diffusion problem is aggravated by counterflow of product gases. Japanese investigators have slightly improved the process by suggesting the use of pulsed reactant flow instead of continuous flow. Los Alamos National Laboratory investigators have used an isothermal forced-flow reactor.

The thermal gradient forced-flow process developed by the Oak Ridge National Laboratory (ORNL) intentionally imposes a steep thermal gradient in the preform. Reactant gases enter the cold region and flow toward the hot face where they react and deposit the ceramic phase. The thermal conductivity increases in that region because of density increase, allowing the higher-temperature deposition front to approach the cold face. While high-quality dense composites can be produced by this approach, regions of high residual porosity are often unavoidable due to the gas flow patterns developed within the substrate. Furthermore, preforms are limited to relatively simple shapes, since it is difficult to maintain thermal gradient in a complex-shaped preform and control gas flow.

A recent development of CVI is microwave-assisted chemical vapor infiltration (MACVI). A number of processing schemes are possible using different combinations of microwave-absorbing and microwave-transparent materials as composite constituents. For example, an absorbing preform (Nicalon SiC fiber) can be combined with a microwave-transparent matrix (Si_3N_4). Composites of 5 cm in diameter and 1 cm thick have been produced to densities of 65% theoretical in less than 20 h by this process. Higher densities require additional microwave power, which is now possible with the new reactors. It is necessary to produce the hot spot at the center of the preform, by appropriate treatment of the materials. Nextel 610 alumina fibers with a carbon coating have shown preferential heating in the interior of the preform. The matrix materials that can be formed by the MACVI process include siliconized SiC, doped SiC, Al_2O_3, and ZrO_2. The inherent nature of microwave heating is that the surface is always cooler than the interior. This is an important advantage to avoid the problem of premature pore closure.

A comparison of CVI and PIP is given in Table 6.5. Both CVI and PIP processes have been used to produce SiC$_f$-reinforced SiC. Either process forms composites with densities of about 85%–90% of theoretical density. The thermal stability of these composites is good. The CVI composite is preferable when the retention of reasonable mechanical properties is required for prolonged periods in oxidizing environments because it gives a coherent matrix, which provides the composite with better integrity that is less achievable from repeated polymer pyrolysis densification steps. A combination of the PIP and CVI processes would be a better option, in which the PIP process can be used initially to get a handleable preform and that can be subsequently densified by CVI.

With the CVI process, a gaseous mixture is passed through fiber preforms where the fiber diameter is typically 10–15 μm, and a very slow deposition condition should be used to get better densification. The Nicalon and

TABLE 6.5

Advantages and Disadvantages of CVI and PIP Processes

Process	Advantages	Disadvantages
CVI/CVD	Even, controlled deposition giving a coherent matrix	Long process times (30–150 h)
	Fiber coatings and matrix built up in single operation	Expensive capital equipment
	Modest reaction temperatures	
	Dopants can be added to the matrix	
PIP	Useful in binding fiber performs	Repetitive infiltration and pyrolysis needed to achieve densification
	Cheaper process than CVI	Long process times with slow heating rates
	Modest process temperatures	Considerable shrinkage on pyrolysis, resulting in matrix cracking

Source: Data from Schwartz, M.M., *Composite Materials*, Vol. II, Prentice Hall, Upper Saddle River, NJ, 1997, p. 230.

Tyranno SiC fibers begin to degrade at about 1200°C; hence, the reaction temperatures need to be lower than this. The onset of thermal degradation of SiC fibers above 1200°C poses limitations on CVI composites (Schwartz 1997b).

SiC_f-reinforced Si_3N_4 composites have also been fabricated by the CVI process (Hoyt and Yang 1995). It is found that the resulting matrix material is amorphous Si_3N_4, which crystallized at temperatures above 1250°C to α-Si_3N_4 with a linear shrinkage of approximately 0.5%. The Vickers hardness of the amorphous deposit is found to be similar to crystalline Si_3N_4, and the amorphous deposit can be a suitable matrix material. Fiber–matrix bonding in the composites heat-treated at 1100°C and 1300°C is found to be strong, and the shrinkage of the matrix around the fiber during devitrification adds to the bonding strength. However, for a damage-tolerant composite, a weak bond between the fiber and matrix is desirable, which is possible by applying an appropriate coating on the fibers.

The CVI process commonly involves the decomposition/chemical reaction of reactants in the vapor state to form the desired ceramic matrix in the preform. For example, methyl trichlorosilane, the precursor for SiC, is decomposed in the presence of H_2 between 1200°C and 1400°C to form SiC:

$$CH_3SiCl_3 \xrightarrow{H_2} SiC_{(s)} + 3HCl_{(g)} \tag{6.25}$$

The vapors of SiC deposit as solid phase within the preform to form the matrix. The CVI process is very slow because the matrix is build up by the deposition of molecular-level layers. The CVI process of forming a ceramic matrix is, indeed, a kind of low-stress and low-temperature process and thus avoids some of the problems associated with high-temperature ceramic processing. It also has the flexibility of forming an interface layer prior to the deposition of matrix in the same reaction chamber. For example, in SiC_f/SiC composites, a carbon interface layer is first deposited, and then the SiC matrix is infiltrated without changing the preform conditions but with a change of feed gases. The graphitic coating on the fibers has a characteristic aligned structure, in which the basal planes are aligned parallel to the fiber direction. When a growing crack front is perpendicular to the fiber direction, the crack gets deflected at the weakly bonded basal planes and thus prevents the crack growing through the fibers. The thermal stresses arising from the process are also accommodated by the easy sliding of basal planes aligned parallel to the fibers. Long columnar grains, perpendicular to the fiber surface, are usually formed during the deposition of SiC. The grains are composed of predominantly β-SiC with small disordered regions of α-SiC.

The nature of porosity in a CMC reinforced with woven fabrics processed by CVI is trimodal. Macropores are present between fiber bundles and between the layers of fabric, with pore sizes less than 100 μm. Micropores are present between fibers in the bundle, and the pore size is usually about 10 μm. Submicron-sized pores are present between columnar grains in the SiC matrix.

The important fact about the CVI process is that it can be carried out at temperatures in the 1000°C–1200°C range. Thus, it is applicable to a broader range of fiber-reinforced composites in view of less fiber degradation at these temperatures. While it is difficult to achieve very low porosities with this method, it can produce several matrices of interest with reasonably limited porosities. However, the residual porosity issue becomes more serious as cross-sectional dimensions increase, and the amount of residual porosity also changes with fiber architecture. Some of the inherent limitations of the CVI are relaxed during forced chemical vapor infiltration (FCVI) of composites; it reduces the requirement for homogeneity and dangers of interrupted deposition and can increase deposition rates. These factors enhance the already reasonable cost potential of the CVI process.

6.3.7.1 Preform Preparation for CVI

The continuous fiber preforms for the CVI process are prepared in three ways. In the first method, woven cloths are stacked and kept together by a specially designed tool giving the desired shape and size to the preform. Better control of the fiber volume fraction and initial porosity is possible in this method. The tool can be removed after sufficient deposition to hold the different layers is formed, and at this stage the preform has a high enough mechanical strength for easy machining.

In another technique, the preform is built from ceramic fiber rods obtained by pultrusion, in which the monofilaments are bound together by an organic binder and then subjected to a pyrolysis treatment. In this way, 3- and 4-D fiber architectures can be produced. Similar preforms can be obtained by braiding of fiber tows.

A third procedure that is gaining importance, particularly in the case of carbon fiber, is needling. This technique binds together cloths or fabric layers by introducing fibers in the third direction using needles with small hooks. The needling is carried out on each layer, which results in an excellent homogeneity in the preform. Another advantage of this technique is that the preform can be directly machined without a binder or matrix deposit.

6.3.7.2 Isothermal Isobaric CVI

One of the commonly used variants of CVI process is isothermal chemical vapor infiltration (ICVI), which relies on diffusion for infiltration. The whole preform is maintained at a uniform temperature in a furnace, while the reactant gases are allowed to flow through the furnace, and they react and deposit the solid materials over the preform. To obtain a uniform deposition around the fibers, the CVI process is carried out at low pressures and low reactant concentrations.

One or several preforms are placed in a hot-wall CVD reactor and the gaseous precursor of the desired matrix is passed into it. One of the main problems with this technique is the formation of infiltration gradient within the preform. This gradient depends on pore distribution, the nature of the precursor, and the infiltration parameters. For a given preform and precursor, it is possible to obtain infiltration homogeneity by varying the temperature, pressure, and, to a lesser extent, total flow rate.

In the case of TiC infiltration, the deposit is limited to the external surface of the preform when the temperature or pressure is too high. Conversely in-depth deposition is favored by a decrease of temperature or pressure. A continuous decrease of residual porosity toward about 10% and a shift of the pore size distribution toward the small diameters could be obtained when the CVI process parameters are optimized. The total flow rate of gas is also an important parameter because it can control the residence time of the gas mixture within the reaction area. The residence time in turn can greatly influence the infiltration process.

The infiltration rate is an important feature of the CVI process, and it should be at the maximum to minimize the cost of fabrication. Measuring the variation of the overall preform mass as a function of time is a simple way of assessing the infiltration rate. Generally, in the CVI process, a compromise must be made between deposit uniformity and infiltration rate. In order to reduce the overall infiltration time, less favorable conditions for uniform densification can be chosen, and intermediate surface machining may be applied to the preforms to reopen the pores. Equally important is the type of precursor (reactant) gas. The parameters optimized for a particular gas may not be applicable to other precursor gases.

Another way to increase the densification rate has been proposed, and that can be practiced effectively in large reactors. As mentioned earlier, at reduced pressure and temperature, deposition in the interior of the preform is more pronounced; hence, the initial infiltration of the preform is carried out by placing the preform in the region of the reactor that is at a lower temperature and/or when it only contacts significantly depleted reactant gases so that deposition is slow and the reactants can penetrate to the finest pores. Once the fine pores are filled, the preform can be moved to regions of higher temperature and/or lower gas depletion to more rapidly fill the coarser pores. Even this can be practiced in smaller reactors in two stages. At the initial stage, the temperature and pressure can be maintained low and later it can be increased to higher levels. Composites produced by this two-stage process exhibit good mechanical strength due to their low porosity, and the time for infiltration by this method is less than 10% of the time needed for the single-step process.

The ICVI technique is now commonly used to form C and SiC matrices, and it has been adapted to plant level for infiltrating multidirectional fiber preforms based on C, SiC, and even Al_2O_3 fibers. It is a slow process and leads to the formation of materials with spatial density gradients and some residual porosity. Usually the process is interrupted several times to remove the surface crust and reopen the external pores for further infiltration. Despite these drawbacks, its use in industry has been justified because of the following advantages: (1) It needs a rather simple technology and procedure; (2) the reinforcements can retain their capability, owing to the rather low process temperatures; (3) the nature of the deposit can be modified easily by changing the precursor gas injected into the reactor (this is very important, especially when there is a need of interface coating [e.g., C and hex-BN coating from hydrocarbon and BF_3–NH_3 precursors, respectively] or a protective coating to improve the resistance to environmental effects [e.g., SiC coating on C/C composites]); (4) a large number of preforms can be infiltrated simultaneously; and (5) preform of different shapes and sizes can be easily infiltrated in the same run.

During this ICVI process, the surface pores tend to close first, restricting the flow of gas to the interior of the preform. This phenomenon is sometimes referred as canning, which necessitates multiple cycles of impregnation and surface machining to obtain a uniform density in the product. The problem of canning can be avoided by using a forced gas flow and a temperature gradient in the preform.

6.3.7.3 Forced CVI (FCVI)

The gases are transported to the interior of the preform mainly by diffusion in the ICVI technique, and the driving force is the concentration gradient occurring between internal and external parts of the preform. This transport mode limits the densification rate, and hence, it is of interest to consider forced convection. The driving force in this case is the total pressure gradient through the preform. This technique is called FCVI, and the precursor gas is injected at a pressure, P_1, through one side of the preform and the exhaust gas is pumped out at a pressure, P_2 ($P_2 < P_1$), on the opposite side. It results in highly increased deposition rates and the infiltration time being reduced from several hundred hours to a few tens of hours. When the FCVI process is carried out under isothermal conditions, there is a depletion of gas phase as it goes through the preform, which favors the deposition at the inlet side of the preform and decreases the permeability. Hence, a combination of pressure gradient and temperature gradient would be preferable to increase the infiltration homogeneity. A low temperature maintained near the inlet is not sufficient for the gaseous species to react/decompose and form the ceramic deposit.

The use of a combination of thermal gradient and forced reactant flow can overcome the problems of slow diffusion and permeability during ICVI. This can eliminate, to some extent, the need for multiple impregnation and surface machining cycles. Thus, the FCVI process takes much shorter infiltration times; still it forms a product with uniform density of matrix and low residual porosity. As a comparison, to infiltrate a 3 mm thick part, the ICVI process could take several weeks, whereas the infiltration of the same part by the FCVI process takes only several hours. Irrespective of the type of CVI process, with increasing densification, a point is reached beyond which the increase in density is not proportional to the infiltration time. The deposition rate in the CVI process is primarily governed by mass and heat transfer as well as chemical kinetics.

A schematic of thermal gradient FCVI process is shown in Figure 6.20. In this process, the preform is placed in a graphite holder, which is in contact with a water-cooled metallic gas distributor. Thus, the bottom and side

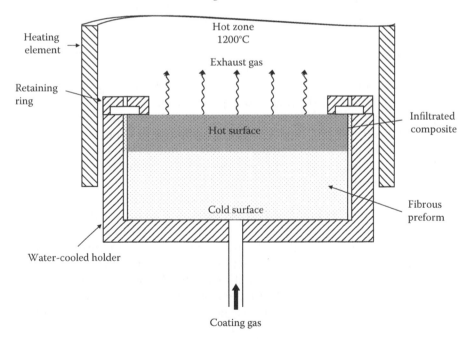

FIGURE 6.20
Schematic of forced and thermal gradient CVI process. (Reprinted from Stinton, D.P. et al., *Am. Ceram. Soc. Bull.*, 65, 347, 1986. With permission.)

surfaces can stay cool, while the top of the preform is exposed to the hot zone, creating a steep thermal gradient. The gaseous reactants pass through the bottom of the preform without undergoing any reaction. The vapor phase moves to the hot zone, where heterogeneous reactions occur on the fibers to produce the matrix. Such a deposition of the matrix material within the hot portion of the preform increases the density and thermal conductivity in the region. Hence, the hot region expands progressively from the top to the bottom of the preform. As a consequence, the deposition also occurs progressively from the top to the bottom of the preform. Once the matrix is deposited completely in the pores at the top region of the preform, then that region is no longer permeable to the gases. The gases then flow radially through the preform and come out through the vented retaining ring.

The initial value of ΔT (temperature difference between hot and cold zones) is an important parameter for the optimization of the FCVI process. It should not be too low or too high; when ΔT is too low, significant deposition occurs in the cold zone, and when ΔT is too high, a rapid deposition in the hot zone can result in an early pore sealing. Another controlling parameter is the gas flow rate, which must be high enough to favor a uniform infiltration profile.

For better deposition, the rate of deposition must be maximized while minimizing density gradients. Since the deposition reaction and mass transport are competing factors (related to each other), high deposition rate results in a well-infiltrated exterior, while the interior of the preform has severe density gradients and a large amount of porosity. On the other hand, very low deposition rates require long infiltration times and are not economically viable. An optimum deposition rate should be used, such that a reasonable quality product is obtained at a reasonable cost.

The FCVI technique has been used mainly to infiltrate 2-D preforms made of carbon or silicon carbide fibers. Such preforms are made by stacking and compressing woven cloth layers in a graphite holder. Alternate layers are oriented at 0°–30°–60°–90° angles yielding a porosity of about 60%.

Circular disks of 240 mm in diameter and 16 mm in thickness have been produced by the infiltration of SiC into the SiC fiber preform using the FCVI process. The residual porosity was 20% after 40 h of infiltration. It is also possible to make tubular shapes by the FCVI process. To prepare this shape, a 3-D preform made by braiding is water-cooled from the inside and heated from the outside using a furnace. The reactant gas is passed along the cooled interior surface, flows through the tubular preform, reacts in the hot external zone, and deposits the matrix. The deposition moves progressively toward the inner surface of the tube. Tubes with an outer diameter of about 100 mm can be made by this process. To prepare SiC/SiC composites by the FCVI process, a mixture of CH_3SiCl_3 and H_2 gases is passed into a SiC preform. The typical process parameters are hot zone temperature between 1100°C and 1200°C, molar ratio of H_2/CH_3SiCl_3 in the range of 5–10, and gas flow rate ranging between 3 and 9 L min^{-1}. In the case of C/C composites, propylene, propane, or methane is passed into a carbon fiber preform with a hot zone temperature of 1200°C–1300°C, a cold zone temperature of 800°C–1100°C, and flow rate between 100 and 400 SCCM (standard cubic centimeters per minute).

The thermal gradient in the FCVI process can also be created by a microwave heating system. This microwave-assisted FCVI technique has been used to infiltrate 1-D fiber preforms consisting of alumina fibers (10–20 μm diameter) laid uniaxially to a cylindrical geometry of 16 mm diameter. The gaseous reactant mixture to form an alumina matrix is $AlCl_3 + H_2 + CO_2 + N_2$. During microwave heating, the temperature difference between the cold outer surface and the hot centerline is in the range of 400°C–600°C. Alumina fiber-reinforced alumina matrix composite with high internal density and good axial uniformity is prepared by this process (Spotz et al. 1993, Skamser et al. 1994) at relatively shorter duration (8–18 h).

Under these modified conditions in the FCVI, 70%–90% dense SiC and Si_3N_4 matrices can be formed in SiC on Si_3N_4 fiber preform in less than a day, whereas the ICVI process would take several weeks to achieve such densities. Binders are not required while preparing the preform in the graphite holder, thus avoiding the problem of incomplete binder removal. The application of moderate pressure to the preform kept in the graphite holder can result in higher fiber volume fraction in the final product. In any bulk infiltration process, such as vapor infiltration, the final density is restricted to about 93%–94% of theoretical density, since beyond this density only closed pores will exist. However, it may be possible to achieve higher densities by the FCVI process, because the pores are closed from top to bottom in a periodic manner.

Although the densification times are significantly reduced and the yield of ceramic matrix is relatively high compared to those of the ICVI, the FCVI process is not suitable for complex-shaped preforms. Unlike ICVI, only one preform can be infiltrated at a time in the reactor and complex graphite fixtures are needed to create the temperature and pressure gradients in the preform.

6.3.7.4 Pulse CVI (PCVI)

In this technique, total pressure cycle is used in order to regenerate the entire gas phase periodically. By using this technique, the infiltration of carbon, silicon carbide, and titanium carbide has been carried out. The pulse chemical vapor infiltration (PCVI) equipment permits the following repeated sequence: (1) rapid injection of the gaseous precursor of the ceramic to be deposited into the reactor up to the desired pressure (it usually takes less than 0.5 s), (2) deposition of the ceramic material by maintaining the pressure for a given time, and (3) evacuation of the chamber down to a residual pressure to pump out gaseous reaction products and unreacted source gas out of the preform. The PCVI process provides a very simple way to control the resident time of the reactant gas mixture in the preform. This is very important when the gaseous precursor undergoes maturation phenomena, that is, the nature and concentration of the gaseous species vary with time. For the deposition of pyrocarbon or SiC, all the pressure pulses are identical in precursor composition and duration. A very attractive application of the PCVI technique is the possibility to fabricate multilayered interphases on a very small scale down to the nanometer level. In this case, a computer-controlled switching is operated to change one precursor to another one and the relative thickness of layers is controlled by the duration and the number of pressure pulses.

Theoretical predictions indicate that the PCVI process can produce a more homogeneous deposit with higher conversion yields at a shorter infiltration time. However, such things are yet to be established experimentally. In addition, upgradation to industrial scale may give rise to technical problems, such as the rapid pumping of a large amount of gases. Conversely, the PCVI technique is a highly flexible method, which permits the control of the microstructure of deposited materials (e.g., the anisotropy of pyrocarbon) and the fabrication of new advanced ceramics with a highly engineered microstructure (e.g., multilayered ceramics at a nanometer scale).

6.3.7.5 Thermal Gradient CVI

In order to overcome the lack of flexibility of the FCVI technique, the use of thermal gradient alone has been suggested and various ways have been investigated to create a temperature gradient within the preform. The principle behind this is to heat preferentially the heart of the preform in a cold-wall reactor, and the radiation heat losses lead to lower temperature in the outer zone and hence the densification front moving progressively from the inside to the outside of the preform.

One of the methods to achieve this type of thermal gradient is microwave heating, which permits internal heating of preforms. This technique is efficient owing to the high yield of ceramics and low production of waste products. This technique has been used to infiltrate SiC-based matrices in SiC fiber preforms (Devlin et al. 1992). It has been found that the temperature within the preform is difficult to control during the densification process. A significant variation of the thermal gradients is observed, which depends on the nature and thermal conductivity of the fibers and the matrix. The partially densified SiC matrix in the internal zone of the preform is heated less due to its higher conductivity compared to that of fibers. As a consequence, the decrease in the thermal gradient results either in cessation of the internal infiltration or in the deposition of a SiC + Si mixture rather than SiC alone.

Another way to produce temperature gradient in a preform is to use inductive heating by direct coupling, provided that the preform and the matrix have sufficient electrical conductivity. This technique is first used for the processing of SiC whisker-reinforced alumina composites and later has been used to form C/C composites in one single run (Golecki et al. 1995). A stack of carbon fiber disk preforms have been infiltrated with cyclopentane to form carbon matrix in one cycle of duration 26–50 h. This high rate of infiltration, more than 10 times faster than in ICVI, is made possible mainly due to the high temperature created inside the regions of the preform. The density variation within a disk is only ±5%–8% and the amount of tar generated is also very small.

A catalytic chemical vapor infiltration (CCVI) technique has been proposed as a potential improvement over the conventional CVI technique for the formation of C/C composites. In this method, the preform is impregnated with a metal catalyst (e.g., Ni) that activates the decomposition reaction thereby increasing the deposition rate. With a propylene precursor, carbon can be deposited within the preform as fine filaments at temperatures as low as 375°C and at very high rates. Nevertheless, this technique cannot produce a dense matrix and must be followed by conventional CVI in order to densify the matrix.

To significantly reduce the infiltration time in the fabrication of fiber-reinforced CMCs, it has been suggested that a combination of liquid and vapor infiltrations would be useful. A liquid route such as PIP is used to fill the large pores in the preform and then the fine pores are filled by the ICVI technique. This mixed route permits the fabrication of CMCs with good mechanical properties in a tenth of the time. In addition to the overall time reductions, the secondary CVI step does not need special tooling to hold the fiber layers.

6.3.7.6 CVI Modeling

As mentioned earlier, CVI is directly related to CVD. The general CVD process can be described as a set of successive steps: (1) transport of the gaseous precursor from the reactor inlet toward the hot reaction area by forced convection; (2) homogeneous gas-phase reactions occurring in the hot zone and forming intermediate reactive species; (3) transport of these intermediate species to the substrate surface; (4) heterogeneous chemical reactions on the surface of the substrate, that is, adsorption of reactive species, reactions either between adsorbed species themselves or between surrounding gas and adsorbed species, and adsorption of gaseous species produced by the surface reactions; (5) transport of the gaseous products from the substrate surface to the bulk of the gas phase; and (6) transport of these gaseous products and unreacted precursor species from the reaction zone to the outlet of the reactor by forced convection. These steps in the CVD process are more or less independent of each other. Some of the steps (steps (1), (3), (5), and (6)) are physical in nature (depends on heat, momentum, and mass transfers), and others are chemical in nature (homogeneous and heterogeneous reactions of steps (2) and (4)). These steps in the CVD process are mainly governed by kinetics and to a lesser extent by thermodynamics. The overall kinetics of the

CVD process can be limited either by mass transfer (steps (3) and (5)) or by chemical reaction (steps (2) and (4)).

In the case of the ICVI process, additional factors must be taken into account because of the different nature of the substrate. Steps (1) and (2) are almost the same, which control the composition of the gaseous phase surrounding the preform. In the ICVI process, the reactive species must diffuse through the pore network of the preform in order to reach its heart. Hence, step (3) becomes a fundamental step of the process. Step (4) proceeds anywhere on the internal surface of the pores depending on the diffusion of the reactive species. The importance of step (5) is also increased in the ICVI process because the gaseous products must diffuse from the core to the external zone of the preform. Step (5) can play an important role if the gaseous products within the preform affect the surface reaction, for example, as inhibitor. In the FCVI technique, the mass transport through the porous preform occurs not only by diffusion but also by forced convection (for which the driving force is the total pressure gradient and the limiting factor is the permeability of the preform). In the case of PCVI, the rate of transport of gaseous reactants increases during pressurization by forced convection; however, the diffusion becomes the main mode of transport once the desired pressure is reached.

6.3.7.6.1 Thermodynamics

The first theoretical approach for obtaining a preliminary knowledge of a CVI chemical system is thermodynamics. Based on thermodynamics, it is possible to calculate the theoretical composition of the deposit and the gaseous product and the yield for any experimental condition such as temperature, pressure, and initial composition of gas phase.

An example of a thermodynamic study is the calculation of both the heterogeneous and homogeneous reaction equilibrium in the CH_3SiCl_3–H_2 system for the CVD of SiC. It is found that in a hot-wall reactor, the precursor molecule, CH_3SiCl_3, is not the actual source for SiC deposition but is decomposed into intermediate species ($SiCl_2$ and CH_4) (Langlais et al. 1995). These intermediate species are the source species for Si and C, respectively, which on reaction deposit SiC. The reactions taking place in the CVD reactor for the deposition of SiC are

$$CH_3SiCl_3 + H_2 \rightarrow CH_4 + SiCl_2 + HCl \quad \text{(homogeneous reaction)} \quad (6.26)$$

$$CH_4 + SiCl_2 \rightarrow SiC + 2HCl + H_2 \quad \text{(heterogeneous reaction)} \quad (6.27)$$

6.3.7.6.2 Kinetics

In a CVD/CVI reactor, owing to the continuous flow of reactant gases, heterogeneous equilibrium is never reached except under very specific conditions, which are decided by kinetic factors. An Arrhenius-type plot is generally used to determine the chemical and mass transfer regimes. When the activation energy is low, the reaction kinetics is controlled by mass transfer, whereas the kinetics is governed by the chemical reactions when the activation energy is high. The transition between these two kinetic regimes can also be found by varying the flow rate. If the deposition rate increases with the total flow rate, the mass transfer is the controlling factor. If the deposition rate is unaltered or decreases with increasing total flow rate, the kinetics is considered to be controlled by the chemical reactions. The chemical reactions giving rise to solid deposit can also include homogeneous reactions. A decrease in deposition rate with increasing total flow rate or decreasing residence time of the gaseous precursor in the reaction area indicates the occurrence of homogeneous reaction. In the case of carbon deposition from various hydrocarbons, an increase of the residence time favors the transformation of the precursor by gas-phase reactions, called "maturation," giving rise to efficient intermediate species for the formation of various types of carbons. In specific cases where the deposition rate does not depend on the total flow rate or residence time of gaseous reactants, the kinetic process is only controlled by heterogeneous reactions. Under such conditions, the activation energy is directly related to the chemical reaction for the deposition, and kinetic laws can be derived by varying the partial pressures of the various precursor species to determine the order of reaction.

6.3.8 Chemical Vapor Composite Deposition

It is a recently developed process in which powders, fibers, or both are mixed in the reactant gas stream and deposited on a substrate. This is a less costly process for the fabrication of high-performance ceramic, metal, and carbon composites. It eliminates the need for a preform and yields net-shape products with a 0.13 mm tolerance. The deposition rate is 100 times faster than CVI, and also this process permits easy tailoring of the density, composition, and microstructure of the composite material. The composite requires only light machining after the deposition. A schematic of this process is shown in Figure 6.21. Deposition rates up to 2 mm h^{-1} have been obtained in this process. An NNS composite product can be deposited on a removable graphite substrate. Any short fibrous materials or particulates that can withstand high temperatures for extended periods can be used in this process. SiC matrix composites utilizing a variety of ceramic powders (SiC, Al_2O_3, and TiB_2) and whiskers (SiC and Si_3N_4) have been deposited. This is also a potential method to deposit CNT-reinforced ceramic composites. CMC tubes of 1.8 m (6 ft) length and 0.2 m (8 in.) diameter have been made by this process (Reagan and Huffman 1989).

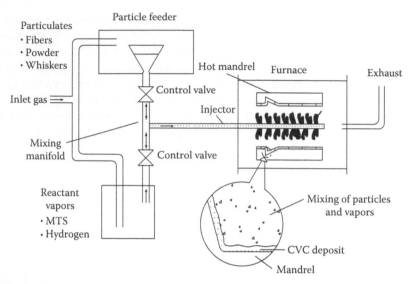

FIGURE 6.21
Schematic of chemical vapor composite deposition. (Reprinted from Reagan, P. and Huffman, F.N., *Fiber-tex 1988*, Buckley, J., Ed., NASA Scientific and Technical Information Division, Washington, DC, 1989, p. 247. With permission.)

6.3.9 Potential Cost-Effective Processing Methods

New in situ composites are being developed by chemical reaction sintering, which include refractory carbide-, nitride-, or boride-reinforced Al_2O_3, Si_3N_4, and SiALON composites. Improved hardness and toughness as well as better sintering capabilities can be achieved by this approach. In addition to that, unique microstructures at low cost can be obtained by this process.

6.3.9.1 Sullivan Process

Another new process developed by Sullivan Mining Corporation, Greenwood, United States, uses a liquid route instead of conventional powder methods to produce CMC parts. For example, Si-based liquid is allowed to react with NH_3, and the by-products are removed using supercritical fluid extraction and then heat-treated, resulting in the formation of Si_3N_4 with high strength and diamond-like hardness. SiC and other carbide ceramics can be produced by this process by changing the reactants to include hydrocarbons.

The Sullivan process is capable of producing complex-shaped components at a fraction of the cost of other processes, including hot pressing and reaction bonding. This method does less damage to the reinforcements than conventional processes. The liquids used in this process completely fill the spaces between fibers and do not shrink like powders, and hence, more complex shapes are possible. A liquid is also much easier to handle and pump into preforms than powder, which makes the process attractive for mass production.

There are two basic steps in this process to form a CMC: (1) reaction of the raw materials and (2) supercritical fluid extraction of the by-products. This leaves an amorphous ceramic matrix ready for heat treatment. The reaction takes place in an autoclave with ammonia solvent. This ammonia supplies N_2 for the formation of Si_3N_4. Introducing H_2 into the autoclave assists in polymerization and facilitates extraction. Other additives, such as metals, react with O_2 contamination, to help the formation of second phases in the matrix.

The extraction step in the Sullivan process is able to remove any undesirable impurities. Hence, low-cost industrial-grade raw materials can be used in the process, from which it is possible to produce high-quality components. The polymeric precursors used as raw materials also instill flexibility and molecular-level control of composition in the final product. After the extraction of by-products, the ceramic parts can be heat-treated at temperatures much lower than those of conventional processes. Further reduction in sintering temperature by the use of microwave furnaces is being explored.

6.3.10 Summary of CMC Processing

CMCs with fully dense matrices offer several advantages as structural materials over the CMCs with less dense matrices, such as better matrix cracking resistance and good stability under environmental attack at high temperatures. In general, powder-based processing methods allow rapid production of CMC green bodies in a single step from low-cost matrix powders. While offering advantages such as easy handling and faster production, the requirement of a high-temperature consolidation step is a major constraint. Fiber and interface properties may be affected by the exposure of CMCs to these extreme temperatures, either through reaction, grain growth, or structure change. Hence, to make the powder route as a versatile method for the processing of CMCs, it is necessary to minimize the densification temperatures to avoid fiber degradation. It is usually assumed that the cost of product is directly related to the processing temperature (although not applicable to all materials); hence, the temperature reduction for consolidation can also have the effect on the cost of product.

Clearly, there are various routes to reaction technology for the formation of ceramics; the important examples are the reaction sintering (or bonding) of SiC and Si_3N_4 from C and Si compacts, respectively. Processing of composites such as Si_3N_4-bonded SiC and SiC-bonded B_4C by infiltrating compacts of SiC + Si and B_4C + C with N_2 or NH_3 gas and molten Si, respectively, at a high temperature are important examples of the industrial application of this technology for refractory, wear, and armor use. Fiber- and whisker-reinforced Si_3N_4 or SiC by reaction forming has also been investigated. The matrix material is formed around SiC fibers/whiskers at high temperatures where it is reactive and hence may also attack fibers or whiskers.

The CVI is a flexible processing technique used for the fabrication of CMCs. In this process, a porous fiber preform is densified with a ceramic matrix, such as carbon, carbide (SiC, TiC, B_4C, etc.), nitride (Si_3N_4, BN, etc.), or oxide (Al_2O_3,

ZrO$_2$, etc.). The deposition of ceramic matrix results from chemical reactions of a gaseous precursor, which involves a set of complex phenomena, for example, the transport of gaseous reactants into the preform by diffusion and convection. During the CVI process, generally low pressures and moderate temperatures are used, which permit the use of fibers of limited thermal stability such as Nicalon SiC fibers and alumina-based fibers. It is an NNS process and can be used to make complex and variable shapes. Another important feature of the CVI process is its capability to deposit successively the interface coating on the fibers and the matrix and eventually an external seal coating in the same reactor.

6.4 Mechanical Properties of CMCs

A mere increase in elastic modulus and strength is not the primary objective in CMCs because many monolithic ceramics themselves have high modulus and appreciable strength. However, an increase in elastic modulus and strength may be desirable for low-modulus and low-strength matrix materials such as glasses, glass–ceramics, MgO, and mullite.

The mechanical behavior of continuous fiber-reinforced ceramics mainly depends on the following three factors: the strength of the fibers, the fiber–matrix interface bonding, and the residual stress present in the composite due to the difference in the CTE of the fiber and matrix. When the matrix fails, the load is transferred from the matrix to the fibers in the wake of the crack, and the fibers must be able to take the load. Hence, high fiber strength is important. Thus, the ultimate load-bearing capability of the composite is determined by the load-bearing capacity of the fibers. The level of toughening depends on how well the interface between the fiber and the matrix is engineered. In an ideal composite, the interface should be strong enough to transfer load from the matrix to the fiber yet weak enough to allow the debonding of the fiber once the crack reaches the interface. Toughening occurs after debonding by crack deflection, crack bridging, and pullout. The influence of fiber–matrix bond strength on the toughness is illustrated by stress–strain curves in Figure 6.22. Unreinforced ceramics show purely elastic behavior and the failure is brittle. Since the strength of fiber is usually higher than that of the matrix, some strengthening occurs in SiC fiber-reinforced SiO$_2$. However, the strong interface does not allow debonding in the presence of matrix crack; the potential toughening mechanisms, such as crack deflection and crack bridging, are not activated; and the composite fails in a brittle manner. In composites with an engineered fiber–matrix interface in which debonding occurs as the crack passes the fiber, leaving the fiber intact and continues to bear the load. On further stress application, the fibers failing slightly away from the crack plane are pulled out from the matrix, thus contributing to further toughening. Thus, by the incorporation of continuous strong fibers with well-engineered fiber–matrix interface (with BN coating) makes the

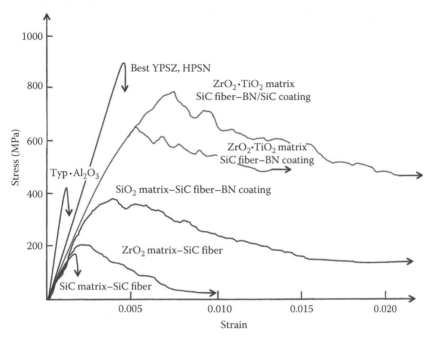

FIGURE 6.22
Stress–strain behavior of ceramic composites with and without interface coating. (Adapted from Lee, S.M., Ed., *Reference Book for Composites Technology*, Vol. 1, Taylor & Francis Group, Boca Raton, FL, 1989, p. 131. With permission.)

otherwise brittle material, to have improved tolerance to strain and reduced probability to catastrophic failure.

The influence of interface coating and intermediate heat treatment on the stress–strain behavior of SiC$_f$-reinforced SiC has been reported by Udayakumar et al. (2011a). Three types of SiC$_f$/SiC composites are prepared through the ICVI process using ceramic-grade Nicalon 8-harness satin fabric layers. The first composite has CVI-derived BN interface (composite A), the second composite also has CVI-derived BN interface (but it is subjected to an intermediate heat treatment at 1200°C in argon (Ar) atmosphere for 5 h after 430 h of SiC matrix infiltration (composite B)), and the third composite is prepared by stabilizing the BN interface at 1200°C in Ar atmosphere for 5 h prior to SiC matrix infiltration (composite C). The tensile stress–displacement curves and scanning electron microscope (SEM) fractographs of these composites are shown in Figure 6.23. Composite A has very low tensile strength and failed in a brittle manner due to the strongest bond existing between the fiber and matrix. The intermediate heat treatment given to composite B has modified the interfacial bonding between the fiber and matrix. The tensile strength of this composite is (233 MPa) found to be higher than that of composite A (70 MPa), and also there is an increase in the strain-to-failure value. It has been reported that the BN deposited at low temperature and low pressure changes to ammonium borate hydrates as a result of

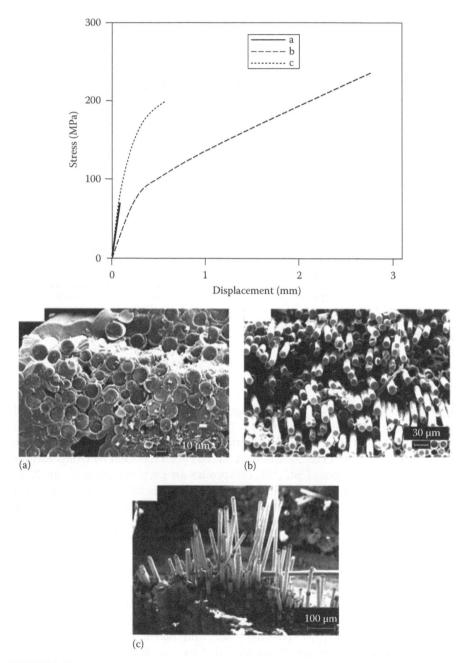

FIGURE 6.23
The influence of interface coating on the stress–strain behavior of SiC_f/SiC composites and the corresponding fractographs: (a) composite A, (b) composite B, and (c) composite C. (Reprinted from Udayakumar, A. et al., *J. Eur. Ceram. Soc.*, 31, 1145, 2011. With permission.)

reaction with moisture (Cholet et al. 1994). This might have happened in composite A during loading/unloading of composites into/from the reactor, when CVI cycles are interrupted for surface machining to open up the pore closures for further infiltration. The intermediate heat treatment has improved the crystallinity and stability of BN in composite B to some extent and modified the bonding strength between the interface and the matrix to a bond weak enough to facilitate good fiber pullout. Composite C also has higher tensile strength (200 MPa) and exhibited nonlinear failure behavior. However, the strain-to-failure value of composite C is lower than composite B. The percentage elongation values for composites A, B, and C are 0.14%, 4.5%, and 1.1%, respectively.

SEM micrograph of composite A shows practically no fiber pullout, thereby supporting the low toughness obtained for this composite. It can be due to the formation of strong interface bonding. It is very clear from the tensile test and fractographs that medium fiber pullout observed for composite C is also supporting the moderate elongation showed by this composite. The structure and bonding nature of interface coating are shown in Figure 6.24. In the case of composite A, the interface is not clearly visible and it is very thin (~100 nm). The interface might have reacted with moisture and subsequently with the matrix causing the formation of the strongest bond between the fiber and matrix. In the case of composite B, the interface is clearly visible and it is significantly thicker (~400 nm) compared to composite A. The interface is constrained to a large extent after 430 h of SiC

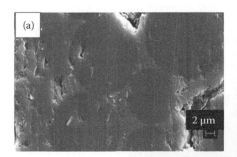
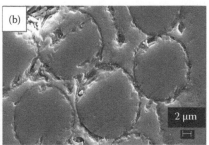
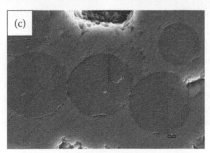

FIGURE 6.24
Microstructure of SiC_f/SiC composites showing the nature of interface: (a) composite A, (b) composite B, and (c) composite C. (Reprinted from Udayakumar, A. et al., *J. Eur. Ceram. Soc.*, 31, 1145, 2011. With permission.)

matrix infiltration, and it resulted in a porous interface and modified the bond between the fiber and the matrix to a weak bond to facilitate the fiber debonding. The interface region is also clearly visible in composite C; however, its thickness (~230 nm) is lesser than the interface thickness in composite B.

In order to evaluate the interfacial shear strength between the fiber and the matrix of the composites, single-fiber push-through test is performed using a nanoindenter with nano-positioning system (Nanoindenter G200-MTS Nano Instruments), which has a diamond indenter. Composite sample (cross section) of thickness 1.5 mm is mounted on the epoxy mold that has a blind hole of diameter ~5 mm at the center to enable the fibers to be pushed out. The sample thickness is reduced to ~350 μm by grinding and then the surface is polished with diamond slurry. An individual fiber surface is selected using the nano-positioning system with appropriate optics, and load is applied at a rate of 10 μN s^{-1}. The fiber bore the load at the beginning and then slipped through the matrix once the force balance between the frictional force and applied force is achieved. After the fiber is pushed, the indenter tip touched the matrix. The sequence of fiber push-through test is shown in Figure 6.25. The force at which the fiber starts slipping is used to calculate the interfacial shear strength using the following equation:

$$\tau_i = \frac{F}{2\pi rt} \tag{6.28}$$

where

τ_i is the interfacial shear strength
F is the force at which the fiber starts to slip through the matrix
r is the radius of the fiber
t is the thickness of the sample

The load–displacement curves of composites obtained during the push-through test are shown in Figure 6.26. In the case of composite A, the fiber

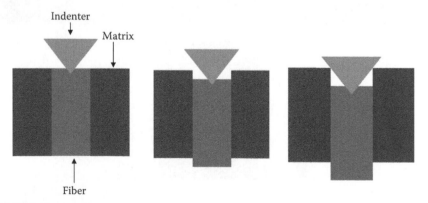

FIGURE 6.25
Schematic representation of the sequences of fiber push-through test using nanoindenter. (Reprinted from Udayakumar, A. et al., *J. Eur. Ceram. Soc.*, 31, 1145, 2011. With permission.)

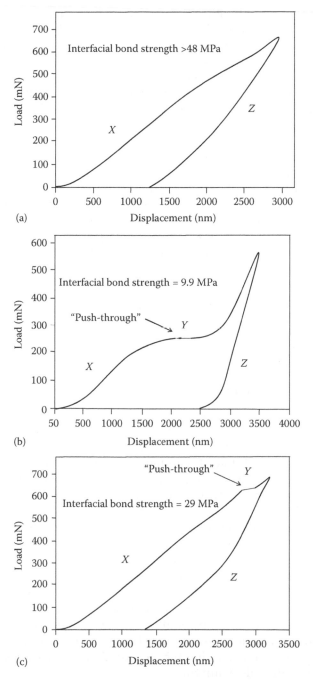

FIGURE 6.26
Load–displacement curves obtained from the nanoindenter: (a) composite A, (b) composite B, and (c) composite C. (Reprinted from Udayakumar, A. et al., *J. Eur. Ceram. Soc.*, 31, 1145, 2011. With permission.)

slipping is not observed within the maximum load capacity of the nanoindenter, indicating a strong bonding between the fiber and matrix (>48 MPa—calculated by taking the maximum load in the indentation curve). The interfacial shear strength calculated for composite B (10 MPa) is weak enough to facilitate debonding. It indicates that the intermediate heat treatment has modified the bond strength. The interfacial shear strength obtained for composite C is 29 MPa, which indicates a relatively stronger bond between the fiber and matrix.

Crack deflection at the interface frequently follows fiber/matrix debonding. This improvement in fracture toughness owing to the weak interface bonding has been confirmed experimentally in many CMCs. In fact, this mechanism is the major source of improving fracture toughness in CMCs. Interface bonding is controlled either by selecting fiber and matrix materials, such that they do not undergo any chemical reaction at the interface during processing, or by applying a diffusion barrier coating to prevent a strong bond between fiber and matrix. The most widely used coating materials at present are carbon and boron nitride, because they are having a weak crystallographic orientation, which can preferentially delaminate. These coatings also prevent damage to the fibers during processing. However, these materials are prone to environmental attack at elevated temperatures. Hence, a variety of environmentally resistant multilayer coatings and oxide-based coatings are under development.

6.4.1 Strength and Modulus

Crivelli and Cooper (1969) have reported flexural strengths as high as 380 MPa for carbon fiber-reinforced vitreous silica. It has been found that this strength is retained up to 800°C. In another study made by Phillips (1974) in Pyrex glass reinforced with high-stiffness and high-strength C_f, he showed that the strength determined by the four-point bend test for the high-stiffness C_f-reinforced composite is 459 MPa and high-strength C_f-reinforced composite is 575 MPa.

Various glass matrix composites reinforced with C_f have shown an increase in strength and modulus with fiber volume fraction, in accordance with the rule of mixtures (Prewo et al. 1986, Dawson et al. 1987). Nicalon SiC fiber-reinforced borosilicate glass is another typical composite in which such a linear increase in strength with fiber volume fraction is observed (Dawson et al. 1987). However, there is a decrease in strength beyond 55 vol.% fibers in C_f/glass composites (Phillips et al. 1972) mainly due to matrix porosity. Young's modulus also follows a similar trend. After making correction for the effect of porosity on modulus, the measured modulus values are in good agreement with the rule of mixtures.

Glass-ceramic matrix composites containing Nicalon SiC fibers are found to be very promising materials. A glass–ceramic matrix has the advantage of processing in the glassy state at low temperature to high density (close to theoretical density) without any damage to the fibers. The flexural strength of this composite is found to increase with increasing temperature until the softening of the matrix (Prewo et al. 1986). It has been suggested that the increase is due to the ability of soft glass–ceramic matrix to distribute the stresses more uniformly. However, the matrix became too soft to sustain any load at very high temperatures.

SiC_w-reinforced Al_2O_3 composites are found to have substantially improved fracture toughness (Becher et al. 1988), strength (Wei and Becher 1985), thermal shock resistance, and high-temperature creep resistance than that of monolithic Al_2O_3 (Porter et al. 1987). Similar behavior has been observed in SiC_w-reinforced Si_3N_4 composites prepared by a combination of reaction bonding and hot pressing (Shih et al. 1992). However, it is found that the stability of SiC_w as well as fracture toughness depends on the amount of additives used during the processing.

Liu and Huang (2001) have reported the impact response, compressive strength, and flexural strength of 3-D C_f-reinforced ceramics prepared by a combination of pressure infiltration and sol–gel processing using a mixture of silica sol and alumina particles. It has been found that when there is less residual silica, the composite shows lower impact energy due to the more brittle nature of mullite than silica. High-density composites formed as a result of high infiltration pressure show better compressive strength. The flexural strength of the composites is found to decrease with sol viscosity, since the low-viscosity sol leads to a dense matrix and strong interface.

Papakonstantinou et al. (2001) have studied the mechanical properties of C_f-reinforced polysialate composites. It is found that the mechanical properties of these composites are comparable to carbon–carbon and CMCs. The polysialate composites are able to retain 63% of their original strength at 800°C mainly because the polysialate matrix protects the carbon fibers from oxidation. It has been reported that these composites have potential applications in aerospace, automobile, and naval sectors, and also they are less expensive.

Brodkin et al. (1999) evaluated the high-temperature mechanical properties of TiC/TiB_2 composites fabricated by transient plastic-phase processing. The flexural strength of these composites is insensitive to temperatures up to 1000°C, whereas a significant reduction in strength is observed at 1200°C. Furthermore, the authors have reported that there is an evidence for plastic deformation of the TiC constituent. It is attributed that a direct contribution to the strength from the borides comes from the fact that at high temperature they act as non-yielding particulate reinforcement in a plastic matrix. Some solid solution strengthening of the TiC_x due to the solubility of boron in titanium carbide has also been suggested.

The effect of zirconia distribution in alumina has been studied by Balasubramanian (1996). One type of zirconia/alumina composite is prepared using mixed alumina and zirconia powders, and another type is

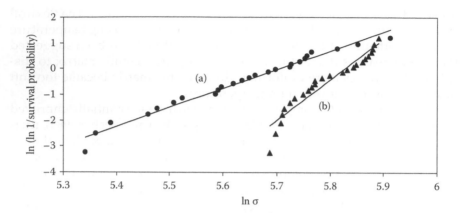

FIGURE 6.27
Weibull moduli of zirconia–alumina composites prepared using (a) mechanically mixed powder and (b) sol–gel-derived powder. (From Balasubramanian, M., Processing and characterization of alumina-zirconia powders and composites, PhD dissertation, IIT Madras, Chennai, India, 1996.)

prepared using sol–gel powder. The mechanical properties of these composites are found to be comparable. However, the composite prepared using sol–gel powder has higher Weibull modulus (15.7) than the other composite (Figure 6.27). The lower Weibull modulus of the mechanically mixed powder composite (7.3) is found to be due to the nonuniform distribution of zirconia grains and the presence of defects around the zirconia grains.

6.4.2 Fracture Toughness

The addition of particles to a ceramic or glass matrix to enhance its properties is well known. The toughening mechanisms in these types of composites may include crack deflection, crack pinning, microcracking, and transformation toughening. However, crack deflection is often the primary toughening mechanism in ceramic particle-reinforced ceramics.

The degree of chemical interaction between particles and matrix is important for easy debonding, and therefore, the level of toughening is achieved by crack deflection. The level of thermal expansion mismatch between the reinforcement and matrix is also important, since it determines the amount of stresses produced at the interface and hence the formation of microcracks. When there is no microcrack formation around the particles, the stress field may deflect the crack.

The incorporation of platelets will improve the fracture toughness of CMCs, but not to the level of whisker-reinforced CMCs. However, the improvement in fracture toughness will be better than particulate CMCs. Crack deflection and bridging by the platelets are the probable toughening mechanisms operating in these CMCs. In some cases, the strength of the composite is reduced by the addition of platelets either due to excessive microcracking or they act as strength-limiting flaws, especially when they are small.

The addition of whiskers to ceramic matrices improves the strength as well as toughness when compared to the addition of particulates and platelets. The presence of whiskers can activate crack bridging, whisker pullout, crack deflection and pinning, and microcracking, which are responsible for the significant improvement in the fracture toughness. Among these mechanisms, the mechanisms operating at any given instance depend on the local stress state and the orientation of whiskers with respect to the crack tip. Toughening by the whiskers, in general, increases with increasing whisker content.

Whisker-reinforced CMCs retain their toughness at high temperatures, generally up to about 1100°C. In fact, the apparent toughness increases partly due to longer pullout and bridging length and partly due to nucleation of creep cracks; however, this increase in toughness is generally offset by the onset of creep at around 1200°C. In addition, non-oxide whiskers, such as SiC and Si_3N_4, are reactive at elevated temperatures. In the presence of oxygen, they form silica at temperatures as low as 1200°C. This silica can react with alumina matrix and form mullite, which further reduces composite toughness.

A combined addition of SiC_w and tetragonal ZrO_2 particles in an Al_2O_3 matrix leads to better toughness than that of either SiC_w/Al_2O_3 composite or ZrO_2-toughened Al_2O_3. However, when this hybrid composite is exposed to high temperatures in an oxidizing atmosphere, the oxidation of SiC is enhanced by the presence of ZrO_2, resulting in the severe degradation of the composite in a relatively short duration (Backhaus-Ricoult 1991).

6.4.3 Fatigue Behavior

The fracture resistance of CMCs under cyclic loading conditions needs to be evaluated for designing the component in a variety of potential applications. For example, it is not unusual to have a design requirement for a ceramic component used in a gas turbine to withstand more than 30,000 cycles of loading.

No significant loss of strength is observed on cyclic loading of C_f/glass composites (Philips 1983). However, the density of matrix cracks is more under cyclic loading than under static loading conditions, and the work of fracture is lower under cyclic loading. Prewo (1987) has studied the tensile fatigue behavior of Nicalon SiC fiber-reinforced lithium aluminosilicate (LAS) glass–ceramics. Two types of LAS matrix materials are used such that one composite shows a linear tensile stress–strain curve till failure and another shows a markedly nonlinear behavior due to extensive microcracking prior to ultimate failure. The residual tensile strength and elastic modulus of the first composite after a certain number of fatigue cycles are same as that of as-fabricated composite. The second composite subjected to cyclic loading below the proportional limit shows similar behavior, whereas on cycling to stress levels above the proportional limit, it results in a second linear stress–strain region with a lower slope. Suresh et al. (1988) and Suresh (1991) found mode I crack growth in SiC_w/Si_3N_4 composites after cyclic compressive loading. Whisker breakage and pullout have also been observed. It has been suggested that the mode I fatigue crack growth could be due to residual tensile stress generated at the crack tip on unloading.

Sintering and other processing aids used in ceramics can form glassy phases at the grain boundaries. The presence of these intergranular glassy phases affects the high-temperature mechanical properties of ceramics, since they can lead to rather conspicuous subcritical crack growth. Such subcritical crack growth is very critical in CMCs because fibers such as SiC can undergo easy oxidation. Han and Suresh (1989) examined the fatigue crack growth in a SiC_w (33 vol.%)/alumina composite subjected to tensile cyclic loading at 1400°C. The subcritical fatigue crack growth has been observed in the composite at stress intensity values far below the fracture toughness. It has also been found that the nucleation and growth of flaws at the interface is the main damage mechanism. The oxidation of SiC_w to a silica-type glassy phase in the crack tip region is also observed. The alumina matrix can react with this SiO_2 to form aluminosilicates (SiO_2-rich or stoichiometric mullite, and the like). The viscous flow of glass can result in interfacial debonding and coalescence of cavities.

In a manner analogous to PMCs, there can be a hysteric heating in CMCs under cyclic loading conditions, but it is due to an interfacial friction. A lubricating layer on the fiber may be beneficial in improving the fatigue life of CMCs. A thicker coating on fibers is expected to provide greater protection to the fibers against abrasion damage, and hence, less frictional heating during fatigue is observed in a Nicalon SiC/C/SiC composite fabricated by the CVI process. However, the composite with a thinner coating exhibits much higher frictional heating (Chawla 1997, Chawla et al. 1998). At higher frequencies, more heating is observed because of the increase in heat generated per unit time. Substantial damage in terms of modulus is observed in Nicalon SiC/C/SiC composite after fatigue loading. At a lower stress, the level of damage is not significantly dependent on frequency. However, at a given frequency, higher stresses induce more damage to the composites.

The sequence of fiber orientation in different layers of a laminate can also affect the high-frequency fatigue behavior. In SiC/Si_3N_4 composites, frictional heating is substantially higher in angle-ply laminates [±45] than in cross-ply laminates [0/90] (Chawla 1997). Temperature rise in these composites correlates very well with stiffness loss as a function of fatigue cycles.

It is of interest to study the behavior of CMCs under isothermal exposure as well as under thermal cycling, since most of the ceramics are intended for high-temperature applications. Wetherhold and Zawada (1991) have studied the behavior of Nicalon SiC/aluminosilicate glass composite under isothermal and thermal cycling conditions. It is found that the samples isothermally exposed and thermally cycled at 650°C–700°C showed rapid oxidation and loss in strength. Oxidation behavior overshadowed any thermal cycling effect in these samples and the embrittlement is attributed to oxygen diffusion from the surface, which destroyed the weak carbon-rich interface. However, less embrittlement is observed at 800°C, and this is attributed to smoothening of the surface by glass flow and low oxidation (Zawada and Wetherhold 1991). Boccaccini et al. (1997, 1998) have studied the thermal fatigue behavior of Nicalon SiC fiber-reinforced glass composites. Even though the thermal mismatch in this system is almost zero, a decrease in Young's modulus and a simultaneous increase in internal friction are observed when increasing the thermal cycles. An interesting finding from that study is crack healing when the composite is cycled to a temperature above the glass transition temperature of the matrix.

In general, one can reduce the thermal fatigue damage by choosing a matrix material that has high yield strength and a large strain-to-failure value. The eventual fiber/matrix debonding can only be avoided by selecting the fiber and matrix such that the difference in CTE values is low.

6.4.4 Creep

The creep resistance in CMCs can be improved by the incorporation of fibers or whiskers. The creep rate (under four-point bending) of SiC_w (20 vol.%)-reinforced alumina is significantly lower compared to that of unreinforced alumina (Lin and Becher 1990). This improvement is attributed to the retardation of grain-boundary sliding by SiC_w present at the grain boundaries. Wiederhorn and Hockey (1991) have studied the creep behavior of particle and whisker-reinforced CMCs under tension and compression. It is found that the creep rate in tension is faster than in compression. This difference is attributed to the ease of cavitation and microcracking during tension rather than in compression.

Although continuous ceramic fibers improve the toughness of ceramics at room temperature, most of these fibers are not sufficiently creep resistant. In fact, creep rate of many fibers is much higher than that of the corresponding monolithic ceramics. Hence, for the creep resistance of CMCs, the intrinsic creep resistance of the fiber, matrix, and interface region should be considered. Oxide fibers have fine grain size and generally contain some glassy phase. Non-oxide fibers are also fine-grained and multiphasic, but they are susceptible to oxidation. Non-oxide fiber/non-oxide matrix composites, such as SiC_f/SiC and SiC_f/Si_3N_4, generally show good low-temperature mechanical properties, but their poor oxidation resistance at high temperature is a major limitation. Non-oxide fiber/oxide matrix composites, such as carbon/glass, SiC/glass, and SiC/alumina, generally do not possess high oxidation resistance because of the rapid permeation of oxygen through the oxide matrix (Prewo et al. 1986).

The creep behavior of monolithic Si_3N_4 and unidirectional and crossplied SiC_f/Si_3N_4 composites at 1200°C has been studied by Yang and Chen (1992). The creep resistance of composite is found to be superior to that of the monolithic Si_3N_4, and among the composites, the creep resistance of the unidirectional composite is superior to that of the cross-plied composite.

From the preceding discussion on the high-temperature behavior of non-oxide-based composites (either non-oxide fiber or non-oxide matrix), one can infer that the oxidation resistance in air at high temperatures is the prime factor for better creep resistance. If this is the case, oxide fiber/oxide matrix composites would be the most promising because of their inherent stability in air. There are many oxide fibers and oxide matrices available to produce oxidation-resistant composites. Alumina-based fibers are the most widely used oxide fibers, while glass, glass–ceramics, alumina, and mullite are the most common oxide matrices. There are two categories of oxide/oxide composites: one with controlled interface coating and the other without coating. In the first category of composites, the toughness characteristics of the composites are substantially improved by the tailored interface coating. Even though the strength and modulus of the second category of composites are better than the unreinforced oxide, the toughness is not substantially improved because of the strong interface bonding. Hence, interface tailoring through coating is essential in order to achieve the desired properties in the CMCs.

6.5 Thermal Conductivity

In addition to toughness, the incorporation of SiC whiskers improves the thermal conductivity of ceramic matrices. Increase in thermal conductivity combined with toughening can resist the coalescence of thermal shock-induced cracks into critical flaws, resulting in improved thermal shock resistance compared to monolithic ceramics. The influence of microstructure on the thermal conductivity of Si/SiC ceramic composites has been studied by Amirthan et al. (2011). The thermal conductivity is calculated by multiplying the thermal diffusivity, specific heat, and room-temperature bulk density. The thermal diffusivity values decrease over the entire temperature range. The inverse temperature dependence of diffusivity of these SiC ceramic composites is suggesting a dominant phonon conduction behavior. The thermal conductivity also decreases over the entire temperature range. This value is much lower than the reported room-temperature thermal conductivity of liquid-phase-sintered (110 W mK^{-1}) and hot pressed SiC (120 W mK^{-1}) ceramics (Liu and Lin 1996). However, this value is much better than the thermal conductivity exhibited by the commercially available CVI–SiC/SiC composite. There are three main reasons for the low thermal conductivity behavior of the previously mentioned Si/SiC composites. First reason is the presence of glassy carbon. The SiC fibers/grains are covered by glassy carbon and it is a poor thermal conductor. The second reason is the direction of the fiber with respect to the heat flux. The absence of fibers in the third direction (i.e., parallel to the heat flux) results in anisotropic thermal conductivity. Third, the phonon velocity varies with respect to the nature of materials and there may be a chance of scattering at the interface of the two different materials. Since the Si/SiC composite materials are composed of three different phases, such as SiC, Si, and glassy carbon, the phonon movement is hindered. Apart from these reasons, the pores present in these composites can also act as a barrier for thermal conductivity. The thermal conductivity of the CVI-treated Si/SiC composite is significantly lower than the untreated composites. Commonly, fine SiC crystallites are produced during CVD. When the crystallite sizes are fine, the mean phonon free path drastically comes down due to the scattering at the boundaries. This significantly affects the thermal conductivity of the materials.

6.6 Wear Behavior

The influence of interface coating on the wear behavior of SiC_f/SiC composites has been studied by Udayakumar et al. (2011a). Three types of SiC_f/SiC composites, that is, SiC_f/SiC composites with carbon interface, boron nitride interface, and boron nitride interface subjected to an intermediate heat treatment, are made by the isothermal and isobaric CVI of ceramic-grade Nicalon SiC fiber preforms. The density achieved in all the three types of composites is around 2.6–2.7 g cm^{-3}. The sliding wear/friction and solid particle erosion experiments are carried out on these composites using a pin-on-disk setup and jet erosion test rig, respectively. The composites with carbon interface show higher hardness, lower sliding wear, and erosion losses compared to the composite with boron nitride interface with and without intermediate heat treatment. The wear rate of composite made with the BN interface coating with intermediate heat treatment is higher when compared to the other composites. This is due to the fact that the bonding between the fiber and matrix in this composite is weaker compared to the other composites. Hence, the interface between the fiber and the matrix failed extensively after matrix fragmentation followed by fiber pullout and fiber fracture. In the case of composite made with carbon interface, the wear damage is the least and the damage is localized in the matrix. This is attributed to the lubricating behavior of the C interface and relatively stronger bonding between the fiber and the matrix.

The reciprocating sliding friction and wear characteristics of Si/SiC ceramic composites have been studied by Amirthan and Balasubramanian (2011). Phenolic composite with cotton fabric is made, pyrolyzed, and infiltrated with molten silicon to form Si/SiC ceramic composites. This composite is further infiltrated with SiC by the CVI process. Reciprocating sliding wear testing of these composites is carried out against alumina balls.

The friction coefficient, wear rate, and SEM micrographs of wear tracks of these composites are shown in Figure 6.28. The composite without SiC infiltration shows low friction coefficient values. The free carbon present in this material plays a major role in controlling friction coefficient. It can be seen that the wear track is covered with the carbon debris. Interestingly, the friction coefficient of CVI-treated sample passes through four transitions. Since

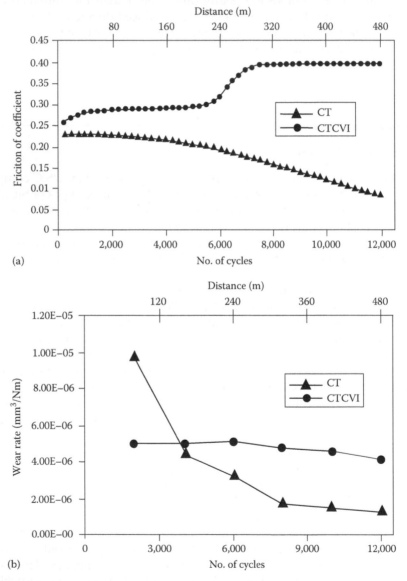

(a)

(b)

FIGURE 6.28
Wear behavior of Si/SiC composites: (a) friction coefficient; (b) wear rate.

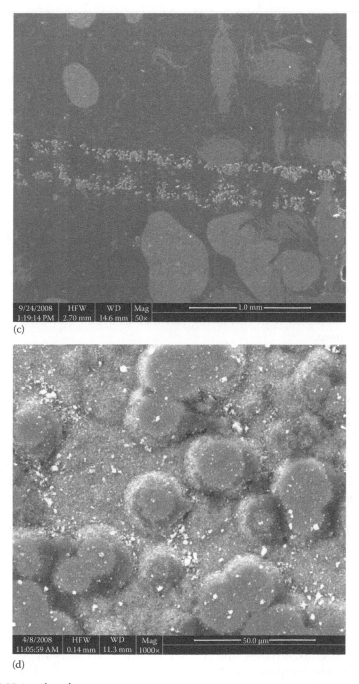

(c)

(d)

FIGURE 6.28 (continued)
Wear behavior of Si/SiC composites: (c) wear surface of the composite without CVI treatment; (d) wear surface of the composite with CVI treatment. (Reprinted from Amirthan, G. and Balasubramanian, M., *Wear*, 271, 1039, 2011. With permission.)

the material surface is coated with the spherical SiC grains of average diameter 20 μm, the surface hardness is significantly improved. Each SiC spherical grain is acting as a solid ball and directly contacting the rubbing alumina balls. The initial increase in the friction coefficient might be due to the resistance offered by this solid SiC grains. The second stable region is due to the blunting or surface smoothening of spherical particles. However, this blunting causes the increased sliding contact surface area. Hence, beyond 6000 cycles, this hard surface may be giving more resistance to the sliding balls and leads to the pullout of alumina grains. These liberated grains may be taken back between the two sliding surfaces and cause the increased friction coefficient. The SEM micrograph also shows the debris on the wear track and it is confirmed as alumina particles by energy dispersive spectroscopy (EDS) analysis. In the fourth stage, the rate of increase of friction coefficient is considerably reduced. This post-transition period is associated with the buildup of a surface layer of fragmented and compacted wear debris that provides lubrication effect under the running-in process. The wear rate of composite without CVI treatment is gradually decreasing. In contrast, in the early stage of the running-in process (i.e., before 2000 cycles), the CVI-treated sample shows very less wear rate. However, in the later stage, its wear rate gradually increases with increasing number of cycles and then there is a slight decrease.

6.7 Applications

CMCs can be used in many applications, where high-temperature mechanical properties are very important. Like MMCs, the applications of CMCs can also be classified into aerospace and non-aerospace applications. The following characteristics are attractive for the use of CMCs in the aerospace sector:

- High specific stiffness and strength leading to a weight reduction
- Higher operating temperatures leading to a greater thermal efficiency
- Longer service life and lower maintenance cost
- Signature reduction

CMCs can lead to improvements in the performance of aerospace vehicles including aircraft, missiles, and reentry vehicles. In future hypersonic aircrafts, the skin temperature may reach as high as 1600°C. The radomes, nose tips, leading edges, and control surfaces will experience slightly higher temperatures. Currently, sacrificial, non-load-bearing thermal protection materials supported by load-bearing components are used.

The CMC thermal protection materials can be load bearing at those high temperatures, and they are reusable.

The non-aerospace applications of CMCs are engine components used at high temperatures and in corrosive environments, cutting tool inserts, high-temperature filters, wear-resistant parts, nozzles, heat exchanger tubes, etc. In such applications, the components can range from simple to complex shapes and tend to be smaller in size. Thus, there are dense, particle- and whisker-reinforced CMCs commercially available for these applications. Again most of these applications warrant high-temperature mechanical properties.

An important application of CMCs is cutting tool inserts. The successful CMCs for cutting tool applications are TiC particle-reinforced Si_3N_4 and Al_2O_3 and SiC_w/Al_2O_3. SiC_w/Al_2O_3 composite is the most common CMC used as cutting tool inserts for high-speed cutting of superalloys. For example, the cutting of Inconel 718 alloy by this composite is three times faster than ceramic tools and eight times faster than cemented carbides. The important characteristics of CMCs, which make them good candidates for cutting tool inserts, are as follows:

- Better wear resistance
- Good thermal shock resistance
- High strength and fracture toughness
- Appreciable thermal conductivity

SiC_w/Al_2O_3 composites are generally made by hot pressing and the volume fraction of SiC_w is usually 0.3–0.45.

As wear-resistant parts, CMCs offer superior abrasion resistance, low friction, good elevated-temperature properties, and good mechanical properties at high speeds. Some of CMC wear-resistant parts are vanes, seals, nozzles, bearings, and wear guides. Toughened zirconia and whisker- and continuous fiber-reinforced ceramics are mainly used in these applications.

In the energy-related applications involving heat transfer, storage, and recovery, monolithic ceramic materials can be used. However, a major drawback of these materials is their poor thermal shock resistance. Hence, fiber-reinforced CMCs would be a better option. For example, continuous fiber-reinforced CMC hot-gas filters for particle separation can be very efficient in hot, coal-derived gas streams. The use of CMCs in these applications makes sense because it will result in higher operating temperatures, higher efficiency, reduced emission of pollutants, longer service life, etc.

Candle-type filters made with Nextel™312 alumina-based fibers and silicon carbide are used to remove particulate matter from gas streams up to a temperature of 1000°C. The high-temperature capability of this CMC filter can eliminate the necessity to cool the gas stream prior to filtration and hence increases the process efficiency and eliminates the cost and complexity of gas dilution, air scrubbers, or heat exchangers. The ceramic fibers increase the toughness of the composite, which results in a filter with excellent resistance to thermal shock and catastrophic failure. The ceramic composite filter developed by 3M is used in advanced coal-fired power generation systems, such as pressurized fluidized bed combustion and integrated gasification combined cycle reactors, and incineration of waste materials.

Nicalon SiC fiber-reinforced glass composites are used as pad inserts and take-out paddles in direct hot glass contact equipment (Beier and Markman 1997). The potential applications of CMCs include heat engine parts, components used in aggressive environments, special electronic/electrical components, energy conversion devices, and military systems.

Radiant burners made from CMC transfer a substantial fraction of the energy input directly to the process load. Increased radiant output and fuel efficiency are the main benefits of using CMC radiant burners.

6.8 Concluding Remarks on CMCs

Although CMCs are being developed for high-temperature applications, the current generation of ceramic composites is susceptible to deformation and creep at temperatures as low as 900°C. It is mainly due to the presence of low-melting phases at the grain boundaries, which might have been introduced either intentionally or unintentionally during processing. These grain-boundary phases soften at a much lower temperature than the softening temperature of matrix material, resulting in the sliding of grains relative to each other and thus initiating creep cracks and deformation of the bulk material. Similarly the current generation of oxide fibers also has restricted use temperature of the order of 1000°C, because of grain-boundary creep. Non-oxide fibers such as SiC fibers also exhibit significant creep in the same temperature range, and the creep rate depends on the oxygen content in the fiber. Hence, the poor thermal stability of fibers is considered as one of the main barriers to the high-temperature applications of fiber-reinforced CMCs. Current research emphasis is mainly focusing on the potential to develop and use single crystal oxide fibers as a possible reinforcement material.

In addition to high-temperature deformation, susceptibility to environmental attack is a problem for many non-oxide matrix and reinforcement materials used in CMCs. Both SiC and Si_3N_4 form a protective surface oxide layer at temperatures approaching 1200°C. However, decomposition reactions taking place at higher temperatures render the oxide layer non-protective. These high-temperature decomposition reactions occur even when the non-oxide fibers are embedded within the matrix material. Cracks formed in the matrix during processing or service can expose the reinforcement to an oxygen-containing environment. Alternatively, the diffusion of oxygen along grain boundaries, or along interphases, can also cause internal volatilization of the reinforcement, resulting in cavitation, bloating, and/or blistering on the reinforcement. Finally, the oxidation of protective boron nitride or carbon coating generally applied on the fibers can lead to the formation of a strong interfacial bond that will embrittle the composite.

Hence, it appears that single crystal oxide fiber-reinforced oxide matrix composites would have the highest potential for structural applications at temperatures higher than 1400°C. However, the grain-boundary creep of the matrix and thermal shock resistance are the main issues yet to be resolved, and therefore, it will take several years to develop and commercialize this type of materials.

Currently, the barriers for large-scale applications of CMCs are the high production costs, non-availability of approved design procedures, and the lack of models for strength and toughness. Most of the CMCs require hot-pressing for better consolidation, but it is very difficult to make complicated shapes by this process. Sintering or sintering followed by hot isostatic pressing is the alternate route for the consolidation of complex shapes.

6.9 Cement-Based Composites

The properties of cement can be improved by incorporating fiber reinforcements. Fibers, such as glass-, asbestos-, steel-, and plant-based fibers, are widely used in cement to improve the mechanical properties.

6.9.1 Asbestos Cement

Once it was very popular fiber-reinforced cement. There has been a rapid decrease since the 1980s in worldwide sales of asbestos cement sheeting products mainly due to the well-publicized health hazards associated with the usage of asbestos fibers. Necessary safety precautions are essential to prevent the inhalation of dust during cutting and drilling of such sheets. Another significant problem is that this composite material is rather brittle, since the fibers are fracturing instead of pullout during the failure of the composite. There were many deaths every year as a result of people falling through roofs when not using the required crawling boards. Even then asbestos-reinforced cement is familiar to engineers as the ubiquitous roofing and cladding material for the past 90 years because of its very low cost and excellent durability. The asbestos fiber content is normally between 9 and 12 wt.% for flat or corrugated sheet, 11 and 14 wt.% for pressure pipes, and 20 and 30 wt.% for fire-resistant boards, and the binder is generally a Portland cement. About 40 wt.% finely ground silica may also be included when the products are made by autoclave process, where the temperature may rise up to 180°C.

6.9.2 Glass Fiber–Reinforced Cement

Glass fiber-reinforced cement (GRC) composite is generally made by combining alkali-resistant glass fibers with a matrix consisting of Portland cement and inorganic fillers. E-glass fibers can be used with a polymer modification to protect the glass against attack by the alkalis in the cement. The most suitable glass fiber is zirconia-based alkali-resistant glass fiber. This type of fiber is normally produced in the form of strands consisting of 204 filaments and the diameter of individual filaments varying from 13 to 20 µm. Several strands may be joined together as a roving, which is cut to a length of 12–38 mm for making GRC. The presence of zirconia in the glass imparts resistance to the alkalis for the cement because the Zr–O bonds are only slightly attacked by the OH⁻ ions, thus improving the stability of the glass network.

Many methods are available for producing GRC, including premixing combined with gravity molding, pressing, injection molding, and extrusion. One of the most common techniques is spray-up, with or without the removal of water. It is almost similar to the spray-up process used for the fiber-reinforced plastics, in which the liquid resin is being replaced by a cement paste. A cement/sand mortar paste is sprayed simultaneously with chopped glass fiber from a dual-spray head gun, which may be handheld or mechanized. The mortar is typically made with cement and sand in the ratio of 1:1 and often contains an acrylic polymer to improve air curing, moisture movement, and durability. In addition to that, pulverized fuel ash and metakaolin clay are often added to improve long-term durability. A chopper unit cuts the glass fiber rovings to predetermined lengths and compressed air is used to inject them into the slurry stream. Roller compaction or vacuum dewatering may be used to remove excess water and consolidate the deposit. Filament winding and hand lay-up processes can also be used, and block work walls can be built by stacking the blocks and applying a rendering premixed glass fiber cement.

6.9.2.1 Properties

The mechanical properties of GRC composite depend on the production process and the fiber content and length. It has been reported that the uniaxial tensile strength and strain-to-failure of GRC composite samples aged for 28 days increase with increasing fiber content and length. Although the mechanical properties are good after 28 days aging, the strength and toughness of GRC may change after long periods of time.

The durability of GRC was found to be strongly influenced by the environment to which it is exposed. There was little change in flexural or tensile strength after 10 years in dry air, whereas there was a decrease in strength of the composite by more than 50% under water or in natural weathering after the same period.

The decrease in impact strength was even more severe, which was reduced by an order of magnitude. Various reasons have been proposed, which include loss in fiber strength and failure strain due to alkali attack and increased bonding between individual glass fibers by the filling of voids in the fiber bundles with lime crystals and calcium silicate hydrates. The calcium hydroxide growth can be eliminated by including metakaolin (calcined China clay), which reacts with the free lime to cause it to disappear within 28 days. Silica fume with a particle size of less than 0.1 μm may not be a good choice, since the lime particles may penetrate into the fiber bundle and result in the formation of hard calcium silicate hydrates that are as damaging to the fibers as lime deposition.

6.9.2.2 Applications

The full potential of GRC in construction is yet to be realized due to concerns regarding durability under damp conditions. However, these concerns are gradually being resolved as more field experience disproves the notion. Cladding panels made of GRC have been used since the 1970s. There is increasing emphasis on the use of single skin cladding panels attached to prefabricated steel frames by flexible anchors, which is known as a GRC Stad Frame system. Another important use is in permanent formwork for bridge decks and the advantages are that no temporary support is required and a dense high-quality cover is provided to the reinforcement.

A German process for producing corrugated sheets uses continuous glass fiber rovings in the main stress direction. The lengthwise reinforcement is laid using unidirectional fiber mat, which is sandwiched between two cement layers to make the sheet. Other applications of GRC include cable ducts, sewer lines, sound barriers, drainage systems, septic tanks, and roofing slabs.

6.9.3 Natural Fiber–Reinforced Cement

The use of natural fibers in cement is common in both developed and developing countries. Cellulose fibers derived from wood are the main natural fibers used in developed countries. Wood is mechanically and chemically pulped to separate the individual fibers of length between 1 and 3 mm and width up to 45 μm. The elastic modulus of individual fibers may vary between 18 and 80 GPa with strength between 350 and 1000 MPa depending on the orientation of cellulose chains. Cellulose fibers produced from wood have several advantages, when used in cement sheets. The cost of the fibers are lower than that of manmade fibers, fibers are derived from renewable resources, these fibers can be used in the existing plants of asbestos cement, and they have adequate tensile strength for cement reinforcement. However, these fibers are sensitive to humidity changes and the elastic modulus of the fibers reduces considerably under wet conditions.

The use of plant fibers in developing countries is generally aimed at producing cheap cement-based roofing sheets often of corrugated or folded plate design. Long fibers that are specific to the locality are used, such as banana, bamboo, coir, elephant grass, flax, jute, palm, pineapple leaf, sisal, and sugarcane. Lengths of fibers may be up to 1 m or more and they are hand placed into a paste of sand and cement. Corrugated sheets of dimension up to 2×1 m^2 and 6–10 mm thickness and tiles may be produced with fibers laid in preferential directions.

6.9.3.1 Manufacture

Wood fiber-reinforced cement sheets are manufactured by modified Hatschek machinery, formerly used in the asbestos cement sheet industry. The fibers are dispersed in a cement slurry with high water content and then the excess water is extracted by vacuum filtration on a large rotating drum.

In Australia, cellulose fibers have completely replaced asbestos fibers in flat sheet products made with calcium silicate in an autoclave. Autoclaved systems are said to have greater dimensional stability in relation to moisture and temperature changes, compared to air-cured hydrated cement binders.

Plant fiber-based composites are made by simple hand lay-up processes that are potentially suitable for low-cost housing applications. Hand lay-up involves the application of a thin mortar layer on a mold, followed by laying up of alternate layers of fiber and mortar. The fibers can be rolled into the mortar or worked onto it manually. In the mixing technique, there is a limit to the fiber content and length, since the workability is reduced with increasing fiber content and length. However, the reduced flow properties may not be deterrent for the production of thin components such as corrugated sheets and shingles, and for these applications, plasticity and fresh strength are sufficient. The short fibers are mixed with cement mortar and spread on a mold surface and then shaped. Corrugated shape can be achieved by pressing between two corrugated sheets.

A wide range of properties are available with wood fiberboards depending on the moisture content and fiber type and content. Typical properties are elastic moduli of 12–15 GPa, tensile strengths of 6–20 MPa, and flexural strengths of 15–30 MPa. These strengths match the roofing requirements of international standards specified for asbestos cement. For handlaid plant fiber composites, the cracking strengths are only 1–3 MPa. Although they may not have enough strength to match the standards, they have better handling characteristics and increased toughness.

The high alkalinity of cement prevents microbiological decay of the fibers, but the calcium hydroxide penetrates the fiber and mineralizes or petrifies it. The natural fibers are not expected to give the composite a long lifetime; however, short wood fibers used as alternatives to asbestos have been shown to be more durable than other plant fibers.

6.9.3.2 Applications

Wood fiber cement composites can replace asbestos cement in roofing, tiling, and cladding applications. Plant fiber-reinforced cement composites are mainly used for local roofing and cladding applications where compliance with international standards is not important and long life is not warranted.

6.9.4 Polypropylene Fiber–Reinforced Concrete

Extensive use has been made of these fibers in concrete to alter the properties of fresh concrete. They are added in small quantities (0.1% by volume) as short (<25 mm length) fibrils or monofilaments. The other way of incorporating polypropylene fiber that has been used for many years is chopped twisted fibrillated twine, typically 0.44% by volume and 40 mm in length. This fiber has only been used in precast products subjected to impact loading such as pile shells.

The concrete mixing technique requires little or no modification from normal practice, since the fiber volume is not very significant in most of the cases. Usually, the fibers are available in water-soluble bags and a bag containing 0.9 kg of fiber is placed in the mixer for each cubic meter of concrete. During mixing the fibers are released and dispersed with the other ingredients.

6.9.4.1 Properties

The main benefit of adding this fiber is a reduction in plastic shrinkage cracking that can frequently occur in badly made concrete slabs. The addition of about 0.44 vol.% polypropylene twine increases the impact strength of concrete by holding cracked sections together to enable them to withstand further impacts.

Polypropylene is extremely resistant to the alkalis in concrete and the cement matrix protects the fibers from ultraviolet radiation. Little change in fiber strength has been observed even after 18 years of exposure in a variety of environments. The accelerated tests carried out on this composite have predicted a lifetime in excess of 30 years.

6.9.4.2 Applications

There is an increasing use as flooring materials in industries for similar reasons as for steel fiber concrete. Nearly 10% of ready-mix concrete used in United States for general construction contains polypropylene fibers.

6.9.5 Steel Fiber–Reinforced Concrete

About 1 vol.% of short steel fiber is incorporated to improve certain mechanical properties of concrete. Even though the fiber content is less than the critical fiber volume (~2 vol.%), useful properties in the composite have been achieved in many practical systems. To improve anchorage efficiency during fiber pullout, bends or crimps are introduced in the fiber. It may not be feasible to incorporate more amount of fibers because of cost considerations and the difficulty in distributing with concrete matrix since it generally contains about 70% by volume of aggregate particles that obviously cannot be penetrated by fibers.

6.9.5.1 Manufacture

In the early development of steel fiber concrete, mixing problems often occurred with balling up of fibers resulting in unsuitable mixes. These problems have now largely been resolved by the suitable selection of fiber type and content and mix proportions. It is generally recommended that the aspect ratio of fiber should be less than 70 and the volume of fibers should be between 0.3% and 0.8%. Generally the fibers are the last ingredient to be added to the concrete mix and care should be taken to ensure that no clumps are added. Gunning or spraying steel fiber concrete is widely used for tunnel linings and rock slope stabilization. Typical fiber lengths range from 25 to 40 mm. As in other composites, the greater the fiber aspect ratio and fiber content, the better the performance of the sprayed concrete, but the more difficult it is to mix and spray.

In a third manufacturing process known as SIFCON process, about 20 vol.% of fiber is placed into a mold and then infiltrated with a fine-grained cement-based slurry. This concrete gives very high strength and toughness and it can be used in localized regions such as beam/column intersections.

6.9.5.2 Properties

Short steel fibers at low volume fractions provide virtually no increase in the compressive or uniaxial tensile strength of concrete. The main benefit of using steel fiber is better crack control. Post-cracking tensile strengths of 0.4–1.0 MPa are possible, and flexural strength may be increased by about 50% of the matrix strength at 1.5 vol.% of fibers.

The main durability problem that is likely to occur in this concrete is corrosion of mild steel fibers, especially when they are exposed across cracked sections. However, the problem is not severe with uncracked concrete or when stainless steel fibers are used.

6.9.5.3 Applications

Steel fiber concretes can be used as a replacement for conventional steel mesh in industrial floor slabs. This is particularly beneficial for large-area pours because they avoid interruptions to the construction process caused by placing mesh.

Mild steel fibers with conventional mixing and compaction techniques have been used in structures such as spillways and highway and airfield pavements. As mentioned earlier, tunnel lining and rock slope stabilization are other major applications of steel fiber concrete.

One of the most successful uses of stainless fibers has been in castable refractories for use at temperatures up to 1600°C. In these products, initial cost is not the prime consideration rather than the performance; hence, stainless fibers are used to increase the product life, typically by 100%.

Questions

State whether the following statements are true or false and give reasons:

6.1 It is very easy to get uniform distribution of whiskers in ceramic composites.

6.2 Modulus value of ceramic materials can be improved significantly by incorporating reinforcements.

6.3 Strong bonding is preferred in ceramic composites.

6.4 Fracture strength of a ceramic composite made with high-modulus ceramic matrix is higher than a ceramic composite made with low-modulus ceramic matrix.

6.5 Carbon fibers can be used as reinforcements for ceramic matrices.

6.6 Liquid infiltration is an easy method for the fabrication of CMCs.

6.7 Non-oxide fiber/oxide matrix composites generally do not possess high oxidation resistance.

Answer the following questions:

6.8 How does the incorporation of fibers modify the stress–strain behavior of ceramics?

6.9 Explain the toughening mechanisms operating in CMCs.

6.10 Explain the stages of fracture behavior in a fiber-reinforced ceramic composite.

6.11 What is the role of fibers in CMCs?

6.12 Explain the phase transformation toughening in alumina–zirconia composites.

6.13 What are the problems with cold pressing and sintering of ceramic composites?

6.14 Compare the directed oxidation process and the CVD process for making ceramic composites.

6.15 What are the advantages and disadvantages of the sol–gel process?

6.16 What are the ways to the control of cracking of gels during drying?

6.17 In a ceramic composite, creep rate in tension is faster than in compression. Why?

References

Amirthan, G. and M. Balasubramanian. 2011. *Wear* 271: 1039.

Amirthan, G., A. Udayakumar, and M. Balasubramanian. 2011. *Ceram. Int.* 37: 423.

Amirthan, G., A. Udayakumar, V.V. Bhanu Prasad, and M. Balasubramanian. 2010. *Wear* 268: 145.

Backhaus-Ricoult, M. 1991. *J. Am. Ceram. Soc.* 74: 1793.

Balasubramanian, M. 1996. Processing and characterization of alumina-zirconia powders and composites. PhD dissertation, IIT Madras, Chennai, India.

Becher, P.F., C.H. Hsueh, P. Angelini, and T.N. Tiegs. 1988. *J. Am. Ceram. Soc.* 71: 1050.

Beier, W. and S. Markman. 1997. *Adv. Mater. Process.* 152(12): 37.

Bhaduri, S.B. and F.H. Froes. 1991. *J. Metals* 43(5): 16.

Bhatt, R.T. and R.E. Phillips. 1990. *J. Mater. Sci.* 25: 3401.

Boccaccini, A.R., D.H. Pearce, J. Janczak, W. Beier, and C.B. Ponton. 1997. *Mater. Sci. Technol.* 13: 852.

Boccaccini, A.R., C.B. Ponton, and K.K. Chawla. 1998. *Mater. Sci. Eng.* A241:142.

Brodkin, D., A. Zavaliangos, S. Kalidindi, and M. Barsoum. 1999. *J. Am. Ceram. Soc.* 82: 665.

Brown, D.R. and F.W. Salt. 1965. *J. Appl. Chem.* 15: 40.

Cao, J.J., W.J.M. Chan, L.C. De Jonghe, C.J. Gilbert, and R.O. Richie. 1996. *J. Am. Ceram. Soc.* 79: 461.

Chawla, N. 1997. *Met. Mater. Trans.* A28: 2423.

Chawla, K.K. 2003. *Ceramic Matrix Composites*. Boston, MA: Kluwer Academic Publishers, p. 301.

Chawla, N., Y.K. Tur, J.W. Holmes, J.R. Barber, and A. Szweda. 1998. *J. Am. Ceram. Soc.* 81: 1221.

Chen, P.L. and I.W. Chen. 1992. *J. Am. Ceram. Soc.* 75: 2610.

Chen, M., P.F. James, F.R. Jones, and J.E. Bailey. 1989. In *Proc. ECCM3*, Bunsell, A.R., Lamaicq, P., and Massiah, A. (Eds.), pp. 87–92. London, U.K.: Elsevier Applied Sciences.

Chen, M., P.F. James, F.R. Jones, and J.E. Bailey. 1990. In *Inst. Phys. Conf.*, Ser. No. 111, Holland, D. (Ed.), pp. 227–237. Bristol, U.K.: IOP Publishing Ltd.

Cholet, V., L. Vandenbulcke, J.P. Rouan, P. Baillif, and R. Erre. 1994. *J. Mater. Sci.* 29: 1417.

Chrysanthou, A. et al. 1994. *J. Alloys Compd.* 203: 127.

Claussen, N. and J. Jahn. 1980. *J. Am. Ceram. Soc.* 63: 228.

Crivelli, V.I. and G. Cooper. 1969. *Nature* 221: 754.

Dawson, D.M., R.F. Preston, and A. Purser. 1987. *Ceram. Eng. Sci. Proc.* 8: 815.

Devlin, D.J., R.P. Currier, R.S. Barbero, B.F. Espinoza, and N. Elliott. 1992. *MRS Symp. Proc.* 250: 245.

Evans, A.G. 1985. *Mater. Sci. Eng.* 71: 3.

Faber, K.T. and A.G. Evans. 1983a. *Acta Metall.* 31: 565.

Faber, K.T. and A.G. Evans. 1983b. *Acta Metall.* 31: 577.

Golecki, I., R.C. Morris, D. Narasimhan, and D. Clements. 1995. *Appl. Phys. Lett.* 66: 2334.

Greil, P. 1995. *J. Am. Ceram. Soc.* 78: 835.

Greil, P., T. Lifka, and A. Kaindl. 1998. *J. Eur. Ceram. Soc.* 18: 1961.

Gupta, R.K. et al. (Eds.). 2010. *Polymer Nanocomposites Handbook*. Boca Raton, FL: CRC Press, p. 47.

Han, L.X. and S. Suresh. 1989. *J. Am. Ceram. Soc.* 72: 1233.

Harris, B. 1980. *Metal Sci.* 14: 351.
Holmquist, M., A. Kristoffersson, R. Lundberg, and J. Adlerborn. 1997. *Key Eng. Mater.* 127–131: 239.
Hoyt, J.T. and J.M. Yang. 1995. *SAMPE J.* 27(2): 11.
Illston, T.J., C.B. Ponton, P.M. Marquis, and E.G. Butler. 1993. *Third Euroceramics*, Vol. 1, Duran, P. and Fernandez, J.F. (Eds.). Madrid, Spain: Faenza Editirice Iberica, pp. 419–424.
Kaya, C., P.A. Trusty, and C.B. Ponton. 1998. *Br. Ceram. Trans.* 97(2): 48.
Kaya, C. et al. 2000. *J. Am. Ceram. Soc.* 83: 1885.
Kovar, D., B.H. King, R.W. Trice, and J.W. Halloran. 1997. *J. Am. Ceram. Soc.* 80: 2471.
Langlais, F., F. Loumagne, D. Lespiaux, S. Schamm, and R. Naslain. 1995. *J. Phys. II.* 5/C5: 105.
Lee, S.M. (Ed.). 1989. *Reference Book for Composites Technology*, Vol. 1. Boca Raton, FL: Taylor & Francis Group, p. 131.
Lin, H.T. and P.F. Becher. 1990. *J. Am. Ceram. Soc.* 73: 1378.
Liu, H.-K. and C.-C. Huang. 2001. *J. Eur. Ceram. Soc.* 21: 251.
Liu, D.M. and B.W. Lin. 1996. *Ceram. Int.* 22: 407.
Lu, T.C., J. Yang, Z. Suo, A.G. Evans, R. Hecht, and R. Mehrabian. 1991. *Acta Metall. Mater.* 39: 1883.
Lundberg, R., R. Pompe, and R. Carlsson. 1986. *Ceram. Eng. Sci. Proc.* 9(7–8): 901.
Manor, E., H. Ni, and C.G. Levi. 1993. *J. Am. Ceram. Soc.* 76: 1777.
Mitomo, M. and S. Uenosono. 1992. *J. Am. Ceram. Soc.* 75: 103.
Munoz, A., J.M. Fernandeza, and M. Singh. 2002. *J. Eur. Ceram. Soc.* 22: 2727.
Naslain, R. 1999. *Adv. Compos. Mater.* 8: 3.
Naslain, R. and F. Langlais. 1986. *Proc. Twenty-First Univ. Conf. Ceram. Sci.* New York: Plenum, p. 145.
Padture, N.P., S.J. Bennison, J.L. Runyan, J. Rodel, H.M. Cahn, and B.R. Lawn. 1991. In *Advanced Composite Materials*, Sacks, M.D. (Eds.). Westerville, OH: American Ceramic Society, pp. 715–720.
Papakonstantinou, C.G., P. Balaguru, and R.E. Lyon. 2001. *Composites* B32: 637.
Petrovic, J.J. et al. 1997. *J. Am. Ceram. Soc.* 80: 3070.
Phillips, D.C. 1974. *J. Mater. Sci.* 9: 1847.
Phillips, D.C. 1983. In *Fabrication of Composites (Handbook of Composites Series, Vol. 4)*, ed. A. Kelly and S.T. Mileiko. Amsterdam, the Netherlands: North-Holland, p. 472.
Phillips, D.C., R.A.J. Sambell, and D.H. Bowen. 1972. *J. Mater. Sci.* 7: 1454.
Porter, J.R., F.F. Lange, and A.H. Chokshi. 1987. *Am. Ceram. Soc. Bull.* 66: 343.
Presas, M., J.Y. Pastor, J. LLorca, A.R. de Arellano-Lopez, J. Martinez-Fernandez, and R.E. Sepulveda. 2005. *Scr. Mater.* 53: 1175.
Prewo, K.M. 1987. *J. Mater. Sci.* 22: 2695.
Prewo, K.M., J.J. Brennan, and G.K. Layden. 1986. *Am. Ceram. Soc. Bull.* 65: 305.
Rabin, B.H., G.E. Korth, and R.L. Williamson. 1990. *J. Am. Ceram. Soc.* 73: 2156.
Reagan, P. and F.N. Huffman. 1989. In *Fiber Tex 1988*, Buckley, J. (Ed.). Washington, DC: NASA Scientific and Technical Information Division, pp. 247–257.
Rice, R. 1990. *Processing of Ceramic Composites*, Vol. 1. *Advanced Ceramics Processing and Technology*, Binner, J.G.P. (Ed.). Park Ridge, NJ: Noyes, pp. 123–210.
Rupo, E.D. and M.R. Anseau. 1980. *J. Mater. Sci.* 15: 114.
Russell-Floyd, R.S. et al. 1993. *Ceramic International Technology 1994*, Birkby, I. (Ed.). London, U.K.: Sterling Publications Ltd., p. 62.

Schwartz, M.M. 1997a. *Composite Materials*, Vol. II. Upper Saddle River, NJ: Prentice Hall, p. 216.
Schwartz, M.M. 1997b. *Composite Materials*, Vol. II. Upper Saddle River, NJ: Prentice Hall, p. 230.
Shih, C.J., J.M. Yang, and A. Ezis. 1992. *Compos. Sci. Technol.* 43: 13.
Skamser, D.J., P.S. Day, H.M. Jennings, D.L. Johnson, and M.S. Spotz. 1994. *Ceram. Eng. Sci. Proc.* 15(5): 916.
Smirnov, B.I., Y.A. Burenko, B.K. Kardashev, F.M.V. Feria, J. Martinez-Fernandez, and A.R.A. Lopez. 2003. *Phys. Solid State* 45: 482.
Spotz, M.S., D.J. Skamser, P.S. Day, H.M. Jennings, and D.L. Johnson. 1993. *Ceram. Eng. Sci. Proc.* 14(9–10 pt.2): 753.
Stinton, D.P. et al. 1986. *Am. Ceram. Soc. Bull.* 65: 347.
Subrahmanyam, J. and R.M. Rao. 1995. *J. Am. Ceram. Soc.* 78: 487.
Suresh, S. 1991. *J. Hard. Mater.* 2: 29.
Suresh, S., L.X. Han, and J.J. Petrovic. 1988. *J. Am. Ceram. Soc.* 71: c158.
Udayakumar, A. et al. 2011a. *Wear* 271: 859.
Udayakumar, A. et al. 2011b. *J. Eur. Ceram. Soc.* 31: 1145.
Varela-Feria, F.M. 2002. *J. Eur. Ceram. Soc.* 22: 2719.
Waku, Y., N. Nakagawa, T. Wakamoto, H. Ohtsubo, K. Shimizu, and Y. Kohtoku. 1998. *J. Mater. Sci.* 33: 1217.
Washburn, E.W. 1921. *Phys. Rev.* 17: 273.
Wei, G.C. and P.F. Becher. 1985. *Am. Ceram. Soc. Bull.* 64: 333.
Wetherhold, R.C. and L.P. Zawada. 1991. In *Fractography of Glasses and Ceramics*, Frechete, V.D. and Varner, J.R. (Eds.). *Ceramic Transactions*, Vol. 17. Westerville, OH: American Ceramic Society, p. 391.
Wiederhorn, S.M. and B.J. Hockey. 1991. *Ceram. Int.* 17: 243.
Yang, J.M. and S.T. Chen. 1992. *Adv. Compos. Lett.* 1: 27.
Yasuoka, M., K. Hirao, M.E. Britto, and S. Kanzaki. 1997. *J. Ceram. Soc. Jpn.* 105: 641.
Zawada, L.P. and R.C. Wetherhold. 1991. *J. Mater. Sci.* 26: 648.
Zeng, J., Y. Miyamoto, and O. Yamada. 1991. *J. Am. Ceram. Soc.* 74: 2197.

Bibliography

ASM Handbook. 2001. *Composites*, Vol. 21. Materials Park, OH: ASM International.
Chawla, K.K. 1998. *Composite Materials: Science and Engineering*, 2nd edn. New York: Springer-Verlag Inc.
Chawla, K.K. 2003. *Ceramic Matrix Composites*. Boston, MA: Kluwer Academic Publishers.
Kingery, W.D., H.K. Bowen, and D.R. Uhlmann. 1976. *Introduction to Ceramics*. New York: John Wiley & Sons.
Krenkel, W. 2008. *Ceramic Matrix Composites*. Weinheim, Germany: Wiley-VCH.
Mallick, P.K. 2008. *Fiber-Reinforced Composites*, 3rd edn. Boca Raton, FL: CRC Press.
Schwartz, M.M. 1997. *Composite Materials*, Vol. II. Upper Saddle River, NJ: Prentice Hall.
Warren, R. (Volume ed.). 2000. *Carbon/Carbon, Cement, and Ceramic Matrix Composites*. Vol. 4. *Comprehensive Composite Materials*, Kelly, A. and Zweben, C. (Eds.). Oxford, U.K.: Elsevier Science Ltd.

7

Carbon–Carbon Composites

Carbon is a truly remarkable element existing in three allotropic forms, namely, diamond, graphite, and fullerenes, and each allotrope is having significant scientific and technological importance. Its most abundant allotrope, graphite, can be produced in different forms: from amorphous to highly crystalline, from highly dense ($\rho = 2.2$) to highly porous ($\rho = 0.5$), and different shapes. Examples include coke, glassy carbon, carbon black, porous carbon, activated charcoal, graphite powder, graphite electrodes, and carbon fibers. Carbon is a preferred material for structural applications, where extreme environmental conditions such as high temperature and corrosive liquids or gases are encountered. This is mainly due to its high strength, chemical inertness, and ability to retain the strength at high temperatures, even above 1500°C. Theoretically, carbon materials with covalently bonded atoms should possess very high strengths in the range of 40–50 GPa. However, the bulk synthetic graphite has less than 2% of the theoretical strength. Therefore, for long there has been a quest by scientists to explore the ways to achieve the maximum possible strength. This, coupled with the search for high-performance reinforcing fibers to produce advanced composites, had led to the development of carbon fibers. The judicious incorporation of these carbon fibers into carbon matrix has led to the birth of carbon–carbon (C/C) composite materials with improved properties.

Carbon-based materials are one of the most promising candidates for high-temperature applications because of their intrinsic high thermal stability (>3000°C), low density (<2.2 g cm^{-3}), and increasing strength with temperature up to 2500°C. In its bulk form, its utility is limited for many applications because of low strain-to-failure, high flaw sensitivity, anisotropy, variations in properties and difficulties in fabricating large size, as well as complex-shaped components. The requirement for high-temperature mechanical properties has led to the development of C/C composites by incorporating carbon fibers. These composites generally consist of graphite fibers and graphite matrix. The attractive properties of monolithic graphite are combined with the high strength and versatility of composites in this new class of materials.

Lack of suitable high-temperature material led to the development of C/C composite materials in the late 1960s. The initial usage was limited to military applications, such as rocket nozzles and reentry parts for missiles, due to the high cost of these materials.

C/C composites refer to a family of complex advanced materials that consist of carbon (mainly graphite form) fibers embedded in a carbon (mainly graphite form) matrix. The presence of carbon fibers in these materials makes them stronger, tougher, and more resistant to thermal shock than conventional graphite. In addition to this, the other important and useful properties of C/C composites are high strength at high temperature (up to 3000°C) under inert atmosphere, low coefficient of thermal expansion (CTE), high thermal conductivity (higher than that of copper and silver), good fatigue and creep resistance, and low recession in high-temperature ablation environments. The combination of low thermal expansion and high thermal conductivity makes these composite materials resistant to thermal shock. C/C composites are made with the densities in the range 1.6–2.0 g cm^{-3}, which are much lower than those of commonly used metals and ceramics, and hence the C/C composite components weight will be very low, which is an important consideration for aerospace applications. Actually, the mechanical strength of C/C composites increases with temperature, in contrast to the strength of most engineering materials, in which the strength decreases with increasing temperature (Figure 7.1). These extraordinary properties of C/C composites have made them extremely useful materials for aerospace and defense applications such as brake disks, rocket nozzles, leading edges of reentry vehicles, and thermal management components in space vehicles. They are also used in biomedical parts and glass and

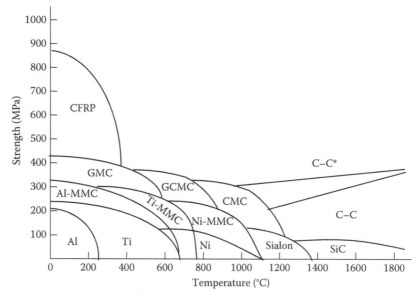

FIGURE 7.1
Variation of strength of some engineering materials with temperature. (Reprinted from Manocha, L.M., *Sadhana*, 28, 349, 2003. With permission.) CFRP, carbon fiber reinforced polymers; GMC, glass matrix composites; GCMC, glass–ceramic matrix composites; CMC, ceramic matrix composites; MMC, metal matrix composites; C–C, carbon/carbon composites; C–C*, advanced carbon/carbon composites.

ceramic industries. A well-known example for the practical application of C/C composites is American space shuttle parts, such as the fuselage nose and the leading edge of the wings, and that space shuttle has successfully completed more than 100 missions under the extreme reentry conditions.

Like most composite systems, well-engineered C/C composites can be made to exhibit graceful failure and show pseudoplastic behavior. They also possess high thermal stability under inert environment. However, in common with all forms of carbon, C/C composites are also prone to oxidation at moderately elevated temperatures (>500°C in air and >700°C in steam). Therefore, there is a necessity to protect them with surface coatings when they are used at elevated temperatures in an oxidizing environment, and this constitutes one of the major disadvantages of C/C composites. The other major drawback inhibiting their wide usage is the high manufacturing costs. C/C composite production is expensive not only due to the high costs of carbon fibers but also the need to carry out repeated processing cycles, irrespective of the manufacturing route.

C/C composites are being made in wide range of configurations, from simple unidirectional or random fiber-reinforced constructions to woven, multidirectional structures in block, hollow cylinder, and other configurations. The variety of carbon fibers and multidirectional weaving techniques now available allow tailoring the properties of C/C composites to meet complex design requirements.

Although both the constituents in a C/C composite are the same, each constituent can have a range of structure from amorphous carbon to graphite structure, that is, structures with varying degrees of crystalline perfection. Solid carbons are generally derived from organic precursors by a pyrolysis process known as carbonization, and they exist as non-graphitic or a mixture of graphitic and non-graphitic forms. The degree of crystalline perfection is obtained from x-ray diffraction (XRD) analysis. It can be characterized in terms of L_a, the mean width of the layer planes, and L_c, the mean height of the ordered layer or stack in the C-direction. Well-graphitized carbon can have L_a and L_c values of the order of several hundred nanometers, whereas poorly graphitized carbon may have values <10 nm. The anisotropy of graphite, the degree of graphitization, the preferred orientation of crystallites, and other microstructural variables can result in a broad range of properties in carbon material. In C/C composites, this range of properties can extend to both the constituents. Coupled with these facts, and the availability of a variety of processing techniques and a wide range of fiber architectures can lead to enormous flexibility in the design of C/C composites and their properties.

The real birth of C/C composites came after the development of carbon fibers with acceptable tensile strength from rayon precursors in 1958–1960. The development of C/C composites since the 1960s has resulted in the production of C-fibers from new precursors such as PAN and pitch, and the development of different methods for the deposition of matrix. However, the C/C composites are still expensive, and their usage is limited to specialized applications where their cost can be justified because of their unique combination of properties needed for those applications. Hence, the majority of applications are found in the aerospace industry, and their application for missile nose tips, reentry vehicle heat shields, and rocket motor nozzles is well documented. Another application that relies on their thermal conductivity and tribological properties is aircraft brakes. Apart from aircrafts, vehicles such as racing cars and high-speed trains also benefit from C/C composite brakes. Other applications include torus walls for fusion reactors, bio-implants, and high-temperature tooling. The specific applications necessitate different requirements, which may be strength, stiffness, high thermal conductivity, low thermal expansion, high wear resistance, or high toughness. These in turn influence the selection of fiber, the geometry of the reinforcement, the choice of matrix material, and the processing method.

7.1 Carbon Fiber Reinforcements

High-performance C/C composites are tailored to the mechanical properties requirement of the final structure. The choice of carbon fiber, its surface morphology and treatment, and the weave patterns are also important and lead to a variety of processing options. The types of carbon fibers commercially available are high-tensile, high-modulus, and intermediate-modulus fibers. These fibers are prepared from pitch or polyacrylonitrile (PAN). The rayon-based carbon fibers are used only in a few specific applications, such as military or space structures.

The important characteristics of some of the commercially available C-fibers are given in Table 2.4. The broad range of mechanical properties of different C-fibers allows the fabrication of composites with optimized strength or stiffness. The maximum mechanical properties with any particular fiber type can be realized in unidirectional C/C composites. The mechanical properties required for any structural component can be adapted by design criteria from finite element calculations. According to the loading conditions in different directions, the C-fibers can be suitably aligned. However, for cost-effective manufacturing, carbon fabrics are often used, since they are now commercially available with different weaving patterns. Although the transverse mechanical properties are low for the composites fabricated using 2-D fabrics and thereby limiting the load-bearing capacity in the transverse direction, they are more commonly used than 3-D fiber architectures. 3-D reinforcement architectures are mainly used in a few critical applications in space or military structures. Nowadays, 3-D fiber preforms are manufactured by many companies worldwide. However, these 3-D preforms are often restricted to internal use and are not commercially available. Moreover, the preform cost increases with the increasing number of desired fiber directions. Therefore, commercially made C/C composites are mostly based on 2-D fiber reinforcements, such as fabrics.

7.2 Matrix Systems

The carbon matrix in C/C composites has to effectively transfer the mechanical loads to the fiber reinforcements, and the mechanical strength of the matrix appears to be unimportant. However, the matrix is also a vitally important element as the reinforcement in the C/C composites. It acts as a binder to maintain the alignment of the fibers and fiber bundles and at the same time isolates the fibers from one another. The literature shows that the matrix is multifunctional and contributes to the physical and chemical, as well as mechanical, properties to the composites. In particular, the failure behavior of the composite strongly depends on the matrix microstructure. As already mentioned, carbon matrix can exhibit a wide range of structures and textures ranging from near amorphous to fully graphitic structure. These are controlled by the nature of the precursor material, processing conditions, and the ultimate heat treatment temperature. The most important criteria for the selection of polymer precursors to form carbon matrix are

1. High carbon yield
2. Minimal shrinkage during pyrolysis
3. Amenable for all types of polymer composite manufacturing routes, such as resin transfer molding, prepreg manufacturing, and filament winding
4. Low solvent content
5. High degree of pre-polymerization with lowest viscosity
6. Availability from multiple sources
7. Low cost
8. High pot and storage lifetimes

A high carbon yield is required so as to obtain a fully densified C/C composite without the need for re-impregnation/re-carbonization cycles. Fully densified C/C composite means that the remaining open porosity is less than 10%. This goal can be reached by using matrix precursors with carbon yields above 84%. Higher carbon yields are always combined with a reduced shrinkage during pyrolysis and thereby cause minimum damage to the C/C composites. Usage of solvents as well as a precursor with a low degree of pre-polymerization reduces the carbon yield. Easy processing is important, because the manufacturing of green bodies (carbon fiber-reinforced polymers) has to be performed as precisely as possible without any defects. All damages occurring during the production of green bodies will be retained in the final C/C component. Considering all these facts, the three types of matrix precursors used for C/C composites are thermosetting resins, thermoplastics, and gaseous hydrocarbons.

7.2.1 Thermosetting Resin Precursors

Thermosetting resins are potentially very attractive carbon matrix precursors because they allow the use of conventional polymer matrix composite fabrication techniques to make different shapes prior to carbonization. The requirements for a suitable thermoset resin precursor are

1. High carbon yield under simple pyrolysis conditions.
2. Shrinkage on carbonization should not damage the carbon fibers.
3. The carbon matrix formed should contain open rather than closed pores.

Although the number of resins that could be used is almost unlimited, when processing variables, such as viscosity, and cure conditions, shrinkage during carbonization, matrix microstructure, but most important of all carbon yield are considered, the choice comes down to a relatively few. Phenolics, phenolic furfuryl alcohols, polyimides (PIs), polyphenylenes, and polyarylacetylenes have been used as thermosetting precursor resins. From the beginning of C/C composite development since 1960s, phenolics and their derivatives are widely used. These resins are cheap, easy to process, and have a reasonable carbon yield. These resins normally tend to have a carbon yield of 45%–60%, but resins with a char yield of over 80% have been developed, and a char yield value of as high as 95% has been reported for an aromatic diacetylene dipolymer (Economy et al. 1992).

Other types of precursor systems such as isocyanates are now being used due to their high carbon yield and better pyrolysis behavior. As these resins tend to be thermally stable up to 400°C, it gives a possibility of manufacturing a more perfect green body with a lower rate of damage and reduced shrinkage.

The thrust for the improvement in thermal stability and increased carbon yield resulted in the development of special types of PIs with a carbon yield above 84%. Another thermosetting resin, polyarylacetylene was also used by the Aerospace Corporation and many researchers in Europe. Both these thermosets enabled the formation of C/C composites with a single pyrolysis cycle and thus led to time and cost savings. However, the high price of the polymer and poor handling qualities do not favor the wide usage. All thermosetting resins form glassy carbon after pyrolysis, which cannot be fully graphitized during the subsequent high-temperature treatment.

7.2.2 Thermoplastic Precursors

Pitches on the other hand are very attractive precursors for carbon matrices because of their high carbon yield and the formation of graphitizable carbon. All pitches belong to the category of thermoplastic matrix precursors. They are derived from petroleum or coal tar residues and are complex mixtures of many organic compounds. The carbon yield depends on the composition of the precursor pitch and the pyrolysis conditions. The pyrolysis of pitch occurs in the liquid state by the evaporation of species of increasing molecular weights as the temperature is increased. However, at temperatures between 385°C and 400°C, cracking reactions take place by releasing low molecular weight aromatic fragments, as the side chains to the polyaromatic molecules are severed. The most significant feature in the liquid phase pyrolysis of pitch is the formation of spheres of liquid crystal or mesophase in the pyrolyzing liquid. These spheres with initial diameters of about 100 nm are the precursors of the graphitic structure. The mesophase development can be affected by contacting surfaces, and there is a strong tendency for the lamellae to align with respect to the surfaces. This effect is very important in C/C composites where the mesophase aligns preferentially with the fiber surface producing a region of preferred orientation, which is maintained after the conversion to carbon. However, when the pitch is carbonized under pressure, the orientation effects are altered even to the extent that the alignment of matrix is normal to the fiber surface. However, the effect of pressure on the alignment is not fully understood.

Among the pitches, isotropic pitches are low-cost precursors. High carbon yields with the pitches are feasible only by high-pressure carbonization, but that increases the carbonization costs and often limits the size and geometry of the product. The carbon yields from pressureless carbonization of isotropic pitches are comparable to that from low-cost phenolics.

The development of mesophase pitches and their industrialization has offered a new class of matrix precursor systems. Their polyaromatic structure with a high carbon/hydrogen ratio is responsible for their superior carbon yield. However, the higher the carbon yield, the higher will be the viscosity and consequently the processing temperature. Therefore, in their early development, mesophase pitches were used as matrices only when hot pressing techniques are used, which limits low-cost industrial applications. Another disadvantage is the thermoplastic behavior, and hence cross-linking of these pitches is only feasible by oxidation or treatment with sulfur. Therefore, manufacturing of C/C composites with pure mesophase pitches is not possible. Pre-oxidized mesophase pitches are commercially available now, which are suitable for manufacturing C/C composites in a one-step pyrolysis. Such a one-step cycle process has been developed for industrial applications. In one of the processes, binders such as sulfur or binder pitches are added during impregnation to form a bulk mesophase pitch. However, a standard method is blending other carbon precursors with mesophase pitch. The basic process involves the coating of carbon fiber bundle with a blend of pitch and coke. A thermoplastic coating surrounds the carbon fiber bundle encased by the powder blend. This coated carbon fiber bundle can be used to make different types of textile preforms and then processed to form C/C composites.

7.2.3 Gaseous Precursors

Gaseous precursors are used in the chemical vapor infiltration (CVI) process. Among the commonly used gaseous precursors (CH_4, C_2H_6, C_2H_4, C_2H_2, etc.) for carbon deposition, methane (CH_4) is the most widely used. Temperature in excess of 550°C is required for the thermodynamically favorable carbon deposition.

7.3 Processing of C/C Composites

C/C composites can be prepared by a variety of processing techniques such as molding of random or oriented-fiber composites with a carbon–char yielding resin binder. Other types of structures having multidirectional (2-D, 3-D, or M-D) fibers are produced by using fiber preforms made by dry weaving, piercing of fabrics, or by modified filament winding. Fully automated computer-controlled equipment for fabricating three-directional cylindrical, conical, and contoured fiber preforms has been developed. The carbon fiber preforms are densified by a process that fills the open volume of the preform with a dense, well-bonded carbon matrix. The actual densification process to be used is dictated by the characteristics of the preform structure and the required properties of the final composite. Methods of introducing the carbon matrix into the preform include impregnation with carbon–char yielding organic liquid followed by pyrolysis and CVI of carbon from a hydrocarbon precursor gas.

In general, C/C composites are processed by any one of the following three routes (sometimes a combination of two routes is used):

1. CVI
2. Impregnation and pyrolysis using thermosetting resin precursors
3. Impregnation and pyrolysis using thermoplastic pitch precursors

Generally the CVI process is carried out in the temperature range of 800°C–1500°C, and the pyrolysis of the resin and pitch precursors is also carried out in the same temperature range. Subsequent (final) heat treatment may involve temperatures up to 3000°C. These three methods lead to different microstructures in the composite, partly because of the difference in the method of formation of carbon from the three types of precursors but mainly because of the formation of carbon forms having different structure and properties from different precursors. A schematic of the pore-filling mechanisms of these three types of precursors is shown in Figure 7.2. The CVI process lays down the carbon directly onto the fibers, whereas the carbon is produced in the void between the fibers after a heat treatment from the polymer precursors.

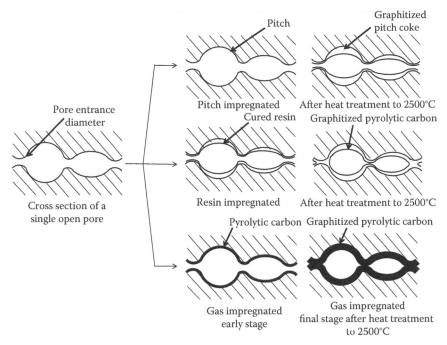

FIGURE 7.2
Schematic mechanisms of pore filling and pore blocking by liquid impregnation and CVD. (Reprinted from Fitzer, E., *Carbon*, 25, 163, 1987. With permission.)

The pitch shrinks during pyrolysis onto the surface of the pore inner wall, whereas thermosetting resins shrink-off from the inner wall surface. The resultant volumetric shrinkage during heat treatment (pyrolysis) leads to significant porosity, and hence pores and shrinkage cracks are common in the matrix. To a certain extent, the choice of processing method depends on the geometry of the final product. During CVI, the inner walls of the pores are directly coated with carbon. Since the CVI processing tends to deposit carbon primarily on the near-surface region, it is more suitable for thin sections, whereas thick sections are normally produced by pitch or resin impregnation. Whatever may be the production route, multiple impregnations are necessary to achieve high density and optimum properties.

Thermosetting polymer resins are generally used for impregnating fibers because they cure at low temperatures and form highly cross-linked non-melting amorphous solids. After pyrolysis, these resins form glassy carbon, which graphitize at temperatures above 3000°C. Another class of liquid impregnants includes molten coal tar and petroleum pitches. Pitches have the advantages of low softening point, low melt viscosity, and high coking value and also they tend to form graphitic coke structures.

The CVI process involves gas-phase pyrolysis of a hydrocarbon, such as methane, propane, or natural gas. In this process, the gaseous hydrocarbon infiltrates into the porous fiber preform heated to a temperature of

1000°C–1400°C and deposits carbon on the fiber surface. This CVI process can be carried out under isothermal and thermal gradient or differential pressure conditions.

At present C/C composites are produced commercially by the CVI process, in which isothermally heated stack of components are impregnated simultaneously in a large furnace. In principle, the gaseous reactants should infiltrate into the interfiber space completely and then crack (decompose). However, these gases have a tendency to crack at the outer surface itself, which blocks the passages for the gases to reach the interior of the preform. In order to produce dense composites, the surfaces are ground after certain time of infiltration and then the components are re-infiltrated. CVI is a very slow process and it takes a couple of months to get dense C/C composites. Often a combination of processing routes is used to produce a fully dense C/C composite at a reduced processing time and cost. For example, a C/C composite is initially produced by pitch or resin route and finally densified by CVI process.

The selection of a process and specific processing conditions depend on the fiber type, matrix precursor, and the geometry, size, and the number of parts, together with the mechanical and thermal property requirements. There are no standard C/C composite properties because the number of materials (fibers and matrix precursors) and process conditions are almost limitless. The carbon fibers and their orientation in the composites determine the mechanical properties. The carbon matrix is responsible for the load transfer to the fiber reinforcements and determines the physical and chemical properties of the C/C composite material.

The selection of matrix precursor has a dominant influence on the manufacturing process as well as on the final properties of C/C composites. CVI methods are more suitable to manufacture near-net-shaped parts, because no shrinkage occurs during P_yC deposition in the preform. In the case of polymer precursors, such as pitches or thermosetting polymers, the shrinkage during pyrolysis can be reduced by using a precursor with high carbon yield.

Industrial manufacturing processes are selected based on the cost-effectiveness, and hence polymer precursors are widely used. All technologies known for fiber-reinforced polymers can be applied with polymer precursors. The special knowledge of individual manufacturer centers mainly on the precursors used, the method to combine the different manufacturing techniques, and the parameters for the final heat treatment.

7.3.1 Thermosetting Resin–Based Processing

The resins are usually dissolved in an organic solvent or furfuryl alcohol with a catalyst/curing agent before infiltration. The resin is allowed to cure after infiltration into the preform. However, resin-based composites are often fabricated from a pre-impregnated woven cloth of carbon fiber, often known as prepreg. In these prepregs the resin is in the partially cured state to retain some tackiness. They are cut to size and then laid up in the desired array. The resin in the

impregnated composite is cured and then pyrolyzed by heating it to a temperature in the range of 350°C–800°C. During the pyrolysis, a large number of volatiles such as H_2O, H_2, CH_4, CO, and CO_2 are expelled. Frequently hot pressing
at pressures up to 10 MPa and temperatures in the range of 150°C–350°C for
periods up to 10 h are used to enhance density during the curing process. The
pyrolyzed composite is subsequently graphitized at temperatures in excess of
1000°C. The flow diagram for the preparation of C/C composites using thermosetting polymer precursor is shown in Figure 7.3.

The carbon matrix produced by the pyrolysis of cross-linked resins is considered to be non-graphitizable, consisting of basic nanostructured units.
Associated with this structure are nanoscale pores, which are however closed
after heat treatment to temperatures over 1000°C. This type of texture means
that the material is essentially isotropic and no evidence of graphite structure could be observed from XRD analysis. However, it has been found that
when graphitization is carried out under stress or in the presence of certain

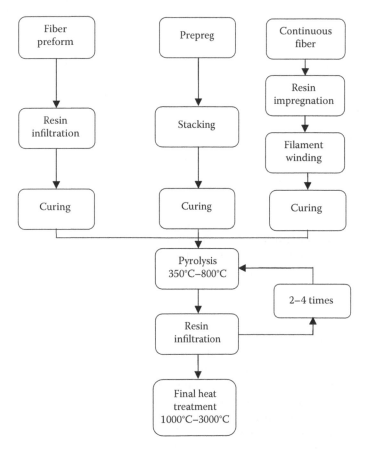

FIGURE 7.3
Flow diagram for the preparation of C/C composites using thermoset resin.

catalysts, anisotropic regions begin to develop in the matrix at temperatures as low as 1000°C. It is believed that there are some interfacial effects occurring between carbon fiber and matrix during this heat treatment. During pyrolysis the resin shrinks by as much as 50%, whereas the dimensions of carbon fibers change very little. It has been postulated that the driving force for graphite formation is the stress accumulation caused by differences in shrinkage between fiber and matrix (Kamiya and Inagaki 1981). It has been reported that when using a polyfurfuryl alcohol precursor and PAN C-fibers, the carbon basal planes tend to align along the fiber axis during heat treatment, and all the matrix carbon between individual fibers is essentially anisotropic after heat treatment at temperatures as low as 1800°C (Hishiyama et al. 1974).

The earlier observations were made in composites, in which carbon fibers were used without any surface treatment. It has been found that if surface-treated C-fibers are used, weak brittle composites is formed because of the formation of strong bonding between the fiber and matrix (Thomas and Walker 1978). However, when the fibers are weakly bonded to the matrix, the shrinkage during pyrolysis and cooling from the heat treatment leads to the separation of matrix from the fibers. Multiple impregnation–pyrolysis cycles are therefore necessary to achieve an acceptable density. Surface machining to open up surface pores is frequently employed to assist further resin impregnation. Due to the relatively low carbon yields of resin precursors also, multiple impregnations (usually four to six times) are needed to achieve an acceptable density. Unfortunately, the density increase during successive impregnation cycles becomes smaller and smaller; hence, a balance needs to be struck between the advantages to be gained from re-impregnation in terms of improved properties and the cost.

7.3.2 Thermoplastic Pitch–Based Processing

The most widely used approach for introducing a carbon matrix into the fibrous preform is through infiltration with a molten pitch. The pitch-impregnated preform is subjected to pyrolysis (carbonization) in an inert atmosphere to convert the organic compound to coke and a high-temperature treatment at 1000°C–2700°C for graphitization of coke. This densification process is carried out at atmospheric or reduced pressure and is usually repeated several times in order to get appreciable density. There is also a variant of this process, known as high-pressure processing, in which high isostatic pressure is used during impregnation and coking stages. It is a more efficient process, and the coke yield for pitch increases from 50 to 85 wt.%, when the pressure is increased to 70 MPa. The flow diagram for the preparation of C/C composites using thermoplastic pitch is shown in Figure 7.4.

The carbonization process of pitch can be characterized by the viscosity changes. Heating to a temperature of 185°C essentially results in the melting of the isotropic pitch and a dramatic reduction in viscosity. After that, there is a negligible change in viscosity until 450°C, when the mesophase begins

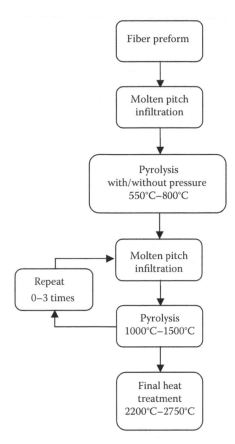

FIGURE 7.4
Flow diagram for the preparation of C/C composites using thermoplastic pitch. (Adapted from Manocha, L.M., *Sadhana*, 28, 349, 2003. With permission.)

to develop. Although the rate of mesophase formation is affected by many factors such as pitch composition and the removal rate of lower molecular weight fraction, the rate increases with increasing temperature and the viscosity also simultaneously increases. Higher temperatures and longer times accelerate the mesophase formation until the pitch becomes a brittle, predominantly crystalline solid.

Mesophase pitches are known to bloat severely upon carbonization, and this has been found to commence with the coalescence of the bulk mesophase. The yield of carbon from pitches can be significantly increased from ~50% to 90% by carbonizing under high pressure. Hence, high-pressure application during carbonization would be beneficial in two ways: it prevents bloating and at the same time increases the yield. As a result, the pitch route for producing C/C composites is normally carried out under high pressure and high temperature in an autoclave, and this process is known as hot isostatic pressure impregnation carbonization (HIPIC). In this process, a carbon fiber preform is vacuum impregnated with molten pitch, placed in a metal container with an excess of pitch. The can is then evaluated, sealed, and placed inside the hot isostatic pressure (HIP) chamber. The temperature is raised above the melting point of the pitch at a programmed rate. The pressure is then increased to about 10 MPa (100 bar). The molten pitch is forced to enter into the pores of the preform by isostatic pressure. The temperature is then gradually increased to carbonize the pitch (650°C–1100°C). After carbonization, the composite is removed from the can and cleaned up by machining the surface.

Irrespective of the densification approach, two or more densification cycles are required to achieve the desired density; however, the number of cycles can be reduced by using HIP process. A new approach of intermediate graphitization has evolved in which the composites after certain impregnation/carbonization cycles are heat treated at a high temperature and then re-impregnated. This intermediate high-temperature treatment opens up the pores for further infiltration, thus resulting in better density. A comparison of densification achieved with a pitch precursor under different processing conditions is shown in Figure 7.5.

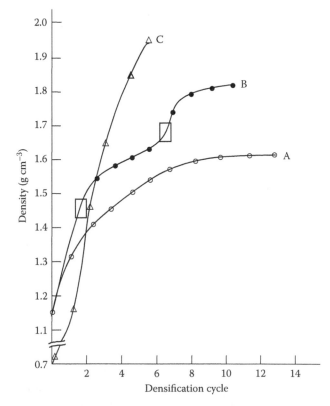

FIGURE 7.5
Density of C/C composites with different densification cycles. (Reprinted from Manocha, L.M., *Sadhana*, 28, 349, 2003. With permission.) A, impregnated carbonization at normal pressure; B, intermediate graphitization; C, pressure impregnation carbonization.

7.3.3 Chemical Vapor Infiltration

C/C composites are also produced by CVI techniques. The CVI process is based on the pyrolysis of gaseous species and deposition on the surface of fibers in the preform. The CVI process for ceramic matrix composites (CMCs) is extensively described in Chapter 6 and the process is similar for the C/C composites also. It is controlled by the deposition parameters and the diffusion rates of gaseous species inside the pores. The homogeneity of pore filling with minimum closed porosity is the key factor for fast densification by CVI process.

The formation of carbon from a hydrocarbon gas within a substrate is a complex process. It is controlled by the preform geometry and infiltration conditions, particularly gas flow rate and concentration, pressure, and temperature. Three types of carbon deposits have been identified, which are smooth laminar, isotropic, and rough laminar. Essentially, smooth laminar deposition is obtained at low temperatures (<1300°C) and intermediate hydrocarbon partial pressures and gas pressures, whereas rough laminar deposition occurs at intermediate deposition temperatures and intermediate partial pressures. Isotropic coatings are obtained at high deposition temperature and low partial pressures. The effect of deposition conditions on the microstructure of carbon matrix formed from propane has been studied and four regions are identified. Essentially, high temperature and low propane concentration favor isotropic deposits, and low temperature and high propane concentration favor a columnar structure. Hence, the CVI process offers considerable scope for the manipulation of microstructure. Each microstructural feature has its own characteristic property; isotropic carbon has low density and smooth laminar carbon is prone to thermal stress microcracking but has better modulus. It is found that the rough laminar structure has optimum properties (Granoff et al. 1973).

Various CVI techniques have been developed to infiltrate carbon fiber preforms. There are four basic CVI processes and they are

1. Isothermal CVI
2. Thermal gradient CVI
3. Pressure gradient CVI
4. Rapid CVI

All these four methods have advantages as well as disadvantages. Depending on the requirements of the final C/C composite part, all methods can be used independently, modified with one another, or combined with polymer impregnation techniques. The four basic methods on the context of C/C composites are described here.

7.3.3.1 Isothermal CVI

The isothermal CVI technique is the most widely used, and most of the C/C composite brakes are made by this technique. It is similar to the fabrication of CMCs, and a schematic of this process is shown in Figure 6.19. In this process, the fiber preforms are heated in the CVI reactor up to the infiltration temperature, which can be in the range between 800°C and 1200°C, and then the reactant gases are passed through the preform. The deposition of carbon on the fiber surface relies on diffusion of gases through the pores. To avoid early sealing of the surface pores, the surface reaction rate needs to be slower than the diffusion rate. A slow deposition rate can be achieved by operating the reactor under a reduced pressure (10–100 mbar) and at a low temperature, typically 1100°C. Unfortunately, under these conditions the rates of weight gain are low and hence processing times are very long. Whatever may be the processing conditions, the early surface pore closure is inevitable and full densification may not be achieved in a single run. Therefore, the C/C composite parts have to be removed from the reactor intermittently and machined to open up the closed surface pores. The number of intermediate machining steps depends on the thickness of the C/C component. For example, the machining step is usually performed up to three times for C/C components with a thickness of 40 mm. The number of these intermediate steps can be reduced by controlling the process parameters. Hence, to build up the required density levels for better mechanical and thermal properties, processing times of several hundred hours and multiple infiltration cycles are essential. Typical resident times for CVI of C/C composite brakes for aircraft are 120–200 h. C/C composite brake manufacturers, such as Dunlop, have a furnace capacity of about 10 ton of brakes per process cycle. The precursor gas used is methane, which is used simultaneously for CVI and furnace heating in the Dunlop industry.

The reproducibility of the isothermal CVI process is excellent. The P_yC matrix produced by this process has a high density, high modulus, and good graphitizing ability. Although the process is time-consuming, the cost is competitive.

7.3.3.2 Thermal Gradient CVI

In order to reduce the processing times, the thermal gradient and pressure gradient CVI processes have been developed. The thermal gradient CVI is normally performed as a cold wall CVI process. The fiber preform is placed around a graphite mandrel, which is inductively heated. Thus, the hottest portion of the preform is the inside surface, which is in contact with the mandrel. The outer surface is exposed to a cooler environment because of the proximity to the water-cooled conduction coils. It is important that the fiber preform should have low thermal conductivity in order to establish thermal gradient. A schematic of thermal gradient CVI process to produce C/C composites is shown in Figure 7.6. Usually, the deposition rate increases with increasing temperature, and hence more deposition can occur near the intersection. The deposited P_yC increases the thermal conductivity of the preform, so the highest temperature region moves toward the cold wall and the deposition front follows the thermal front. The main advantage of this process is its faster infiltration rate and the fact that it can operate at atmospheric pressure. However, there is a major disadvantage with this process,

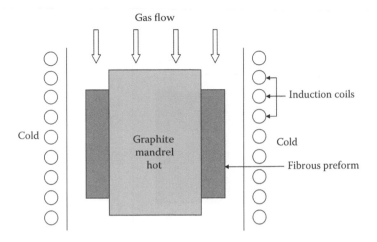

FIGURE 7.6
Schematic of thermal gradient CVI process. (Reprinted from Krenkel, W., *Ceramic Matrix Composites*, Wiley-VCH, Weinheim, Germany, p. 77, 2008. With permission.)

which is the need of separate susceptor for each composite part. Hence, this process is limited to the densification of individual parts. This process is more suitable to produce large C/C composite parts, such as rocket nozzles.

7.3.3.3 Pressure Gradient CVI

This method is a modified version of the isothermal CVI process. It relies on the forced flow of gaseous reactants through the pores in the fiber preform, thus removing the surface reaction limitation (early closure of surface pores) of the isothermal CVI method. A C-fiber preform (often carbon felt) is placed in an isothermally heated furnace with an outer graphite tool, which only leaves a small gap for the forced gas flow of the precursor gas. The gaseous precursor, usually methane, is forced to flow through the preform through the small opening. A schematic of pressure gradient CVI process is shown in Figure 7.7. A pressure difference that forces the gas flow through the pores is created across the wall of the structure, depositing carbon at the farthest surface from the gas entry. While passing through the porous fiber preform, there is a pressure drop in the flow direction and that increases with increasing deposition. Unlike the isothermal CVI process, the deposition rate in this process is independent of diffusion effects due to the forced flow of the precursor gas. Once the matrix is deposited completely in the pores at the top region of the preform, then that region is no longer permeable to the gases. The gases then flow radially through the preform and come out through the vented retaining ring (not shown in the figure). Although high deposition rates may be achieved, there are severe drawbacks in this technique such as the dependency on robust high-temperature pressure seals. This process is also limited to the production of single, simple-shaped component at a time. Hence, this method may not be suitable for the commercial production of C/C composite parts in large numbers.

Gas out

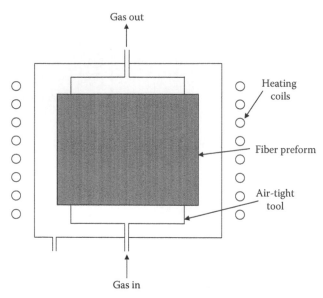

Gas in

FIGURE 7.7
Schematic of pressure gradient CVI process.

7.3.3.4 Rapid CVI Process

A rapid CVI technique developed by CEA in France can be applied industrially for full the densification of C-fiber preforms and it is known as film boiling
method. In this process, the porous C-fiber preform acts as a carbon susceptor
and is fully immersed in a liquid hydrocarbon, such as cyclohexane or toluene.
Out of these two hydrocarbons, the carbon yield is higher with toluene. The
liquid boils during inductive heating and the vapor penetrates into the pores.
On thermal decomposition, these hydrocarbon vapors form a carbon deposit
on the inner surfaces of the preform and densify the preform. The hydrocarbon vapor, which is not decomposed, is cooled and can be used again. A complete densification of C/C brakes can be achieved within 10 h by this process
(Rovillain et al. 2001). Another advantage is that the carbon matrix is highly
oriented and can easily be graphitized. The CVI reactor used in this process is
simple and can be made with glass, because the C-fiber preform is cooled over
the complete densification cycle by the surrounding liquid precursor.

This rapid CVI method is controlled by kinetics, allowing the gaseous species to diffuse deep into the porous preform. However, reactions inside the
preform structure can occur even before the gas diffusion comes to an end.

7.3.4 Redensification Cycles

The mechanical properties of C/C composites increase with increasing degree
of densification. Therefore, it is important to control the pore-filling behavior
during re-impregnation and re-carbonization to avoid the formation of high

amounts of closed pores. During re-impregnations, the pore filling depends on the following parameters:

- Impregnation pressure
- The viscosity of polymer
- Carbonization pressure
- The type of polymer

Low impregnation pressures, in combination with low polymer viscosities, fill larger pores easier than smaller ones. However, during the curing process, or carbonization, the infiltrated polymers may flow out from the pores. This can be prevented by applying high pressures (up to 100 bar) during the carbonization of thermoplastic pitches, whereas the curing of the thermosetting polymers is performed under a pressure of up to 40 bar. The disadvantage is the encapsulation of the condensates (from thermosets) or pyrolysis gases (from pitch) at high pressures, which can result in the damage to the composite. In principle, re-impregnations with pitches are better for composites containing large pores, because the pitch shrinks during pyrolysis onto the surface of the pore inner wall. Carbonized and graphitized thermosetting resins shrink-off from the inner wall surface, therefore influencing the graphitization behavior during the high-temperature treatment. During CVI densification, the inner walls of the pores are directly coated with carbon. The CVI impregnations also have the ability to fill up large pores. However, the premature closing of small pore-mouth results in closed porosity.

7.3.5 Graphitization

The final graphitization treatment determines the chemical and physical properties of C/C composites. This graphitization treatment is performed from 1700°C to more than 3000°C. In general, increasing temperatures result in increasing graphitization of the matrix. Pitch- or CVI-derived carbon shows better graphitization than that derived from phenolic resin. Although phenolic resins form a glassy carbon, it can be graphitized by the application of mechanical stress during graphitization treatment. Increasing the degree of graphitization improves the electrical and thermal conductivity, which is important for heating elements or first-wall materials in fusion reactors. However, the failure behavior is also affected by the increasing degree of graphitization. Highly graphitized carbon matrix results in a pure shear failure of the matrix, with the loss of ductility (Krenkel 2008a). However, the creep resistance increases with the increasing degree of graphitization.

In the case of C/C composites made with thermoset precursors, the final graphitization treatment determines the mechanical properties as well as the fracture behavior. The failure behavior can be modified or tailored from pure brittle failure to a pure shear failure by suitably selecting the graphitization temperature.

7.4 Microstructure of Carbon–Carbon Composites

The important factors to be considered for achieving high toughness and thermal conductivity in C/C composites are proper choice of carbon fibers, carbon matrix microstructure, density, and macrostructure of the composites. The microstructures of the various types of C/C composites are quite complex and differ depending on the type of fiber and matrix precursor and the processing conditions. The important factors that control the microstructure are

1. Fiber type
2. Fiber architecture (1-D, 2-D, 3-D, etc.)
3. Fiber surface treatment, if any
4. Number of impregnation/carbonization cycles
5. Incorporation of oxidation protection system
6. Combined densification routes (e.g., pitch and CVI)

Since the number of factors is very large, it may not be possible to produce a catalogue of the structures of C/C composites associated with the different variables. The two important features in the microstructure of C/C composites are the orientations of the graphite layers relative to the fiber directions and the geometry of pores and voids within the composite. The largest scale at which one might view the microstructure of C/C composites is the submillimeter scale where the fiber architecture is observed. The way of arrangement of fiber bundles defines the overall geometry of the matrix. Observation on increasingly smaller scales reveals fiber and matrix texture. At the nanometer scale, the internal structure of the fiber and the structure of the fiber–matrix interface are revealed. In addition to this complex range of microstructural features, there is a wide range of internal pores that also profoundly affects the properties. These range from equiaxed pores to high aspect ratio cracks at interfaces. Pores and cracks also vary in size from the millimeter scale (such as the cracks between bundles) to the nanometer scale (such as the pores between the basic structural units).

Processing routes and the choice of precursor significantly influence the density and matrix microstructure (crystallinity, orientation of graphitic planes, and type, size, and quantity of defects). Some of these striking structural features and their effect on C/C composite properties are listed in Table 7.1. The voids and microcracks are very less in pitch-derived C/C composites processed through HIP route than in those composites made by CVI process or phenolic resin impregnation. The latter composites have the lowest densities. In order to have desired fiber/matrix bonding, pyrolytic carbon coatings are generally applied to the fibers.

TABLE 7.1

Micro-/Macrostructural Features of C/C Composites and Their Effect on Composite Properties

Nature	Scale	Position	Effect
Micromechanical cracking	Fiber diameter scale	Cracking in matrix	Load transfer among fibers
		Fiber–matrix interface	Transverse properties of fiber bundles
Minimechanical cracking	Cloth layer thickness scale	Interface between fiber bundles	Porosity
		Interface between fiber bundle and matrix	Load transfer among fiber bundles and laminates
		Matrix–matrix interface	Major influence on mechanical and thermal properties

Source: Data from Manocha, L.M., *Sadhana*, 28, 349, 2003.

The degree of graphitization of carbon matrix is also an important factor governing the properties of C/C composites. A semicrystalline or randomly oriented carbon matrix is desired from the strength point of view, since crack propagation is difficult in this type of matrices. However, a highly graphitic matrix is favorable for good thermal and electrical conductivities. The crystalline carbon matrix is generally formed in C/C composites prepared from CVI route or pitch route, whereas amorphous carbon matrix is formed in thermosetting resin route. However, in composites made with high strength, surface-treated carbon fibers, a graphitic carbon matrix can form from the thermosetting resin precursors. Due to strong fiber–matrix bonding, shrinkage during pyrolysis results in the generation of thermal stresses in the matrix, which cause stress graphitization of the matrix during heat treatment at or above 2000°C.

As mentioned earlier, the two methods of carbon deposition, either by using liquid phase or by infiltration with a hydrocarbon gas, generally lead to two different microstructures in the composites. This is due to the difference in the formation of carbon with different structure and properties from different precursors. Perhaps the most significant difference is that the CVI process lays down carbon directly on to the fibers, whereas the liquid precursors fill the spaces between the fibers, and carbon is produced after the carbonization treatment, which involves substantial volumetric shrinkage. Thus, the formation of carbon matrix in the composite is entirely different, and pores and shrinkage cracks are very common in the matrix produced from liquid precursors. These are filled by subsequent re-infiltrations followed by pyrolysis/heat treatment operations.

The microstructure of carbon matrix produced by CVI depends on the deposition conditions such as concentration of gaseous reactants, pressure, temperature, as well as preform geometry. As mentioned earlier, the carbon matrix formed by CVI can have three types of microstructures around the fibers.

The major difficulty in CVI is to achieve a uniform deposition of carbon within the preform. To realize this, the rate of deposition must be much slower than the rate of diffusion of the gas into the porous preform. This puts constraints on the deposition conditions, which should be low temperature and low hydrocarbon partial pressure, thus favoring the rough laminar structure. However, in the thermal gradient CVI technique, the temperature gradient across the sample allows the use of higher deposition temperature for uniform deposition, so other types of microstructures are possible. A characteristic feature in the microstructure of CVI processed composite is the presence of elliptical closed pores, which are formed when the CVI carbon deposit closes narrow pore necks and prevents further infiltration.

The microstructure of thermoset resin-based C/C composites is dominated by large-scale porosity and shrinkage cracks. When carbon matrices are produced from thermosetting resins, the resin is normally used in the solution form. Hence, the evaporation of solvent will take place and the condensation reactions leading to cross-linking will also release water vapor, resulting in a significant volumetric shrinkage. The microstructure of a C/C composite produced using thermoset resin varies significantly according to the final heat treatment temperature. When carbonized alone, a thermoset resin forms a glassy, isotropic carbon that is considered non-graphitizable. This is made up of basic structural units of small dimensions with no significant ordering. Such a structure can be found in matrix-rich regions of C/C composite, far away from fiber bundles. However, when the resin is heat treated in a carbon fiber preform, graphitic material is observed near the fiber–matrix interface region. This is believed to be due to the restraint exerted by the fibers on the matrix during shrinkage, the resultant stresses cause graphitization of the carbon matrix during heat treatment. The graphite structure in thermoset resin-derived matrices is oriented in such a way that the graphite layers encircle the fibers. Irrespective of the fiber type (rayon-, pitch-, and PAN-derived carbon fibers), similar effects have been observed for phenolic resin-based C/C composites after graphitization at 3000°C (Manocha 1986). However, the anisotropy was discontinuous in the vicinity of rayon-based fibers, in the form of long areas with pitch-based fibers and intermediate for PAN-based fibers.

The interaction of the resin with the fiber surface is also important, since it determines the location of the cracks. The first carbonization cycle is critical for determining the location of the cracks. If the precursor is bonded well to the fibers (e.g., as with surface-treated, high strength (HT) carbon fibers), then the composite exhibits high shrinkage, extensive cracking in the body of the matrix, and possible fiber damage. As a result of this, the composite microstructure consists of clumps of fibers, which are bonded together but separated by large cracks. However, if the fiber–matrix bonding is poor (e.g., as with C-fibers without surface treatment), the matrix shrinks away from the fibers leaving fissures or annular gaps at the interface.

Further shrinkage of carbon matrix occurs during heat treatment to temperatures above 2000°C because of gradual transformation to graphitic structure.

This also modifies the form and distribution of the porosity developed on carbonization. The pores generated during this treatment facilitate further resin penetration on re-infiltration. Hence, an intermediate graphitization treatment would be beneficial to achieve full densification in the composite.

The liquid infiltration using pitch has many features similar to thermoset resin infiltration. The major difference is that the carbonization treatment is usually carried out under pressure. This reduces volatilization of carbon-based compounds and shrinkage and increases carbon yield from 50% to 85%. Matrix carbon resulting from mesophase development within the pitch is considerably influenced by the fiber architecture. In C/C composites, the mesophase tends to align parallel to fiber surfaces producing a region of preferred orientation, which extends to surrounding matrix. Within a fiber bundle, where the volume fraction of fibers is high, the interlamellar matrix is well oriented. This orientation significantly affects the modulus and thermal conductivity of C/C composites. However, variations in processing conditions can affect the matrix orientation, particularly the pressure application during carbonization (Stover et al. 1977). When the carbonization is carried out under atmospheric pressure, the graphitic layer planes are aligned because of the viscous flow of the mesophase along the fiber surface, whereas the high-pressure application modifies the flow and hence the orientation. High pressures usually favor the formation of a transversely aligned matrix, that is, the layer planes are normal to the fiber surface, but this effect is not clearly understood. It appears that the surface characteristics of the fibers are important since this orientation effect have not been observed in fibers coated with a thin amorphous layer of carbon. Other factors do, however, seem to be important because it has been reported that both parallel and perpendicularly oriented graphite are produced in the carbon matrix carbonized under atmospheric pressure (Murdie et al. 1988).

During the conversion of pitch from a viscous solid to a polycrystalline solid, there is extensive bulk shrinkage. This leads to the formation cracks, which run parallel to the lamellar structure of the solid. Cracks will also form due to the difference in contraction between matrix and fiber during cooling from the processing temperature. The location of these cracks depends on the fiber–matrix bond strength. In the case of a strong bond, shrinkage cracks are formed within the matrix, whereas they occur predominantly at the fiber–matrix interface when the bonding is weak.

All C/C composites contain pores and cracks, which are extremely important microstructural features in determining mechanical and thermal properties. The presence of a network of pores and cracks becomes advantageous, since they provide conduits for further impregnation to achieve densification. This is particularly important since the majority of commercially produced composites need 3–6 re-impregnations to build up the desired density. A comparison of the characteristic features and properties of carbon matrices from the different precursors is given in Table 7.2.

TABLE 7.2

Comparison of Characteristic Features and Properties of Carbon Matrices from Vapor Phase, Pitch, and Resin Precursors

Characteristic Feature or Property	Type of Carbon Matrix		
	CVD	Pitch	Resin
Density	High, ~2000 kg m⁻³	High increases with HTT up to the value for graphite	Low, 1300–1600 kg m⁻³
Carbon yield	C directly deposited, no further thermal degradation	Varies according to pitch composition (50%–80% w/w)	About 50% for phenolics, increasing to 85% for polyphenylenes
Porosity	Low, except for isotropic form. Laminar fissures	Macro-sized gas entrapment pores plus shrinkage and thermal stress fissures	High microporosity (pore diameter <1.0 nm), becomes closed above 1000°C. Macro-voids may be evident due to vapor evolution during curing of resins
Microstructure (in the bulk state)	Varies from isotropic to highly oriented laminar forms	Macro-domains (1–100 μm) showing preferred orientation developed from the mesophase state	Isotropic—except on the nanoscale, i.e., the basic structural units are randomly oriented
Orientational effects within fiber preforms	Strong orientation of laminar matrix with fiber surfaces. Orientation also on crack surfaces within other matrix types if subsequently treated to CVI	Preferred orientation of lamellae with fiber surfaces, increases as HTT develops the graphite structure. Modified by pressure pyrolysis	Preferred orientation at fiber surfaces but to a much lower extent than other precursors, increases with HTT
Graphitizability/ crystallinity	Laminar forms highly graphitizable; L_c and L_a > 200 nm after 3000°C HTT	Highly graphitizable; L_c and L_a usually less than 100 nm	Normally non-graphitizing
Purity/ composition control	Controlled by gas-phase composition, enables other elements to be incorporated into deposit (e.g., Si, B)	Controlled by source, can be of high purity. Not easy to incorporate other elements except as powders	Similar to pitch

TABLE 7.2 (continued)

Comparison of Characteristic Features and Properties of Carbon Matrices from Vapor Phase, Pitch, and Resin Precursors

Characteristic Feature or Property	Type of Carbon Matrix		
	CVD	Pitch	Resin
Reactivity to oxidizing gases	Very low reactivity for highly oriented pyrographite	Low reactivity, decreases with HTT	Usually high reactivity due to micropore network, decreases with HTT but still relatively high
Thermal expansion	Depends on preferred orientation, can be highly anisotropic, approaches values for the crystal in the two major directions	Depends on domain orientation; expansion partially accommodated by lamellar cracks $(1–5 \times 10^{-6}\,K^{-1})$	Isotropic in bulk $(\sim 3 \times 10^{-6}\,K^{-1})$
Thermal and electrical conductivity	Determined by preferred orientation, approaching single crystal graphite values	Depends on domain orientation, HTT and internal porosity, increasing with HTT	Isotropic in bulk
Young's modulus	7–40 GPa, depending upon structure	5–10 GPa, depending on grain size, porosity, and degree of graphitization but up to 14 GPa for very fine (1 μm) grain size materials	10–30 GPa
Strength	Depends on microstructure and degree of graphitization; 10–500 MPa	Depends on porosity and pore geometry; 10–50 MPa for most polycrystalline carbons/graphites, rising to 120 MPa for very fine-grain (1 μm) graphites	Approximately 80–150 MPa for glassy carbons. Lower for resin carbons, depending on porosity
Failure strain	0.3%–2.0% depending on structure; higher values at highest deposition/HT temperatures	Up to about 0.3% depending on grain size and degree of graphitization	Up to about 0.4%; largest values for glassy carbons at HTT

Source: Data from Rand, B., in *Essentials of Carbon-Carbon Composites*, ed. G.R. Thomas, The Royal Society of Chemistry, London, U.K., p. 164, 1993.
HTT, High temperature treatment.

7.5 Properties of C/C Composites

C/C composites in the true sense cover a large range of materials with different properties. The main properties of interest in C/C composites are strength, stiffness, fracture toughness, wear resistance, thermal conductivity, and resistance to oxidation at high temperatures. The factors controlling these properties are quite different and extensive especially in such multiphase materials. The mechanical properties of the constituents and their volume fraction, bonding, fiber orientation, microstructure, and crack propagation mechanism control the mechanical properties, whereas the thermal properties are mainly governed by thermal transport phenomena. Moreover, the fiber reinforcements are likely to undergo a change in properties during processing at high temperatures. This change will also influence the ultimate properties of C/C composites.

7.5.1 Mechanical Properties

The type of fiber and matrix precursor, fiber architecture, and processing method will all determine the mechanical properties of C/C composites. As a result of the large number of variables involved, C/C composites have a wide range of properties. The stiffness and strength of C/C composites are dominated by the fibers, and consequently the specific arrangement of fibers in the composite has a strong influence on these properties. However, the C/C composites are very complex as a result of the physical and chemical changes that can occur during processing. Hence, the mechanical properties are dependent on a whole lot of factors, among which the most important are

1. Carbon fiber type (rayon, PAN, pitch, high strength, high modulus)
2. Fiber architecture (weaved fabric stack, pierced fabric, knitted preform, braided preform)
3. Fiber directions (1-D, 2-D, 3-D, etc.)
4. Fiber volume fraction
5. Matrix precursor (resin, pitch, gaseous hydrocarbons)
6. Final density of the composite

There are some other factors, which will also influence the mechanical properties, but they are difficult to quantify. They are change in fiber properties during heat treatment, fiber damage during construction and densification, fiber–matrix bonding, and difference in thermal expansion between fiber and matrix. Hence, the comparison between different sets of published data must be treated very cautiously because of the need to consider so many factors. Nevertheless, some valid comparisons can be made.

TABLE 7.3

Room-Temperature Mechanical Properties of Graphite and C/C Composites

Material	Elastic Modulus (GPa)	Tensile Strength (MPa)	Compressive Strength (MPa)	Shear Strength (MPa)	CTE (10⁻⁶/°C)
Graphite	7.5–11	20–30	83	—	2.8
C/C with unidirectional, continuous	240–280	650–1000	620	7–14	1.1
C/C with 2-D fabric	110–125	300–350	150	7–14	1.3
C/C with 3-D (woven orthogonal fibers)	55	170	140	21–27	1.3

Source: Adapted from Sheehan, J.E. et al., *Annu. Rev. Mater. Sci.*, 24, 19, 1994.

A comparison of room-temperature mechanical properties of fine-grained monolithic graphite with unidirectional and three-directional fiber-reinforced C/C composites clearly indicates the benefits of incorporating carbon fibers in the carbon matrix (Table 7.3). It also indicates the dependence of mechanical properties on fiber architecture. The incorporation of unidirectional fibers considerably increases the strength in the principal fiber direction, while the 3-D network of fibers lead to more modest increase in room-temperature strength and tremendous increase in fracture toughness. It is found that the room-temperature mechanical properties are significantly higher in the fiber direction than in the transverse direction for both the unidirectional fiber and fabric-reinforced composites. The 3-D, woven orthogonal fiber-reinforced composite, on the other hand, has more isotropic mechanical properties. Strength utilization of fibers in unidirectional fiber composites (UDCs) is between 25% and 50% depending on the processing conditions. However, as in other fiber-reinforced composites, the shear properties of C/C composites are relatively low. The same is true for the tensile properties of 1-D and 2-D composites in directions normal to the fiber direction. These low shear and transverse tensile strengths are mainly due to weak fiber–matrix interface that is designed to reduce composite brittleness and improve fracture strength in the fiber direction. The modulus values of C/C composites in the fiber direction are almost the same as those predicted by the rule of mixtures.

The strength and fracture of C/C composites are governed by the Cook–Gordon theory, which states that if the ratio of the adhesive strength of the interface to the cohesive strength of the solids is in the right range, a large increase in strength and toughness is achieved in a brittle material. Extensive work has been carried out to achieve the highest possible fiber properties in C/C composites. Composites with strong interface bonding fail catastrophically without fiber pullout at a lower stress while those with controlled interface fail in a mixed tensile cum shear mode exhibiting high strength. A generalized view of the effect of heat treatment temperature on the fracture mode and strength of C/C composites is shown in Figure 7.8.

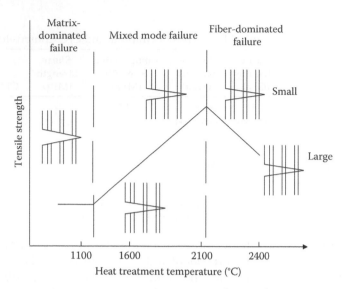

FIGURE 7.8
Effect of heat treatment temperature on strength and fracture mode of C/C composites—a general view. (Reprinted from Manocha, L.M., *Sadhana*, 28, 349, 2003. With permission.)

The tensile strength of a unidirectional C/C composite in the longitudinal direction may be expressed by the rule of mixtures, as

$$\sigma_c = \sigma_f V_f + \sigma_m (1 - V_f - V_p) \tag{7.1}$$

where
σ_f is the tensile strength of fiber
V_f is the volume fraction of fibers
σ_m is the tensile strength of matrix
V_p is the volume fraction of pores

Similarly, the elastic modulus of the composite may be expressed as

$$E_c = E_f V_f + E_m (1 - V_f - V_p) \tag{7.2}$$

where E_f and E_m are the elastic moduli of fiber and matrix, respectively. The failure strain of the matrix (ε_m) in C/C composites is lower than that of the fibers (ε_f), so in a well-bonded composite, the failure of the composite occurs when the matrix fails. In this condition, catastrophic brittle failure will occur and the full reinforcing potential of the fibers will not be realized. The principal factor in avoiding brittle failure is to have poor adhesion between fiber and matrix. There is considerable evidence to show that the longitudinal strength of weakly bonded composites is higher than that of strongly bonded composites. In another study, it has been reported that, for composites that fail at the matrix failure strain, the strength can be greater when stiffer and weaker fibers are used as the reinforcement (Fitzer and Huttner 1981). The difference in thermal expansion between fiber and matrix also influence the failure strength. This difference can cause the matrix to be prestressed in tension on cooling from the final heat treatment temperature. When the failure is matrix dominated, the prestressing leads to failure at lower strains and reduced strengths are expected. However, the matrix-dominated properties of unidirectional composites, such as transverse modulus and strength, and interlaminar shear strength (ILSS) of well-bonded C/C composites are superior to those of weakly bonded composites.

When the modulus value of the matrix is much less than that of the fibers, Equation 7.2 reduces to

$$E_c = E_f V_f \qquad (7.3)$$

This is applicable to most of the polymer matrix composites; however, the measured values of many C/C composites were found to be much larger than this. A substantial body of evidence is beginning to emerge to show that an appreciable degree of preferred orientation develops in carbon matrix on heat treatment. This directionality of carbon matrix significantly influences the modulus values of the resultant composites.

The results reported in the literature are difficult to compare because investigating the influence of one parameter while maintaining all other parameters constant is difficult if not impossible. While it is possible to optimize the matrix microstructure to improve transverse properties, the greatest improvement in properties can be obtained by placing fibers in the appropriate directions. The simplest arrangement is the use of simple 2-D reinforcement such as woven carbon fabrics. However, the maximum fiber volume fraction (V_f) in any given direction is reduced, which is only 0.39, a very low value compared to unidirectional composites, where it can be 0.78. While this results in a composite with virtually isotropic properties in the two orthogonal directions, the strength and modulus values are lower than that of 1-D composite. Most bidirectional composites are in fact made from woven cloth, which contains undulating fibers, responsible for some loss of strength compared to simple cross-ply lay-up of unidirectional fibers. It has been demonstrated convincingly that the strength of a C/C composite made with an eight-harness satin weave fabric is much greater than that of a composite made using the same fibers but with a plain weave (Manocha and Bahl 1988). Although the properties in the X- and Y-directions may be adequate while using the 2-D fibers, the properties through the thickness are still inadequate.

Three-directional fiber (3-D) composites have been designed to improve the poor through-thickness properties of 1- and 2-D fiber composites. Since the maximum possible V_f in any given direction is only 0.2 in the 3-D composites, this means a reduced effectiveness of strength improvement by the fibers. Even though the maximum strength is lower than that of 1- and 2-D composites, 3-D composites show significant pseudoplastic behavior.

C/C composites also exhibit high fracture toughness (20–100 MPa m$^{1/2}$) and good creep and fatigue resistance. When the C/C composites are tested at 2000°C under inert atmosphere, they have been found to exhibit about 10%–20% increase in strength. However, the strength drops down by 10%–20% when tested in air, due to oxidation. Carbon is a highly preferred material for nuclear applications because of its stability under nuclear radiation and so are C/C composites. Under low levels of neutron irradiation (10^{21} n cm^{-2}), these composites have been found to exhibit an increase in strength and fracture toughness by 20%–30% and Young's modulus by about 30%.

C/C composites are unique that they retain the room-temperature mechanical properties up to about 2200°C. In fact the tensile strength of these composites in the fiber direction was found to increase with increasing temperature. The shear, transverse tensile, and compressive strength of C/C composites with respect to temperature are mainly influenced by the matrix and fiber–matrix interface properties. Generally, these properties also improve with increase in temperature mainly attributed to the closing of microcracks as the temperature is increased. The steady-state creep rates of most C/C composites are at least four orders of magnitude lower than that of most advanced ceramics.

As mentioned earlier, the structure of the carbon matrix is very much influenced by processing variables, which in turn affect the mechanical properties of the C/C composites. A study has been made to find the influence of CVI temperature on the microstructure of C/C composites prepared using carbon fabric and a reactant gas concentration of 10% propane (Oh and Lee 1988a,b). When the CVI process is carried out at 1100°C, the matrix is well infiltrated and well bonded to the fibers, whereas at 1400°C, the matrix is loosely bonded to the fibers and high porosity is observed. The bulk density is found to decrease from 1.79 to 1.37 g cm^{-3} with increasing deposition temperature. Increasing the concentration of reactant gas at constant temperature resulted in increased composite bulk density and the corresponding stress–strain curves changed from catastrophic failure to a stepwise pseudoplastic failure. A higher degree of preferred orientation of the matrix is observed as the propane concentration is increased, resulting in a weaker fiber–matrix bond.

The mechanical properties of thermoset resin-derived C/C composites depend very much on the type of fiber and whether or not the fiber has been surface treated. The highest mechanical properties are achieved by using non-surface-treated high-modulus PAN or mesophase pitch-based C-fibers. The selection of resin also has a major influence on the mechanical properties.

The usually employed thermosetting resins give a carbon yield of 50%–60% by weight and a single impregnation would result in a density of 1.3–1.4 g cm^{-3}. Typically, 4–6 impregnation–pyrolysis cycles would be required to raise the densities to 1.7–1.8 g cm^{-3}. Another factor that has a major influence on the mechanical properties is the heat treatment temperature. The strength is low, and brittle fracture predominates when the heat treatment temperature is below 2000°C. At heat treatment temperatures above 2400°C, the graphitic phase can form resulting in pseudoplastic behavior. The heat treatment at 2800°C and above tends to reduce the strength due to thermal damage of fibers.

The effect of processing conditions on the mechanical properties of pitch-derived C/C composites is difficult to assess since no results from systematic studies are available in the open literature. The mechanical and other properties are controlled not only by the carbon yield but also by the resulting microstructure. Pitch carbonized under atmospheric pressure results in a well-graphitized carbon with sheath layers parallel to the fiber surface, whereas a transversely oriented matrix structure, possessing more isotropic properties, has been observed when high pressures are applied during carbonization. In general, multiple impregnation–pyrolysis cycles improve the strength of a pitch-based C/C composite. However, higher graphitization temperatures reduce flexural strength. This is due to the difference in thermal contraction between carbon matrix and fiber generating shrinkage stresses resulting in crack formation between fiber and matrix. Hence, graphitization aids densification by generating such cracks for infiltration but results in a lower composite strength.

Another important factor, which influences the strength of C/C composites, is the porosity. The strength of polycrystalline ceramics, particularly graphite and glassy carbons, may be expressed in terms of porosity by a simple empirical Knudsen equation:

$$\sigma = \sigma_o \exp^{-\beta p} \tag{7.4}$$

where
σ_o is the strength at zero porosity
β is a constant
p is the porosity

A similar relationship has been observed in C/C composites. Moreover, the geometry of pores is also very important.

One property that is particularly affected by heat treatment temperature is the modulus of the composite. An increase in heat treatment temperature causes a rapid increase in modulus in the fiber direction of a UDC, whereas slightly decreases the modulus perpendicular to the fiber axis (Yasuda et al. 1980).

One of the greatest advantages of C/C composites is their ability to retain high strength and stiffness at high temperatures far in excess of the maximum useful temperatures of other ceramic materials. Unfortunately, only a few papers have been published on the high-temperature mechanical properties, probably due to the sensitive nature of these materials for defense applications. The Young's modulus of a C/C composite heat treated at 1800°C showed a progressive decrease with increasing test temperature up to 1400°C and a rapid decrease above this temperature (Hill et al. 1974). However, with a more graphitic carbon matrix (heat treated at 2600°C), the modulus increased with test temperature up to 1000°C and then it decreased. However, there is no clear trend of strength with test temperature.

As a consequence of the fabrication process, carbon matrices usually have a large number of internal cracks or voids, and hence the fracture toughness values are very important for any specific application. However, a very few data have been published for this property also. Fracture toughness can be defined simply as the resistance to propagation of a crack through a body. In C/C composites, it strongly depends on the type of carbon fiber used and the orientation of the initial crack with respect to fiber direction. In a C/C composite made with 2-D fabric, severe crack blunting and delamination are observed when crack propagation is perpendicular to the fibers. However, very low fracture toughness values are obtained when crack propagation is parallel to the plane of the cloth. The fracture toughness value of C/C composite is increased to 25 MPa m$^{1/2}$ from 7 MPa m$^{1/2}$ (corresponding to graphite). The occurrence of progressive cracking will be indicated by stepwise unloading of the stress–strain curve, which results from a high degree of fiber pullout and that will make the material extremely tough. This type of stress–strain curves has been observed in many C/C composites, indicating high toughness (Fitzer 1987).

It is generally accepted that the maximum service temperature of carbon is limited to ~2000°C because of the onset of creep. C/C composites were developed for defense/aerospace applications to push this limit to higher temperatures. It has been reported that the high-temperature creep of C/C composites is characterized by an initial transient creep followed by steady-state creep rate, which increases progressively with increasing temperature in the range of 2060°C–2600°C.

7.5.2 Thermal Properties

The bulk thermal expansion and thermal conductivity are highly anisotropic in the UDCs but approximately isotropic in the 3-D fiber composites. The CTE of C/C composites tends to decrease with increasing temperature; nevertheless, it is relatively lower than the other ionic- and covalently bonded materials.

There is considerable interest in developing C/C composites with high thermal conductivity. These composites can be used in many important applications such as first-wall tiles of nuclear fusion reactors, hypersonic aircraft parts, thermal radiator panels, and electronic heat sinks. A very large number of processing permutations are possible with the availability of a wide variety of fiber and matrix materials. This wide range of options means that a wide range of thermal conductivities is possible, and also there are opportunities to tailor the thermal conductivity of carbon-based materials.

It is generally accepted that graphite is a phonon conductor, and >99% of heat is transported by phonons or quantized lattice vibrations. These phonons may be scattered by other lattice vibrations or by any defects present in the crystal lattice. Graphite is a unique material having an extremely high anisotropy of thermal conductivity. The thermal conductivity within the layer planes is more than 200 times higher than the out-of-plane thermal conductivity, which means that graphite is very a good thermal conductor in two directions and virtually an insulator in the third direction. Hence, the thermal conductivity of any graphite assembly, such as C/C composite, is critically dependent on the orientation of the layer planes. The conduction in polycrystalline graphite will be affected by the presence of voids and pores, and the tortuous heat conduction path, which depends on the orientation of layer planes. The most significant contribution to scattering is grain boundary scattering. As the crystallite size decreases, the thermal conductivity decreases.

When carbon fibers are incorporated into a carbon matrix, the resultant thermal conductivity is not only influenced by the fiber architecture but also by the geometry of pores. A C/C composite is therefore a ternary system consisting of fibers, matrix, and pores. The latter act as a barrier to heat flow, and the geometry and orientation of pores are crucial, which are clearly influenced by the fiber architecture. There is a misconception that the thermal conductivity of a C/C composite is dominated by the fiber conductivity, and it is not possible to achieve thermal conductivity values more than that of fibers. However, it has been reported that the thermal conductivity of the matrix parallel to the fibers is significantly higher than that of fibers themselves in a C/C composite fabricated by the CVI process (Whittaker and Taylor 1990). Surprisingly, the matrix thermal conductivity perpendicular to the fiber is found to be more than twice that of the fiber. Clearly, the orientation effects are important, and it is possible to manipulate the thermal conductivity by the appropriate selection of matrix precursor and processing conditions, particularly the graphitization temperature, since more graphitic carbon is associated with higher thermal conductivity.

To find the microstructure and thermal conductivity relationship, C/C composites with isotropic, smooth laminar, and rough laminar microstructures have been prepared by the CVI process. It was found that the rough laminar matrix exhibited the highest conductivity because of more graphitic carbon.

For thermoset resin-based C/C composites, it is generally considered that fibers will dominate the thermal conductivity because of the low thermal conductivity of the non-graphitizable carbon matrix. Pitch-derived matrices are considered to be highly graphitizable and expected to show better thermal conductivity. Depending on the fiber architecture also, the thermal conductivity of C/C composites can vary. For 2- and 3-D composites, because of the complex geometry of pores and the orthogonal fiber orientations, the thermal conductivity in the fiber directions will be lower than that of 1-D composites. It has been reported that the thermal conductivity of 2-D composites is 190 W m^{-1} K^{-1} at 225°C parallel to the fibers, whereas it is 300 W m^{-1} K^{-1} for the 1-D composites in the fiber direction (Whittaker et al. 1990). In another study, it was found that the thermal conductivity of 3-D pierced fabric composite made using low-modulus fibers was lower than that of similar composite made with high-modulus fibers. It has been reported that the addition of coke to either pitch or thermoset resin precursor slightly decreases the longitudinal thermal conductivity but significantly increases the transverse thermal conductivity.

The CTE of graphite is unique. The coefficient is very low parallel to the layer planes and it is even negative at 20°C. However, the CTE normal to the layer planes is very high, even up to 26×10^{-6} K^{-1} is found at high temperatures. Hence, for the carbon fibers, in which the layer planes are arranged parallel to the fiber axis, the CTE is nearly zero along the fiber axis; however, it is ~10×10^{-6} K^{-1} normal to the fiber axis. The thermal expansion of a 1-D C/C composite will be restrained by the fiber but also influenced by the layer plane orientations within the matrix. The differential thermal contraction between fiber and matrix on cooling from the high processing temperatures is responsible for the formation of a network of cracks parallel to the fibers. They may be beneficial in providing access during repeated impregnations to improve density. A considerable amount of thermal expansion on heating from ambient temperature will also be useful initially for filling voids. However, in composites made using preforms with different fiber orientations, the expansion will be strongly affected by the fiber orientations.

C/C composite, being a heterogeneous material consisting of fibers, matrix, and pores with a variety of microstructures in the constituents, estimation of its thermal properties becomes complex. However, C/C composites with tailored thermal conductivities can be made by proper choice of fibers, matrix precursors, fiber configuration, and processing conditions. Typical thermal conductivities of C/C composites with different fiber/matrix combinations are shown in Figure 7.9. Composites having highly oriented graphitic fibers or matrix or their combination such as PAN fiber and CVI matrix or mesophase pitch-based carbon fibers and matrix exhibit very high thermal conductivities of the order of 250–350 W m^{-1} K^{-1} in the fiber direction. Although these composites exhibit highly anisotropic behavior with low conductivities in the transverse directions, always there is a scope of improvement by varying the fiber architecture. The CTE also depends on the fiber orientation, and it is $0-1 \times 10^{-6}$ K^{-1} in the fiber direction and $6-8 \times 10^{-6}$ K^{-1} in a direction perpendicular to the fibers. The combination of good strength, relatively high thermal conductivity, and low thermal expansion makes C/C composites resistant to severe thermal shock.

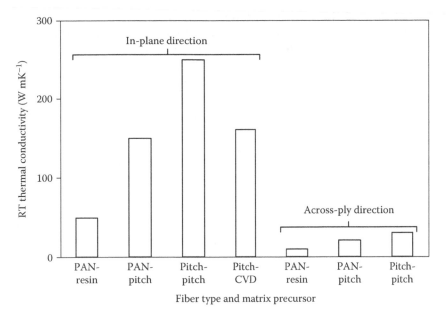

FIGURE 7.9
Thermal conductivity of C/C composites with different fiber–matrix combinations. (Reprinted from Manocha, L.M., *Sadhana*, 28, 349, 2003. With permission.)

7.5.3 Tribological Properties

C/C composites have widened the scope of application of carbon-based materials for wear-resistant components from bearing seals and electrical brushes to brake pads for military and civilian aircrafts, sports cars, and high-speed trains. This has been due to the inherent tribological properties of graphite with additional strength and thermal conductivity contribution from the reinforcing carbon fibers. C/C composites exhibit low coefficient of friction in the fiber direction (~0.3) and moderate in the perpendicular direction (0.5–0.8). The friction and wear mechanism of C/C composites is quite complex because of the complex microstructural features. Various factors like peak temperature, and the formation of debris and transfer films on surfaces further affect the frictional properties.

7.6 Oxidation Protection of C/C Composites

The most severe drawback inhibiting the widespread use of C/C composites is their susceptibility to oxidation above 500°C under normal atmospheric conditions. This becomes progressively more severe with increasing temperature and the rate of oxidation is limited only by the diffusion of oxygen through the surrounding gas to the specimen surface. Carbon is oxidized according to the following chemical reactions:

$$C_{(s)} + O_{2(g)} \rightarrow CO_{2(g)} \tag{7.5}$$

$$2C_{(s)} + O_{2(g)} \rightarrow 2CO_{(g)} \tag{7.6}$$

The second reaction is predominant at higher temperatures, whereas CO_2 is formed at lower temperatures. The oxidation attack starts on the atoms present at the edge of the material, which means crystalline defects in the carbon structure. Therefore, the carbon matrix is more sensitive to oxidation than the reinforcing C-fibers, since the defects are more common in the matrix. Oxidation rate at low temperatures can be reduced by increasing the crystallinity and purity of the carbon matrix by subjecting it to higher graphitization temperatures.

The oxidation mechanism depends on the temperature at lower oxidation temperatures (<800°C), that is, the oxidation rate is controlled by the rate of chemical reactions (kinetics), whereas at higher temperatures, the burn-off rate is limited by diffusion effects. The active sites for oxidation in the carbon matrix can be reduced by increasing the final graphitization temperature or blocked by contaminants applied as salt impregnations.

The oxidation behavior of a number of C/C composites has been studied. It is found that the matrix is more reactive and the fibers are oxidizing at a slower rate. The reaction rate is faster for pyrolytic graphite than isotropic carbon (Yasuda et al. 1980). The oxidation of composites preferentially occurs at the sites of high energy such as fiber–matrix interfaces. The rate of oxidation increases with increasing operating temperature but reduces by the increase in heat treatment temperature of the C/C composite. The latter effect is due to a reduction in the retained impurities, relaxation of the carbonization stress, and a reduction of reactive edge sites, despite an increase in the fraction of open pores at higher heat treatment temperatures.

There are many industrial applications, in which oxidation protection is not at all required, because the C/C composites are used under inert gas or vacuum. However, oxidation protection of C/C composites has to be applied when they are used in an oxidative atmosphere above 420°C. The selection of oxidation protection systems depends on the oxidation conditions, such as temperature and surrounding atmosphere, fiber architecture, matrix microstructure, and the required lifetimes. Two main approaches are being utilized to protect C/C composites against oxidation. One is to use internal inhibitors to slow down the rate of reaction between carbon and oxygen. Another approach is to use diffusion barriers to prevent oxygen from reaching the carbon surface.

7.6.1 Bulk Protection Systems

C/C composites used in oxygen-containing atmospheres below 650°C can be protected by bulk protection systems. C/C composite grades with phosphoric salt impregnations are commercially available, which are used in the manufacturing of hollow glasses at temperatures of the order of 700°C without any further oxidation protection. The salt impregnations decrease the oxygen flow rates and thereby oxidation attack. It has been reported that the phosphoric salt impregnated C/C composites can be used in hollow glass manufacturing up to 2000 h, if the contact times and contact areas are sufficiently low. The extent of oxidation depends on temperature, time, and the surface area attacked by oxygen (Krenkel 2008b).

Bulk treatments are often performed with silicon, which reacts with carbon to form SiC. This SiC exhibits superior oxidation protection compared to salt impregnations. Silicon treatments of the bulk C/C composites are well known, and different industrial methods are available. Silicon treatments can be performed with Si vapor, by melt impregnation under vacuum and/or pressure, by capillary impregnation of liquid silicon, by pack cementation, or by a combination of cementation and liquid impregnation. The characteristics of different methods are shown in Figure 7.10.

The silicon treatment by a combination of cementation and liquid impregnation is the only procedure that can be used for fully densified C/C composites without any dimensional change. The degree of conversion from carbon to SiC depends on the reactivity of the carbon, open porosity, and density. A broad range of C/C composites from C/C–SiC to C/SiC can be realized according to the final requirements. The higher the amount of SiC formed, the better the oxidation resistance. However, increasing the degree of

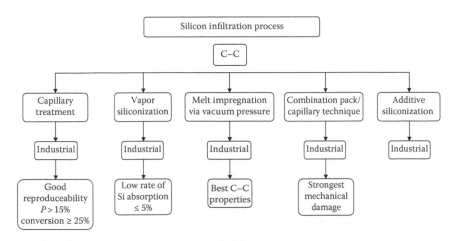

FIGURE 7.10
Bulk protection methods via silicon treatments. (Reprinted from Krenkel, W., *Ceramic Matrix Composites*, Wiley-VCH, Weinheim, Germany, p. 77, 2008. With permission.)

conversion is always combined with increasing brittleness of the composites, including fiber damage by chemical reactions between C-fibers and silicon.

Silicon treatments of C/C composites are used not only for oxidation protection but also for the improvement of wear resistance for tribological applications. The best properties are obtained by multilayer CVI coatings on the inner surface of bulk materials. Such methods have been used, in particular for C/SiC and SiC/SiC components applied in space or aircraft structures. A combination of nanoscale multilayer coating of P_yC, SiC, BN, or B_4C directly applied on the internal surface of fiber preforms improves the oxidation resistance as well as the ductile failure behavior of the high-temperature composites. Boron-containing layers are able to form boron oxide readily, even at low oxidation temperatures, resulting in self-healing behavior of the matrix during oxygen attack. Hence, the bulk oxidation protection of C/C composites can also be obtained by using boron additives.

Another method for bulk protection is powder immersion reaction-assisted coating (PIRAC), a method developed by Prof. Gutmanas, Technion, Haifa (Eliezer 2003). PIRAC with chromium or titanium can be used as bulk protection for C/C composites. The formed carbides fill the intralaminar cracks in the composites and thereby prevent strong oxidation attack. However, for long-term oxidation resistance, it is necessary to give outer chemical vapor deposition (CVD) coatings. The PIRAC method is a simple and cost-effective CVI technique. The components to be protected are packed in a powder mixture of metal and its halogenide. The package is sealed to vacuum-tight by using metal foil and heated in a conventional furnace to the CVI reaction temperature, preferably below 1100°C. The furnace chamber is filled with an inert gas. Enclosed in the metal foil under low pressure, a metal subhalogenoid vapor is formed that diffused inside the C/C composite to form the metal carbide.

7.6.2 Multilayer Surface Coatings

While selecting a prospective thermal protection system, a number of important factors and associated system requirements need to be considered. The primary task is to apply a coating to isolate the composite from environment. In order to achieve this, the coating system must have at least one component that can act as an efficient barrier to oxygen. The primary oxygen barrier should have low oxygen permeability, and its main function is to totally encapsulate the carbon, ideally without any defects through which the oxidizing species can ingress. Optimally a material, which forms a protective oxide in situ, can be used. It is also equally important to minimize the diffusion of carbon outward to avoid carbothermic reduction of oxides that may be present. The mechanical compatibility between the protective layer and substrate is also an important factor to be considered to overcome the problem of coating spallation.

The CTE of carbon is very low compared to that of the refractory ceramics that may form part of the coating system. Hence, any applied coating is likely to contain microcracks because of the difference in thermal contraction from the high coating temperatures. Moreover, the composite will undergo thermal cycle from ambient to the working temperature, thereby generating more thermomechanical stresses. Therefore, it is imperative that the coating system should possess a self-healing capability. The most successful solution to date has been the incorporation of a glass or glass-forming compound as the outermost layer, which can flow into and seal any cracks in the primary coating. The seal should work in the temperature range from 400°C (the oxidation threshold of the coating) to the microcracking temperature of the primary oxygen barrier. It is also necessary to establish good adhesion between the C/C composite and the coating and between the different layers of the coating system. Since the coating must be able to withstand any stresses generated at the surface of the component, it is essential to consider the mechanical properties of the coating. Ideally, a low-modulus coating is desirable to accommodate expansion mismatch strain during thermal cycling.

This combination of properties cannot be met by any single material, and hence, a multilayer system consisting of different materials is generally used. The main advantage of this multilayer coating technology is associating the specific advantages of each layer while limiting their drawbacks. Individual layers are strategically stacked on the substrate to provide protection over the whole temperature range of interest. While a wide range of coating systems have been developed and patented, the most successful systems are composed of three basic layers: bond layer, functionally active layer, and primary, erosion-resistant, oxygen barriers (Figure 7.11).

The most commonly used primary oxygen barrier materials are SiC and Si_3N_4. Both of them are refractory materials and oxidation resistant due to the formation of protective skin of SiO_2 on oxidation. The vapor pressure of SiO_2 is relatively low even at temperatures as high as 1650°C and oxygen diffusivity in SiO_2 is also very low. Although the viscosity of SiO_2 is low enough at high temperatures to flow and seal any cracks, it is very high below 1150°C to afford any self-healing properties.

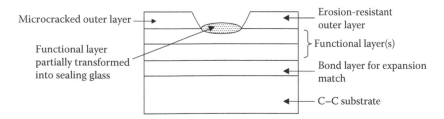

FIGURE 7.11
Schematic of a multilayer thermal protection system. (Reprinted from Warren, R., Carbon/carbon, cement, and ceramic matrix composites, in *Comprehensive Composite Materials*, Vol. 4, eds. A. Kelly and C. Zweben, Elsevier Science Ltd., Oxford, U.K., p. 419, 2000. With permission.)

SiC coatings formed by CVD process are normally used as surface coatings. The thermal expansion mismatch of bulk C/C and SiC coatings leads to the formation of thermally induced cracks during cooling down from CVD temperature to room temperature. An increase in SiC layer thickness results in a decrease of the crack distance, which means increased crack densities. Oxygen can diffuse through these cracks to cause oxidation. However, these cracks can be almost completely closed by heating the samples to the CVD temperatures. Therefore, the oxidation rate of surface-coated C/C composites depends not only on physical parameters such as reaction kinetics and diffusion rates but also on crack formation and closing effects, which are influenced by the coating parameters and the heat treatment temperature of the coating. It has been reported that the oxidation rate of surface-coated C/C composite is maximum at about 800°C (Weiss and Lauer 2004). This maximum oxidation rate is caused by the high ratio of internal surface area to external surface area available for oxidation (by crack formation) and highest oxidation kinetics. For long-term applications, therefore, imparting self-healing effects would be beneficial to close these cracks. The self-healing capacity can be tailored according to the oxidation temperatures and atmospheres.

SiC forms a silica layer on oxidation, with a parabolic growth of the layer thickness over a period of time. Since the SiO_2 layers possess the lowest diffusion rate of oxygen, SiC is the most favored oxidation protection system. An additional advantage of the formation of silica is that it can also induce self-healing because of volume increase during oxidation. One limitation of such a layer is the fact that a completely covered SiO_2 layer can be obtained only at temperatures above 1000°C, whereas at lower oxidation temperatures (700°C–1000°C) only isles of SiO_2 are formed. Hence, glass formers are needed for low-temperature protection, and the boron-derived additives are the common glass formers. These boride additives can be incorporated in the bulk C/C composite itself. Another possibility is to include the boron oxide former, such as B_4C in the multilayer surface coating.

To prevent oxygen reaching the substrate and to be capable of sealing cracks formed during service, the optimum solution appears to be the introduction of functionally active layers. Glasses have been considered for such use for many years. The use of borate glasses has been studied extensively because the viscosity of B_2O_3 in the temperature range of 600°C–1100°C is more appropriate and it has a tendency to wet SiC and Si_3N_4. However, the usefulness of borate glasses is limited by their tendency to vaporize above 1000°C. Moisture sensitivity and volatilization can be reduced by increasing the viscosity of borate glasses by adding a refractory oxide up to 25%. For example, ZrO_2 or HfO_2 may be added, which is useful in the temperature range of 1200°C–1600°C. To provide protection at temperatures, >1100°C silica glass may also be employed. The diffusion rate of oxygen through SiO_2 layer depends on the layer thickness, which increases with increasing oxidation temperature and time. The diffusivity of oxygen in SiO_2 is seven orders of magnitude lower to than that in B_2O_3, resulting in a superior oxidation protection. Nevertheless, sealing based on silica glass have its own limitations. The first one is the incompatibility with the SiC layer underneath. Because of this, the thermal shock resistance of SiO_2 is limited. The thermal shock resistance decreases with increasing glass layer thickness, and spallations are very common when the layer thicknesses are more than 8 µm. Moreover, the lifetime of the glass sealing is reduced by the crystallization occurring on high-temperature exposure. The viscosity of SiO_2 is one or more order of magnitude higher than B_2O_3, which also reduces the sealing efficiency. A nonstoichiometric SiC coating containing an excessive amount of Si leads to the formation of pores in the SiO_2 glass sealing, due to the following chemical reaction:

$$Si_{(s)} + SiO_{2(s)} \rightarrow SiO_{(g)} \tag{7.7}$$

The SiO_2 glass sealing is thermally unstable, at temperatures above 1800°C. A further disadvantage of SiO_2 is poor corrosion resistance against water vapor with high flow rates, which is encountered in gas turbine applications. Silica glass sealing can be replaced by $MoSi_2$-based glass sealing, which has a lower viscosity and so a better sealing effect. For high-temperature applications above 1800°C, different high-temperature refractory oxides such as ZrO_2 or Y_2O_3 have to be applied. However, oxygen diffusivity in these oxides is some orders of magnitude higher than SiO_2 glass sealing. As mentioned earlier, silica glass viscosity is too high at temperatures below 1100°C to close cracks effectively. To overcome the deficiencies of these systems, more complex systems such as TiO_2–SiO_2–B_2O_3 (Gray 1988) and P_2O_5–SiO_2–Al_2O_3 have been developed (Tawil et al. 1993).

A more recent approach is to use functionally active layers, which form glass on oxidation. The main advantage of such layers is that they actively absorb oxygen. Several boron- and silicon-containing compounds in various combinations have been studied as the functionally active layers. These include B_4C, TiB_2, and $MoSi_2$, although the latter is prone to spallation, known as $MoSi_2$ pest, and undergoes ductile to brittle transformation at ~1000°C. Boron carbide is able to form boron oxide at relatively low temperatures and that can close the cracks in the multilayer coating. In principle, borides can be used up to 1600°C, but the boron oxide is neither stable under water flow nor thermally stable at temperatures above 1200°C–1250°C (due to the evaporation of borosilicates). Therefore, additional sealing layers with superior high-temperature properties are needed.

Bond layers are used primarily to accommodate the thermal expansion mismatch between the composite substrate and functional layers but also to prevent the outward diffusion of carbon from the substrate. SiC and Si_3N_4 are the most commonly used bond coat materials, and they can be applied either by CVD or using a slurry technique. In principle, the lower CTE of Si_3N_4 would be an advantage for bond coat, but it has been found that there is no significant improvement over SiC.

The emphasis of thermal protection system has now shifted to the development of new multilayer systems, understanding the principles of thermal protection and predicting the behavior. An early notable success of thermal protection was the protection of nose cone and leading edges of the space shuttle. The coating was formed by encapsulating the C/C composite part in a powder pack comprising SiC, silicon, and Al_2O_3 and then heated to 1750°C–1850°C for 4–7 h (Shuford 1984a,b). A coating of thickness 125–170 μm was produced by this process. The compounds present in the powder mixture reacted at high temperature and produced reactive species, which then reacted with C in the substrate and formed SiC. The following reactions occurred during the formation of coating:

$$4Al_2O_3 + 4SiC + Si \rightarrow Al_4C_3 + 5SiO + CO \tag{7.8}$$

$$SiO + 2C \rightarrow SiC + CO \tag{7.9}$$

The cracks and pores were filled with a mixture of glazes derived from tetraethyl orthosilicate, SiC powder, Al_2O_3 powder, and aluminum phosphate. This approach is based on the dramatic improvement of the protective properties of refractory coatings by covering with a glassy layer of a B_2O_3-based glass, SiO_2, or more complex systems such as a borate glass containing $ZrSiO_4$ particles.

Several multilayer protection systems with a number of combinations have been developed. A complex four-layer system has been patented (Bentson et al. 1989), which consists of an inner sealant layer comprising a boron-rich layer with a zirconia source, an outer sealant layer consisting precursors to form a complex borate glass, and a granular refractory material (A preferred composition on weight basis is 30% B_4C, 5% SiO_2, 15% Li_2ZrO_3, 30% SiC, and 20% pitch), an inner coating of B_4C (5–25 μm thick), and an outer coating of SiC to a thickness of 100–300 μm applied by CVD. This coating system was found to be successful on a C/C composite thermally cycled up to 1460°C.

Another multilayer coating system comprises an initial boride sealant layer applied by slurry painting or CVD, middle oxygen barrier layer (stoichiometric or siliconized SiC) applied by CVD, and a final borosilicate overglaze layer (Dietrich 1991). A multilayer system has been developed for C/SiC composites, which is also applicable to C/C composites (Goujard et al. 1994). It consists of three layers deposited by CVD, which are an inner SiC layer (120–140 μm thick), a thin B_4C middle layer (10–15 μm thick), and a SiC outer layer (40–60 μm thick). In another system, the inner layer is SiC, the intermediate layer is AIN, and the outer layer is Al_2O_3. It has been suggested that alternative materials such as HfO_2 or ZrO_2 for the outer layer and TiB_2, HfN, ZrC, Pt, or Ir for the intermediate layer can also be used.

An idealized four-layer coating system has been proposed for protecting C/C composites used above 1800°C. It consists of an inner layer of refractory carbide, such as TaC, TiC, HfC, or ZrC. Above this layer is an inner refractory oxide layer, a modified silica glass layer, and an outer layer of refractory oxide. Candidate materials for the refractory oxide layers include ZrO_2, HfO_2, Y_2O_3, and ThO_2.

A method for introducing an additional protection to C/C composites is to incorporate oxygen inhibitors or getters into the carbon matrix during lay-up or densification cycles. The most successful of these inhibitors is boron, although silicon and titanium compounds such as SiC, Ti_5Si_3, and TiB_2 can also be used. While these reduce the reactivity of the C/C composites in air, they become effective only after a significant fraction of the composite is gasified.

7.7 Applications of C/C Composites

C/C composites developed to meet the needs of the space program are nowadays considered as high-performance engineering materials with potential applications in many non-aerospace industries. Accordingly, the applications of these materials are also steadily growing in other engineering sectors. Nevertheless, the main applications of C/C composites in terms of mass consumption are still in aerospace sector as high-performance brake systems in aircrafts. New innovations in these materials still owe to requirements from aerospace industries. In other engineering sectors, C/C composites are used as engine components, refractory materials, hot-press dies and heating elements, high-temperature fasteners, liners and protection tubes, and guides in glass industries. Major applications of C/C composites involve exposure to high temperatures, for example, heat shields of reentry vehicles, aircraft brakes, hot-pressing dies, and rocket nozzles. Hence, the potential applications of C/C composites may call for use at temperatures exceeding 1000°C and even approaching 2200°C for times ranging from a few hours to a few thousand hours.

C/C composite brakes are being used in some aircrafts and automobiles. C/C composite brake systems offer significant weight savings and increase in service life over steel and cermet brakes for both military and commercial aircrafts. Carbon with a density of 2.26 g cm^{-3} and specific heat nearly two and a half times that of steel offers a 60% weight saving for a similar brake. For Concorde aircraft about 600 kg is saved by using C/C composite brake. The number of landings has increased by two to four times as compared to cermet/steel brakes because of the low wear rate of C/C composites.

The properties required for brake disk friction materials are

1. Good strength and impact resistance
2. Adequate and consistent friction characteristics
3. Low wear rate
4. High thermal conductivity

C/C composites meet all these requirements and they are superior to high strength graphite for brake applications because of their higher strength and toughness but with almost the same density and thermal conductivity.

The aircraft brake design involves the following considerations: Brake disks used in aircrafts are required to provide the frictional torque to stop the aircraft quickly and to absorb the several hundred mega joules of heat generated while applying brake. The aircraft brake system consists of multiple brake disks having alternate rotors and stators forced against each other by hydraulic pressure. Friction between the rotating and stationary disks causes them to heat up to 1500°C and the surface temperature can reach as high as 3000°C; hence, good thermal shock resistance is also required. With its high thermal conductivity and low CTE, the C/C composite is an ideal material to withstand that thermal shock conditions. In view of these requirements, a brake material must be a good structural material with good mechanical and thermal properties, an efficient heat sink, and have excellent abrasion resistance.

Consider a brake system for a Boeing 767 aircraft. The aircraft has a mass about 170,000 kg and the usual takeoff velocity is about 320 km h^{-1}. This will give a kinetic energy during takeoff as 670 MJ. In the event of an aborted takeoff, this energy must be dissipated in about 30 s by the eight brakes in the aircraft. An aborted takeoff is, indeed, a worst scenario, but the braking material must be able to withstand such conditions, and C/C composite brakes can meet these requirements. Using C/C composite brake systems in Boeing 767 airliner reduces weight by 400 kg over the conventional metal brake systems and the increased durability of C/C composites permits 3000 aircraft landings, which is almost double compared to the landings possible with metal rotor brake systems.

Out of the total quantity of C/C composites produced in the world, about 60% by volume is used in aircraft braking systems. Even though the cost of C/C composite brakes was high initially, the advances in processing technology since 1970s have now reduced the cost of these materials from \$350 to \$65 kg^{-1}. Therefore, now it has become commercially advantageous to employ C/C composite brakes in civil aircrafts, and they are now being used in Boeing 747-400, 757, 767, and 777 airliners and all the Airbus family aircrafts.

The aerospace field continues to be the primary area of application for the C/C composites. In addition to the leading edges of space shuttle, C/C composites have been used in solid propellant rocket nozzles and exit cones and as ablative nose tips and heat shields for reentry vehicles. Since on an average a rocket motor fuel burns for about 30 s, the property demands for nozzles and exit cones are short lived but intense. Hence, oxidation is generally not a problem in this application, and a controlled ablation should be built into the material structure. Dense C/C composite is preferred for this application because of its superior ablation resistance. Initially, exit cones were made with 2-D fiber architecture, but after the development of 3-D weave technology, the exit cones are made with 3-D fiber architecture. Although the space shuttle has the best-known C/C composite reentry heat shield, the greatest number of C/C parts (11% by volume or 37% by value) is used in the nose cones of ballistic missiles. For the level of thermal and mechanical stresses generated during reentry, 3-D C/C composite is the most appropriate material. The major advantages of C/C composites are the high thermal conductivity, which avoids surface cracking caused by thermomechanical overload and the high heat capacity, which means that the component can effectively operate as a heat sink.

The use of C/C composites for reentry vehicle heat shield was successfully demonstrated in 1971, and the satisfactory performance of C/C composites in missile nose tip applications was reported at about the same time. They are also being considered for hypersonic aircraft parts such as the leading edges and for gas turbine components.

The applications in advanced rocket propulsion systems include nozzles, thrust chambers, and ramjet combustion liners. The advantages of C/C composites for these applications are simpler designs and reduced weight and volume. A total weight saving of 30%–50% has been predicted, if a rocket nozzle is completely made using C/C composites.

Some non-aerospace applications for C/C composites are beginning to develop. C/C composites are also being used in Formula 1 race cars as brakes and clutches. The capacity to dissipate large quantities of heat has led to their use in high-speed passenger train brake systems. However, it is unlikely that C/C composites would be used in commercial and personal automobiles in the near future because of their high costs.

Various C/C composites have been evaluated as bone replacement. Carbon is the best biocompatible material among the known biomaterials and is compatible with human bone, blood, and soft tissue. With this excellent biocompatibility and the ability to tailor the modulus to match that of bone, C/C composites have become an attractive material of choice for medical implants. However, the medical applications of these materials are limited due to the long elapsed time to obtain license for using such materials and the availability of thermoplastic matrix composites at a cheaper rate with good workability.

Other applications of C/C composites include hot-press dies, furnace heating elements, kiln furniture, and "gob" interceptors used in glass-making. C/C hot pressing dies are now commercially produced, and such dies can withstand higher pressure and offer longer life than monolithic graphite die. Future applications that are being actively researched are hypersonic vehicle airframe structures, space structures, and engine components for gas turbines. C/C composites also have great potential in energy sectors as polar plates for fuel cells and storage batteries. Other potential applications include high-temperature ducting systems, components for nuclear reactors, electrical contacts, seals and bearings, and components for advanced turbine engines.

C/C composite tiles are being used in the first wall of fusion reactors as multidirectional first-wall bumper limiter tiles, RF limiter tiles, and first-wall diverters. Some of these are required to withstand a continuous temperature of 2200°C with occasional spikes to 3300°C. The next generation plasma fusion reactors will also require advanced C/C composites possessing very high thermal conductivities to cope with the anticipated severe heat loads.

C/C composites have great potential as high-performance engineering materials. However, the high cost and the availability of limited database have restricted their use. Another limitation is that their susceptibility to oxidation at high temperatures. Ceramic coatings that protect C/C composites from oxidation for hundreds of hours at temperatures up to 1400°C and for short durations up to 1750°C have been developed. As the production techniques become refined and scaled up, C/C composites will become more economically acceptable materials for many applications. The availability of low-cost carbon fiber will also have a major impact on the future of C/C composite products. As the technology becomes more economical and viable, more and more applications for C/C composites can evolve.

Questions

State whether the following statements are true or false and give reasons:

7.1 SiC may not be a suitable material for surface coating on C/C composites to prevent low-temperature oxidation.

7.2 Pitch infiltration and pyrolysis followed by CVI is an economic option to prepare C/C composites.

7.3 An intermediate high-temperature treatment during pitch infiltration and pyrolysis process improves the density.

7.4 The strength of C/C composites is better when produced using thermosetting resin.

7.5 C/C composites produced using pitch have better electrical conductivity.

7.6 C/C composites are better materials for nuclear applications.

Answer the following questions:

7.7 Describe the processing methods and properties of C/C composites.

7.8 Compare the pitch infiltration process and CVI process to produce C/C composites.

7.9 What is the main limitation of C/C composites?

7.10 What are the materials added to C/C composites for bulk oxidation protection.

7.11 What are the layers in a surface coating of C/C composites used for high-temperature applications? Explain the role of each layer and mention the preferred material for each layer.

7.12 What are the advantages of using C/C composite aircraft brakes?

References

Bentson, L.D., R.J. Price, M.J. Purdy, and E.R. Stover. 1989. US Pat. 5 298 311.

Dietrich, H. 1991. *Mater. Eng.* 108(8): 32.

Economy, J., I.L. Jungand, and T. Gogeva. 1992. *Carbon* 30: 81.

Eliezer, R. 2003. Coating of graphite and C/C composites via reaction with Cr powder. MSc dissertation, Technion, Haifa, Israel.

Fitzer, E. 1987. *Carbon* 25: 163.

Fitzer, E. and W. Huttner. 1981. *J. Phys. D*, 14: 347.

Goujard, S., L. Vandenbulcke, and H. Tawil. 1994. *J. Mater. Sci.* 29: 6212.

Granoff, B., H.O. Piersonand, and D.M. Schuster. 1973. *Carbon* 11: 177.

Gray, P.E. 1988. US Pat. 4 894 286.

Hill, J., C.R. Thomas, and E.J. Walker. 1974. In *Carbon fibres: their place in modern technology*, ed. P.C. Oliver. London, U.K.: The Plastics Institute, p. 122.

Hishiyama, Y., M. Inagaki, S. Kimura, and S. Yamada. 1974. *Carbon* 12: 249.

Kamiya, K. and M. Inagaki. 1981. *Carbon* 19: 45.

Krenkel, W. 2008a. *Ceramic Matrix Composites*. Weinheim, Germany: Wiley-VCH, p. 81.

Krenkel, W. 2008b. *Ceramic Matrix Composites*, Weinheim, Germany: Wiley-VCH, p. 84.

Manocha, L.M. 1986. In *Carbon 86*. Baden, Germany: Deutsch Kenamisch gesellschaft, p. 671.

Manocha, L.M. and O.P. Bahl. 1988. *Carbon* 26: 13.

Murdie, N., J. Don, W. Kowbel, P. Shpik, and M.A. Wright. 1988. In *International SAMPE Technical Conference Series*, pp. 73–82. Covina, CA: SAMPE.

Oh, S.-M. and J.-Y. Lee. 1988a. *Carbon* 26: 769.

Oh, S.-M. and J.-Y. Lee. 1988b. *Carbon* 26: 763.

Rand, B. 1993. In *Essentials of Carbon-Carbon Composites*. ed. G.R. Thomas, p. 164. London, U.K.: The Royal Society of Chemistry

Rovillain, D., M. Trinquecoste, E. Bruneton, A. Derre, P. David, and P. Delhaes. 2001. *Carbon* 39: 1355.

Sheehan, J.E. et al. 1994. *Annu. Rev. Mater. Sci.* 24: 19.

Shuford, D.M. 1984a. US Pat. 4 465 777.

Shuford, D.M. 1984b. US Pat. 4 471 023.

Stover, E.R., J.F. D'Andrea, P.N. Bolinger, and J.J. Gebhardt. 1977. In *Extended Abstracts of the 13th Conference on Carbon*, pp. 166–167. Irvine, CA: American Carbon Society.

Tawil, H., X. Bernard, and J.-C. Cavalier. 1993. UK Pat. WO93/13033.

Thomas, C.R. and E.J. Walker. 1978. *High Temp-High Press.* 10: 79

Weiss, R. and A. Lauer. 2004. Project report on Gradierte CVD- und PIRAC multibeschichtungen auf C/C als Korrosions-und oxidationschutz durch innovative hoch temperatur prozesse, FKZ 03N5039B.

Whittaker, A.J. and R. Taylor. 1990. *Proc. R. Soc. London* A431: 199.
Whittaker, A.J., R. Taylor, and H. Tawil. 1990. *Proc. R. Soc. London* 430: 167.
Yasuda, E., S. Kimura, and Y. Silsusa. 1980. *Trans. Jpn. Soc. Compos. Mater.* 6: 14.

Bibliography

ASM Handbook. 2001. *Composites*, Vol. 21. Materials Park, OH: ASM International.

Chawla, K.K. 2003. *Ceramic Matrix Composites.* Boston, MA: Kluwer Academic Publishers.

Chawla, K.K. 2012. *Composite Materials: Science and Engineering*, 3rd edn. New York: Springer-Verlag, Inc.

Krenkel, W. 2008. *Ceramic Matrix Composites.* Weinheim, Germany: Wiley-VCH.

Mallick, P.K. 2008. *Fiber-Reinforced Composites*, 3rd edn. Boca Raton, FL: CRC Press.

Manocha, L.M. 2003. High performance carbon–carbon composites, *Sadhana* 28: 349.

Schwartz, M.M. 1997. *Composite Materials*, Vol. II. Upper Saddle River, NJ: Prentice Hall.

Warren, R. 2000. Carbon/carbon, cement, and ceramic matrix composites, Vol. 4, in *Comprehensive Composite Materials*, eds. A. Kelly and C. Zweben. Oxford, U.K.: Elsevier Science Ltd.

8

Nanocomposites

Nanomaterials and, in particular, nano-reinforcements have in recent years been the subject of intense research, development, and commercialization. The nanomaterials not only have very small physical dimensions but also exhibit some unusual properties by virtue of their small size. The use of nanomaterials makes it possible to design and create new materials with unprecedented flexibility and improvements in their physical properties. The ability to tailor properties by using heterogeneous nanomaterials has been demonstrated in several fields. The most convincing examples of such designs are natural structures, such as bone (a hierarchical nanocomposite built from ceramic platelets and organic binders). Since the constituents of a nanocomposite have different chemical compositions and structures and hence properties, they can serve various functions. The nanocomposites are a class of composites, containing a material having at least one dimension below 100 nm, wherein the small size offers some level of controllable, enhanced performance that is different from the micro-/macrocomposites. Taking some clues from nature and based on the demands that emerging technologies put on building new materials to satisfy several functions at the same time for many applications, scientists have been devising synthetic strategies for producing high-performance nanocomposites. These strategies have clear advantages over those used to produce homogeneous micro-grained materials. Besides, the push for nanocomposites comes from the fact that they offer useful new properties compared to conventional composite materials.

The challenge to produce stronger, tougher, and lightweight materials is driven by demands for property improvements at an affordable cost. The main advantage of nanocomposites is that it is possible to achieve improved properties at significantly lower filler loading levels and hence the cost may not increase substantially.

A tire made of rubber compounded with carbon black is one of the earliest primitive nanocomposites. As early as the 1860s, the ability of carbon black to improve the mechanical properties of vulcanized rubber was recognized by researchers. By virtue of its high surface area and mechanical properties, carbon black is able to significantly enhance the properties of rubber. Other well-known nano-reinforcements available in the early twentieth century were fumed silica and precipitated calcium carbonate.

The development of nanoclay–polyamide system by Toyota in 1988 can be regarded as a key milestone in the modern nanocomposite era. Following this development, Toyota launched the first commercial product in 1993, made of polymer nanocomposite (PNC) that was nanoclay–polyamide 6 timing belt cover. From 2001, Toyota is producing body panels and bumpers using PNCs. Similarly, General Motors began using nanoclay–polymer composites from 2002 for step assists on its Safari and Chevrolet Astro models.

Nanocomposites can also be considered as solid structures with nanometer-scale repeat distances between the different phases that constitute the structure. They consist of two or more inorganic/organic phases in some combinatorial form. In general, nanocomposite materials can demonstrate different mechanical, electric, optical, and electrochemical properties than those of each individual component. Any specific property of the nanocomposite material is often more than the summation of the properties of the individual components, since the interfaces play an important role in controlling the overall properties. Due to the high surface area of nano-reinforcements, nanocomposites have very large interfacial area. Hence, the special properties of nanocomposites often arise from the interaction of its phases at the interfaces. In contrast, the interfaces in conventional composites constitute a much smaller volume fraction of the bulk materials.

Both simple and complex approaches to prepare nanocomposites exist. A practical nanocomposite system, such as supported catalysts consisting of metal nanoparticles placed on ceramic support, can be prepared simply by the evaporation of metal onto chosen substrate or dispersal through solvent chemistry. On the other hand, nanocomposite material such as bone, which has a complex hierarchical structure, is difficult to entirely duplicate by the existing processing techniques.

The notion is that there must be some advantage in using nanomaterials whether it is for the enhancement of mechanical properties or another desirable property. If the nanocomposite is to have identical properties to its macroscopic counterpart, then there is no point in developing nanocomposites. Many nanocomposites have been developed over the years. Although the full potential of many nanocomposites is not realized, the difficulties are being identified and the properties are getting improved.

There are many types of nanocomposites that have received significant research and development, which include polymer/polymer, ceramic/polymer, ceramic/metal, and ceramic/ceramic nanocomposites. The details of the common nanocomposites are described in the following sections.

8.1 Polymer Nanocomposites

Polymers reinforced with a second organic or inorganic phase are common in the production of modern plastics. PNCs are a new class of polymer composites consisting at least one nanomaterial. The most noteworthy effort in the last 25 years has demonstrated a doubling of the tensile modulus and strength without sacrificing impact resistance for nanocomposites containing as little as 2 vol.% nanometer level reinforcement. In addition, the heat distortion temperature (HDT) of the polymer matrix is increased up to 100°C, which extends the use of it to higher temperatures.

Besides their improved properties, these nanocomposite materials can be processed using the conventional polymer-processing equipment to near-net shape. Since high degrees of stiffness and strength are realized with far less quantity of inorganic material, they are much lighter compared to conventional polymer composites. The weight advantage has a significant impact on environmental concerns among many other potential benefits. In addition, their outstanding combination of barrier and mechanical properties may eliminate the need for a multilayer design in packaging materials.

Materials formed by combining inorganic nanomaterials and organic polymers offer properties that are representative of both components, and they have often been referred as hybrid materials. The traditional composite materials often exhibit localized heterogeneity that can lead to the loss of desired properties. For example, even with inorganic particles of the order of a cubic micron in size, the concept of true hybridization at the molecular level is not realized. With the advent of nano-reinforcement materials, there is a possibility of increasing the interaction between the organic and inorganic phases by several orders of magnitude. To put this statement in perspective, consider a nanomaterial of size 1 nm, and this is considerably smaller than any linear dimension associated with a polymer molecule. Hence, the interaction of nano-sized particles with a polymer molecule is truly a molecular-level interaction.

The nano-reinforcements are available as essentially isotropic to highly anisotropic needlelike or sheetlike elements. Uniform dispersion of these nanomaterials can lead to ultrahigh interfacial area between the constituents. To illustrate this, consider a 1 μm cube and the surface area is 6 μm^3 or 6 × 10^6 nm^2. If the cube is broken into cubes of 1 nm length, then the surface area of each cube is 6 nm^2. The total number of cubes generated would be 10^9. Therefore, the total surface area of 10^9 cubes would be 6 × 10^9 nm^2, representing an increase of the total surface area by a factor of 10^3. The factor for the real situation may be somewhat less than 10^3, but the important point is that the available surface area increases by orders of magnitude, as we move from a micron-sized material to an equivalent weight of nano-sized material. For example, interfacial area in a dispersion of layered silicates in polymers is about 700 m^2 cm^{-3}. This large interfacial area differentiates nanocomposites from conventional composites and filled plastics. The dominance of interfacial regions implies that the behavior of PNCs cannot be understood by simple scaling arguments that begin with the behavior of conventional polymer composites.

Three major characteristics that define and form the basis of PNC performance are nanoscopically confined matrix polymer, nanoscale inorganic constituents, and nanoscale arrangement of these constituents. The driver for current research is to develop the tools for processing and characterization, and theory to optimize and enable full exploitation of the potential of these unique characteristics.

The proliferation of reinforcement–polymer matrix interfaces means the majority of polymer chains reside near the reinforcement surface. Since an interface limits the conformations that a polymer molecule can adopt, the free energy of the polymer in this region is basically different from those polymer molecules far away from the interface. Thus, in PNCs with only a small amount of nanoparticles, the entire polymer matrix may be considered as nanoscopically confined interfacial polymer. The restrictions in chain conformations will affect molecular mobility, relaxation behavior, and thermal transitions such as glass transition temperature of the polymer. Still more complications arise in semicrystalline and mesostructured liquid-crystalline polymers, since the interface alters the degree of ordering and packing that affects crystallite growth structure and organization.

The second major characteristic of PNCs is related to the dimensions of the nano-reinforcements. When the dimensions of the particle approach the fundamental length scale of a physical property, new mechanical, optical, and electric properties arise, which are not present in the macroscopic counterpart. Dispersions of nanomaterials exhibiting unique properties create bulk materials with the properties dominated by that of nanomaterials.

Finally, as with any composite, the arrangement of constituents, especially the reinforcements, determines the nanocomposite behavior. Conceptually, the spatial ordering of spherical, rodlike, or platelike nanoparticles with varying degrees of orientation order (if any) will manifest a wide variety of systems. Thus, the final properties of a PNC will depend as much on the individual properties of the constituents as on the relative arrangements and subsequent synergy between the constituents. Ultimately, PNCs offer the possibility of developing a new class of materials with their own manifold of structure–property relationships.

Several benefits of PNCs have been identified that allow them to compete with traditional materials, which include

- Enhancement of stiffness and strength with minimal loss of ductility and impact strength
- Better thermal stability
- Better flame retardancy
- Better barrier properties
- Good abrasion resistance
- Reduced shrinkage and residual stress
- Tailored electric, electronic, and optical properties

Enhancements in mechanical properties, thermal stability, and barrier properties make these materials prime candidates for packaging and automotive applications.

The advent and increasing availability of nanomaterials offers the potential of developing new polymer composite materials with hitherto unseen properties. A short list of potential nano-reinforcements for structural nanocomposites includes carbon nanotubes (CNTs), clay layers, and metal oxide, nitride, and carbide clusters. Nevertheless, the development of these nanocomposites has been frustrated because of the difficulties in incorporation and uniform dispersion of these materials into polymer media.

The CNTs offer the prospect of significantly increasing the mechanical properties of matrix materials. Initial attempts on the fabrication of PNCs with CNTs, however, did not result in the expected level of performance. Indeed, the nanocomposite properties were often inferior to that of the polymer matrix, and the reasons could be due the problems at the interface as well as agglomeration of the CNTs. A proper bonding of polymer matrix with CNTs could improve the properties. However, the surface energy of CNTs is usually very low, and the functionalization of the surface may provide a means to improve bonding, although the functional groups would presumably damage the graphene lattice. Consequently, the mechanical properties of CNT-based nanocomposites have not been proven exceptional as on the first decade of the twenty-first century. Although the prohibitively high cost and poor mechanical properties may not permit the use of CNT-based nanocomposites for any structural applications at present, the other properties may be of interest in some other applications, for example, electrically and thermally conductive polymers. Imparting some electric conductivity to polymers can render them suitable for electrostatic paint spraying, resulting in a less-expensive and an environment friendly process due to less wastage of paint and the elimination of the need for a primer coat. The addition of short, conductive microfibers may not be suitable for this purpose because of the resulting poor surface finish. Since surface finish is often very important for automotive applications, CNTs may be a suitable choice. Hyperion Catalysis, Inc., brought CNT composites to commercial status by producing CNT-based paints for electrostatic spraying, and Ford Motor Company introduced this multiwalled carbon nanotube (MWCNT) nanocomposites in mirror housings on the 1998 Ford Taurus.

It is reasonable to expect that there would be significant increase in the interfacial drag upon incorporation of nanomaterials into polymer media, resulting in the difficulties in mixing. This is, in fact, what is observed when precursor clay is added to a polymer matrix. As the clay is wet out and separate into tactoids (nanolayers), there is an increase in viscosity. As these tactoids separate into individual platelets (exfoliation), the viscosity can increase by orders of magnitude.

Chemical surface modification of conventional reinforcements to achieve compatibility with a polymer matrix is well documented. The same concepts can be applied to nano-reinforcements, but there are some challenges that are unique to these materials because of high surface area. It is necessary to look for surface chemical reactions that are thermodynamically and kinetically favored in order to accomplish functionalization.

Out of the PNCs, only layered clay mineral and CNT-based nanocomposites have attained commercial significance. The layered clay minerals impart improved mechanical properties, fire resistance, and barrier properties to the polymers, and these composites are widely used in packaging applications. The CNTs impart electric conductivity, in addition to improved mechanical properties to the polymers, and these composites have been commercially exploited to assist in the electrostatic painting of automotive components. These two types of nanocomposites comprise almost all of the commercial activity to date on PNCs.

8.1.1 Clay–Polymer Nanocomposites

Clay–PNCs are the materials comprising nanometer-sized clay platelets that are dispersed in a polymer matrix. A clay mineral is a potential nanoscale additive for polymeric materials. It comprises silicate layers in which the fundamental unit is 1 nm thick platelets. Various organic molecules can enter into the layers (intercalation). The intercalation causes an increase in the space between silicate layers, and the increase depends on the size of organic molecule. The term nanoclay is often loosely used to mean a nanoclay precursor, which under appropriate treatment will delaminate into the individual constituent plates. The term nanoclay should be applied to those platelets, which have at least one dimension in the nanoscale. In the case of montmorillonite clay, the platelets are approximately 1 nm in thickness. The other dimensions are of the order of 150 nm and sometimes as large as 1 or 2 μm.

8.1.1.1 Types of Clay Composite Structures

Depending on the level of interfacial interactions between polymer matrix and layered silicates, four types of clay–polymer composites can form (Figure 8.1). They are conventional composites, intercalated nanocomposites, exfoliated nanocomposites, and intermediate nanocomposites:

Conventional composites: In the conventional polymer composite, the layered silicates (clays) act as conventional, micron-sized fillers.

Intercalated nanocomposites: Intercalated nanocomposites are formed by the insertion of polymer chains in between the silicate layers while maintaining their regular arrangement in galleries.

Exfoliated nanocomposite: In the exfoliated clay polymeric nanocomposites, the layered clay platelets are delaminated and dispersed as single platelets in the polymer matrix.

Intermediate nanocomposites: Clay–PNCs, in which the clays are partially intercalated and partially exfoliated, are intermediate nanocomposites. This type of nanocomposite is often formed.

The ordered structure of layered silicate is lost during the exfoliated nanocomposite formation, and the average distance between the exfoliated layers

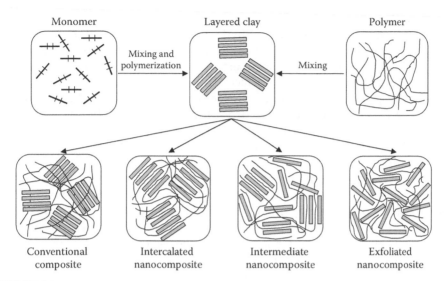

FIGURE 8.1
Four types of clay–polymer composites.

depends on clay loading. In the fully exfoliated state, the individual clay platelets will have the highest aspect ratio possible, and hence the highest improvement in barrier properties is expected. Moreover, exfoliating the clay into individual platelets leads to the smallest particle size, thereby providing the maximum clarity to the polymer films. In addition, exfoliated clay–PNCs have higher strength, stiffness, dimensional stability, and heat resistance than the corresponding unfilled polymer. Therefore, nanocomposite technology has the potential to expand the use of polymers into a wide variety of applications including those where higher barrier properties are needed. It may be possible to calculate the effective aspect ratio from the permeability measurement. Typical values of effective aspect ratios found for amorphous, unoriented nanocomposites are 150–200. Hence, the effective aspect ratio has proven to be good indicator of the extent of clay exfoliation.

Achieving high degree of exfoliation requires the selection and optimization of many variables such as the process of incorporating clay, the choice of clay, the clay treatment, and the optional use of dispersing aids. Achieving exfoliation of the clay by balancing many variables requires a significant amount of research. Natural Na-montmorillonite clay is hydrophilic and hence not compatible with most organic molecules. The sodium cations present in the interlayer space of montmorillonite can be exchanged with organic cations to yield organophilic (hydrophobic) montmorillonite.

8.1.1.2 Clay Surface Modification

Clay minerals usually contain hydrated sodium or potassium ions, which render them hydrophilic. Hence, the inorganic clays are only miscible with hydrophilic polymers, such as polyethylene oxide or polyvinyl alcohol (PVA). Mixing these clays with a hydrophobic polymer is like mixing oil with water. In order to render the silicate layers miscible with hydrophobic polymers, it is essential to convert the normally hydrophilic surface to a hydrophobic (organophilic) surface. This conversion can be achieved by the exchange of metallic cations by organic cations. Usually, the exchangeable organic cations contain a hydrophilic group (which bonds with clay surface) and an organophilic group (which is compatible with the polymers). These organic compounds are known as coupling agents or compatibilizing agents.

The first compatibilizing agents used in the preparation of clay–PNCs were amino acids (Usuki et al. 1993a). After this, numerous compatibilizing agents have been used for the preparation of nanocomposites; among them, the most popular are alkylammonium ions because they can easily exchange the ions situated between the clay layers. Silanes can also be used because of their ability to react with the hydroxyl groups present at the surface and edges of the clay layers.

Amino acids are molecules containing a basic amino group ($-NH_2$) and an acidic carboxyl group ($-COOH$). In an acidic medium, a proton from the $-COOH$ group is transferred to the $-NH_2$ group. This $-NH_3^+$ cation can exchange the cations (Na^+, K^+) present between the clay layers so that the clay becomes organophilic. A wide range of ω-amino acids $\left(COOH-(CH_2)_{n-1}-NH_3^+\right)$ have been used to exchange the interlayer cations of montmorillonite clay. Amino acids were widely used in the preparation of clay–polyamide 6 nanocomposites, because the acid functional group has the ability to polymerize caprolactam (Usuki et al. 1993a). This interlayer polymerization delaminates the clay layers and favors the formation of an exfoliated nanocomposite.

Alkylammonium ions are better alternatives to amino acids for the preparation of clay–PNCs, since they can be easily intercalated between the clay layers. Their basic formula is $CH_3-(CH_2)_n-NH_3^+$, where n varies between 1 and 18. The length of the ammonium ions (n values) has a strong influence on the final structure of nanocomposites. It is found that the alkylammonium ions containing more than eight carbon atoms are favoring the formation of exfoliated nanocomposites, whereas shorter alkylammonium ions are leading to the formation of intercalated nanocomposites.

Maharaphan et al. (2001) modified sodium-activated montmorillonite clay using dodecylamine ($CH_3(CH_2)_{11}NH_2$). The dodecylamine is dissolved to form dodecylamine hydrochloride ($CH_3(CH_2)_{11}NH_2$ HCl) (in concentrated hydrochloric acid medium), which ionized to dodecylammonium ions $\left(CH_3(CH_2)_{11}NH_3^+\right)$. The sodium-activated montmorillonite clay is dispersed in the solution containing dodecylammonium ions, to exchange the sodium ions in montmorillonite with dodecylammonium ions, and thus the hydrophilic clay is converted into organophilic clay (organomodified montmorillonite clay [OMMT]). During this process, the spacing between the clay layers is increased to about 10 Å. Alkylammonium ions can also lower the surface charge of clay, and hence organic species with different polarities can enter into the layers.

In order to make the clay amenable for exchange with organic cations, the clay should be sodium activated first. Sodium activation can be carried out by the following procedure (Kornmann and Berglund 1998). Twenty-five grams of bentonite clay is dispersed in 2 L of 1 M sodium chloride solution and stirred overnight to allow the exchange of natural cations by sodium ions. The suspension is centrifuged for 20 min at 10,000 rpm, and the sedimentation is redispersed in deionized water to remove the excess sodium ions. The filtered and dried clay is ready for organic modification.

The sodium-activated clay is treated with dodecylamine to get organoclay (OMC) as reported elsewhere (Maharaphan et al. 2001). The detailed steps involved in the treatment are shown in Figure 8.2. The x-ray diffraction (XRD) patterns of the unmodified and modified clay are shown in Figure 8.3.

Expanded OMCs: To expand the basal spacing of OMCs, another compound can be incorporated. Many organic, oligomeric, and polymeric materials are available for this purpose. Further expansion of clay layers will improve exfoliation during melt compounding with polymers. This provides yet

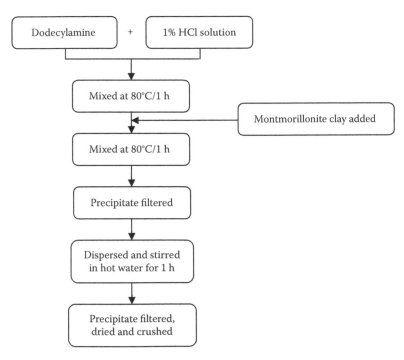

FIGURE 8.2
Flow diagram for the treatment of clay to make it organophilic.

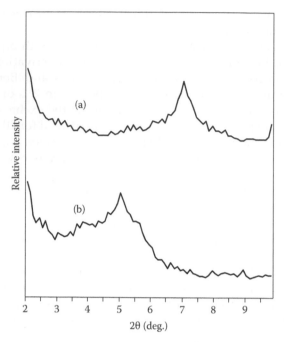

FIGURE 8.3
XRD patterns of the (a) unmodified clay and (b) modified clay. (Reprinted from Jawahar, P. et al., *Wear*, 261, 835, 2006. With permission.)

another method of fine tuning a clay for exfoliation into a polymer matrix. The expanding agent can be incorporated by the following processes:

1. Dispersion of the OMC and expanding agent in water, organic solvents, or a mixture of solvents.
2. Melt compounding of the OMC and expanding agent, either simultaneous or subsequent melt compounding with polymer. (Basal spacing up to 4.5 nm has been achieved by using expanding agents, and these expanded OMCs have been found to be useful for the preparation of PNCs by melt compounding. Examples of expanding agents include carbinol terminated polydimethylsiloxane, polyethylene glycol distearate, polycaprolactone, and polyvinylpyrrolidone [PVP].)

8.1.1.3 Processing of Clay–Polymer Nanocomposites

There are several methods by which clay can be incorporated into a polymeric material. These methods are broadly classified into two categories, and they are in situ polymerization and melt compounding. In first method, the clay layers are dispersed in the monomer or monomer solution, and then polymerization is carried out. In the second method, the clay layers are incorporated into the molten polymer during the conventional processing of polymers, such as extrusion. Clay–PNCs can also be prepared by using polymer solution, in which the clay layers are dispersed in the polymer solution.

It is desirable to use melt compounding as a means of incorporating clays into polymers for several reasons. Firstly, many polymers are produced on a large scale by melt processing, and it is convenient to use these materials as they are currently produced. Secondly, melt compounding is a versatile process capable of producing a variety of formulations at different volume scales. Thirdly, the high-shear forces exerted by the melt compounding equipment may permit the incorporation of significantly higher concentrations of clay compared to that can be achieved by a commercial in situ polymerization process. The presence of exfoliated clay significantly increases the viscosity of polymerizing medium, and this may cause limitations for commercial-scale in situ processes, which typically use low-shear forces. Interestingly, it has been reported that the high polymerizing-medium viscosity becomes an advantage for the injection molding of nanocomposites, such as polyamide 6,6 nanocomposites, because they can be injection molded at faster cycle times by using higher injection speeds and pressures without flashing.

8.1.1.3.1 Monomer Intercalation Method

Polyamide 6 is a commonly used engineering polymer because of its good mechanical properties. The first clay–PNC synthesized was based on this matrix by monomer intercalation method. The insertion of ε-caprolactam into the clay gallery is difficult because of the lack of affinity between the silicate layers of the clay and the ε-caprolactam monomer. Organic modification of the clays will render them compatible with organic molecules. The Toyota group has chosen the ammonium cation of ω-amino acids as the exchange cations, since these acids can catalyze the ring opening polymerization of ε-caprolactam. Various ω-amino acids containing different numbers of carbon atoms have been tried, and basal spacing after swollen by ε-caprolactam is measured. It is found that the basal spacing of the montmorillonites swollen by ε-caprolactam is equal at 25°C and 100°C when the number of carbon atoms is less than 8. The basal spacing is equivalent to the sum of the molecular length of the ω-amino acid and silicate layer thickness. However, the basal spacing exceeded this sum at 100°C for montmorillonites modified with ω-amino acids having more number of carbon atoms. It has been suggested that for better swelling by ω-amino acids, the number of carbons should be more than 11 (Usuki et al. 1993b).

The conceptual approach to the monomer intercalation method is illustrated in Figure 8.4. The monomers enter into the layers of modified clay, and the polymerization of the monomers occurs within the layers, resulting in an expanded interlayer distance. When the separation exceeds certain limit, then the individual silicate layers are separated and homogeneously dispersed in the polymer matrix.

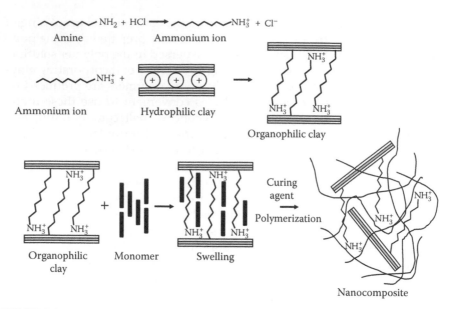

FIGURE 8.4
The conceptual approach to the monomer intercalation method to form clay–PNCs.

The Na-montmorillonite clay ion exchanged with 12-aminododecanoic acid allows the insertion of ε-caprolactam molecules into the clay gallery. After the intercalation of ε-caprolactam molecules into the gallery, the polymerization occurs to form polyamide 6. Gradually, the silicate layers are separated and uniformly dispersed in the polyamide 6 matrix. This nanocomposite showed significant improvements in mechanical, thermal, and gas barrier properties at clay loadings of only 2–5 wt.%. The HDT of nanocomposite containing 5 wt.% montmorillonite is found to be 152°C, which is 87°C higher than that of polyamide 6 (Kojima et al. 1993).

The following sequence of mechanism has been proposed for the exfoliation in clay–polyamide 6 nanocomposites (Okada et al. 1995):

1. Swelling of the OMC by caprolactam monomer by the intercalation of the monomer
2. Polymerization initiated by the addition of caprolactam to the end of the dodecanoic acid through the ammonium group
3. Swelling of the OMC by more caprolactam monomer
4. Polymerization of the caprolactam within the gallery accompanied by further intercalation of the monomer
5. Near complete consumption of the caprolactam with exfoliation

Similar mechanisms have been proposed for exfoliation of clay during caprolactone polymerization and epoxy polymerization. There are several implications if this proposed mechanism is generally applied to many polymer systems. For the swelling of the clay by monomer, it is necessary to select appropriate clay treatments, and if necessary low molecular weight dispersing aids should be used. The polymerization of the monomer in the gallery of the clay requires that the tether should participate in polymerization, as either an initiator or a monomer.

8.1.1.3.2 Monomer Modification Method

Acrylic resins are used in coating materials and paints, and they are synthesized by copolymerization of acrylic monomer. Acrylamide is one of the important monomers for water-based acrylic paints because of the good water solubility of quaternary ammonium salt of acrylamide. The quaternary ammonium salt (N-[3-(dimethylamino) propyl] acrylamide) can exchange the sodium cations of montmorillonite and bond to the silicate layers. This modified acrylamide is copolymerized with ethyl acrylate and acrylic acid by free-radical polymerization to form acrylic resin. A nanocomposite film formed by curing this acrylic resin is found to be transparent, and the gas permeability of the films decreased to about 50% by the addition of 3 wt.% clay (Usuki et al. 1995). A clay–polymethyl methacrylate nanocomposite was also synthesized using modified organophilic clay in the same manner (Biasci et al. 1994).

8.1.1.3.3 Covulcanization Method

Vulcanized rubbers are usually reinforced with carbon black to improve mechanical properties. Carbon black is an excellent reinforcement owing to its strong interaction with rubbers, but it often decreases the processability of rubber compounds at high loadings. Therefore, using silicate layers in rubber would be a better option, but rubbers are more hydrophobic than polyamide 6; consequently, it is more difficult to achieve dispersion of silicate layers in a rubber matrix with alkylammonium-treated montmorillonite.

Toyota group modified the montmorillonite clay with amine-terminated butadiene-acrylonitrile oligomer in a solvent mixture of N,N¹-dimethyl sulfoxide, ethanol, and water. This modified clay is blended with nitrile-butadiene rubber (NBR) by roll milling, and then the rubber is vulcanized with sulfur. The tensile stress at 100% elongation of nanocomposite containing 10 phr (parts per hundred parts of rubber) montmorillonite is found to be equal to that of the rubber containing 40 phr of carbon black. The permeability of hydrogen and water is found to be decreased by 70% in the clay–rubber nanocomposite.

8.1.1.3.4 Common Solvent Method

Polyimides are widely used in microelectronic applications because of their thermal and chemical stability and superior electrical properties. It would be desirable to reduce the thermal expansion coefficient, moisture absorption, and dielectric constant of polyamides, since these properties of currently available polyimides do not meet the requirement of advanced microelectronics.

The Toyota group has also reported the synthesis and properties of clay–polyimide nanocomposites. The montmorillonite clay is modified by dodecyl ammonium ion and dispersed homogeneously in dimethylacrylamide (DMAC) solvent. Polyamic acid, the precursor of polyimide, is also dissolved in DMAC. After polymerization reaction, the DMAC solvent is removed, and clay–polyimide nanocomposite is obtained.

It is found that the permeability of water vapor decreased markedly with increasing amount of montmorillonite in the nanocomposite. A 2 wt.% addition of clay brought the permeability coefficient of water vapor to a value less than half that of polyimide. However, the water absorption of polyimide is not decreased by the incorporation of nanoclay.

8.1.1.3.5 *Polymer Melt-Intercalation Method*

Polypropylene (PP) is one of the most widely used polyolefin. The performance level of this polymer can be improved by forming nanocomposites. However, there are many problems encountered during the preparation of PP nanocomposites. There is a difficulty in the intercalation of the monomer and then exfoliating the clay to a nanometer level by subsequent polymerization. Intercalation using a polymer solution is also difficult, because of the highly hydrophobic nature of the polymer. Hence, melt intercalation is considered for the fabrication of PP nanocomposites. A PP nanocomposite could be formed by blending PP with organophilic montmorillonite above the melting temperature of PP (about 160°C). However, the problem is that PP does not have any polar groups in its backbone chain and is one of the most hydrophobic polymers. Hence, the nano-level dispersion of silicate layers in a PP matrix was not realized even by using a montmorillonite clay treated with dioctadecyl dimethyl ammonium ion, in which the polar surfaces of the clay could be covered with one of the most hydrophobic ions. A solution to this problem has been found, that is, the use of a polyolefin oligomer having polar groups as a compatibilizer.

While preparing PP nanocomposites using a compatibilizer, two important factors in terms of the digomer should be considered. Firstly, the oligomer should contain a certain amount of polar groups to intercalate between silicate layers. Secondly, the oligomer should be miscible with PP. Since the presence of polar groups in the oligomer will affect the miscibility with PP, there must be an optimum number of polar groups in the compatibilizer. The Toyota group has examined two types of maleic anhydride-modified PP (MA-PP) oligomers containing different amounts of maleic anhydride groups to prepare a PP nanocomposite. A mixture of MA-PP and octadecylammonium-treated montmorillonite is prepared by melt processing. The MA-PP molecules intercalated into the silicate layers during this process. This mixture is then melt blended with PP at 210°C in a twin-screw extruder to obtain the PP nanocomposite. From the XRD analysis, it is found that the PP nanocomposite prepared with the compatibilizer containing less maleic anhydride groups has more exfoliation compared to that prepared with the compatibilizer containing more maleic anhydride groups. Hence, the miscibility of the oligomers is also as important as the compatibility, and the compatibilizer with an optimum number of polar groups should be used for better dispersion of silicate layers. The process of forming PP nanocomposite is schematically shown in Figure 8.5. The driving force for the intercalation of MA-PP originates from the strong hydrogen-bonding tendency between the maleic anhydride group and the oxygen ions on the silicate layers. The interlayer spacing of the montmorillonite increases because of this, and the interaction between the silicate layers weakens. During strong shear mixing with molten PP, the silicate layers can separate and disperse uniformly in the matrix. It is found that there is a significant increase in storage modulus of the nanocomposites, probably due to the dispersion of silicate layers on a nanometer level.

The polymer melt-intercalation technique can also be used to prepare a polystyrene (PS) nanocomposite with styrene–methylvinyloxazoline copolymer as a compatibilizer. In this process, powdered PS, the compatibilizer, and montmorillonite clay treated with octadecyltrimethylammonium are melt blended at 180°C using a twin-screw extruder. Several clay–PNCs have also been prepared by this melt-intercalation technique without a compatibilizer, which include polyamide 6, polyamide 6,6, polyamide 11, and polyamide 12 nanocomposites.

8.1.1.4 Clay–Epoxy Nanocomposites

The in situ polymerization of prepolymers in OMC galleries has been successful for the preparation of exfoliated clay nanocomposites of polyimide, polyether, acrylonitrile rubber, unsaturated polyester (UP), epoxy, PS, and polysiloxane matrices. In addition to being a compatibilizing agent for the intercalation of the polymer precursor, the exchanged organic cation is potentially capable of functioning as a polymerization catalyst, and it is best demonstrated in the synthesis of clay nanocomposites of amine-cured epoxies. Exchange cations derived from protonated amines are shown to cause the polymerization rate in the spatially restrictive galleries of the clay to be competitive with the extragallery polymerization rates, resulting in the formation exfoliated clay–PNCs.

The preparation of amine-cured epoxy nanocomposites through the curing of prepolymers in the galleries of protonated onium ion (alkylammonium ion) exchanged clays depends on two crucial factors: firstly, the ability of the onium ion to serve as a compatibilizing agent, which allows the co-intercalation of the resin and curing agent, and, secondly, the ability of the onium ion to catalyze the polymerization reaction within the gallery. The catalytic function of the onium ion is very important because, as discussed earlier, it allows the intragallery polymerization at a competitive rate to that of extragallery polymerization. Consequently, the interlayer spacing is very much increased to facilitate exfoliation, resulting in the complete dispersion

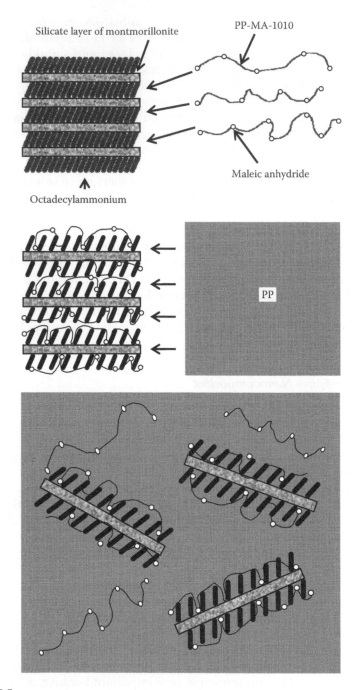

FIGURE 8.5
The scheme for the processing of clay–PP nanocomposite. (Reprinted from Pinnavaia, T.J. and Beall, G.W., *Polymer-Clay Nanocomposites*, John Wiley & Sons, Chichester, U.K., 2000, p. 107. With permission.)

of silicate layers in the polymer matrix, and thus they can contribute fully to the reinforcement mechanism. The dispersion of montmorillonite nanolayers in any epoxy matrix is readily achieved with acidic alkylammonium ions such as $C_{18}H_{37}NH_3^+$.

Two types of clay–epoxy nanocomposites have been prepared using two different types of epoxy resins, more specifically, Shell EPON 828 and 826 for flexible and glassy epoxy matrices, respectively (Lan et al. 1995). The curing agents used are Huntsman Chemical Jeffamine D-2000 and D-230 (polypropylene glycol bis(2-aminopropyl) ethers) for flexible and glassy epoxy matrices, respectively, with molecular weight of ~2000 and 230. The respective curing agents can form a rubbery epoxy matrix (T_g of −40°C) or a glassy epoxy polymer matrix (T_g of ~82°C). The methodologies used for the preparation of clay–epoxy nanocomposites using rubbery and glassy epoxy matrices differed slightly. For the rubbery system, the required amount of OMC is added directly to a mixture of epoxy resin and curing agent. For the glassy system, the OMC is pre-intercalated with the epoxy resin at 50°C overnight, followed by the addition of the curing agent. The mixtures of OMC, resin, and curing agent are degassed and poured in molds. The mixtures are pourable when the loading of OMC is below 20 wt.%. Both the nanocomposites showed exfoliated structure.

The key to achieving an exfoliated clay structure in an epoxy matrix is first to treat the clay with a hydrophobic onium ion and then disperse the modified clay in the epoxy resin. The curing agent is added after allowing sufficient time for the epoxy to intercalate. As the precursor molecules diffuse into the gallery, the onium ions catalyze intragallery polymerization at a rate that is competitive with extragallery polymerization. This leads to initial gel formation in the gallery and to the exfoliation of clay nanolayers. Complete network cross-linking can then proceed with the retention of exfoliated clay state. The relative rates of reagent intercalation, chain formation, and network formation can be controlled by a judicious choice of temperature for each process.

8.1.1.5 Clay–Unsaturated Polyester Nanocomposites

UP resin is amenable for the formation nanocomposites by the in situ polymerization technique, since it is available as low-viscosity liquid prepolymer. The OMCs can be dispersed uniformly in this liquid resin, and then on curing a nanocomposite is formed. Many investigations on this nanocomposite system showed significant improvement in properties because of the presence of clay nanolayers.

Kornman and Berglund (1998) have reported a 32% increase in tensile modulus by adding 5 wt.% montmorillonite. The fracture toughness is found to be doubled by dispersing 3 wt.% of nanoclays. In another study, the properties of nanocomposite prepared by mixing the OMMT directly into polyester–styrene solution are compared with those of another nanocomposite formed by dispersing the OMMT in polyester resin and then adding styrene (Suh et al. 2000). It is found that the degree of clay dispersion altered on changing the mixing pattern.

In contrast, Bharadwaj et al. (2002) found a decrease in tensile modulus at higher clay contents, mainly attributed to the hindrance of cross-linking by the presence of clay platelets. Inceoglu and Yilmazer (2003) compared the nanocomposites formed using modified and unmodified montmorillonite clay and found that the tensile, flexural, and impact strengths of composites formed with modified clay are higher than those formed with unmodified clay.

8.1.1.5.1 In Situ Polymerization

The flow diagram for the preparation of clay–polyester nanocomposite is shown in Figure 8.6. The required amount of OMC is dispersed in the polyester resin and stirred for 1 h using a high-shear mechanical stirrer rotating at a speed of 1000 rpm. After mixing, it is kept undisturbed for a few hours, so that air bubbles present inside the mix come out slowly. Cobalt naphthanate accelerator followed by methyl ethyl ketone peroxide catalyst are added sequentially and then stirred gently. It is then cast in a mold to get the nanocomposite. The castings are allowed to cure at room temperature for 24 h and post-cured at 70°C for 3 h.

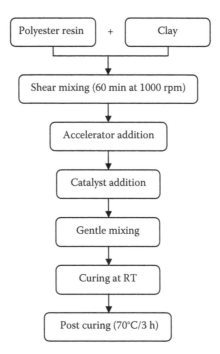

FIGURE 8.6
Flow diagram for the preparation of clay–polyester nanocomposite.

8.1.1.6 Clay–PET Nanocomposites

Polyethylene terephthalate (PET) is one of the plastics produced in large volume that has been used in many years as fibers and films. Another fastest growing application area for PET is food and beverage packaging. This growth is mainly due to the important properties of PET, such as gas barrier, processability in high-throughput blow molding, clarity, color, and resistance to flavor scalping. Thus, PET has become the preferred packaging material for a wide variety of products such as carbonated soft drinks, water, sport drinks, and fruit juices. However, the barrier properties of PET to oxygen and carbon dioxide are not sufficient to package certain large-volume products that are sensitive to oxygen and loss of carbon dioxide such as beer, wine, and tomato-based products. Even small soft drink containers are sometimes problematic owing to the high surface–volume ratio, which leads to unacceptable loss of carbon dioxide and consequently loss of shelf life. The discovery of enhanced barrier properties in polymeric materials by using exfoliated platelets of aluminosilicate clays has led to considerable research on applying this technology to improve the barrier properties of PET. The conceptual advantage of this approach is based on the use of conventional PET as the polymer backbone, with barrier improvement arising from low levels of relatively inexpensive clay minerals. This, coupled with the potential to process this material in conventional stretch blow molding equipment, makes this approach very attractive.

Two general routes exist for the incorporation of clay during polymerization. The first is based on the dispersion of clay into ethylene glycol, one of the monomers used to produce PET. This technology was developed by Nanocor, Inc. The key steps in this approach include finding a clay treatment that gives exfoliation of the clay into ethylene glycol and finding polymerization conditions that permit polymerization while maintaining the exfoliation of the clay. One of the methods of modifying the clay involves the treatment of sodium montmorillonite clay with a polar polymer, such as PVP or PVA. The montmorillonite intercalated with PVP or PVA can exfoliate in a wide variety of media, including ethylene glycol. This clay-dispersed ethylene glycol is mixed with molten dimethyl terephthalate to form PET nanocomposite. In this polymerization reaction, a significant increase in the low-shear melt viscosity can occur with increasing clay content. This limited the molecular weight of the PET that is formed directly in the melt phase polymerization, but it is found that the PET nanocomposites undergo solid-state polymerization to a molecular weight suitable for conventional polymer processing, such as stretch blow molding.

The second route involves the in situ incorporation of a readily available OMC. By the proper selection of the alkylammonium tether, the OMC can be tuned for PET. The selection of alkylammonium compound has a major influence on the exfoliation of the OMC, which will be reflected in the resulting barrier properties and the visual appearance of the PNCs.

Although the PET nanocomposites prepared by melt compounding process exhibit significantly improved barrier properties, none of the PET nanocomposites prepared by this process have shown gas barrier properties as high as those exhibited by PET nanocomposites prepared by in situ process. This indicates that the exfoliation level achieved in the in situ process exceeds the level achieved by melt compounding.

It is found that the mechanism proposed for polyamide 6 nanocomposite is not applicable to PET nanocomposites. Instead of the polymerization of monomer in the gallery, a low molecular weight PET causes the exfoliation of the clay. It suggests that an oligomer concentrate process to prepare PET nanocomposites would be possible. This may be considered as a specific case of the expanded OMC approach in which the expanding agent is a digomeric version of the matrix polymer. Based on this approach, a PET nanocomposite can be prepared by the following steps:

1. Melt compounding of oligomeric PET with OMC
2. Increase the molecular weight of PET by solid-state polymerization, chain extension, or compounding or a combination and variation of these methods

It is found that the PET nanocomposites prepared by the oligomer process have significantly improved barrier properties compared to those of nanocomposites prepared by the melt blending of OMC with high molecular weight PET and, in some cases, have achieved equivalent barrier properties to those of PET nanocomposites prepared by the in situ process (Barbee et al. 1999).

8.1.1.6.1 Multilayer Nanocomposite Containers

In addition to monolayer PET nanocomposite containers, PNC multilayer containers are of considerable research and commercial interest in the packaging industry. There are several potential approaches to use the nanocomposite technology in multilayer containers. In one approach, the PET nanocomposite layer is used as a barrier layer in multilayer containers. The second approach involves the use of a high-barrier matrix PNC, such as polyamide nanocomposite, as the barrier layer. The third approach involves the application of a PNC coating on the preformed parts.

Incorporating a PET nanocomposite layer as a middle layer of a multilayer PET container has some advantages. One advantage is that the multilayer containers have a neat PET as the food contacting surface. Another potential advantage is that the recycling becomes easy when same polymers are used in all layers. The third advantage is that multilayer containers having PNC internal layer have lower haze than that of monolayer PNC containers. It is believed that sandwiching the nanocomposite between two layers of unfilled polymer masks light scattering caused by clay platelets present at or near the surface of the polymer upon stretching. One disadvantage is that the incorporation of PET nanocomposite layer inside a multilayer structure dilutes the barrier improvement achieved by the clay platelets.

Clay–polyamide nanocomposites are ideal for PET-based multilayer containers because they have high gas barrier properties, processing is similar to that of PET, and they adhere well to PET. In addition, the polyamide nanocomposites are amenable to existing multilayer technologies, including preform injection molding machines and stretch blow molding machines. The actual performance of containers depends on many factors including the container weight and shape, the thickness and placement of the barrier layer, and the type of the nanocomposites and the clay loading. Containers to meet the barrier needs of a wide variety of targeted applications can be made by suitably varying these factors. Still, haze remains as a critical issue for the use of PNCs in stretch blow molded containers. Although the level of haze exhibited by the multilayer nanocomposites is acceptable for colored containers, it may not be suitable for clear, colorless containers. Considerable research is needed to understand and reduce haze in order to use these materials commercially.

Another approach of using multilayer technology for improving the gas barrier properties of PET is to apply a very thin nanocomposite coating after forming the article. A highly clay loaded, thermosetting melamine formaldehyde nanocomposite coating that imparts very high-barrier properties to PET film has been developed (Harrison et al. 1996). Another coating material for PET is epoxy nanocomposite made with clay or other platelet materials. It has been reported that these highly loaded nanocomposite coatings have very high gas barrier properties. Hence, a considerable amount of research is being devoted to develop these nanocomposite coatings to commercial scale (Carblom and Seiner 1998).

8.1.1.7 Clay–Rubber Nanocomposites

Since the early days of the rubber processing, fine particulate fillers have been used in rubber compounding. Particulate fillers are generally divided into two groups: inert fillers and reinforcing fillers. Inert fillers are added to the rubber to increase the bulk volume and reduce the cost of product. In contrast, reinforcing fillers such as carbon black and silica are added to enhance the mechanical properties, to alter the electric conductivity, to improve the barrier properties, and/or to increase the fire and ignition resistance of rubber. However, a minimum of 20 wt.% of filler is usually needed for a significant property improvement. This high loading of fillers may reduce the processability of the rubber compounds. The expansion of the polymer industry and the demand for new, low-cost composites with improved properties at lower particle content are some of the key challenges for the rubber industry.

In recent years, clay–rubber nanocomposites have attracted much attention and interest of many industrial and academic researchers, since they often exhibit outstanding properties at low levels of clay loading. Elastomers with their low modulus stand to gain much in terms of modulus and strength from the addition of nanoclays. Research on the clay–rubber nanocomposites has focused mainly on four common elastomers, which are natural rubber (NR), ethylene propylene diene (EPDM) rubber, styrene butadiene rubber, and NBR. However, there are some reports on clay–rubber nanocomposites based on other types of elastomers, such as silicone rubber, polybutadiene rubber, and ethylene propylene rubber.

8.1.1.7.1 Clay–Rubber Nanocomposite Preparation

Clay–rubber nanocomposite preparation methods can be generally divided into four major groups:

1. In situ polymerization
2. Solution intercalation
3. Direct-melt intercalation
4. Intercalation through latex compounding

In situ polymerization is similar to the preparation of plastic-based nanocomposites. The organically modified clay is swollen within the monomer solution (or liquid monomer). The polymerization is then initiated either by incorporating curing agent or initiator or by increasing the temperature, if it is sufficiently reactive at high temperature.

In the solution intercalation technique, a solvent system is selected such that it can dissolve the rubber and the modified clay can swell in the solvent. The organically modified clay is first allowed to be swollen, and the layers come apart in the solvent. The rubber is then dissolved in the solvent separately, and this solution is mixed with the clay-dispersed solvent. Upon solvent removal, the clay layers are surrounded by the rubber molecules, resulting in clay–rubber nanocomposite.

Direct-melt intercalation is a most promising method, and it has great advantages over the earlier two methods by being compatible with current industrial processes and environmentally benign due to the absence of solvents. In this method, molten rubber and modified clay are blended under shear. The rubber chains penetrate the silicate galleries to form either intercalated or exfoliated nanocomposites.

Intercalation through latex compounding is also a promising method to prepare clay–rubber nanocomposites. The latex compounding starts with the dispersion of layered silicates in water, and the clay particles swell owing to the hydration of interlayer cations. Rubber latex is then added and mixed with the clay dispersed in water for a period of time and later coagulated.

8.1.1.7.2 Clay–Natural Rubber Nanocomposites

Rubber nanocomposites based on NR and organically modified layered silicate have mainly been prepared in a two-roll mill. It is found that the OMC nanolayers are largely exfoliated in the NR matrix and there is no difference in distribution of nanolayers before and after vulcanization. It is also found that the cure characteristic of NR is affected by the presence of OMC and the optimum curing time of NR is sharply reduced with the incorporation of OMC, since the presence of amine groups in the clay facilitates the sulfur curing reaction of NR. It has been reported that the thermal decomposition temperature of NR is increased by the incorporation of OMC, since the exfoliated nanolayers prevent the volatile decomposition products from diffusing out during the thermal degradation process (Varghese and Karger-Kocsis 2003).

In another study, it is found that the compatibility between the OMC layer and NR matrix is much higher when a nanocomposite is prepared by the solution mixing method instead of mechanical mixing method. However, both the methods are suitable for obtaining a uniform dispersion of OMC nanolayers in the NR matrix (López-Manchado et al. 2004).

8.1.1.7.3 Clay–EPDM Nanocomposites

EPDM rubber is a very common general-purpose rubber with a significant commercial importance. It has been one of the main rubbers investigated to find the effects of nanoclays on the properties. Clay–EPDM nanocomposites have been prepared by simple static mixing in confined chambers such as Haake and solution blending and on a laboratory two-roll mill. However, most clay–EPDM nanocomposites have been produced by conventional melt blending process. The advantages of melt blending process are its compatibility with present polymer-processing methods and its environmental friendly characteristics due to the nonusage of organic solvents in this process. The nanocomposites prepared by the melt blending process were found to have better physical properties than those prepared by the other methods. A comparative study has been made on the properties of nanocomposites prepared by melt blending and using two-roll mill (Gatos et al. 2004). It was found that the nanocomposite prepared by melt blending has higher tensile strength and elongation at break than the nanocomposites prepared by a two-roll mill. This is attributed to the higher shear rates experienced by the clay, leading to better dispersion of clay layers, and consequently to higher mechanical properties.

The type of curing agent (peroxide or sulfur) has an effect on the internal structure of OMMT–EPDM nanocomposite. It is found that the peroxide curing system leads to the formation of intercalated OMMT–EPDM nanocomposite, whereas the sulfur curing system causes the formation of an exfoliated nanocomposite (Zheng et al. 2004).

The intercalation of EPDM chains into clay galleries, as well as the exfoliation of clay layers in EPDM matrix, leads to outstanding properties for clay–EPDM nanocomposites (Gatos et al. 2004). The tensile strength of OMMT–EPDM nanocomposites is enhanced three to four times than the EPDM compound without nanoclay. The exfoliated silicate layers in EPDM increased the elongation at break by 140%, compared to EPDM compound without nanoclay. The mechanism for the improvement of elongation in nanocomposite is not clear at present but could be due to better physical bonding between silicate layers and EPDM matrix. The tear strength is significantly enhanced by the incorporation of OMCs in EPDM. This increase in tear resistance is attributed to the uniform dispersion of silicate nanolayers, which could act as a physical barrier to the growing crack. Another important characteristic of clay–EPDM nanocomposites is their excellent gas barrier properties. The permeability of oxygen in the nanocomposite is reduced by 60% compared to EPDM rubber compound due to the uniform dispersion of silicate nanolayers, which could make a tortuous path for oxygen molecules (Chang et al. 2002b).

8.1.1.7.4 Clay–SBR Nanocomposites

Clay–SBR nanocomposites have been prepared, and they have better tensile strength, hardness, and tear strength than those of SBR composites reinforced with carbon black. These nanocomposites have been prepared by solution blending using toluene as solvent, but other methods such as latex compounding can also be used. When comparing the nanocomposites prepared by solution blending and latex compounding, the nanocomposites prepared by the second method showed better tensile strength, tear strength, and hardness. However, the clay–SBR nanocomposites prepared by the solution method had higher elongation at break and higher permanent set (Wang et al. 2000b).

In another study, the effect of peroxide and sulfur curing systems on the mechanical properties of clay–SBR nanocomposites has been investigated (Sadhu and Bhowmick 2004). It is found that the clay–SBR nanocomposites with very similar cross-link densities cured by peroxide and sulfur systems exhibited comparable modulus and tensile strength. However, the elongation at break of sulfur-cured clay–SBR nanocomposite is much higher than that of the peroxide-cured nanocomposites due to the flexible nature of sulfide linkages.

8.1.1.7.5 Clay–Nitrile Rubber Nanocomposites

Clay–NBR nanocomposites have been prepared mainly with exfoliated structure (Kim et al. 2003; Wu et al. 2003). These nanocomposites have been prepared using different processing techniques such as melt intercalation, solution blending, ball milling of layered silicate in emulsified solution followed by latex shear blending, and by co-coagulation of nitrile rubber latex and layered silicate aqueous suspension followed by two-roll milling.

8.1.1.7.6 Clay–Silicone Rubber Nanocomposites

OMC–silicone rubber nanocomposites can be prepared by melt-intercalation process (Wang et al. 1998). Nanocomposites having intermediate-type structure have been prepared by this process. The mechanical properties are found to be improved with the incorporation of OMMT, and those properties were close to the mechanical properties of nanosilica-filled silicone rubber composites. The thermal stability of silicone rubber was also found to be increased from 381°C to 433°C by the incorporation of OMMT. The increase in decomposition temperature can be attributed to the increases in physical and chemical links between OMMT and silicone rubber matrix, which prevent the silicone rubber chains from degradation.

8.1.1.7.7 Clay–Polybutadiene Nanocomposites

Clay–polybutadiene rubber nanocomposites are mainly prepared by the solution method. The organically modified clay is dispersed in toluene with continuous stirring, and a polybutadiene–toluene solution is added to this dispersion and vigorously stirred for about 12 h. The nanocomposite is obtained by removing the solvent. By using this method, a nanocomposite has been prepared with intermediate structure having intercalated and exfoliated clay layers (Wang et al. 2000b). Some of the mechanical properties such as tensile strength and elongation at break are found to be better than those of carbon black nanocomposites. This layered silicate could be used as a promising reinforcement in the polybutadiene rubber, provided that the clay layers are uniformly distributed in the rubber matrix at the nano-level.

8.1.1.8 Characterization

The structure of clay–PNCs is generally established using wide angle x-ray diffraction (WAXD) analysis and transmission electron microscopy (TEM) observation. Due to its easiness (no special sample preparation is needed) and availability, WAXD analysis is most commonly used. Based on the position, shape, and intensity of the basal reflections from the clay, the nanocomposite structure (intercalated or exfoliated) may be identified. For example, the extensive layer separation in an exfoliated nanocomposite results in the eventual disappearance of any coherent XRD peak from the clay layers. On the other hand, the finite layer expansion associated with polymer intercalation in the intercalated nanocomposite results in the shifting of basal reflection peak to lower diffraction angles corresponding to the gallery height. Although WAXD analysis is a convenient method, no information about the spatial distribution of silicate layers or any structural inhomogeneities is available. Moreover, at very low level of clay loadings (less than 5%), the intensity of the basal peak may not be sufficient to show up, which gives wrong information. Hence, WAXD analysis cannot be used as a conclusive evidence for the structure of clay–PNCs. On the other hand, TEM allows the observation of internal structure, spatial distribution of clay layers, and defects through direct visualization. However, special care must be exercised to select a representative cross section of the sample. Moreover, sample preparation for TEM observation is a tedious and time-intensive process, and TEM observation gives only the qualitative information of the whole sample.

The XRD patterns and TEM micrographs of OMC-filled polyester composites are shown in Figure 8.7. The XRD pattern of OMC shows a sharp peak at 2θ value of 5° corresponding to the basal plane diffraction. The interlayer distance (d-spacing) of OMC obtained from Bragg's law is 17.7 Å. The basal plane diffraction peak of OMC is absent in nanocomposites containing 2 wt.% OMC. This reveals that the interlayer distance of OMC is

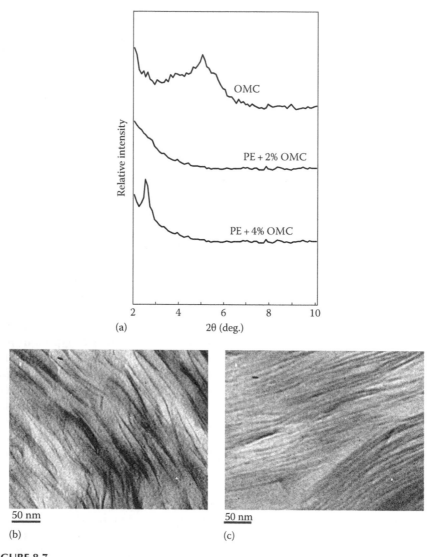

FIGURE 8.7
XRD patterns (a) and TEM micrographs of (b) PE+ 2% OMC and (c) PE+ 4% OMC nanocomposites. (Adapted from Jawahar, P. and Balasubramanian, M., *Int. J. Plastics Technol.*, 9, 472, 2005. With permission.)

more than 70 Å or the layers are randomly dispersed in the polymer matrix. The corresponding TEM micrograph also reveals the random distribution of clay platelets. From the TEM micrograph of nanocomposite containing 4 wt.% OMC, it can be observed that the silicate platelets are closely packed than the nanocomposites with 2 wt.% OMC. The XRD pattern and the parallel arrangement of the clay platelets indicate the formation of intercalated nanocomposites.

8.1.1.9 Properties of Clay–Polymer Nanocomposites

PNCs made with clay frequently exhibit remarkably improved physical and mechanical properties compared to those of pristine polymers. Improved properties include modulus, strength and heat resistance, and resistance to gas permeability and flammability. The main reason for these improved properties in nanocomposites is more interfacial interaction between the polymer and clay layers compared to conventional filled polymer systems.

8.1.1.9.1 Mechanical Properties

Findings from many investigations on clay–PNCs showed dramatic improvements in tensile properties. Lan and Pinnavaia (1994) have reported more than a 10-fold increase in strength and modulus on incorporating 7.5 vol.% OMC in a rubbery epoxy matrix. It is also found that the performance of nanocomposites is related to the degree of delamination of clay layers in the polymer matrix, since an increase in the degree of delamination can increase the interaction between reinforcement and matrix. Fornes and Paul (2003) have reported the better effectiveness of nanoclay on increasing the modulus compared to glass fibers. It is found that for doubling the modulus value of polyamide 6, only about 6.5 wt.% of clay is needed, whereas the amount of glass fiber needed is almost three times. Kojima et al. (1993) suggested that the improvement of tensile modulus for clay–polyamide 6 nanocomposites is due to the constraints exerted by clay layers for the polymer chains mobility.

The fracture surfaces of cured polyester resin with and without OMC are shown in Figure 8.8. The fracture surface of nanocomposite containing OMC is very rough, indicating that the resistance to crack propagation is very high. It has been reported that the rough surface is an indication of the resistance to crack propagation and/or the deflection of the propagating crack by the clay layers (Bernd et al. 2003).

8.1.1.9.2 Gas Barrier Property

Clay layers are believed to increase the barrier properties by creating a tortuous path for gas molecules to travel through a polymer matrix (Figure 8.9). A dramatic improvement of barrier properties with a simultaneous decrease in the thermal expansion coefficient is observed in

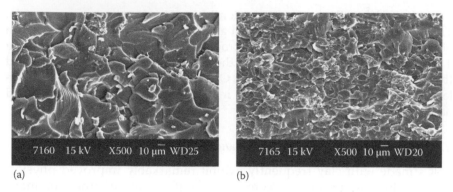

(a)   (b)

FIGURE 8.8
Fractographs of (a) cured polyester resin and (b) clay–polyester nanocomposite.

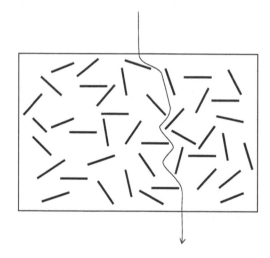

FIGURE 8.9
Tortuous path of gas molecules in a clay–PNC is schematically shown.

clay–polyimide nanocomposites (Yano et al. 1991, 1993, 1997). The addition of a small fraction of OMC reduced the permeability of small gaseous molecules, such as O_2, H_2O, He, and CO_2. For example, the permeability coefficient of water vapor had decreased 10-fold with 2 wt.% clay loading in polyimide. The permeability is found to decrease with increasing aspect ratio of silicate platelets. Such excellent barrier properties have generated considerable interest in developing clay nanocomposites for food packaging applications.

Platelet-shaped particles enhance the barrier of polymers according to a tortuous path model, developed by Neilson, in which the platelets present in the polymer obstruct the passage of gases and other permeants through the polymer. According to this model, the barrier improvement is a function of the volume fraction of platelets (V_p) and a function of the aspect ratio of the platelets (α), which is given as

$$P_{NC} = \frac{(1-V_p)P_m}{1+\alpha V_p/2} \tag{8.1}$$

where
P_{NC} is the permeability of the nanocomposite
P_m is permeability of the matrix polymer

The barrier properties can be improved by increasing the volume fraction and aspect ratio of platelets.

8.1.1.9.3 Water Absorption

The weight increase of the nanocomposite samples kept in water for a period of 7 days is shown in Figure 8.10. It is observed that the addition of OMC has brought down the water absorption. For the period of 7 days, the percentage increase in weight due to water absorption for pristine polyester is 0.48%, whereas it has come down to the value of 0.19% for the OMC content of 3 wt.%.

8.1.1.9.4 Glass Transition Temperature

The variation of the damping factor for pristine polyester and OMC-filled composites is shown in Figure 8.11. The damping factor (tan δ) increases on the addition of OMC. The stiffness variation between the nanoclay and

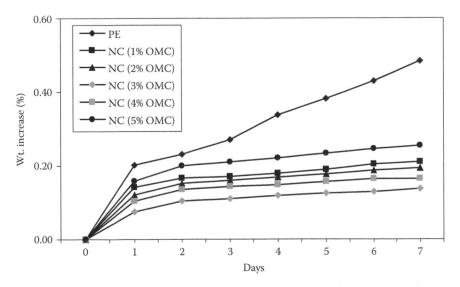

FIGURE 8.10
Water absorption behavior of polyester and clay–polyester nanocomposites. (Reprinted from Jawahar, P. and Balasubramanian, M., *Int. J. Plastics Technol.*, 9, 472, 2005. With permission.)

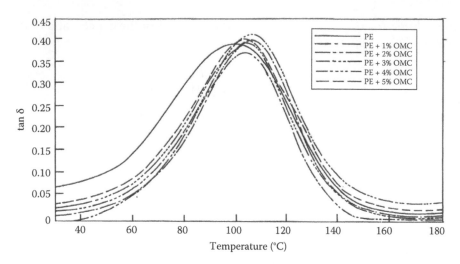

FIGURE 8.11

Dynamic mechanical analysis of polyester and clay–polyester nanocomposites. (Reprinted from Jawahar, P. and Balasubramanian, M., *Int. J. Plastics Technol.*, 9, 472, 2005. With permission.)

polymer matrix might have caused internal damping, resulting in increased damping factor for nanocomposites.

The glass transition temperature can be obtained from peak tan δ value of damping curves. The glass transition temperature (T_g) value of pristine polyester is 99.6°C. It increases on addition of OMC, and it is 105.3°C for the clay content of 3 wt.%; thereafter, it decreases to the value of 102°C for the OMC content of 5 wt.%. This improvement in T_g suggests that the clay layers stiffen the polymer matrix at high temperatures. Due to the large surface area of the nanoclays, large amount of polymer chains has strong contact with the clay surfaces that prevents segmental motion of the polymeric chains (Chang et al. 2002a).

8.1.1.9.5 Thermal Stability

The thermogravimetric analysis of the phenolic resin and the SiO_2–phenolic nanocomposite is shown in Figure 8.12. Both the materials show three stages of decomposition. In the case of SiO_2–phenolic nanocomposite, the decomposition temperature is shifted to higher temperature. The first stage of decomposition for the phenolic resin starts at 220°C and ends at 330°C, whereas the first stage of decomposition for the nanocomposite starts at 300°C and ends at 400°C. The second stage of decomposition for the phenolic resin ends at 450°C and for the nanocomposite ends at 500°C. After that, there is gradual loss in both the cases. It is clear that the completion of the first stage of decomposition of the nanocomposite takes place about 80°C higher than the phenolic resin. It is attributed that the increased char formation in the nanocomposites reduces the production of combustible gases and decreases the exothermic heat generated during real combustion situations (Pearce and Liepins 1975).

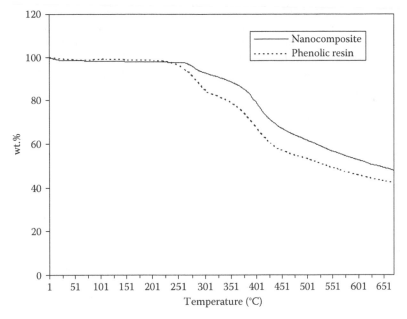

FIGURE 8.12

Thermogravimetric analysis of phenolic resin and its nanocomposite. (Reprinted from Periadurai, T. et al., *J. Anal. Appl. Pyrolysis*, 89, 244, 2010. With permission.)

8.1.1.9.6 Flame Retardancy

The potential of clay–polyamide nanocomposites for flame-retardant applications was first reported by Unitika Ltd., Japan, in 1976 (Fujiwara and Sakamot 1976). Gilman (1999) has reported flame-retardant properties of nanocomposites in detail. According to him, the clay must be nano-dispersed to improve the flame retardance but need not be completely delaminated. The flame-retardant characteristic of nanocomposites comes from the formation of a carbonaceous–silicate char. This char insulates the underlying material and slows down the mass loss rate.

8.1.1.9.7 Heat Distortion Temperature

HDT of a polymeric material is an index of heat resistance under an applied load. It is found that there is a marked increase in HDT from 65.8°C for the neat polyamide to 152.8°C for a polyamide nanocomposite containing 4.7 wt.% OMC (OMMT). The nano-dispersion of OMMT also increased the HDT of PP (Nam et al. 2001) and polylactic acid (Sinha et al. 2003).

8.1.1.9.8 Optical Properties

Traditional composites tend to be largely opaque because of light scattering by the micrometer level particles or fibers. Although the lateral dimensions of silicate layers are in the micrometer range, the thickness is only ~1 nm.

Hence, the nanocomposites formed by dispersing single layers of clay will be transparent to visible light. The domain sizes in nanocomposites are reduced to a level such that true "molecular composites" are formed. As a result, these composites are often highly transparent, which renders them useful for applications outside the boundaries of traditional composites (Wang et al. 1996; Wang and Pinnavaia 1998).

8.1.1.9.9 *Tribological Properties of Clay–Polyester Nanocomposites*

The tribological properties of clay–polyester nanocomposites containing different amounts of OMC have been reported by Jawahar et al. (2006). The wear rate of polyester resin slid against steel is found to decrease with increasing OMC content up to 3 wt.%. After that, there is an increase in wear rate; even then, the wear rate of nanocomposite containing 5 wt.% clay is lower than that of pristine polyester resin. On the other hand, the wear rate of composites made with unmodified clay is found to increase with increasing amount of clay (Figure 8.13). The improvement in wear resistance in nanocomposites is attributed to the presence of exfoliated and intercalated nano-sized clay platelets. These nano-sized clay platelets prevented the large-scale fragmentation of polyester matrix. The wear surface of nanocomposites shows finer wear tracks compared to polyester resin wear surface because of this (Figure 8.14). The wear tracks of unmodified clay–polyester composites are very rough with delamination throughout the contact surface. Hardness also plays a vital role in improving wear resistance (Khedkar et al. 2002). The uniform distribution of nano-sized clay platelets improved the hardness of polyester resin, whereas the unmodified clay did not improve the hardness, since it remained as microtactoids, which may not increase the hardness at lower concentrations. In this composite, the clay particles remained as microtactoids, which came out easily and acted as an abrasive. The same trend is also observed for the friction coefficient.

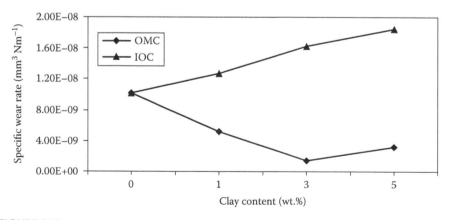

FIGURE 8.13
Wear behavior of polyester and clay–polyester nanocomposites. (Reprinted from Jawahar, P. et al., *Wear*, 261, 835, 2006. With permission.)

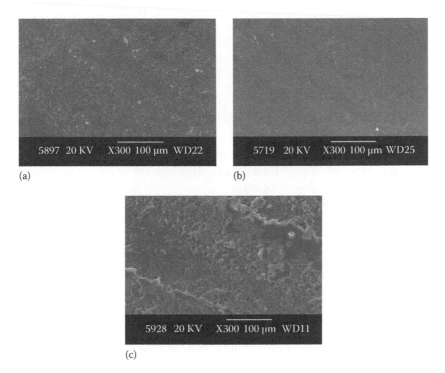

FIGURE 8.14
Wear surface of polyester (a) and clay–polyester composites containing (b) 1% OMC and (c) 1% unmodified clay. (Reprinted from Jawahar, P. et al., *Wear*, 261, 835, 2006. With permission.)

The nanoclay dispersion decreased the coefficient of friction, whereas the micron-level clay particles increased the friction coefficient.

8.1.2 Graphite–Polymer Nanocomposites

Natural flake graphite is a layered material with good electric conductivity ($\sim10^4$ S m^{-1}). Graphite intercalated compounds can be prepared by intercalating small compounds between graphene layers because of the weak van der Waals forces that exist between graphene layers. Expanded graphite (EG) can be formed when these intercalated compounds are subjected to thermal treatment. During this process, the intercalated compounds vaporize and separate the layers apart, leading to an expansion in the C-direction, and the resulting low-density puffed-up material is called EG. The layered structure of parent graphite is maintained in the EG, whereas the volume expansion is as high as 200–300. Consequently, various graphite–PNCs can be prepared using this EG by different methods, including in situ polymerization, melt intercalation, exfoliation–adsorption, and direct solution intercalation. The incorporation of layered graphite can significantly increase the electric conductivity of polymers. These types of conducting nanocomposites have attracted considerable attention because of their potential applications in advanced technologies, for example, in antistatic coatings, electrochemical displays, sensors, redox capacitors, electromagnetic shielding, and secondary batteries.

8.1.2.1 Synthesis of Graphite Nanosheets and Graphite Oxide

Exfoliated graphite consists of a large number of delaminated graphite nanosheets. The electric conductivity of this material is comparable to the original graphite. The exfoliated graphite can be broken down to yield graphite nanosheets using sonication.

The first step in the preparation of graphite nanosheets is the formation of graphite intercalated compound, which is readily formed by heating graphite powder with potassium metal under vacuum at 200°C. The C-axis expands from 3.4 to 5.4 Å after this intercalation. It is then treated with ethanol, which leads to the following reaction:

$$2KC_8 + 2C_2H_5OH \rightarrow 16C + 2KOC_2H_5 + H_2 \tag{8.2}$$

The resulting hydrogen gas evolution aids the separation of graphite sheets into nanosheets with a thickness of 40 ± 15 graphene layers.

Unlike layered silicates, graphite does not undergo any ion exchange reaction with alkylammonium ions, but these cations can intercalate the interlayers of graphite oxide (GO), which is obtained by the oxidation of graphite. The preparation of GO involves the oxidation of graphite using an oxidizer in the presence of nitric acid. When the bonds between the graphite layers are broken by strong oxidation, hydrophilic GO with lamellar structure is formed. It readily absorbs water or other polar liquids. Similar to layered silicates, the swelling of GO allows the intercalation of various compounds, such as alkylammonium salts and polymers. On the formation of GO, the layered structure of graphite is preserved, but the aromatic character is partially lost. The structure contains various amounts of hydroxyl groups, ether groups, double bonds, and carboxyl groups. The suggested formula for GO is $C_7O_4H_2$, and it is known to be thermally unstable. GO is reduced by heating or by reducing agents and becomes a graphite-like material, which contains few or no oxygen and is electrically conductive.

8.1.2.2 Preparation of Graphite–Polymer Nanocomposites

Several approaches have been considered to prepare graphite–PNCs, and the three main techniques are described in the succeeding text.

8.1.2.2.1 In Situ Intercalative and/or Exfoliation Polymerization

In this process, the EG or exfoliated graphite nanosheets are dispersed in the liquid monomer, or a monomer solution and the monomers enter in between the intercalated graphite sheets or graphene layers. The polymerization can be initiated by either heating or radiation after the diffusion of a suitable initiator. The main advantage of this technique is that a well-dispersed intercalated or exfoliated graphite–PNC is generally obtained. With this process, the nanocomposites are also prepared through emulsion polymerization where the EG is well dispersed in the aqueous phase.

8.1.2.2.2 Melt Intercalation and/or Exfoliation

The EG or exfoliated graphite nanosheets are mixed with the molten polymer. Due to the neutral nature of graphite, the polymer molecules can diffuse into the interlayer space or galleries of EG and form either an intercalated or an exfoliated graphite–PNC.

8.1.2.2.3 Exfoliation–Adsorption

The EG is broken down into graphite nanosheets and graphene layers in a solvent, in which the polymer is soluble. In this process, the graphene layers held together by very weak forces are dispersed in an appropriate solvent. The polymer is then dissolved in this dispersion, and the dissolved polymer is adsorbed onto the delaminated graphite sheets. When the solvent is evaporated or the polymer is precipitated out, the graphite sheets or graphene layers reassemble by sandwiching the polymer chains, resulting in the formation of an ordered multilayer nanostructure or an exfoliated nanostructure. It should be noted that intensive stirring such as sonication is needed to separate the graphite sheets from the EG worm and help the polymer chains intercalate into graphite nanosheets or graphene layers.

8.1.3 Nanofiber-Reinforced Composites

The availability of high-performance fibers, such as the variety of the recently developed nanofibers, with very high strength and ultralight-weight is enough to intrigue the composite material enthusiasts to investigate opportunities for new advanced nanocomposites. The advent of several novel forms of nano-carbons has led to attempts to develop nanocomposites with these nano-sized reinforcements. The mechanical properties of CNTs are exciting, since the nanotubes are seen as the ultimate carbon fiber ever mode. These materials are considered as the most important reinforcements in composite materials because of their excellent mechanical properties. The fibrous nanomaterials of carbon, such as carbon nanofibers (CNFs), single-walled carbon nanotubes (SWCNTs), double-walled carbon nanotubes (DWCNTs), and MWCNTs, make up the general varieties of nanofibers for composite materials use.

CNFs are discontinuous hollow filaments having a diameter of ~100 nm. These filaments are produced by a catalytic process in which metal particles are exposed to free carbon at elevated temperatures. They show physical properties approaching those of single-crystal graphite, including high thermal and electric conductivity values and high tensile strength and modulus. Hence, the incorporation of these filaments into polymers renders them suitable for structural and conducting applications, such as static discharge, electrostatic painting, and radio-frequency interference shielding. These discontinuous filaments are advantageous from the processing point of view also, because short-fiber-reinforced composites can be fabricated without the limitations of textile processing and incorporation into the matrix as faced by continuous fibers. For example, CNF–polymer composites may be fabricated by extrusion or injection molding, allowing for both high volume production and easy recycling.

Fascination with CNTs led to the development of a range of advanced nanocomposites but not without the developed understanding of how to manipulate the nanotubes for property improvement. The factors to be considered while developing CNT-based nanocomposites are nanotube starting conditions, dispersion, interaction with the matrices, and alignment. Generally, where development has been achieved, nanotube functionalization has been a key contributor either to help dispersion or to provide complete linkage to the matrix. The full potential of CNTs may well rest with more self-assembly and synergistic behavior between nanotubes and other nano-species. The development of PNCs with nanotubes involves the following five basic approaches:

1. Proper dispersion of nanotubes in polymers by various mixing techniques
2. Approaches to improve shear strength and, in particular, z-axis property enhancement in laminate systems
3. New resin formulations to form hybrid polymer
4. Forming interpenetrating polymer networks (IPNs)
5. Complete integration that yields paradigm changes in the ways composites are made

It has been reported that the addition of CNFs increases the mechanical properties as high as 400%, reduces the resistivity by 14–15 orders of magnitude, and increases the shielding effectiveness up to 40 dB. While multifunctionality is expected in these nanocomposite systems, none of these systems possesses multifunctional advanced properties to date, although this is expected to occur in the near future.

To realize the full potential of CNFs, an appropriate interphase between the CNFs and the matrix polymer should be available. The optimal interphase would provide fiber–matrix adhesion to maximize mechanical properties, such as tensile strength, and would allow for ease of dispersion and good electron transport between the fibers. The science and technology of surface modification for CNFs is slowly emerging. The surface modification CNFs can be accomplished by altering the surface area/surface energy, by the modification of the degree of graphitization of the CNF surface, or by the addition of functional groups intended to enhance covalent bonding.

8.1.3.1 Surface Modification of Carbon Nanofibers

Good fiber distribution and adequate bonding between fibers and matrix are essential for the better performance of CNF-based composites. The fiber surface morphology and chemistry are deciding factors for better adhesion with matrix. Methods of surface treatments include the modification of surface area, hence surface energy to improve wetting and physical bonding, incorporation of functional groups to enhance chemical bonding, and the control of the graphitization of the CNF surface. The incorporation of partially graphitized CNFs by high-temperature heat treatment can give improved composite properties. The intrinsic electric conductivity of the CNFs is maximized by giving a heat treatment at 1500°C. It has been reported that this high-temperature heat treatment of the fibers alters the exterior planes from continuous, coaxial, and poorly crystallized to discontinuous, nested, and conical crystallites inclined at about 25° to the fiber axis (Lake et al. 2010).

Air oxidation is the simplest procedure, while more complex procedures can be taken from organic synthesis methodology used for the oxidation of aromatic compounds. Air oxidation has been accomplished by passing air through the fiber contained in a horizontal tube at 400°C. The extent of oxidation can be varied by varying the time of oxidation. Surface oxygen levels of approximately 4–5 atom.% are obtained after 30 min oxidation.

Sulfuric acid/nitric acid mixture has also been used for oxidation. In this process, 1 part of fiber is added to 20 parts by volume of 1:1 mixture of concentrated sulfuric acid and nitric acid mixture. The mixture is refluxed for 1 h, cooled, and filtered through a sintered glass filter to isolate the fiber. The fiber on the filter is washed with copious amounts of deionized water to remove the remaining acid and then dried at 120°C. Surface oxygen levels as high as 25 atom.% has been obtained by this process.

Yet another process has been reported for the oxidation of nanofiber, in which peracetic acid is used as the oxidizing agent. In this process, 1 part fiber by weight is added to 200 parts by volume of a mixture of 3:1 glacial acetic acid and 30% hydrogen peroxide. It is allowed to stand at room temperature for 5 days. The fiber is separated by filtration and washed with copious amounts of deionized water and then dried at 120°C. Surface oxygen levels up to 10 atom.% have been obtained by this process.

8.1.3.2 Alignment of CNF

The alignment of CNTs or nanofibers in polymer matrix is of great importance in order to exploit their anisotropic electric, thermal, mechanical, and optical properties in various industrial applications. Specifically, the alignment of CNT in polymer by applying magnetic or electric field has received considerable attention due to the capability of alignment in any direction, and this cannot be achieved in any conventional polymer-processing methods.

Magnetic alignment of diamagnetic fiber is caused by diamagnetic anisotropic susceptibility. Hence, the alignment CNTs along the direction of magnetic field utilizing the strong anisotropic diamagnetic property has been attempted. Magnetically aligned CNT thin films were produced from CNT aqueous suspension by applying a magnetic field of 7–25 T. Magnetic alignment of CNTs in polymer has also been carried out to enhance the electric and thermal conductivities of composites. A CNT (3 wt.%)/epoxy nanocomposite was prepared under a magnetic field of 25 T, and CNT (1, 2, and 5 wt.%)/UP composites were prepared under a magnetic field of 10 T (Kimura et al. 2002; Choi et al. 2003). However, the necessary alignment time is not reported, which could depend on CNT structure, matrix viscosity, and magnetic strength.

Many reports have been published about the alignment of CNTs in solution by applying direct current (DC) or alternating current (AC) electric field. In addition, electric field has also been used to make connected structures of CNT by dropping a trace amount of CNT suspension between comblike electrodes having several micron distances. There are three forces acting on each CNT due to the electric field, which are torque, Coulombic, and electrophoretic forces. In the presence of an electric field, each CNT experiences polarization that leads to a torque force. Coulombic attraction is generated among oppositely charged ends of different CNTs. The electrophoretic force is induced by the presence of charged surfaces. Depending on CNTs, functionalization, and matrices, electrophoresis is sometimes observed. Thus, the torque force, Coulombic attraction, and electrophoretic force govern the formation of the aligned ramified network, that is, uniform density between electrodes, and high degree of linearity in connected network structures. The cause of the formation of a ramified network structure, rather than a straight network structure, could be due to the connection of more than two fibers having opposite charges to a single fiber.

8.1.3.3 Processing of Nanofiber-Reinforced Composites

Several processing approaches are available (Table 8.1) for preparing CNT-based nanocomposites. These approaches can lead to the development of advanced composites with better properties. Generally, combinations of these approaches are used based on the starting condition of the nanotubes and the properties required.

By far, the easiest processing route to prepare CNT-based nanocomposites is the mixing of CNTs directly into the liquid polymer. Dry mixing of nanofibers with polymer powder is not preferable because of handling issues, specifically the potential health hazard of airborne material and the difficulty in getting complete and uniform mixing. The use of solvents with surfactants is another possibility to disperse nanofibers and mix with the polymer resin.

TABLE 8.1

A List of Choices for the Processing of Nanocomposites Based on Nanotubes

Starting Nanotubes	Dispersion	Interface	Alignment
Dry as-received	Spreading	Polymer selection	High-shear mixing
In solvent as-received	Bench mixing	Tangle reduction	Extrusion
Purified	Homogenization	Dispersion	Elongational flow
Functionalized	Sonication	Alignment	Fiber spinning
Substrate grown	High shear	Purification	Magnetic fields
Dry aggregate	Stretch drying	Unwrapping	Electric fields
Pearls	Ball milling	Polymerization	Stretch drying
Powder	Polymerization	In situ	Dip/spin coating
In water	In situ	Stabilization for temperature	Gel/wet spinning
Wrapped	Incipient wetting	Functionalization	Solid free-form fabrication
Masterbatched	Extrusion	Metallization	Tape casting
Bucky tubes	Functionalized	Synergism	Substrate growth
Separated	Defunctionalized	B-stage	Self-assembly
Fluff	Wrapping	Sized	Aligned growth
Bucky paper	Separated	Curing agent advancement	Robocasting
Cut	B-stage	Hybridization	Fused deposition molding
Prepreg	Spraying		Extrusion
Sized	Prepreg		Substrate growth
Metallization	Metallization		
Decanted	Viscosity control		

Source: Data from Gupta, R.K. et al., *Polymer Nanocomposites Handbook*, CRC Press, Boca Raton, FL, 2010, p. 211.

However, the retained solvent or surfactant can cause some problems in the final composite unless chemical matching to the system is accomplished. Another problem is the increases in viscosity with the increase in CNT content, and this limits the ability to use higher amount of CNTs, even greater than a few weight percentage.

The percolation threshold content of CNTs for several conducting polymers has been found to be less than 1%. However, the range can vary based on dispersion and segregation level, but, ultimately, conducting polymers can be produced with low concentration of CNTs. For CNT nanocomposites in general, the electric properties are expected to be the first to market. For mechanical property enhancement, a great deal of focus has been on low-concentration CNT systems, because of cost concerns. One approach that has the potential for immediate application is the use of CNTs to improve shear strength. In this case, the nanotubes are placed in the midplane between the plies to improve resistance against shear failure. In many composites, the strength in the in-plane direction is usually very high, but the out-of-plane strength is poor because the regions between plies do not contain reinforcing fiber. One way to improve the out-of-plane strength is to incorporate nanotubes in the regions between fiber plies so that (i) the matrix layer thickness is increased or (ii) the resin–fiber interface region is enhanced. A recent research has shown a 45% increase in shear strength by adding only 0.1 wt.% SWCNTs, which are sprayed on the fiber plies. It has been reported that the strength of a composite determined by double cantilever beam test increased over 250% by growing multiwalled CNTs on the plies (Antonucci et al. 2003). To process a carbon fiber-reinforced epoxy composite with nanotubes for shear strengthening, the following steps could be used:

1. The fluff form of as-prepared nanotubes is purified and functionalized for epoxy resin.
2. The functionalized nanotubes are dispersed in a solvent and subjected to sonication to achieve a well-dispersed condition (more recent approaches include decanting the suspension of CNT to remove large agglomerates).
3. The suspension is sprayed onto carbon fibers to produce uniform coverage of CNTs on carbon fiber plies.
4. These coated plies are stacked face-to-face in the midplane.
5. The composite is processed by vacuum-assisted resin transfer molding using the stacked plies.

Composites in large-scale can be produced by this method because only a small quantity of CNTs is used in these composites. Alignment of CNTs could be employed by self-assembly through functionalization or through substrate growth. These avenues for achieving CNT alignment are currently being explored.

Many conventional methods utilized for producing polymer composites such as resin transfer molding, vacuum-assisted resin transfer molding, and vacuum bagging can also be used to produce CNT-based nanocomposites. However, the main problem is the increase of viscosity of the resin when nanofibers are added directly into the resin.

As reinforcements, the initial potential for CNTs seems to be high because the tensile strength and stiffness are the highest compared to all known fibers, the aspect ratios are in the useful range, and they are light in weight. Still, the highly defective nature of multiwalled nanotubes, the roping of the single-walled nanotubes, and the agglomeration in both the CNTs decreased the opportunity to form high-performance advanced composites. It is found that the viscosity of the polymer mixture increased rapidly when higher amounts of CNTs are added, and this viscosity increase led to poor mixing conditions. Moreover, the CNTs tend to stay agglomerated no matter whatever may be the shear processing condition is. Approaches such as incipient wetting create a dispersion of nanotubes on or in a polymer powder before mixing, which tends to facilitate better dispersion. Better purification and the use of functionalization also aid better dispersion of CNTs and improve properties. While advances have been made for better distribution, it is clear that preparing larger-sized samples still has processing issues.

8.1.3.4 Nanofiber–Polymer Interface

The diverse results obtained by various researchers on the mechanical properties of CNT–PNCs imply that further research in terms of processing methods, CNT treatment, its dispersion and alignment in the matrix, as well as a better understanding and control of CNT–polymer interface is necessary to obtain CNT–PNCs with desirable and predictable properties and performance. In order to study the interfacial characteristics, two CNT nanocomposites with PS and epoxy matrices have been fabricated and characterized at micro- and nanometer scale using field emission scanning electron microscope (FESEM) and TEM (Xu et al. 2002; Wong et al. 2003). The tensile modulus is found to increase by about 10%, while there is only a slight increase in tensile strength (about 5%). Although the extrusion process used for the fabrication of CNT/PS nanocomposite is believed to align CNTs to some extent in the flow direction, the FESEM images showed the presence of CNT agglomerates, ranging from 5 to 20 µm in diameter. The reinforcing effect of CNTs is offset by the agglomerates, since they are acting as flaws or stress concentrators in the composite. Hence, the effective dispersion of CNTs in a polymer matrix still poses a challenge in the processing of CNT nanocomposites. It is found that the agglomerates are coated by PS, suggesting good wetting of CNTs by PS. Failure is found to occur around the CNT agglomerate within the PS matrix but not at the interface, suggesting strong adhesion of CNTs with PS.

Studies carried out on CNT/epoxy nanocomposite have also revealed good bonding between CNT and epoxy matrix. From the fracture surface, it is found that there is a failure in the epoxy matrix but not at the CNT–epoxy interface. However, CNT agglomerates are found at places where clean CNTs are found directly separated from the matrix as a result of poor local dispersion of CNTs. These regions are probably the places of failure initiation.

Mismatch in the coefficients of thermal expansion (CTE) between CNT and polymer can result in the formation of thermal radial stress and deformation along the tube when the polymer is cooled from the melt. Compressive radial stress results in closer CNT–polymer contact and local CNT deformation, which promotes mechanical interlocking. Thus, thermal residual stress from CTE mismatch could be a significant factor contributing to CNT–polymer adhesion (Liao et al. 2006).

Local nonuniformity along a CNT, such as varying diameter and bends and kinks, can contribute to CNT–polymer adhesion by mechanical interlocking. For instance, extra mechanical work is needed for the CNTs to pull out since the polymer should deform at rough contacts in order for them to slip past each other, compared to CNT–polymer contact along smooth CNT surfaces.

Better adhesion between CNT and polymer as well as nonuniformity in CNTs suggests that mechanical interlocking could be an important contributor for CNT–polymer adhesion.

As a summary, CNT–polymer interactions have the following contributions at the nanometer scale for better adhesion of CNT with polymer:

1. When there is no chemical bonding between CNT and polymer, the origins of CNT–polymer interactions are electrostatic and van der Waals forces.
2. Mismatch in CTEs is a significant factor contributing to both non-bond interactions and mechanical interlocking, because compressive residual stress may increase the contact area between CNT and polymer.
3. Local nonuniformity of a CNT embedded in polymer matrix may lead to nano-mechanical interlocking.

8.1.3.5 New Polymer Formulation

Because CNTs are highly polymeric and have molecular structures, it is expected that new polymer formulations may lead to another way of nano-composite development. The new formulations can integrate the nanotubes into the polymer, that is, the nanotubes become the integral part of the polymer chains. There are numerous polymers, in which nanotubes can be part of the polymeric formulation, but the length of the polymer linkage may limit the ability of each nanotube to be integrated. The chains may either react with themselves or react with the nanotube, thereby limiting the development of the new formulation. Figure 8.15 shows schematically the formation of cross-links through the walls of nanotubes, and this could be a fully integrated system, which is called a hybrid polymer. Integration relies on covalent bonding that provides mechanical strength. The composite should be designed to eliminate secondary bonding and form only network, as seen in thermosets.

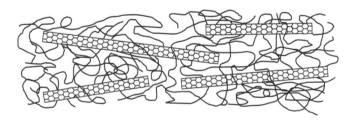

FIGURE 8.15
Schematic of the formation of cross-links in a polymeric material through the walls of nanotubes. (Reprinted from Gupta, R.K. et al., *Polymer Nanocomposites Handbook*, CRC Press, Boca Raton, FL, 2010, p. 15. With permission.)

8.1.3.6 Interpenetrating Polymer Network

Another promising approach to develop nanocomposite is by the use of entangled nanotubes. Generally, most forms of nanofibers tend to agglomerate and entangle. Much of the research on CNT-based nanocomposites is focusing on untangling and dispersing these nanotube systems. Instead, there are opportunities for using the nanotubes in the more natural tangled form. In this approach, the agglomerated or entangled nanotubes are dispersed in a monomer liquid, and the polymerization proceeds with the formation of IPN. In this form, the nanotubes and polymer chains are interlocked that provide secondary strengthening to the composite. Researchers are using this approach more and more, since the monomers can penetrate the agglomerates more easily than the polymer chains (McIntosh et al. 2006).

8.1.3.7 Properties of CNF–Polymer Nanocomposites

The increases in mechanical properties of CNT nanocomposites are not yet at extraordinary levels. The probable reasons could be insufficient bonding between CNT and matrix and the presence of relatively thick matrix between CNTs. The extent of bonding needed between the nanotube and the resin is still not understood. Various methods of functionalization have been developed and bonding between 1 in every 20 carbon atoms is not uncommon now. On considering the micron-sized reinforcements, though they are not bonded by every available surface atom, more bonds are present on them because of their larger size. Although an individual CNT has a high surface-to-volume ratio, an individual micron-sized fiber has more surface area compared to a CNT. Even though both of them are not completely bonded, the number of bonding on the individual reinforcement matters. Therefore, some degree of bonding is needed for a rule of mixture condition that must be fully assessed for CNTs.

One very interesting aspect of composites is that the load transfer from matrix to fiber becomes more difficult as the fibers get further apart. The fact is that as the distance between fibers increases to dimensions comparable to the fiber diameter, the matrix cannot transfer the load. It means that the distances between micron-sized fibers should be in microns or less, and the distances between nanofibers should be in nanometers or less. When agglomerated nanotubes are present in a composite, the distance between them are larger than nanometer sizes, and weak regions of the composite are created. Further investigation of this phenomenon is needed, to prove beyond doubt.

8.1.4 Particulate Nanocomposites

Among the various nano-reinforcements, two of the most extensively investigated nano-reinforcements are layered silicates (clay) and CNTs, which are 2- and 1-D elements, respectively, thus with appreciable aspect ratio (usually >100). Increasing attention has also been paid to low aspect ratio nanoparticles, such as carbon black, silica, and calcium carbonate due to their wide range of potential applications and low cost. Academic interest in this area has also arisen as PNCs containing low aspect ratio nanoparticles are a critical bridge between the conventional microcomposites and the ones filled with high aspect ratio nanoparticles where from the nanoparticle perspective, size is reduced and number density increased, but additional complexity is introduced by the orientational correlations of high aspect ratio particles.

Nanoparticles are used as reinforcing fillers in elastomers and plastics, which include synthetically prepared carbon black, silica, and zinc oxide, as well as naturally occurring calcite, clay, and talc minerals. These particles are commercially available in various sizes, from 10 to 100 nm. In the early days, zinc oxide had been used as primary reinforcing filler in the tire industry and then started using carbon black. Currently, carbon blacks have been widely used in tire manufacturing industry because of their good reinforcing effect. However, small amounts of zinc oxide are added in rubber compounds even today, because of their ability to enhance vulcanization/cross-linking processes involving sulfur. Recently, attention has been shifted to silica because it lowers rolling resistance and thus reduces fuel consumption.

In recent years, the focus of tire industry has shifted toward the better performance tires, such as low rolling resistance, snow traction, and wet traction tires. Silica-filled rubber tires have been known as "green tires" because of the lower rolling resistance, which decreases fuel consumption of automobiles. Silica particles have polar characteristics, while carbon black has nonpolar characteristics. Hence, it is difficult to disperse silica particles compared to carbon black particles due to the formation of agglomerates induced by polar or van der Waals forces.

Small particles tend to agglomerate due to interparticle attractive forces. As the particle size decreases, the surface area-to-volume ratio increases, which facilitate agglomeration. The agglomeration tendency is much greater for polar particles such as oxides and carbonates than for nonpolar particles such as carbon black.

Generally, there are two kinds of interactions that exist in particulate-filled composites. On the one hand, a polymer matrix adheres well to the surface of the particles forming an interface; on the other hand, the particles may also interact with each other and form aggregates. The homogeneous dispersion of nanoparticles in a polymer is usually very difficult due to the strong tendency of the ultrafine particles to agglomerate and the high melt viscosity of the polymer matrix. The two key factors controlling the performance of particulate nanocomposite are (i) uniform dispersion of particles in the matrix and (ii) the particulate–polymer interface. Fine dispersion without significant particle aggregation and adequate interfacial adhesion are essential to achieve high performance from the composite.

The attractive forces between small particles are also important while using these particles in rubbers and plastics. Nonpolar particles such as carbon black with minimal surface oxidation have attractive van der Waals forces, which cause them to agglomerate. Polar particles such as silica, calcium carbonate, titanium dioxide, and zinc oxide have dipole–dipole attractive forces, which are much stronger and lead to the formation of more difficult-to-disperse agglomerates.

8.1.4.1 Surface Modification

Very small polar particles are difficult to disperse by the conventional polymer-processing equipment. This is especially true with SiO_2 particles of size between 10 and 50 nm. Silica is a polar compound due to the differences in electronegativity between the silicon (1.90) and oxygen (3.44) atoms, while carbon black is nonpolar due to the same electronegativity between carbon atoms. Polar compound particles are more difficult to disperse than nonpolar compound particles. This is because polar particles are agglomerated not only by van der Waals forces but also from dipole–dipole forces. As the silica particle size decreases to nanometer dimensions, its surface area per unit volume increases, which increases both the van der Waals force and the hydrogen-bonding force between the silica particles. The silica particle in suspension is negatively charged at certain pH because of the loss of protons from H_3O^+ ions located in the spaces between the oxygen ions of the SiO_2 structure.

A rubber compound with high level of wear resistance with low rolling resistance can be made using silica particles. However, rubber compounds with silica particles were well known to be hard to process and consume significant amount of energy during processing due to their polar character. Silica particles tend to agglomerate easily even after mixing. To prevent agglomeration of silica particles, the interparticle forces should be reduced by coating the particle surfaces with additives, which also make the surfaces more compatible with the surrounding polymer.

Aliphatic carboxylic acids are known when technologists hydrolyzed fats and used the products as soaps and for the formation of emulsions. The chemical structures of these fatty acids were known in the nineteenth century. In the early twentieth century, it was found that the fatty acids would form monolayer on the surface of water and reduce its surface tension. The –COOH ends of the fatty acid would be immersed in water, and the hydrocarbon ends would extend vertically upward on the water surface. Later, it was found that by introducing small amounts of aliphatic carboxylic acids, such as stearic acid and propionic acid with fine zinc oxide particles, the incorporation of zinc oxide into NR could be enhanced. Similar behavior was found with other polar particles such as calcium carbonate and titanium dioxide. By the 1970s, various researchers found that the particulates modified with these carboxylic acids significantly reduced the compound viscosity.

Various coupling agents, such as bis(triethoxysilylpropyl) tetrasulfide, hexadecyltrimethoxysilane, propyltriethoxysilane, and octyltriethoxysilane, are available to treat silica-based materials. These coupling agents contain ethoxy groups or methoxy groups, which react chemically with the hydroxyl groups present on silica surface. The bis(triethoxysilylpropyl) tetrasulfide molecule contains four sulfur atoms located in between the triethoxysilyl-propyl groups. The sulfur atoms can chemically bond with double bonds present in rubber molecular chains, and the alkoxy groups chemically couple with the hydroxyl groups on silica surfaces through a hydrolysis reaction. Thus, this coupling agent chemically bridges the silica surface with rubber chains.

The silane modification helps silica dispersion in two ways: (i) The chemical coupling of silane on the silica surface changes the hydrophilic silica surface to a hydrophobic surface, which makes the silica surface compatible with hydrophobic polymer chains, and (ii) the silane treatment reduces the number of hydroxyl groups on the silica surface, which implies the reduction of hydrogen-bonding forces between silica particles. Hence, the silane-treated silica particle agglomerates can be easily broken during the mixing stage. In addition to that, the silane-bonded silica surface will have improved compatibility with polymer chains, and hence the polymer easily wet the silica surface. Thus, the coupling of silane on the silica surface improves silica particle dispersion as well as the bonding between the polymer and silica particles. The hydroxyl group in the hydrolyzed silane interacts with the hydroxyl group on the silica surface and forms a bond. It also reduces the polarity of the silica surface.

Alkoxysilanes undergo hydrolysis in the presence of an acid as well as a base. The hydrolyzed silanes undergo condensation reaction with hydroxyl groups on the silica surface. The coupling mechanisms between silane and silica surface are shown in Figure 8.16. The rate of acid hydrolysis is significantly higher than that of base hydrolysis and is less affected by other substituents on carbon. The hydrolysis and condensation reactions are also affected by the pH of the solution, and the rate of hydrolysis is slowest at approximately neutral pH. The condensation reaction rate is minimum at pH 4. Alcohols reverse the hydrolysis reaction, and hence the removal of alcohol during the reaction eliminates the possible reverse reaction.

Moisture affects the coupling reaction time and mechanical properties of the silane-treated silica compound. Enough water molecules should be available to achieve complete hydrolysis reaction of an alkoxysilane. When there is insufficient number of water molecules (i.e., less than 3 mol of water molecules for 1 mol of alkoxysilane), the hydrolysis reaction stops before all the alkoxy groups are hydrolyzed. The fully hydrolyzed silane would have 100% coupling on silica surface by the condensation reaction, whereas the partially hydrolyzed silane would react only with one or two hydroxyl groups in the silane depending on the hydrolysis condition, and one or two alkoxy groups per molecule remain in the silane.

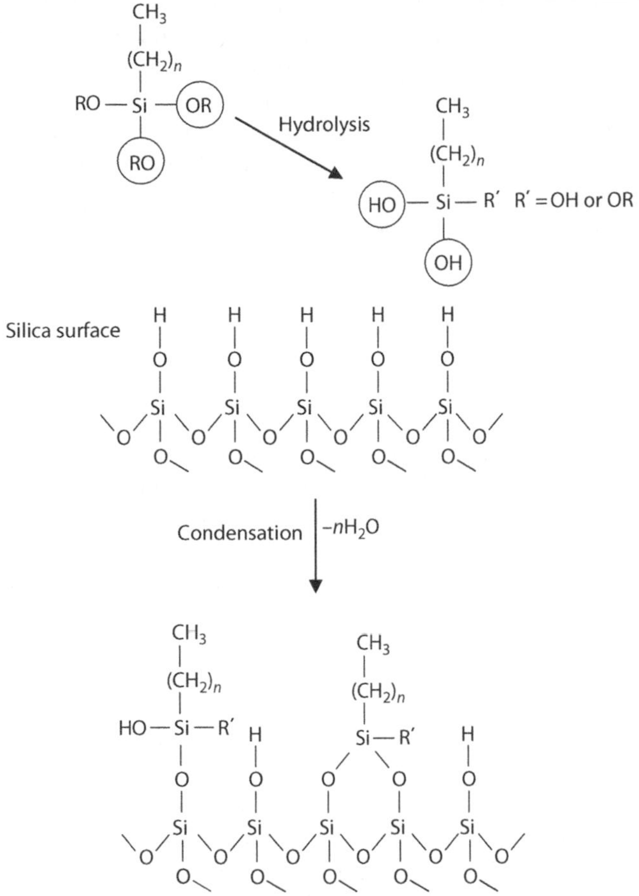

FIGURE 8.16
The coupling mechanisms between silane and silica surface. (Reprinted from Kim, K.J. and White, J.L., *J. Ind. Eng. Chem.*, 7, 50, 2001. With permission.)

Silica particles exist as normal, condensed, or water molecule emulsified form depending on the surrounding conditions. In the normal form, the silica particles are agglomerated with hydrogen/polar bonding. Under hot storage or shipping conditions, the nano-sized silica particles are chemically bonded due to a self-condensation reaction, and this is condensed silica. Once they are chemically bonded to each other, it is very difficult to separate them during processing. Sufficient moisture adsorbed on silica particles causes the water molecules to react with alkoxy groups of silane. Thus, when an alkoxysilane molecule approaches the silica surface, the water molecules present on the surface react with the alkoxy group. Even the partially hydrolyzed silanes undergo further hydrolysis that improves the bonding between silane and silica particles.

During mixing, the polar-bonded silica agglomerates disperse easily in the presence of large amounts of water molecules, because hydrogen bonding is about five times stronger than polar bonding. This facilitates the better dispersion of the silica particles in a polymer matrix.

It has been found that because of steric hindrance, the third alkoxy group usually does not hydrolyze. The unreacted alkoxy group may undergo hydrolysis reaction at a later stage and produces alcohols. However, water-treated silica systems have shown additional hydrolyzation of the alkoxy group remained in the silane. Hence, it generates less alcohol and improves silica particle dispersion as well as the mechanical properties of the composite. Once the silica particles are treated with silanes, the agglomeration tendency of silica is decreased due to reduced interparticle reactions and polarity.

The addition of bifunctional silanes such as bis(triethoxysilylpropyl) tetrasulfide and bis(triethoxysilylpropyl) disulfide improves the chemical bonding between silica particles and an unsaturated elastomer. The tire manufacturing industry has made extensive use of these bifunctional silanes as a chemical bonding agent as well as a processing aid. The silane-treated silica particles significantly improve the hardness and toughness of the elastomer, resulting in improved mileage, and traction on the road surface for automobiles. The temperature and moisture level are very important parameters in determining the degree of coupling during the chemical reaction between silane and silica surface.

Zinc ions not only improve the reaction kinetics of vulcanization but also improve the mechanical properties of elastomers. It has been found that the addition of a zinc ion containing fatty acid into bifunctional organosilane containing elastomers improved their mechanical properties. The zinc ion containing fatty acids aided the formation of cross-links between the sulfur groups in a silane and the double bonds in an elastomer chain. This increased the cross-link density led to a lower tan δ, increased modulus, lower heat buildup, and significantly increased blowout time.

$CaCO_3$ particles can be modified with organic agents to form a hydrophobic surface, which increases the interfacial interactions with hydrophobic polymers. Nonreactive surface treatment can also be tried that decreases the particle–particle and particle–polymer interactions. As a result, the aggregation of particulates decreases, and homogeneity, surface quality, and processability improve, but the strength of the composite decreases due to poor particle–polymer adhesion. The nonreactive coating materials include calcium and magnesium stearates, silicone oils, waxes, and ionomers. Reactive treatment leads to the formation of coupling between particulates and polymer matrix. Various reactive surface modifiers, such as alkylalkoxysilanes, alkylsilyl chlorides, and dialkyl titanates, have been used with many inorganic reinforcements. However, $CaCO_3$ does not have active –OH groups on its surface for the reaction with silanes. Thermoplastics, especially apolar polyolefins, are also inactive to silanes since their polymer chains do not contain any reactive groups. Reactive coupling is thus not expected in $CaCO_3$–polymer systems with alkyl silanes. However, the reactive coupling of $CaCO_3$ to PP has been achieved with the use of amino-functional silane coupling agents. Detailed studies have proved that amino-functional silane coupling agents adhered strongly to the surface of $CaCO_3$ and formed a polysiloxane layer (Demjen et al. 1999).

An alternative approach to modify the $CaCO_3$ for apolar polyolefin systems is grafting, that is, introducing polymer chains onto nano-$CaCO_3$ by irradiation graft polymerization or incorporating grafted polyolefins, such as maleic anhydride- or acrylic acid-grafted PP. In the case of graft polymerization, the low molecular weight monomers can easily penetrate into the agglomerated nanoparticles and react with the activated sites of the nanoparticles inside as well as outside the agglomerates. Introducing a third component with good compatibility with nano-$CaCO_3$ can lead to encapsulation of the nanoparticles by that component and hence a finer dispersion of the nanoparticles. Clearly, the third component can improve both the degree of nanoparticle dispersion and the interfacial interactions between nano-$CaCO_3$ and polymer matrices.

The significant influences of surface treatment on the morphology and properties of $CaCO_3$–PNCs have been reported for the $CaCO_3$–PP and $CaCO_3$–PET nanocomposites (Lorenzo et al. 2002; Wang et al. 2002). In the first case, nano-$CaCO_3$ fillers are treated with stearic acid under high-shear forces and then melt blended with PP. It is found that higher-shear force breaks up the agglomerates of nanoparticles more effectively and uniform surface coverage of stearic acid on nano-$CaCO_3$ decreases filler surface energy, thus improving the nanoparticle dispersion in the matrix. As a consequence, the mechanical properties have been found to be superior to those composites prepared using untreated nano-$CaCO_3$. In the second case, $CaCO_3$–PET nanocomposites are prepared by in situ polymerization using surface-treated $CaCO_3$. It is found that the surface-treated nano-$CaCO_3$ dispersed well in the polymer matrix and also they have good adhesion with the PET matrix.

8.1.4.2 Processing of Particulate Nanocomposites

Several processing techniques, such as in situ polymerization, melt bending, and solution casting, have been used to prepare particulate nanocomposites. Each of these methods has some distinct advantages and disadvantages.

8.1.4.2.1 In Situ Polymerization

In the in situ polymerization approach, PNCs are prepared by polymerizing the monomers or precursors in the presence of nanoparticles. Usually, the nanoparticles are first dispersed into the monomer or precursor, and the mixture is then polymerized by adding the appropriate catalyst. The main advantage of this approach is that a direct and easier dispersion of the nanoparticles into the liquid monomer or precursors is possible that avoids agglomeration of the nanoparticles in the polymer matrix. In situ polymerization method has been proved to be a successful approach for the preparation of $CaCO_3$–PNCs (Avella et al. 2001).

Another novel approach based on in situ polymerization is reverse microemulsion approach. A microemulsion of water in oil generally consists of small water droplets surrounded by a surfactant monolayer dispersed in an oil-rich continuous phase. The size of water droplets is usually less than 100 nm, which allows the controlled synthesis of inorganic nanoparticles within this small volume. Hence, each water droplet is called a microreactor and can be used for the chemical preparation of relatively monodispersed nanoparticles of various inorganic materials. The good dispersion of such microreactors in an oily phase can be taken as an advantage to modify nano-$CaCO_3$ with methyl methacrylate (MMA) and to polymerize MMA subsequently. The advantage of this approach is that the modification of nano-$CaCO_3$ particles and dispersion into the precursors are combined together. It has been reported that the glass transition temperature of the nanocomposite is higher than that of the matrix because of the reduced mobility of polymer chains by the formation of cross-links through the nanoparticles (Xi et al. 2004).

8.1.4.2.2 Melt Compounding

The melt compounding method has been extensively used to fabricate PNCs because of several advantages. Firstly, it is environmentally benign due to the nonutilization of organic solvents during processing. Secondly, it is compatible with conventional industrial processes, such as extrusion, injection molding, and other polymer-processing techniques, and hence can be easily commercialized. Thirdly, the high-shear forces experienced in some of these processing methods may permit the incorporation of significantly higher loadings of nanoparticles in comparison with the nanoparticle loadings achievable in an in situ polymerization process. Finally, this process may allow the use of polymers that are not suitable for in situ polymerization. Melt compounding of polymers with nanoparticles achieved only limited success for most polymer systems because of the high tendency of nanoparticles to form larger aggregates during this process. In addition, polymer degradation during melt compounding is sometimes severe, which cannot be avoided easily. These limitations should be considered while selecting the melt compounding method. In addition to that, the size and loading level of particles and melt-processing conditions such as processing temperature and time, shear force, and configuration of the processing equipment should also be controlled in order to achieve good dispersion of nanoparticles in polymer matrices.

To facilitate uniform dispersion of nanoparticles and avoid the degradation polymer, a master-batch method has been developed. In this method, nanoparticles are melt blended with a polymer at relatively high filler content and then compounded with the raw polymer again. A better dispersion of nano-$CaCO_3$ in a PP matrix by utilizing this method has been reported (Ren et al. 2000).

8.1.4.2.3 *Solution Casting*

It is used only when in situ polymerization and melt compounding processes are unviable due to the nature of a polymer matrix. An example is the preparation of biodegradable poly (L-lactide) composite films containing nano-$CaCO_3$ particles using methylene chloride as a solvent. The surface-treated nano-$CaCO_3$ particles are dispersed into a polymer solution made with methylene chloride using ultrasonication. After that, the suspension is cast in a Petri dish, and the solvent is allowed to evaporate at room temperature. The trapped solvent in the resulting solid is extracted with methanol and then dried under vacuum for a few days.

Finally, the as-cast films are compression molded. It is found that the composites exhibited better performance than pure poly (L-lactide) film during enzymatic hydrolysis due to the good dispersion of nano-$CaCO_3$ particles resulting in enormously large interfacial area susceptible to enzymatic hydrolysis (Fukuda et al. 2002).

8.1.5 Organic–Inorganic Hybrids (Nanocomposites)

The organic–inorganic hybrid materials are new types of nanocomposites, which have attracted much interest in recent years. These materials possess the advantages of both organic polymers and inorganic ceramics. Of the several kinds of inorganic materials, metal oxides such as silica, alumina, and titania are generally used because they are readily prepared in situ by the sol–gel process. The sol–gel process has provided promising opportunities for the preparation of a variety of organic–inorganic hybrid materials. The inorganic phase is formed within the polymer matrix by the sol–gel process.

8.1.6 Applications of Polymer Nanocomposites

In application where better structural integrity is required, nanocomposites with increased tensile strength, heat deflection temperature, stiffness, and toughness can be used. Nanocomposites can slow down the transmission of gases and moist vapor because of the presence of nano-reinforcements with high surface area-to-volume ratio. This renders the gas molecules to travel in a tortuous path. Nanoclays and nano-talcs are favored in such cases because of their platy structure, which effectively prevents the passage of gas molecules in large area. Applications such as food packaging, beer bottles, and tire inner-liners have made use of such nanocomposites with improved impermeability to gases. There is also potential exists to improve vapor barrier properties further, leading to use these nanocomposites in applications such as automotive fuel tanks, hoses, seals and gaskets, and PET bottles.

The timing belt covers of automotive engines are usually made with glass fiber-reinforced polyamide or PP. The timing belt covers made with nanocomposites showed good rigidity and excellent thermal stability, and at the same time there is a weight saving of up to 25%, owing to less inorganic content compared to the conventional composites.

Polyamide 6 films are also used for food packaging applications. It is found that a polyamide 6 nanocomposite containing 2 wt.% montmorillonite clay has only half of the oxygen permeability to that of polyamide 6 (Messersmith and Giannelis 1995). The influence of modifying agents on the mechanical properties of clay–epoxy nanocomposites has been studied (Lan and Pinnavaia 1994). Three types of organic modifying agents ($CH_3(CH_2)_{n-1} NH_3^+$, where $n = 7, 11, 18$) are examined, and it is found that the longer linear alkyl chains facilitated the formation of exfoliated nanocomposites, and there is significant increase in the tensile strength and modulus values.

The use of nanomaterials is becoming increasingly more important in fire-retardant applications. While there are a relatively small number of commercialized fire-resistant materials containing nanomaterials, there is a considerable amount of work going on to develop these kinds of materials. Work is underway to develop materials with unidirectional heat properties by taking the advantage of the thermal conductivity of CNTs, which is at least three orders of magnitude greater along the tube axis than other materials used to make protective clothing. Moreover, CNT materials are highly reflective, and hence they will absorb only a small fraction of the radiation from a fire.

A new class of nano-level hydrotalcite-type additives for polymers has been found to act as an effective fire retardant at levels that do not adversely affect the polymer properties to the same degree as currently used additives. The nano-level flame-retardant, layered double hydroxide gives intumescent, self-extinguishing properties when added with conventional flame retardants. This material can increase the volume of intumescent char formed when added with ammonium polyphosphate in epoxy coatings. This new nanomaterial is also useful in fire-resistant coatings on steel and in flame-retarding thermoplastics. Due to environmental concerns related to the halogenated flame retardants, these nanomaterials are gaining great interest in industry, because they are inherently nonhalogenated. In addition, they improve the modulus of the polymers in which they are added, in contrast to most nonhalogenated flame retardants that usually reduce the modulus and flexibility of the polymer.

Another area of research interest is to reduce the flammability of flexible foams used in furniture with nano-sized flame retardants. The use of penta-bromodiphenyl ether (penta-BDE) flame retardant in polyurethane (PU) foam is banned in many countries. As a consequence, there is a need to develop new user-friendly flame-retardant systems for PUs. It is a challenging task to introduce CNTs and layered double hydroxides in PU foams. The manufacturing of PU foams is a complex process in which reaction rates, bubble formation, coalescence to form cell structures, and phase separation are carefully controlled. Those nano-sized additives can affect bubble nucleation, viscosity, and surface properties.

8.2 Metal Matrix Nanocomposites

It is very well known that the properties of metal matrix composites (MMCs) are controlled by the size and volume fraction of the reinforcements as well as by the nature of the matrix/reinforcement interfaces. An optimum set of mechanical properties can be obtained when fine, thermally stable ceramic reinforcements are dispersed uniformly in the metal matrix.

8.2.1 Processing of Metallic Nanocomposites

The main types of nano-reinforcements used in metallic nanocomposites are short nanofibers, nanotubes, and nanoparticles. The major challenge in processing of nanocomposites with these types of nano-reinforcements is to disperse them uniformly in the metal or alloy matrix, since they are prone to easy agglomeration because of their high surface energy. In addition to that, the processing method should be selected in such a way that it ensures minimal damage to fibrous reinforcements due to applied stresses or due to reaction with the matrix material at elevated temperature. Stress applied during processing may damage nanofibers or align them in the matrix. The chemical reactions may lead to the degradation reinforcement properties and hence affect the properties of the composites. The processing methods used for the fabrication of conventional MMCs are very well applicable to the metallic nanocomposites also, but extra care should be exercised in handling and dispersing nano-reinforcements. Some processes might not be amenable for bulk production, and some other processes are restricted to a specific application. Hence, the right processing technique should be selected after considering all these factors. In the following sections, main emphasis is given for the processing of CNT-based metallic nanocomposites. Those processing methods can also be used for the fabrication of nanofiber as well as nanoparticle-reinforced metallic nanocomposites.

8.2.1.1 Powder Metallurgy Techniques

Powder metallurgy techniques have been used for the fabrication of metal matrix and ceramic matrix composites for a long time. The details have already been described under the processing of MMCs and CMCs. These techniques have been extensively used for the fabrication of CNT–Al and CNT–Cu composites. They are also used for the fabrication of a number of other MMCs with CNTs, such as Mg-, Ni-, and Ti-based alloys; some intermetallics; and Ag and Sn alloys. Dispersing CNTs is easy in this technique, and the level of dispersion depends on the size of starting metal powder. Smaller powders are preferable for better dispersion of CNTs, but they oxidize easily. Mixing in a liquid medium is generally practiced to prevent oxidation and heat generation during mixing in a mill. In most cases, high-energy milling has been employed to achieve better bonding as well as dispersion of CNTs. CNT clusters are broken down during this milling and attached to metal particles by mechanical interlocking. After thorough mixing, the powder mixture is consolidated by cold pressing and sintering, hot pressing, or spark plasma sintering (SPS).

Cold compaction and sintering have been used mainly for the fabrication of CNT–Al and CNT–Cu composites. This process met limited success due to the poor densities of the final composite. As described earlier, the second phase (CNTs) might act as obstacles for diffusion that leads to poor removal of pores during sintering. Furthermore, any CNT clusters remaining in the powder mixture will also be retained in the final product. Therefore, there is a need to densify the product further using some secondary deformation processing.

Hot pressing has the advantage of producing high density (>95%) with fine grained matrix. CNT–Al composites have been prepared by hot pressing a powder mixture at 520°C and 25 MPa pressure (Xu et al. 1999). There are CNT agglomerates mainly present at the Al grain boundaries and a few single CNTs within the matrix grains. Some carbide formation is also noticed. In another study, SWCNT–Al composites have been prepared by hot pressing a mixture of Al nanoparticles (50 nm) and SWCNT at a pressure of 1000 MPa. Even though ultrasonication is used to disperse the powders in alcohol medium, the composite is found to have CNT agglomerates (Zhong et al. 2003). Hot pressing has also been used to prepare CNT–Cu composites at 1100°C in argon atmosphere with a pressure of 32 MPa (Chen et al. 2005). Composites with Ni-coated CNTs showed better densification and good dispersion, whereas uncoated CNTs segregated at the grain boundaries.

SPS is a variation of hot pressing in which the heat source is a pulsed DC that is passed through the die or powder depending on the conductivity of the powder. A schematic of the SPS setup is shown in Figure 8.17. Spark discharges (produced due to pulsed current) produce rapid heating, which enhances the sintering rate. The heating rates achieved in this technique are quite high (up to 1000 K s^{-1}) compared to hot pressing. This method is more suitable for consolidation of nanopowders, since sufficient time may not be available for grain growth. Moreover, it helps in restricting the reactions between the metal matrix and CNTs.

CNT–Al and CNT–Cu composites have been prepared by SPS process. CNT–Cu composites are prepared using 100 nm and 20 µm size Cu powders and CNTs by consolidating at 750°C in SPS equipment for 1 min at a pressure of 40 MPa (Kim et al. 2004). Even though the dispersion is better with nano-Cu powder, the final density is almost similar to the composites made with micron-sized powder. In another study, the yield strength of CNT (10 vol.%)–Cu composite is found to be almost three times that of unreinforced metal, when molecular-level mixed powder is sintered using SPS (Kim et al. 2007). A comparative study made on hot-pressed and SPS

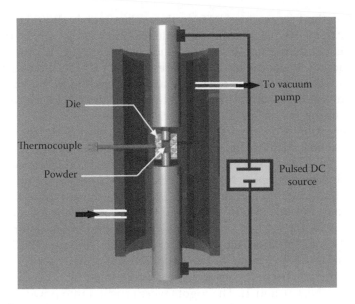

FIGURE 8.17
A schematic of the SPS setup. (Reprinted from Agarwal, A. et al., *Carbon Nanotubes: Reinforced Metal Matrix Composites*, CRC Press, Boca Raton, FL, 2010, p. 24. With permission.)

samples of CNT (1 wt.%)–Al composites showed that the SPS samples had slightly improved hardness, improved strength, and reduced wear loss over the hot-pressed samples (Kim et al. 2009). SPS has also been used in the fabrication of CNT–Fe_3Al composite, and this composite is found to have better compressive yield strength than the CNT–Fe_3Al composite fabricated by hot isostatic pressing (Pang et al. 2007). As mentioned earlier, the distribution of CNT is not affected during densification by these methods. Only shear forces applied during deformation processing can break the CNT clusters, which improve the dispersion and at the same time align them. This can also improve the density of the composite. The deformation processing has been used in many CNT–Cu and CNT–Al composites to improve the properties.

8.2.1.2 Melt Processing

The liquid-phase fabrication methods are efficient processes for the fabrication of MMCs because of their simplicity. As the conventional discontinuously reinforced MMCs, CNT-reinforced MMCs can also be fabricated by the melt-processing methods, such as casting and melt infiltration. If CNTs could remain stable during liquid-phase processing of high-strength metal matrix, outstanding nanocomposites may be obtained. The melt-processing methods are ideally suitable for large-scale manufacturing of CNT–MMCs, because bulk components can be made easily by these methods. However, the dispersion of low-density CNTs in the melt will be a real challenge.

A proper method for mixing CNTs in the molten matrix should be selected. Mixing with CNTs will be easy if the molten metal wets the CNTs. It is found that metals exhibiting good wetting, such as Cs and Se, fill the CNTs by capillary action. If the molten metal does not wet CNTs, then surface tension forces will tend to bring the CNTs together leading to the formation of clusters. In that case, some surface treatments to the CNTs are required to improve wetting.

Melt processing is mainly used for fabricating composites from low melting point metals, such as Mg and Al, and some bulk metallic glass (BMG) matrices. CNT–Mg and CNT–BMG composites have been fabricated by casting of CNT-dispersed melt. Ni-coated CNTs are dispersed in molten Mg at 700°C and then cast into ingots (Li et al. 2005a). The tensile strength and ductility are found to increase by 150% and 30%, respectively, for a composite containing 0.67 wt.% CNT, but the mechanical properties deteriorated at higher CNT concentration due to cluster formation. Zr-based BMG composites containing CNTs are made by using Zr(52.5)–Cu(17.9)–Ni(14.6)–Al(10)–Ti(5) BMG ingots. These ingots are made by arc melting a mixture of high purity Zr, Al, Ni, Cu, and Ti. The ingots were crushed into powders to a particle size of about 200 μm and mechanically mixed with CNTs. The mixture is compacted into a cylinder and melted rapidly in a quartz tube by induction heating under a high purity argon atmosphere and then cast in a copper mold. The segregation of CNTs is avoided by rapid heating and solidification (Bian et al. 2002, 2004). However, CNT clustering is observed in the composites, and there is an increase crystallinity of the BMG due to the formation of a ZrC phase.

Melt infiltration technique has a higher chance to produce composites with a uniform distribution of CNTs even at high loadings, but at the same time complete filling of pores is a real challenge. CNT preforms were prepared by compacting a mixture made with an organic binder (50 wt.%), followed by heating at 500°C–600°C. Composites are made by infiltrating these preforms with molten Al–Mg and Al–Si alloys using high pressure (Uozumi et al. 2008).

The rapid solidification processing (RSP) may help CNTs to survive the process because of the controlled melting and solidification steps. RSP has been used for the successful synthesis of $CNT/Fe_{80}P_{20}$ metallic glass nanocomposites (Wei et al. 2003). The addition of CNTs greatly enhanced the thermal stability of the glassy matrix and decreased the low-temperature electric resistivity by 70%. The activation energy of crystallization for this nanocomposite is found to be higher than that of $Fe_{80}P_{20}$ metallic glass. The onset temperature of crystallization for this nanocomposite is found to be almost 75°C higher than that of metallic glass.

Although melt processing is an economical method to produce CNT-based composites, it is not widely used compared to powder metallurgy techniques. This is due to the difficulties in dispersing CNTs in the melt and the formation of interfacial compounds with molten metal.

8.2.1.3 Thermal Spraying

Based on the source of heat, thermal spray techniques can be classified as wire arc, flame spray, plasma spray, high-velocity oxy-fuel (HVOF) spray, and detonation gun spray processes. A generalized thermal spray process with different process variables is shown in Figure 8.18. To form the starting nanocrystalline powders, conventional powders can be cryo-milled. Pure metals (except aluminum) require some alloying for the retention of nanocrystalline structure at elevated temperatures. For example, under the right conditions, precipitates form in Fe alloyed with Al, and these precipitates stabilize the nanoscale grain structure to 75% of the melting temperature of the pure Fe. To form WC/Co and Cr_3C_2/NiCr nanocomposites, the hard particles are broken into nanometer size, and then they are embedded in the binder phase. In all cases, the contamination of some nitrogen or oxygen is unavoidable. The nanoscale powders must be agglomerated so that grains on the order of 50 μm can be introduced into the thermal spray gun. Unlike sintering, this agglomeration does not prevent full densification during the thermal spray process. A reasonably narrow particle size distribution ensures uniform heating of all particles. Nanocrystalline feedstock is generally injected internally (inside the torch), but agglomerated powders can be injected externally. The type of flame or jet produced depends on the thermal spray technique; however, in any technique, the gas heating and gas flow parameters can

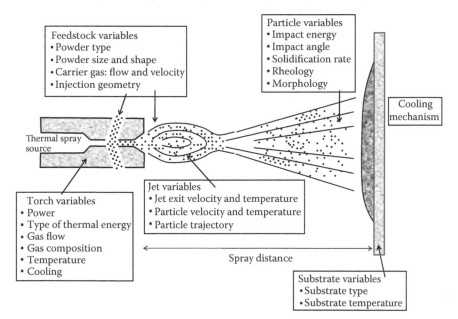

FIGURE 8.18
A generalized thermal spray process with different process variables. (Reprinted from Ajayan, P.M. et al., *Nanocomposite Science and Technology*, Wiley-VCH, Morlenbach, Germany, 2003, p. 12. With permission.)

control the velocity and temperature profile. The temperature and velocity profile, combined with the spray distance (powder travel distance) determine the temperature that the powders reach. Successive impact of particles in a molten or viscous state on the substrate forms a coating.

The ability to maintain the nanocrystalline structure during processing is critical to improve its properties because it is the nanoscale microstructure that is responsible for the unique properties. Several parameters are critical to maintain the nanocrystalline structure: (i) the thermal stability of the powders (nanocrystalline materials experience grain growth at temperatures well below than that observed for conventional materials, since the high surface area drives the early growth), (ii) the degree of melting during flight (this can be controlled by the spray distance, the temperature of the jet, and the velocity of the jet), and (iii) the cooling rate (a high cooling rate leads to high nucleation and slow grain growth rate, which promotes the formation of nanocrystalline structure). The systems that need to maintain their nanocrystalline structure even at elevated temperature are apt to have impurities or a second phase to stabilize the grain structure.

Agarwal and his group have reported the preparation of many CNT–MMCs by thermal spray process (Laha et al. 2004, 2005, 2007; Bakshi et al. 2008, 2009). Free-standing structures of 62 mm in diameter, 100 mm in length, and 0.3 mm wall thickness from Al–Si alloys reinforced with CNTs have been produced by plasma spraying. In plasma spraying, an electric arc is used to ionize an inert gas to produce a plasma jet with gas temperatures and velocities of ~10,000°C and 2,000 m s^{-1}, respectively. Al–Si powder is blended with 10 wt.% CNT for 48 h in a ball mill. The blending in a ball mill is found to be inefficient, since the blended mixture contained some CNT clusters. This mixed powder is used for plasma spraying onto a cryogenically cooled, rotating aluminum alloy mandrel. The presence of CNTs in the plasma-sprayed composites proved the ability of CNTs to withstand the harsh temperature environment of plasma. To improve the flowability of powder during plasma spraying and to improve the dispersion of CNTs in the powder, a spray-drying process has been tried. An aqueous suspension is made by dispersing Al–Si powder and CNTs in water containing a small amount of PVA binder. This slurry is spray dried to form spherical agglomerates containing fine metal particles with uniform distribution of CNTs. This powder is found to have good flowability, and it is possible to make cylinders of 5 mm wall thickness by the plasma spraying process.

HVOF spraying equipment consists of an internal combustion chamber in which a fuel (hydrogen, propylene, acetylene, and propane) is burnt in the presence of oxygen or air, which results in a hypersonic gas velocity. The particle velocities are higher than that achieved with plasma spray, but at the same time, the thermal energy produced is lower (it may reach only 2725°C), which reduces superheating and particle vaporization. HVOF spraying process has been used to produce relatively large-sized CNT-reinforced Al–Si hypereutectic alloy composites (Laha et al. 2005, 2007). Porosity in the composites prepared by this process is found to be less than that in the composites prepared by plasma spraying. The primary silicon size is also found to be lower in HVOF composites compared to plasma-sprayed composites, mainly attributed to the rapid heat dissipation and high-impact fracture in HVOF composite. The elastic modulus and hardness of the HVOF composites are found to be higher than the plasma-sprayed composites by 49% and 17%, respectively. Hence, HVOF could be a potential method for the preparation of CNT–MMCs.

8.2.1.4 In Situ Processing

The homogeneity of composite materials is very crucial for high-performance engineering applications, such as the automotive and aerospace applications. Nonuniform distribution of reinforcement can decrease the ductility, strength, and toughness of the composites. Nanoscale ceramic reinforcements formed in situ are generally dispersed more uniformly in the matrices of MMCs, leading to significant improvements in yield strength, stiffness, and resistance to creep and wear. A variety of processing techniques have been developed for the production of these composites. Particularly attractive among the several techniques available for producing in situ MMCs are the solidification processes in which the reinforcements are formed in situ in the molten metallic phase prior to its solidification. What make these processes so attractive are their simplicity, economy, and flexibility. The judicious selection of solidification processing techniques, matrix alloy compositions, and dispersoids can produce new structures with a unique set of useful engineering properties that are difficult to obtain in conventional monolithic materials. Specifically, the solidification conditions play an important role in dictating the microstructure and the mechanical properties of these materials. Microstructure refinement arising from RSP offers a potential avenue for alleviating reinforcement segregation and growth, thus resulting in enhanced dispersion hardening. For example, RSP of Ti/B alloys are effective in producing in situ Ti-based nanocomposites containing large-volume fractions of reinforcing particles. The reinforcement particles are formed in situ in Ti/B alloys either upon solidification or subsequently during the controlled decomposition of the resulting supersaturated solid solutions. The in situ formed TiC reinforcement in Al, Al/Si, and Al/Fe/V/Si matrices by the RSP route is more effective in improving the tensile properties of these matrices. The in situ composites exhibit excellent strength at room temperature as well as elevated temperatures.

Several workers have used the RSP route to fabricate TiC particulate-reinforced Al-based MMCs. In their work, master materials are prepared by melting a mixture of Al, Ti, and graphite powder in a graphite-lined crucible in an induction furnace under argon atmosphere. Rapidly solidified samples in ribbon form are prepared by chill block melt spinning. These ribbons are further milled into powders (100–250 μm), canned, degassed, and then extruded into rods. TiC particles of 40–80 nm are reported to be formed in situ and distributed uniformly in the aluminum matrix of grain size 0.3–0.85 μm (Tong and Fang 1998). On the other hand, the in situ TiC/Al composites prepared without RSP are found to have agglomerated TiC particles. These particles of size 0.2–1.0 μm accumulate at the aluminum subgrain or the grain boundaries. Another important advantage of RSP is its ability to produce alloy compositions not obtainable by conventional processing methods. In general, materials produced by RSP have excellent compositional homogeneity, small grain sizes, and homogeneously distributed fine precipitates or dispersoids.

8.2.1.5 Mechanical Alloying

In this process, alloying occurs from two or more elements as a result of repeated breaking up and joining of particles in a high-energy ball mill. The process can be used to prepare highly metastable structures, such as amorphous alloys and nanocomposite structures with high flexibility. Scaling up to industrial quantities is easily achieved in this process, but purity and homogeneous distribution of materials remain as a challenge. In addition to agglomeration, high-energy milling can provoke chemical reactions, which can influence the milling process and the properties of the product. However, this can be used to prepare magnetic oxide–metal nanocomposites by mechanically induced displacement reaction between a metal oxide and a more reactive metal. The great interest in mechanical alloying to produce nanocrystalline materials is due to its simplicity and the possibility to scale up.

Nanocrystalline materials in general are single- or multiphase polycrystals with individual grain sizes that are in the nanometer range. Owing to the extremely small dimensions, many properties of nanocrystalline materials are fundamentally different from, and often superior to, those of microcrystalline and amorphous solids. For example, nanocrystalline materials exhibit increased strength and hardness, improved ductility and toughness, reduced elastic modulus, enhanced diffusivity, higher specific heat, and enhanced thermal expansion coefficient compared to conventional polycrystalline materials.

Nanocrystalline materials can be formed by completely crystallizing amorphous solids by proper heat treatment. However, partial crystallization of amorphous alloys can lead to the formation of structure with nano-sized crystallites embedded within the residual amorphous matrix. This special nanocomposite exhibits excellent mechanical or magnetic properties. Zirconium-based nanostructured composites with amorphous matrices have been produced by die casting or mechanical alloying and subsequent consolidation at elevated temperatures. The distribution of finely dispersed nanocrystals increases the flow stress significantly without affecting the ductility much. For example, a Zr(57)–Al(10)–Cu(20)–Ni(8)–Ti(5) alloy containing 40 vol.% nanocrystals has better strength and ductility (Eckert et al. 2001). The Fe-/Si-/B-based amorphous alloys on proper annealing treatment have 10 nm diameter crystals embedded in the residual amorphous matrix and show excellent soft magnetic properties (Yoshizawa et al. 1988).

8.2.2 Metallic Nanocomposite Coatings

Metallic nanocomposite coatings can be formed by thermal spraying or electrochemical deposition. Both the techniques have certain advantages and disadvantages. Thermal spray can be used to form thick coatings; sometimes free-standing structures are also possible. There is no limitation on the substrate to be coated by thermal spray process.

8.2.2.1 Thermal Spraying

Thermal spraying is a commercially relevant, proven technique for producing nanostructured coatings. In this technique, agglomerated nanocrystalline powders are melted, impacted against a substrate at high velocity, and quenched very rapidly in a single step. This rapid melting and solidification results in the retention of a nanocrystalline phase or sometimes leads to the formation of amorphous structure. The retention of the nanocrystalline structure leads to enhanced wear resistance, higher hardness, and sometimes a reduced coefficient of friction compared to conventional coatings.

Plasma spraying and HVOF processes are the most widely used thermal spray methods for producing nanocomposite coatings. Vacuum plasma spraying and low-pressure plasma spraying have been used to produce WC/Co nanocomposite coatings. The high speed and low temperatures result in more strongly adhering and more homogeneous coatings with lower oxide content by the HVOF process.

8.2.2.2 Electrodeposition of Nanocomposite Coatings

Electrodeposition technique has been used to produce a variety of nanomaterials, which include nanocrystalline deposits, nanowires, nanotubes, and multilayer and nanocomposite coatings. However, it is mainly utilized for the preparation of coatings, and it is not feasible to prepare free-standing composites of larger thickness, because of the restriction in the maximum thickness that can be achieved by this technique (<200 µm). Nanocomposite coatings are formed by the co-deposition of nano-level reinforcements with a metallic matrix material. Two types of processes are involved for the incorporation of nanoparticles into metallic coatings, namely, the dispersion of particles in the electrolyte and electrophoretic migration of particles.

Figure 8.19 shows schematically the common sequence of events involved in the co-deposition of particles within metallic layers. It has been postulated that the co-deposition process occurs in five consecutive steps (Roos et al. 1990), which are (1) the formation of ionic clouds on the particles, (2) convection toward the cathode, (3) diffusion through a hydrodynamic

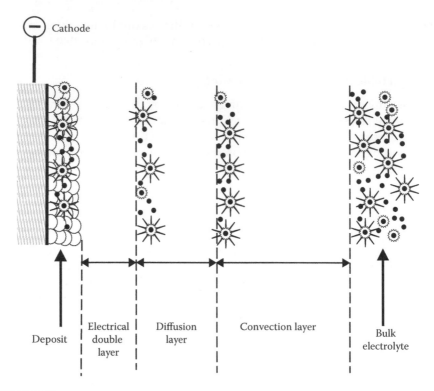

FIGURE 8.19

The common sequence of events involved in the co-deposition of particles within metallic layers to form nanocomposite coating. (Reprinted from Roos, J.R. et al., *J. Met.*, 42(11), 60, 1990. With permission.) ●, Particles (conductive or nonconductive); ○, Surfactant (ionic, non ionic, or organic); ✳, "Clouding" of particles by surfactants; ◉, "Clouding" of particles by cations; ◯, Deposited metal.

boundary layer, (4) diffusion through a concentration boundary layer, and (5) adsorption at the cathode. The simultaneous deposition of metal at the cathode entraps these particles within the coating.

A variety of nanoparticles have been successfully incorporated into metallic materials by electrodeposition, which includes Al_2O_3, diamond, SiC, SiO_2, ZrO_2, TiO_2, and Si_3N_4. Among many metallic deposits, copper and nickel are very common. The characteristics of nanocomposite coatings are dependent on many process parameters, including particle characteristics, such as type, morphology, and surface charge of particles. Other process parameters include particle concentration, electrolyte concentration, temperature, pH, type and concentration of surfactant, current density, pulse time (if applicable), hydrodynamics, and electrode geometry. Among these process parameters, the three important factors influencing the co-deposition process are the applied current density, the type and concentration of particles, and the bath agitation or electrode movement.

A variety of galvanostatic techniques can be utilized for the electrode-position of nanocomposite coatings, which includes DC, pulsed DC, and pulsed reverse current techniques. The DC technique has most commonly been used for the deposition of nanocomposite coatings. In this technique, the incorporation of nanoparticles occurs simultaneously with the reduction reaction of an ionic species to form the metal deposit.

In the pulsed DC technique, cathodic current is applied as pulses, and the pulse duration can be varied to control the deposition. This technique has enabled the incorporation of higher concentration of nanoparticles as well as to produce a wider range of deposit compositions with different properties. The pulsed reverse current technique imposes a cathodic current during the on-time and an anodic current during the off-time. This method has been the most successful for incorporating higher concentration of nanoparticles since a fraction of the deposited metal is removed (dissolved) during the off-time. This method also helps in producing a finer deposit with less agglomeration of particles because larger particles are preferentially removed during the reverse pulse period. The other advantages of this method are the possibility to use a lower concentration of nanoparticles in the electrolytic solution and the selective entrapment of similar size nanoparticles.

A combination of pulsed current and ultrasound has led to the formation of well-dispersed alumina nanoparticles in a nickel deposit (Qu et al. 2004). The introduction of ultrasonic vibration during electrodeposition can reduce nanoparticle agglomeration in the electrolyte solution as well as remove the adsorbed large nanoparticles from the metal deposit.

The current density is also found to influence the amount of nanoparticles incorporated in the metal deposit. An increase in current density during the deposition of α-Al_2O_3 nanoparticles with nickel using a sulfamate bath has resulted in a rough surface microstructure with fewer nanoparticles (Banovic et al. 1999). In another study, it has been demonstrated that the degree of nanoparticle incorporation is also dependent on the electrolyte type (Vidrine and Podlaha 2001). When a chloride electrolyte is used, the nanoparticle content was higher at low current density, whereas higher nanoparticle content is obtained at high current density while using a citrate electrolyte.

The amount of nanoparticles in the electrolyte solution has a significant effect on the quantity of nanoparticles incorporated into the metal deposit. In general, the volume fraction of nanoparticles in the metal deposit increases substantially on increasing the quantity of nanoparticles in the electrolyte solution, but there is a saturation limit, beyond that the increase in nanoparticle concentration may not increase the amount of nanoparticles in the deposit.

The size of particles affects the rate of incorporation of particles into the deposit. Studies on this aspect have revealed that as the size of nanoparticles becomes smaller, more number of particles is incorporated for the same duration. For example, the volume fraction of 300 nm Al_2O_3 particles in nickel deposit is found to be much higher compared to that of 50 nm Al_2O_2 particles. However, the number of 50 nm Al_2O_3 particles in the nickel deposit is much larger than that of 300 nm Al_2O_3 particles (Shao et al. 2002a,b). Another study has demonstrated that the distribution of particles is more uniform while using 20 nm SiC particles instead of 1 μm SiC particles (Kuo et al. 2004).

It is not necessary that the particles are always present at the boundaries in the nanocomposite coatings; they could also be embedded within the grain (crystal) itself (Gyftou et al. 2002). These embedded particles can disturb/ perturb the crystalline growth of a metal deposit, resulting in an increased number of defects in the crystals, and thus lead to a nanocrystalline structure with increased hardness.

Electrolyte bath agitation keeps the particles suspended in the electrolyte and transports the particles to the cathode surface. It is found from many investigations that increased agitation generally increases the amount of particles incorporated in the metal deposit. However, excessive agitation may lead to a lower quantity of particle incorporation in the deposit as it could remove the particles from the cathode surface by the vigorous hydro-dynamic forces before they are entrapped. For laboratory scale depositions, magnetic stirrers, rotating disk or cylinder electrodes, and parallel plate channel flow can be employed to achieve agitation.

Apart from bath agitation, the use of surfactants has been found to be effective in keeping the particles suspended in the electrolyte. Cationic and nonionic surfactants can be used to control surface charge on the particles for a stable suspension. The volume fraction of 100–400 nm SiC particles in a nickel deposit is found to increase with increasing concentration of cetyltri-methylammonium bromide surfactant. This is attributed to the availability of more surfactant molecules for adsorption onto the particle surface, which increase the surface charge on the particles leading to stronger attraction toward the cathode (Kuo et al. 2004). Other important parameters controlling electrodeposition include pH, bath temperature, and type and concentration of additives. The rate of deposition is found to increase with small increase in pH and increase in temperature (Wang et al. 2004).

Electrodeposited CNT–MMC coatings can be divided broadly into three categories based on the fabrication method and microstructure of the composite:

1. Composite coatings formed by co-deposition of CNTs and metal ions from the electrochemical bath on a substrate. This is the commonly used method to prepare CNT–MMC coatings by electrodeposition.

2. Metal is deposited on a uniform, aligned array of CNTs present on the substrate. A uniform distribution of CNTs in the metal matrix is achievable through this process.

3. Individual CNTs are coated by the electrodeposition of metal, and they are known as 1-D composites. They have potential applications as nano-sensors, electrodes, interconnects, and magnetic recording lead in computer applications and also can be used as the precursor powder to fabricate larger CNT–MMCs, through powder metallurgy technique.

The main challenge for electrodeposition of CNT–MMCs is also the uniform dispersion of CNTs in the metal matrix. The natural tendency of agglomeration hinders the uniform dispersion of CNTs in the electrolyte bath. This ultimately results in inhomogeneous distribution of CNTs in the coating. The most commonly used technique to improve CNT dispersion in the electrolyte bath is introducing agitation by means of mechanical stirrer, ultrasonication, or magnetic forces. Prior ball milling of CNTs also helps to improve dispersion, since it makes the CNTs shorter, thus reducing the tendency to agglomeration. Surface treatment of CNTs can modify their surface charge and reduce the formation of agglomerates. Various surface treatment methods have been tried, which include acid treatment and sensitization, acid cleaning, adding surfactants, and pre-coating the CNTs with a metal. Many CNT–MMC coatings have been produced through co-deposition route (Chen et al. 2002; Tan et al. 2006; Sun et al. 2007; Guo et al. 2008; Yang et al. 2008; Popp et al. 2009).

The second type of CNT–MMC coatings is prepared by the electrodeposition of metal in the voids between the CNTs in an array or a network. A CNT array on nonmetallic substrate, such as silicon, alumina, or glassy carbon, can be prepared by two routes: (1) growing aligned CNTs on a substrate by CVD process (Ngo et al. 2004; Popp et al. 2009) or (2) dropping CNT suspension on a substrate followed by drying (Kang et al. 2007). The presence of CNTs on the substrate prior to deposition ensures uniform distribution in the composite. However, the coating thickness by this technique is restricted to ~50 μm, which is much lower than the thickness achieved by co-deposition (~180 μm).

The third type of CNT–MMC formed by electrodeposition is metal-coated individual CNTs. They can be obtained by co-deposition on a substrate and subsequent separation from the substrate by ultrasonication. This process often results in a nonuniform coating over the CNTs. A better way to control the thickness and uniformity of coating is by coating the vertically aligned array of CNTs inside a porous substrate (e.g., porous alumina template), which can be leached away after the coating (Xu et al. 2003). Electroless deposition can also be used to prepare such 1-D composites. The CNTs are suspended in an electroless plating bath, and the reducing agent present in the bath reduces the metallic salt to metal, which coats the CNTs. Prior to coating, the CNTs are functionalized by surface treatment to create nucleation sites on the CNT surface. The preparation of 1-D CNT composites with Ni, Cu, Ag, and Co has been reported by many authors (Chen et al. 2000; Dingsheng and Fingliang 2006; Zeng et al. 2007).

There appear to be no investigations on the influence of nanoparticles shape (e.g., nano-rods, platelets, angular, or equiaxial) on the properties of nanocomposite coatings. The different shapes may have different effects during the incorporation and can affect the final microstructure.

The effects of particle inclusion on the crystallization behavior of metallic matrix during electrodeposition have not been studied. This aspect is also important, since the inclusion of particles may cause a reduction of cathode surface area for deposition (if the particles are nonconductive), an increase of cathode surface area (if the particles are conductive), or a modification of deposit microstructure.

8.2.3 Properties and Applications

Metal carbide/ductile metal systems are used as cutting tool materials, because the carbide provides the necessary hardness and the metal provides toughness. For example, composites such as WC/Co and WC/TiC/Co are commonly used in cutting and forming tools. When the particle or grain sizes of the components in these nanocomposites are in the nanometer range, then they have much better mechanical properties. Typically, reductive decomposition of W- and Co-containing salts, followed by gas-phase carburization, is used to prepare WC/Co nanocomposites. However, this process produces carbon-deficient metastable carbide phases with inferior mechanical properties. In an alternative approach, a polymer precursor, such as polyacrylonitrile, is used as the carbon source to obtain WC/Co nanocomposites with good mechanical properties (Zhu and Manthiram 1994). Another interesting nanocomposite system is TaC/Ni, because of its excellent thermal stability and outstanding mechanical properties. These nanocomposites can be prepared by devitrification of sputtered amorphous films of Ni/Ta/C. The TaC particles of size ~10–15 nm are uniformly distributed in the Ni matrix with the grain size of about 10–30 nm even at 700°C. No grain growth is observed in these nanocrystalline duplex-phase composites suggesting excellent thermal stability. The measured hardness value of this composite matches to that of conventional WC/Co nanocomposites at a much reduced volume fraction (0.35) of particles.

WC/Co coatings are of great interest because their excellent wear resistance is already known. It is expected that the nanostructuring could further increase the wear resistance and decrease the coefficient of friction. However, thermally sprayed WC/Co coatings do not always exhibit improved wear properties. It has been reported that WC/Co coatings formed by HVOF exhibit decreased wear resistance due to the decomposition of the carbide during spraying (Stewart et al. 1999). Nanoparticles attain temperatures almost 500°C higher than the micron-sized particles. Vacuum plasma spraying, however, can produce coatings with significantly improved wear resistance and lower coefficient of friction, presumably because of the prevention of the oxidation of the carbide phase.

In general, the nanocomposite coatings are found to have better microhardness and corrosion resistance than the respective metallic coatings. For example, the presence of TiO_2 nanoparticles has increased the corrosion resistance of nickel deposit (Li et al. 2005b). The wear resistance of ZrO_2/Ni nanocomposite coatings is found to be better than a pure nickel coating (Wang et al. 2005).

Cr_3C_2/NiCr composite coatings are also used in applications where wear resistance is required. They have an added advantage over WC/Co, which is their excellent corrosion resistance. Nanostructuring of these coatings has also resulted in improved hardness, as well as reduced coefficient of friction. The improved homogeneity at the nano-level, as well as a high density of Cr_2O_3 nanoparticles formed by oxidation during spraying, causes the improved properties compared to conventional materials. Key to achieving excellent properties in the coating is minimizing the degree of melting, so as to maintain the nanostructure in the final coating. However, significant deformation or splatting of particles is required during spraying to ensure a large contact area between the particles and substrate. Thus, some melted particles led to continuous, good-quality coatings.

8.3 Ceramic Nanocomposites

Many efforts are under way to develop high-performance ceramics; those can have potential applications such as highly efficient gas turbine, aircraft, spacecraft, and automobile components. Even the best processed ceramic materials pose many unsolved problems such as relatively low fracture toughness, degradation of mechanical properties at high temperatures, and poor resistance to creep, fatigue, and thermal shock. Attempts to overcome these problems have involved incorporating second phases such as particulates, platelets, whiskers, and fibers in the micron-sized range in the matrix. Even though there is some improvement in those properties, the improvement is not meeting the goals. Recently, the concept of nanocomposites has been considered, in which the microstructures of ceramic matrices are controlled by the incorporation of nanometer-sized second phases. The dispersed phase can go into the grains or grain boundaries, that is, intragranular or intergranular (Figure 8.20). These nanocomposite materials can be produced by adding a small amount of additive. The additive segregates at the grain boundary or precipitates as molecular- or cluster-sized particles within the grains or at the grain boundaries. Optimized processing condition can lead to excellent structural control at the molecular level in most nanocomposite materials. Intragranular dispersions aim to generate and fix dislocations during the processing and/or to control the size and shape of the matrix grains. The intergranular nano-dispersoids play an important role in the control of the grain boundary structure of oxide and nonoxide ceramics, which improves their high-temperature mechanical properties.

Significant scientific effort has been directed toward making ceramics more flaw tolerant through the control of their microstructure. A best example of tough ceramics through microstructural design is self-reinforced silicon nitrides, which were developed during the 1970s. These ceramics have high

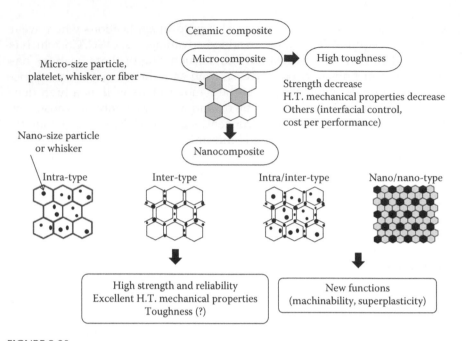

FIGURE 8.20
Microcomposites and new concept of nanocomposites with inter- and intragranular designs and their characteristics. (Reprinted from Niihara, K., *New Design Concept for Advanced Materials*, http://www.sanken.osaka-u.ac.jp/labs/scm/about_resE.html, 1996. With permission.)

toughness and room-temperature strength, along with good corrosion and oxidation resistance. However, their high-temperature (>1000°C) properties are compromised by low creep resistance and the occurrence of subcritical crack growth. These are caused by the softening of the glassy phase present at the grain boundaries as the temperature is increased. A possible way to overcome this problem is the fabrication of SiC/Si_3N_4 nanocomposites by incorporating SiC nanocrystals.

Considerable attention has been paid to develop "functionally graded nanocomposite materials," in which a gradually varying dispersoid (nanoparticles)-to-matrix ratio in chosen directions continuously changes the material properties. An example of such a material is SiC-dispersed graphite, which has served well as thermal barriers of the space shuttle due to its excellent resistance to oxidation and thermal shock. One route to prepare these composites is chemical vapor deposition using multicomponent gas reactions. For example, 10–100 nm sized SiC dispersions in pyrographite can be prepared using precursors of $SiCl_4/C_3H_8/H_2$ or $SiCl_4/CH_4$ by CVD. The entire range of compositions from 100% carbon to 100% SiC has been prepared by this method (Wang et al. 1990). Changing the deposition conditions can control the morphology of the second phase and hence the microstructure of the composite, which in turn influences the mechanical properties.

Significant interest in ceramic nanocomposites was generated in 1991 when Niihara reported large improvements in both the fracture toughness and the strength of ceramic when nanometer range particles are embedded within a matrix of larger grains and at their grain boundaries. These ceramics reportedly showed up to 200% improvement in both strength and fracture toughness, better retention of strength at high temperatures, and better creep resistance (Niihara 1991).

Nanocrystalline carbide-embedded composites, particularly those with amorphous or diamond-like carbon (DLC) matrices are more suitable for tribological applications. The challenges here are to minimize the formation of soft graphite-like phase during synthesis and to retain a high fraction of the DLC matrix. Pulsed-laser deposition has been used successfully to prepare ~10 nm TiC/TiCN nanocrystals embedded uniformly in DLC. Hardness values as high as 60 GPa, coefficient of friction values as low as 0.1, and high toughness values have been achieved in these composites. The two-phase structure in the nanocomposites provides crack deflection mechanisms, which reduce the tendency toward easy brittle failure in these hard composites (Voevodin et al. 1997).

Dispersing metallic particles into ceramics improves their mechanical properties, especially the fracture toughness. A wide variety of properties, including magnetic, electric, and optical properties can also be tailored in the nanocomposites due to the size effect of nano-sized metal dispersions. Conventional powder metallurgical methods and solution-based chemical processes such as sol–gel and co-precipitation methods have been used to prepare composite powders for metal/ceramic nanocomposites. Some of the composite powders are W, Mo, Ni, Cu, Co, or Fe/Al_2O_3; Ni or Mo/ZrO_2, Fe, Ni, or Co/MgO; and so on. The composite powders are sintered in a reducing atmosphere to give homogeneous dispersions of metallic particles within the ceramics. Fracture strength and toughness are enhanced due to microstructural refinement by the metallic nanoparticles and their plasticity. Ferromagnetism is a value-added supplement to the excellent mechanical properties, when transition-metal nanoparticles are dispersed. A good magnetic response to applied stress is found in these ferromagnetic–metal/ceramic nanocomposites, allowing the possibility of remote sensing of initiation of fractures or deformations.

Nanocomposite technology is also applicable to functional ceramics such as ferroelectrics, piezoelectrics, varistors, and solid ionic conductors. Incorporating a small amount of nanoparticles into $BaTiO_3$, ZnO, or ZrO_2 can significantly improve its strength, hardness, and toughness, which are very important in creating highly reliable devices operating under severe conditions. In addition, the electric conductivity of ceramics can be enhanced by dispersing metallic nanoparticles or nanowires. Dispersion of soft materials into a hard ceramic material generally decreases its mechanical properties, especially the hardness. However, the addition of soft nanoparticles into several kinds of ceramics can improve their mechanical properties. For example, adding soft hexagonal BN to Si_3N_4 can enhance its fracture strength not only at room temperature but also at very high temperatures up to 1500°C (Ajayan et al. 2003a). Some of the nanocomposite materials also exhibit superior thermal shock resistance and machinability because of the characteristic plasticity of second phase and the interface regions between that phase and the hard ceramics.

Advanced ceramic materials that can withstand high temperatures (>1500°C) without degradation are needed for applications, such as structural parts of engines, gas turbines, heat exchangers, and combustion systems. Such hard, thermally stable, oxidation-resistant ceramic composites are also in demand for aerospace applications. Silicon carbide–silicon nitride nanocomposites can perform best in adverse high-temperature oxidizing conditions. Such nanocomposites are optimally produced from amorphous silicon carbonitride, which forms the nanocomposite consisting of microcrystals of Si_3N_4 and nanocrystals of SiC. This silicon carbonitride is obtained by the pyrolysis of polyhydridomethylsilazane $[CH_3SiH–NH]_m$ $[(CH_3)_2 Si–NH]_n$ at about 1000°C. The oxidation resistance of this nanocomposite arises from the formation of a thin silicon oxide layer (Riedel et al. 1995).

8.3.1 Processing of Ceramic Nanocomposites

Processing is the key to the fabrication of nanocomposites with optimized properties. Several processing techniques have been used to prepare ceramic nanocomposites, including conventional powder processing, sol–gel processing, and polymer precursor processing. Some of the commonly used processes for creating ceramic nanocomposites are described in the succeeding text.

8.3.1.1 Mechanochemical Synthesis

High-energy ball milling can also induce chemical changes in nonmetallurgical systems, including silicates, ferrites, and organic compounds. Ceramics like SiC, Al_2O_3, ZrO_2, and Si_3N_4 are excellent candidates for demanding structural applications because of their better mechanical and thermomechanical properties. The incorporation of fine SiC and Si_3N_4 particles (<300 nm) in an alumina matrix improved the fracture toughness from 3 to 4.8 MPa $m^{1/2}$ and strength from 350 to 1050 MPa (Sawaguchi et al. 1991). These high strength values are retained to about 1000°C. Co-milling of SiC and Si_3N_4 powders followed by hot pressing produces nanocomposites possessing both inter- and intragranular SiC particles. Intragranular particles are more effective in toughening, because they are mainly responsible for crack deflection and crack impediment. Intergranular particles are often detrimental since they may initiate cracks, but they could provide some advantages such as to refine the grains during processing. The toughness of the composite depends on the volume fractions of these two types of dispersoids, and precisely controlling these fractions is a challenging task (Xu et al. 1997).

A metal–metal oxide nanocomposite can be produced by a displacement reaction between a metal oxide and a more reactive metal in a high-energy ball mill. The reaction may progress gradually, producing a nanocomposite powder, which on consolidation forms the nanocomposite product. For example, the reduction of metal oxides with aluminum during reactive balling can produce nanocomposites of alumina and a metal or metallic alloy. Such ceramics with ductile metallic inclusions produce toughened materials with superior mechanical properties. These nanocomposites also have better thermal shock resistance due to better metal–ceramic interfacial strength (Osso et al. 1998).

Direct milling of a mixture of iron and alumina powders has been used to prepare a nanocomposite consisting of nanoparticles of iron embedded in an insulating alumina matrix. The average particle size of iron particles can be reduced to the 10 nm range. The magnetic properties (saturation magnetization and coercivity) of this system can be tailored by changing the iron content, particle size, and the internal stresses accumulated during milling (Pardavi-Horvath and Takacs 1992). Systems containing smaller ceramic magnetic particles embedded in a nonmagnetic metal matrix can also be prepared by high-energy ball milling. For example, nanocomposites of Fe_3O_4 particles dispersed in Cu have been prepared by ball milling a mixture of Fe_3O_4 and Cu powders, as well as by reactive milling of CuO and metallic iron. Nanocomposites produced by both the processes have a significant super paramagnetic fraction due to the very small particle size of the dispersed magnetic phase. In situ chemical reactions influence the microstructure and magnetic properties of the product. The metastable nanocrystalline/amorphous structures are inherently formed in mechanical alloying due to repeated deformation and fracture events during collisions of particles with the balls. Deformation-induced defect density and the local changes in temperature due to impacts affect the diffusion coefficients of several species. In fact, the final microstructure and stoichiometry of mechanically milled samples often reflect the competing processes of milling-induced disorder and diffusion-controlled recovery, rather than being solely dependent on the starting material characteristics (Jang and Koch 1990).

8.3.1.2 Sol–Gel Processing

Aerogels, due to their highly porous structure, are clearly an ideal starting material to prepare nanocomposites. Aerogels are generally made by sol–gel process using selected metal alkoxides. They are extremely light with densities range from ~0.5 to 0.001 g cm⁻³ and highly porous, having nano-sized pores. Nanocomposites are made by using these aerogels as substrates and introducing one or more additional phases. The nanocomposites most commonly made are silica-based nanocomposites using silica aerogel, but this can be extended to other aerogel systems.

The aerogel-based nanocomposites can be fabricated in various ways: The second phase can be incorporated during the sol–gel processing of aerogel, through the vapor phase after supercritical drying or by the chemical modification of the aerogel backbone through reactive gas treatment. Many varieties of nanocomposites can be produced by these general approaches.

In the first approach, a nonsilica material is added to the silica sol. The added material may be a soluble organic or inorganic compound or an insoluble powder. It should be noted that the additional components must withstand the subsequent processing steps used to form the aerogel. The conditions encountered during the drying process using CO_2 are milder than those using alcohol and are more amenable to forming composites. If the added components are insoluble materials, steps must be taken to prevent them settling. The addition of soluble inorganic or organic compounds provides the possibility to prepare unlimited number of composites. However, two criteria must be met to prepare a composite by this route. First, the added component must not interfere with the chemical reactions taking place during the sol–gel process. The advance prediction of possible interference maybe difficult, but it is rarely a problem if the added component is reasonably inert. The second problem is the loss of the added phases during alcohol soaking or supercritical drying steps. This can be a significant impediment especially when a high loading of the second phase is desired in the final composite. It is often useful to use a chemical binding agent when a metal complex is added, which can bind to the silica backbone and chelate the metal complex. Many people use this process to prepare nanocomposites based on silica aerogels or xerogels. The resulting nanocomposites consist of a dried gel with metal atoms or ions uniformly dispersed throughout the material. Thermal post-processing creates nano-sized metal particles within the ceramic matrix. Such composites can have many applications like catalysts for gas-phase reactions and substrate for the catalyzed growth of nanostructures.

Vapor infiltration through the open pore network of aerogels provides another route to prepare various aerogel-based nanocomposites (Hunt et al. 1995). In fact, the materials adsorbed in the aerogels can be modified into solid phases by thermal or chemical decomposition. Some other materials that have a porous interior structure, such as zeolites, can also be used instead of aerogels. Recently, SWCNTs have been deposited within the pores of zeolites by vapor phase infiltration. These nanocomposites have unique properties, such as superconductivity (Wang et al. 2000a).

Carbon–silica aerogel nanocomposites have been made by the decomposition of hydrocarbons at high temperatures (Song et al. 1995). The fine structure of aerogel allows the decomposition to take place at a low temperature (200°C–450°C). Carbon loading ranges from 1% to 80% have been obtained. The carbon deposition is uniform throughout the aerogel at lower loadings, but the carbon deposition is localized at the exterior surface of the aerogel at higher loadings. These nanocomposites have interesting properties, such as appreciable electric conductivity and higher strength compared to the aerogel.

Silicon–silica aerogel nanocomposites can be made by the thermal decomposition of organosilanes in the silica aerogel. The thermal decomposition of various organosilanes adsorbed in the silica aerogel forms deposits of elemental silicon. The rapid decomposition of the silane leads to localized deposition near the exterior surface of the aerogel. The nanocomposite containing 20–30 nm diameter silicon particles exhibits strong visible photoluminescence at 600 nm.

Transition-metal–silica aerogel nanocomposites have been prepared using transition-metal complexes. Transition-metal complexes deposit metal compounds uniformly throughout the aerogel. The metal compounds can be thermally decomposed to the respective metals. Due to the disperse nature of the metallic phase, it has high reactivity and can be readily converted into metal oxides, sulfides, or halides. The loading of the metallic phase can be increased by repeated deposition steps. These nanocomposites contain the desired metal species with sizes in the range of 5–100 nm in diameter.

8.3.1.3 Polymer-Derived Ceramic Nanocomposites

In polymer-based processing, an amorphous Si/N/C powder is prepared by cross-linking and pyrolysis of polymethylsilazane. While the conventional powder-based process leads to a micro-nanostructure with nano-sized SiC particles mainly present inside Si_3N_4 grains, the polymer-processing route results in nano–nano structures. Hybrid polymer/powder processing can be used to prepare SiC/Al_2O_3 nanocomposites. In this process, alumina powder is coated with a silicon-containing polymer and then pyrolyzed to form SiC nanophase.

8.3.1.4 Novel Processing Method

A novel processing route has been developed to prepare well-dispersed SiC/Al_2O_3 nanocomposites (Sakka et al. 1995). In this process, micron-sized particles of the two ceramics are colloidally dispersed and consolidated to form a compact with uniform distribution. The SiC particles are then oxidized to reduce the size of cores to nanometer range. Subsequently, the interfacial reaction between the oxidized SiC (silica) particles and the alumina matrix produces mullite. The advantage of this process is that the particle size of SiC need not be brought down to nanoscale by milling and can be reduced by controlled oxidation. Moreover, the shrinkage during sintering is small due to the volume increase that occurs during reaction sintering.

8.3.2 Carbon Nanotube–Ceramic Nanocomposites

The nanotubes promise to increase the fracture toughness of composites by absorbing energy during deformation, which will be important, especially for ceramic matrix composites. Possible applications of these nanocomposites are in lightweight armor and conductive ceramic coatings. An increase in fracture toughness (about 10%) has been observed in CNT/nanocrystalline SiC ceramic composite fabricated by hot pressing at 2000°C (Ma et al. 1998).

Nano-sized ceramic powders with CNTs provide another opportunity for developing ceramic–matrix composites with enhanced mechanical properties. Addition of CNTs to alumina can result in lightweight composites with greater strength and fracture toughness. However, the mechanical properties of such composites strongly depend on the processing method and surface treatment of the CNTs. An exciting possibility, as well as a processing challenge, is incorporating CNTs into alumina matrix to improve its mechanical properties. Alumina with 5–20 vol.% MWCNT nanocomposites have been fabricated by dispersing CNTs in γ-alumina powder. The γ-alumina transformed to α-alumina (with mean grain size close to 60 nm) during sintering at 1300°C. The CNTs are lightly oxidized at 640°C in air, to remove the disordered carbonaceous material for easy dispersion. The alumina powder and CNT mixtures are dispersed in organic solvent using ultrasonication, dried, and then hot pressed in a graphite die at 1300°C and a pressure of 60 MPa. The structure of the CNT is found to remain the same before and after processing, without any visible degradation, and the sintered density is >97% of the theoretical density (Chang et al. 2000).

Carbon–carbon nanocomposites have been prepared by dispersing SWCNTs in isotropic petroleum pitch, followed by pyrolysis. These nanocomposites showed better electric and mechanical properties than isotropic pitch-based carbon. The advent of new carbon nanostructures suggests that novel carbon–carbon nanocomposites can be prepared with properties far superior to those of conventional carbon fiber-reinforced carbon composites. The challenges lie in achieving proper dispersion of these nano-reinforcements and tailoring the interface between the nano-carbons and the matrix.

8.3.3 Ceramic Nanocomposite Coatings

Improved wear resistance and friction properties and high-temperature stability are important characteristics of coating materials used in applications such as cutting tools. Most widely used hard coating materials are TiN, TiC, TiAlN, CrN, DLC, WC, $MoSi_2$, Al_2O_3, etc. For improved coatings with lower friction, increased lifetime, increased toughness, and higher thermal stability, new types of materials are being considered, including nanocomposite materials. Nanocomposite structures such as multilayer or isotropic coatings made from nanoscale entities can have properties superior to single-phase materials.

Nanocomposite coatings usually consist of two or more phases combined as multiple layers or as isotropic multiphase mixtures. Multilayer coatings typically have a total thickness of several micrometers, and the different layers provide toughness and other properties, such as oxidation resistance, apart from better tribological properties. Gradient layers are also often introduced to overcome the problem of vast differences in the thermal expansion coefficients of adjacent layers, which would otherwise cause internal stresses and delamination at the interfaces between layers. In multilayer nanocomposite coatings, when the thickness of each layer is in the nanometer range, superlattice effects can increase the hardness and other properties of the coatings (Shinn and Barnett 1994). The hardness of such coatings (e.g., TiN/VN) can be orders of magnitude higher than that of the corresponding base materials. The hardness increase in these multi-layer films arises mainly from the hindrance of dislocation movement across the sharp interfaces between two materials having vastly different elastic properties and lattice mismatch. It should be noted that if the individual layers are very thin (typically 3–5 nm), the hardness increase cannot be obtained because sufficient strain field may not be developed. The variation of composition at the interface (sharp or gradual) also affects the hardening mechanisms, because high-temperature interdiffusion between layers can decrease the sharp variation in shear modulus. Hence, for practical applica-tions, the thickness of the layers in these coatings should be designed so as to obtain the optimum hardness values and wear properties with appro-priate high-temperature stability. It should be noted that when multilayer nanoscale coatings are deposited by the periodic variation of the deposi-tion conditions, a templating effect is often observed. That is, during build-ing superlattice structures from different materials, the layer first deposited can force the next layer of a different material to adopt the crystallographic structure of the first layer. For example, this occurs in TiN/CrN multilayer coatings, the Cr_2N is forced to assume the structure of the TiN underlayer, and in TiN/AlN multilayer coatings, the AlN (originally Wurtzite structure) is forced to form the NaCl structure of the TiN.

In addition to nanoscale multilayer coatings, it is also possible to fabri-cate isotropic nanocomposite coatings consisting of nano-sized crystallites embedded in an amorphous matrix. These coatings generally have a hard phase (transition-metal carbides and nitrides) and a second phase that acts as a binder and provides structural flexibility (amorphous silicon nitride and amorphous carbon). The formation of these nanocomposite structures involves phase separation between two materials, that is, they show complete immiscibility in solid solution, which are often co-deposited by sputtering or plasma deposition. Unlike multilayer systems, the possible material com-positions and particle sizes in these nanocomposite coatings are restricted by material properties and deposition conditions. These nanocomposite coat-ings are generally deposited by plasma-assisted chemical vapor deposition (PACVD) or physical vapor deposition. Very hard (~50–60 GPa) TiN/α-Si_3N_4 nanocomposite coatings have been made using PACVD from $TiCl_4$, $SiCl_4$/ SiH_4, and H_2 at about 600°C (Veprek et al. 1995). The gas-phase nucleation (uncontrollable rates), chloride precursors (the unreacted species can con-taminate the coating), and high processing temperatures are disadvantages in this process. These coatings can also be prepared by sputtering Ti and Si targets under nitrogen atmosphere at room temperature. The disadvantages are that the quality of the films is inferior and the hardness is lower than the PACVD nanocomposite coatings. In addition to improved hardness, the nano-composite coatings also have better oxidation resistance than TiN coatings.

In a single-phase nanocrystalline material, the hardness increase comes from the lack of plastic deformation because the dislocations face far more barriers to mobility. However, this hindered dislocation alone cannot explain the superior hardness of these nanocomposite coatings, since the nanocrystallites within the amorphous matrices are only a few nanometers large and dislocation formation simply does not occur. In this material, plastic deformation occurs through a pseudoplastic deformation, in which the nanocrystals move against each other. Since this process requires higher energy, resistance to plastic deformation is relatively high. Nanocrystalline carbides in amorphous carbon matrices are good example for these categories of nanocomposites, and they can provide hard, low-friction coatings. TiC/C systems are more promising in this regard. Advances in laser-based deposition techniques (e.g., magnetron sputtering-assisted pulsed-laser deposition) have facilitated the fabrication of these nanocomposite coatings. The volume fraction and size of the particles in the coating can be adjusted to obtain optimum toughness and hardness. The fabrication of a nanocomposite with appropriate particle size allows the optimum generation of dislocations and micro- and nano-cracks, which result in a self-adjustment in composite deformation from elastic to plastic at loads exceeding the elastic limit. Thus, the compliance of the coatings is improved, and catastrophic brittle failure can be avoided. Such load-adaptive nanocomposites with optimal design of microstructure are extremely useful for wear applications.

Hard nanocomposite coatings such as TiN and Ti–C–N in DLC matrix have significant advantages in aerospace systems. Apart from their high hardness and toughness and low-friction properties, these coatings also resist attack by corrosive fluids such as engine oils and lubricants used in aircraft engines and parts. Magnetron sputter-assisted pulsed-laser deposition is a more suitable technique for depositing such coatings in different configurations, such as functionally gradient, multilayer, and granular nanocomposites.

The tailoring of the different architectures in creating appropriate coatings with desirable tribological properties is the key to the development of high-performance hard coatings. Nanostructured coatings provide several pathways for reducing stress in coatings and terminating cross-sectional dislocations, with possibilities for controlling dislocation movement by dispersion strengthening or lattice mismatching.

Other DLC matrix-based nanocomposite systems contain carbides based on Ta, Nb, etc. To deposit these coatings, several techniques such as PVD, pulsed-laser deposition, and reactive magnetron sputtering are used. The particle sizes in these coatings are large enough (10–50 nm) to allow dislocations pileup but are too small for crack propagation. The larger grain separation allows the development of incoherency strains and cracks to originate between crystallites under loading, which allow pseudoplastic deformation. Thus, their hardness is much higher, and these coatings also have much higher toughness. These coating systems can also be modified by introducing other elements, for example, W or Cr for producing optically absorbing coatings for solar energy converters and MoS_2 for lubricating coatings.

Chemical and physical vapor deposition and laser ablation methods have been used to prepare a variety of superhard nanocomposites consisting of nitrides, borides, and carbides. In these systems, the hardness of the nanocomposite significantly exceeds that predicted by the rule of mixtures based on the hardness of bulk materials. For example, the hardness of M_nN/α–Si_3N_4 (M = Ti, W, V, etc.) nanocomposites with the optimum content of Si_3N_4 (close to the percolation threshold) reaches 50 GPa, although that of the individual phases does not exceed 21 GPa. For a binary solid solution, such as $TiN_{1-x}C_x$, the hardness increases with increasing x, following the rule of mixtures. Superhardness has also been achieved in coatings consisting of transition-metal nitride and a relatively soft metal that does not form thermodynamically stable nitrides, such as M_nN–M′ (M = Ti, Cr, Zr and M′ = Cu, Ni). However, these systems have low thermal stability, and their hardness decreases upon annealing above 400°C (Choi et al. 1997).

8.3.4 Properties of Ceramic Nanocomposites

Nanocomposites fabricated by hot pressing of polymer-derived powders have yielded the best mechanical properties. However, this technique is more expensive and has limitations on the shapes that can be fabricated. Pressureless sintering and gas-pressure sintering are the most attractive consolidation techniques in terms of cost and flexibility on product shapes. However, attempts on gas-pressure sintering of mixed powders have resulted in poor dispersion, agglomeration of SiC, and changes in glass-phase chemistry due to reaction with SiC. Thus, the focus for future processing rests on the routes that produce commercially viable powders with uniform dispersion of particles, which can be fabricated into dense components by pressureless or gas-pressure sintering. Moreover, the effect of reinforcement on the mechanical properties of silicon nitride nanocomposites is not well understood. As in the SiC/Al_2O_3 nanocomposites, the presence of intergranular SiC grains restricts grain growth, resulting in the formation of Si_3N_4 matrix with finer grains. In fact, at higher reinforcement contents, the grain growth is reported to be severely restricted, resulting in fine α-Si_3N_4 grains, and the composite can show superplastic behavior. The "nano-sized" strengthening and toughening effect due to thermal expansion mismatch proposed for SiC/Al_2O_3 nanocomposites needs to be critically revived for the SiC/Si_3N_4 composites, since the difference between the thermal expansion coefficients of SiC and Si_3N_4 is small and in the opposite sense (CTE of SiC > Si_3N_4). The formation of a liquid phase in Si_3N_4 during sintering further complicates the matter, since the reactions between the reinforcements and the liquid phase could alter its composition and quantity, thus changing the sintering behavior and creep resistance. Thus, systematic studies on the effects of reinforcement size and volume fraction on the microstructure, processing, and properties of Si_3N_4 nanocomposites are needed for a more fundamental understanding of the effect of nano-sized reinforcement (Ajayan et al. 2003b).

Creep in the liquid-phase sintered Si_3N_4 ceramics is controlled by grain boundary transport and sliding due to the presence of glassy phase at the boundaries. Hence, the incorporation of fine reinforcements at the grain boundaries may improve creep resistance. The reduction in amount and hence an increase in viscosity of the intergranular phase due to chemical reactions and accelerated crystallization, the hindrance of grain boundary sliding due to binding by SiC particles, and the obstruction of easy diffusion by the SiC nanoparticles explain the better creep resistance of the nanocomposites (Dusza et al. 1999). Similarly, better creep resistance in SiC/Al_2O_3 nanocomposites containing 5 vol.% SiC nanoparticles has also been reported (Descamps et al. 1999). The presence of nonbridged grain boundaries causes a rearrangement of grains due to initial grain sliding. However, this initial sliding process is stopped by the presence of SiC nanoparticles at the boundaries; hence, only dislocation motion and, to a lesser extent, viscoelastic mechanism contribute to creep. The exact mechanisms of creep resistance in nanocomposites have not been entirely conclusive, and systematically investigating the influence of particle size, oxidation, and the volume fraction of the dispersed phase is desirable. An additional factor that needs careful consideration is the volume fraction of reinforcements on the grain boundaries as opposed to within the grains.

Crack-tip bridging by particles is a primary mechanism of strengthening in ceramic nanocomposites. For example, small brittle SiC particles in SiC/Al_2O_3 composites cause crack-tip bridging at small distances behind the cracks. Residual stresses present around the particles cause the strengthening mechanism to operate effectively even at small volume fraction of SiC. The strength of nanocomposite is increased due to the reduction in the critical flaw size by the fine dispersion of nanoparticles (Perez-Riguero et al. 1998). Intergranular fracture, as observed in monolithic alumina, is suppressed in the nanocomposite, because the crack extension along grain boundaries is suppressed by the particles that strongly bond the matrix grains. Hence, transgranular fracture is a common mode of fracture in nanocomposites.

In summary, the advantage of ceramic nanocomposites lies not only on the strength but also on other mechanical properties such as fracture toughness, hardness, and creep resistance. The degree of improvement in these properties depends on the composite system, that is, the type of reinforcement and matrix. Although there are some generalities in the strengthening and toughening mechanisms in such composites, the actual size, location, and volume fraction of particles strongly influence the final outcome of the mechanical behavior.

Questions

Answer the following questions:

8.1 What are the approaches to create nanocomposites?

8.2 What are the major challenges in nanocomposite formation?

8.3 Explain the role of interface in the context of nanocomposites?

8.4 What are the problems in preparing clay–PNCs?

8.5 What are the advantages and disadvantages of nano-reinforcements?

8.6 What are the peculiar properties of nanocomposites?

8.7 What are the benefits of polymer–clay nanocomposites?

8.8 What is the main problem of clay dispersion in polymer? What are the ways to overcome that problem?

8.9 What are the methods of polymer incorporation in clay layers?

8.10 Explain the process to prepare clay–PP nanocomposites.

8.11 What is the method used to prepare nanocomposite acrylic coatings?

8.12 What are the advantages of nanoclay over carbon black as reinforcement for rubber?

8.13 What are the general characteristics of CNTs?

8.14 What are the potential advantages of CNT-reinforced composites?

8.15 Explain the novel method to form PNCs using CNT aggregates.

8.16 Compare the thermal spray and electrodeposition processes for the formation of nanocomposite coatings.

References

Ajayan, P.M., L.S. Schadler, and P.V. Braun. 2003a. *Nanocomposite Science and Technology*, Morlenbach, Germany: Wiley-VCH, p. 4.

Ajayan, P.M., L.S. Schadler, and P.V. Braun. 2003b. *Nanocomposite Science and Technology*, Morlenbach, Germany: Wiley-VCH, pp. 20–21.

Antonucci, V., K.T. Hsiao, and S.G. Advani. 2003. In *Advanced Polymeric Materials: Structure Property Relationships*, eds. G.O. Shonaike and S.G. Advani. Boca Raton, FL: CRC Press.

Avella, M., M.E. Errico, and E. Martuscelliand. 2001. *Nano Lett.* 1: 213.

Bakshi, S.R., V. Singh, K. Balani, D.G. McCartney, S. Seal, and A. Agarwal. 2008. *Surf. Coat. Technol.* 202: 5162.

Bakshi, S.R., V. Singh, S. Seal, and A. Agarwal. 2009. *Surf. Coat. Technol.* 203: 1544.

Banovic, S.W., K. Barmak, and A.R. Marder. 1999. *J. Mater. Sci.* 34: 3203.

Barbee, R.B., J.C. Matayabas Jr., J.W. Trexler Jr., and R.L. Piner. 1999. Polyester nano-composites for high barrier applications. PCT Int. Appl. WO 9932403 (to Eastman Chemical Company).

Bernd, W., F. Haupert, and M.Q. Zhang. 2003. Compos. Sci. Technol. 63: 2055.

Bharadwaj, R.K., A.R. Mehrabi, C. Hamilton et al. 2002. Polymer 43: 3699.

Bian, Z., M.X. Pan, Y. Zhang, and W.H. Wang. 2002. Appl. Phys. Lett. 81: 4739.

Bian, Z., R.J. Wang, W.H. Wang, T. Zhang, and A. Inoue. 2004. Adv. Funct. Mater. 14: 55.

Biasci, L., M. Aglietto, G. Ruggeri, and F. Ciardelli. 1994. Polymer 35: 3296.

Carblom, L.H. and J.A. Seiner. 1998. PCT Int. Appl. WO 98/24839 (to PPG Industries, Inc.) (6/11/98)

Chang, S., R.H. Doremus, P.M. Ajayan, and R.W. Siegel. 2000. Ceram. Eng. Proc. 21: 653.

Chang, J.H., B.S. Seo, and D.H. Hwang. 2002a. Polymer 43: 2969.

Chang, Y.-W., Y. Yang, S. Ryu, and C. Nah. 2002b. Polym. Int. 51: 319.

Chen, X.H., F.Q. Cheng, S.L. Li, L.P. Zhou, and D.Y. Li. 2002. Surf. Coat. Technol. 155: 274.

Chen, X., J. Jintong Xia, J. Peng, W. Li, and S. Xie. 2000. Compos. Sci. Technol. 60: 301.

Chen, X.-H., Li, W.-H., Chen, C.-S., Xu, L.-S., Yang, Z., and J. Hu. 2005. Trans. Nonferrous Met. Soc. China 15: 314.

Choi, E.S., J.S. Brooks, D.L. Eaton et al. 2003. J. Appl. Phys. 94: 6034.

Choi, H.J., K.S. Cho, and J.G. Lee. 1997. J. Am. Ceram. Soc. 80: 2681.

Demjen, Z., B. Pukanszky, and J. Nagy. 1999. Polymer 40: 1763.

Descamps, P., D. O'Sullivan, M. Poorteman, J.C. Descamps, A. Leriche, and F. Cambier. 1999. J. Eur. Ceram. Soc. 19: 2475.

Dingsheng, Y. and L. Fingliang. 2006. Rare Metals 25: 237.

Dusza, J., P. Sajgalik, and M. Steen. 1999. Key. Eng. Mater. 175: 311.

Eckert, J., A. Reger-Leonhard, B. Weiss, and M. Heimaier. 2001. Mater. Sci. Eng. A301: 1.

Fornes, T.D. and D.R. Paul. 2003. Polymer 44: 4993.

Fujiwara, S. and T. Sakamot. 1976. Kokai patent application no. SHO51, 1976-109998.

Fukuda, N., H. Tsuji, and Y. Ohnishi. 2002. Polym. Degrad. Stabil. 78: 119.

Gatos, K.G., R. Thomann, and J. Karger-Kocsis. 2004. Polym. Int. 53: 1191.

Gilman, J.W. 1999. Appl. Clay Sci. 15: 31.

Guo, C., Y. Zuo, X. Zhao, J. Zhao, and J. Xiong. 2008. Surf. Coat. Technol. 202: 3246.

Gyftou, P., M. Stoumbouli, E.A. Pavlatou, and N. Spyrellis. 2002. Trans. Inst. Met. Finish. 80(3): 88.

Harrison, A.G., W.N.E. Meredith, and D.E. Higgins. 1996. US Pat. 5 571 614 (11/5/96)

Hunt, A.J., M.R. Ayers, and W.Q. Cao. 1995. J. Non-Cryst. Solids 185: 227.

Inceoglu, A.B. and U. Yilmazer. 2003. Polym. Eng. Sci. 43: 661.

Jang, J.S.C. and C.C. Koch. 1990. J. Mater. Res. 5: 498.

Jawahar, P. and M. Balasubramanian. 2005. Int. J. Plastics Technol. 9: 472.

Jawahar, P., R. Gnanamoorthy, and M. Balasubramanian. 2006. Wear 261: 835.

Kang, X., Z. Mai, X. Zou, P. Cai, and J. Mo. 2007. Anal. Biochem. 363: 143.

Khedkar, J., I. Negulescu, and E.I. Meletis. 2002. Wear 252: 361.

Kim, K.T., S.I. Cha, and S.H. Hong. 2007. Mater. Sci. Eng. A449–A451: 46.

Kim, K.T., K.H. Lee, S.I. Cha, C.B. Mo, and S.H. Hong. 2004. Mater. Res. Soc. Symp. Proc. 821: P3.25.1.

Kim, I.-Y., J.-H. Lee, G.-S. Lee, S.-H. Baik, Y.-J. Kim, and Y.-Z. Lee. 2009. Wear 267: 593.

Kim, J.-T., T.-S. Ohand, and D.-H. Lee. 2003. Polym. Int. 52: 1058.

Kim, K.J. and J.L. White. 2001. J. Indust. Eng. Chem. 7: 50.

Kimura, T., H. Ago, M. Tobita, S. Ohshima, M. Kyotani, and M. Yamura. 2002. Adv. Mater. 14: 1380.

Kojima, Y., A. Usuki, M. Kawasumi et al. 1993. *J. Mater. Res.* 8: 1185.

Kornmann, X. and L.A. Berglund. 1998. *Polym. Eng. Sci.* 38: 1351.

Kuo, S., Y. Chen, M. Ger, and W. Hwu. 2004. *Mater. Chem. Phys.* 86: 5.

Laha, T., A. Agarwal, and T. McKechnie. 2005. In *Surface Engineering in Materials Science-III*. eds. A. Agarwal, N.B. Dahotre, S. Seal, J.J. Moore, and C. Blue. Warrendale, PA: The Minerals, Metals and Materials Society.

Laha, T., A. Agarwal, T. McKechnie, and S. Seal. 2004. *Mater. Sci. Eng.* A381: 249.

Laha, T., Y. Liu, and A. Agarwal. 2007. *J. Nanosci. Nanotechnol.* 7: 515.

Lake, M.L., D.G. Glasgow, G.G. Tibbetts, and D.J. Burton. 2010. In *Polymer Nanocomposites Handbook*. eds. R.K. Gupta, E. Kennel, and K.J. Kim. Boca Raton, FL: CRC Press, p. 78.

Lan, T., P.D. Kaviratna, and T.J. Pinnavaia. 1995. *Chem. Mater.* 7: 2144.

Lan, T. and T.J. Pinnavaia. 1994. *Chem. Mater.* 6: 2216.

Li, S., S. Song, T. Yu, H. Chen, Y. Zhang, and H. Chen. 2005a. *Mater. Sci. Forum* 488–489: 893.

Li, J., Y. Sun, X. Sun, and J. Qiao. 2005b. *Surf. Coat. Technol.* 192: 331.

Liao, K., Y. Ren, and T. Xiao. 2006. In *Polymer Nanocomposites*. eds. Y.W. Mai and Z.Z. Yu, pp. 343–344. Cambridge, U.K.: Woodhead Publishing Limited.

López-Manchado, M.A., B. Herrero, and M. Arroya. 2004. *Polym. Int.* 53: 1766.

Lorenzo, M.L., M.E. Errico, and M. Avella. 2002. *J. Mater. Sci.* 37: 2351.

Ma, R.Z., J. Wu, B.Q. Wei, J. Liang, and D.H. Wu. 1998. *J. Mater. Sci.* 33: 5243.

Maharaphan, R., W. Lilayuthalert, A. Sirivat, and J.W. Schwank. 2001. *Compos. Sci. Technol.* 61: 1253.

McIntosh, D., V.N. Khabashesku, and E.V. Barrera. 2006. *Chem. Mater.* 18: 4561.

Messersmith, P.B. and E.P. Giannelis. 1995. *J. Polym. Sci. A: Polym. Chem.* 33: 1047.

Nam, P.H., P. Maiti, M. Okamaoto, and T. Kotaka. 2001. In *Proc. Nanocomposites*, Chicago, IL: ECM Publication.

Ngo, Q., B.A. Cruden, A.M. Cassell et al. 2004. *Mat. Res. Soc. Symp. Proc.*, 812: 3.18.1.

Niihara, K. 1991. *J. Ceram. Soc. Jpn.* 99: 945.

Niihara, K. 1996. *New Design Concept for Advanced Materials*. Niihara Lab., ISIR, Osaka University, Japan. http://www.sanken.osaka-u.ac.jp/labs/scm/about_resE.html.

Okada, A., A. Usuki, T. Kurauchi, and O. Kamigaito. 1995. In *Hybrid Organic-Inorganic Composites*, ed. J.E. Mark, C.Y.-C. Lee, and P.A. Biancon, Chapters 6, 55 through 65. ACS Symposium Series 585.

Osso, D., O. Tillerment, G. Le Caer, and A. Mocellin. 1998. *J. Mater. Sci.* 33: 3109.

Pang, L-X., K.-N. Sun, S. Ren, C. Sun, R.-H. Fan, and Z.-H. Lu. 2007. *Mater. Sci. Eng.* A447: 146.

Pardavi-Horvath, M. and L. Takacs. 1992. *IEEE Trans. Magn.* 28: 3186.

Pearce, E.M. and R. Liepins. 1975. *Environ. Health Persp.* 11: 59.

Perez-Riguero, J., J.Y. Pastor, J. Llorca, M. Elices, P. Miranzo, and J.S Moya. 1998. *Acta. Mater.* 46: 5399.

Periadurai, T. et al. 2010. *J. Anal. Appl. Pyrolysis* 89: 244.

Popp, A., J. Engstler, and J.J. Schneider. 2009. *Carbon* 47: 3208.

Qu, N.S., K.C. Chan, and D. Zhu. 2004. *Scr. Mater.* 50: 1131.

Ren, X.C., L.Y. Bai, G.H. Wang, and B.L. Zhang. 2000. *China Plastics* 14: 22.

Riedel, R., H.J. Kleebe, H. Schonfelder, and F. Aldinger. 1995. *Nature* 374: 526.

Roos, J.R., J.P. Celis, J. Fransaer, and C. Buelens. 1990. *J. Met.* 42(11): 60.

Sadhu, S. and A.K. Bhowmick. 2004. *J. Appl. Polym. Sci.* 92: 698.

Sakka, Y., D.D. Bidinger, and I.A. Aksay. 1995. *J. Am. Ceram. Soc.* 78: 479.
Sawaguchi, A., K. Toda, and K. Niihara. 1991. *J. Ceram. Soc. Jpn.* 99: 510.
Shao, I., P.M. Vereecken, R.C. Cammarata, and P.C. Searson. 2002a. *J. Electrochem. Soc.* 149: C610.
Shao, I., P.M. Vereecken, C.L. Chien, P.C. Searson, and R.C. Cammarata. 2002b. *J. Mater. Res.* 17: 1412.
Shinn, M. and S.A. Barnett. 1994. *Appl. Phys. Lett.* 64: 61.
Sinha, R.S., K. Yamada, M. Okamoto, and K. Ueda. 2003. *Polymer* 44: 857.
Song, X.Y., W.Q. Cao, M.R. Ayers, and A.J. Hunt. 1995. *J. Mater. Res.* 10: 251.
Stewart, D.A., P.H. Shipway, and D.G. McCartney. 1999. *Wear* 225–229: 789.
Suh, D.J., Y.J. Lim, and O.O. Park. 2000. *Polym. Eng. Sci.* 41: 8553.
Sun, Y., J. Sun, M. Liu, and Q. Chen. 2007. *Nanotechnology* 18: 505704.
Tan, J., T. Yu, B. Xu, and Q. Yao. 2006. *Tribol. Lett.* 21(2): 107–111.
Tong, X.C. and H.S. Fang. 1998. *Metall. Mater. Trans.* A29: 893.
Uozumi, H., K. Kobayashi, K. Nakanishi et al. 2008. *Mater. Sci. Eng.* A495: 282.
Usuki, A., M. Kawasumi, Y. Kojima, A. Okada, T. Kurauchi, and O. Kamigaito. 1993a. *J. Mater. Res.* 8: 1174.
Usuki, A., M. Kawasumi, Y. Kojima et al. 1993b. *J. Mater. Res.* 8: 1179.
Usuki, A., K. Okamoto, A. Okada, and T. Kurauchi. 1995. *Kobunsi Ponbunshu* 52: 728.
Varghese, S. and J. Karger-Kocsis. 2003. *Polymer* 44: 4921.
Veprek, S., S. Reiprich, L. Shizhi. 1995. *Appl. Phys. Lett.* 66: 2640.
Vidrine, A.B. and E.J. Podlaha. 2001. *J. Appl. Electrochem.* 31: 461.
Voevodin, A.A., S.V. Prasad, and J.S. Zabinski. 1997. *J. Appl. Phys.* 82: 855.
Wang, G., X.Y. Chen, R. Huang, and L. Zhang. 2002. *J. Mater. Sci. Lett.* 21: 985.
Wang, W., F.-Y. Hou, H. Wang, and H.-T. Guo. 2005. *Scr. Mater.* 53: 613.
Wang, Z., T. Lan, and T.J. Pinnavaia. 1996. *Chem. Mater.* 8: 2200.
Wang, S., C. Long, X. Wang, Q. Liand, and Z. Qi. 1998. *J. Appl. Polym. Sci.* 69: 1557.
Wang, Z. and T.J. Pinnavaia. 1998. *Chem. Mater.* 10: 3769.
Wang, Y., M. Sasaki, T. Goto, and T. Hirai. 1990. *J. Mater. Sci.* 25: 4607.
Wang, N., Z.K. Tang, G.D. Li, and J.S. Chen. 2000a. *Nature* 408: 50.
Wang, H., S. Yao, and S. Matsumura. 2004. *J. Mater. Process. Technol.* 145: 299.
Wang, Y., L. Zhang, C. Tang, and D. Yu. 2000b. *J. Appl. Polym. Sci.* 78: 1879.
Wei, B.Q., Y. Li, Ph. Kohler-redlich, R. Lück, and S. Xie. 2003. *J. Appl. Phys.* 93: 1748.
Wong, M.H., M. Paramsothy, Y. Ren, X.J. Xu, S. Li, and K. Liao. 2003. *Polymer* 44: 7757.
Wu, Y.-P., Q.-X. Jia, D.-S. Yu, and L.-Q. Zhang. 2003. *J. Appl. Polym. Sci.* 89: 3855.
Xi, Q., C.F. Zhao, J.Z. Yuan, and S.Y. Cheng. 2004. *J. Appl. Polym. Sci.* 91: 2739.
Xu, X.J., M.M. Thwe, C. Shearwood, and K. Liao. 2002. *Appl. Phys. Lett.* 81: 2833.
Xu, C. L., B.Q. Wei, R.Z. Ma, J. Liang, X.K. Ma, and D.H. Wu. 1999. *Carbon* 37: 855.
Xu, Y., A. Zangvil, and A. Kerber. 1997. *J. Eur. Ceram. Soc.* 17: 921.
Xu, Q., L. Zhang, and J. Zhu. 2003. *J. Phys. Chem.* B107: 8294.
Yang, Y.L., Y.D. Wang, Y. Ren et al. 2008. *Mater. Lett.* 62: 47.
Yano, K, A. Usuki, and A. Okada. 1997. *J. Polym. Sci. A: Polym. Chem.* 35: 2289.
Yano, K, A. Usuki, A. Okada, T. Kurauchi, and O. Kamigaito. 1991. *Polymeric Preprints,* 32: 65.
Yano, K., A. Usuki, A. Okada, T. Kurauchi, and O. Kamigaito. 1993. *J. Polym. Sci. A: Polym. Chem.* 31: 2493.
Yoshizawa, Y., S. Oguma, and K. Yamauchi. 1988. *J. Appl. Phys.* 64: 6044.
Zeng, Q., J. Luna, Y. Bayazitoglu, K. Wilson, M.A. Imam, and E.V. Barrera. 2007. *Mater. Sci. Forum* 561–565: 655.

Zheng, H., Y. Zhang, Z. Peng, and Y. Zhang. 2004. *Polym. Polym. Compos.* 12: 197.
Zhong, R., H. Cong, and P. Hou. 2003. *Carbon* 41: 848.
Zhu, Y.T. and A. Manthiram. 1994. *J. Am. Ceram. Soc.* 77: 2777.

Bibliography

Agarwal, A., S.R. Bakshi, and D. Lahiri. 2010. *Carbon Nanotubes: Reinforced Metal Matrix Composites.* Boca Raton, FL: CRC Press.
Ajayan, P.M. 2003. *Nanocomposite Science and Technology.* Morlenbach, Germany: Wiley-VCH.
Gupta, R.K., E. Kennel, and K.J. Kim (Eds.). 2010. *Polymer Nanocomposites Handbook.* Boca Raton, FL: CRC Press.
Krishnamoorti, R. and R.A. Vaia (Eds.). 2002. *Polymer Nanocomposites.* Washington, DC: American Chemical Society.
Mai, Y.W. and Z.Z. Yu (Eds.). 2006. *Polymer Nanocomposites.* Cambridge, U.K.: Woodhead Publishing Limited.
Pinnavaia, T.J. and G.W. Beall. 2000. *Polymer-Clay Nanocomposites.* Chichester, England: John Wiley & Sons.

Further Reading

1. Ashok Kumar (Ed.). 2010. *Nanofibers.* Vukovar, Croatia: Intech.
2. Reddy, B.S.R. (Ed.). 2011. *Advances in Nanocomposites—Synthesis, Characterization and Industrial Applications.* Rijeka, Croatia: InTech.
3. Biswas, M. and R.S. Sinha. 2001. Recent progress in synthesis and evaluation of polymer–montmorillonite nanocomposites, *Adv. Polym. Sci.* 155: 167–221.
4. Hussain, F., M. Hojjati, M. Oamoto, and R.E. Gorga. 2006. Polymer matrix nanocomposites, processing, manufacturing and application: An overview, *J. Comp. Mater.* 40: 1511–1575.
5. Ray, S.S. and M. Okamoto. 2003. Polymer/layered silicate nanocomposites: A review from preparation to processing, *Prog. Polym. Sci.* 28: 1539–1641.
6. Vaia, R.A. and E.P. Giannelis. 2001. Polymer nanocomposites: Status and opportunities, *MRS Bull.* 26: 394–401.

Appendix: Laboratory Practice

A.1 Determination of Viscosity of a Thermosetting Resin

Take about 400 mL of liquid resin in a beaker of 500 mL. The viscosity of resin is generally determined using Brookfield-type viscometer (ASTM-D2393). The type of spindle and its rotational speed is variable, and select the adequate ones suitable to the sample used. Immerse spindle slowly into the sample liquid up to the mark in the spindle. Switch on the viscometer to rotate the spindle. Indicator reading starts to move from zero. Stop the spindle either when the reading stabilizes or after a specified time (2–3 min), and take reading. To obtain the viscosity of the sample, multiply the reading by the appropriate multiplying factor given in the table, which is determined from the spindle number used and the rotational speed selected. Modern equipments can give the viscosity value directly. The viscosity of different resins can be determined or varying quantities of solvent can be added to a particular resin and the viscosity can be determined after each addition of solvent. The unit of viscosity is milli-Pascal second (mPa s).

A.2 Determination of Gel Time of a Thermosetting Resin

Attach the reciprocating (up and down movement) spindle to the gel timer. Mix 100 mL of resin with curing agent in a 150 mL disposable plastic beaker. Start the gel timer. Place the cup with the liquid resin into the gel timer retaining ring/on the base. Adjust the cup position to prevent the spindle from touching the sides or bottom of the cup. When the resin is in the liquid state, the spindle freely goes up and down. After gelation, the movement of spindle is arrested and the timer stops. Note down the reading on the time counter. This time corresponds to gel time of the particular resin and curing agent combination (ASTM-D2471). The gel time of the resin for varying amount of curing agent can be determined.

A.3 Determination of Peak Exotherm and Cure Time of Polyester Resin

Take the resin in a cup and mix the curing agent. Start the stopwatch immediately and immerse the thermocouple lead (sensor) into the resin. The temperature indicator connected to the thermocouple shows the temperature. Continuously monitor the temperature. The temperature will reach to a maximum and then start decreasing. Note down the maximum rise in temperature and the corresponding time. Those represent the peak exotherm temperature and cure time, respectively.

A.4 Determination of Heat Deflection Temperature

Heat deflection temperature (HDT) is determined to assess the softening characteristics of polymeric materials. In this test, a rectangular bar of the polymeric material is loaded in three-point bending mode inside a suitable nonreacting liquid medium, such as silicone or transformer oil. The load on the bar is adjusted to create a maximum bending stress of either 1.82 or 0.455 MPa (ASTM-D648). The deflection of the bar at the center is monitored as the temperature of the liquid is increased at a uniform rate of $2 \pm 0.2°C/min$. The temperature at which the deflection increases by 0.25 mm from its initial room temperature position is called the HDT of the material at the specific stress.

A.5 Determination of Binder Solubility of Chopped Strand MAT

Cut a strip (400 × 70 mm) of chopped strand mat. Insert the mat below the rollers in a resin tray. Attach one end of the strip with a permanent clip. Attach the other end with a movable clip. Tie a thread with this clip and suspend a weight corresponding to 1/3 of the surface density of the mat through a roller. Pour the required quantity of resin to completely immerse the mat and start the stopwatch immediately. Once the binder is dissolved, the mat tears off due to the weight. Stop the watch and note down the time. This time duration corresponds to the solubility of binder (ASTM-D2558). The value should be less than 50 s.

A.6 Determination of Wetting Time of Chopped Strand Mat

Cut a strip (300 × 300 mm^2) of chopped strand mat. Take the resin (three times to that of the weight of the mat) in a tray and spread uniformly. Place the mat over the resin and start the stopwatch immediately. The clear white regions start disappearing. Once 70% of the white regions are disappeared, stop the watch and note down the time. This time duration corresponds to the wetting time of chopped strand mat (ASTM-D3374).

A.7 Laminate Preparation by Hand Lay-Up Process

Laminate can be made on any smooth surface, preferably over a thick glass plate. Clean the surface thoroughly, apply a releasing agent, such as wax or PVA (a solution of 10% polyvinyl alcohol in water), and allow it to dry. Mix the required amount of activator with the polyester resin. Apply a gel coat (optional) and allow it to gel. Cut the required size (generally 30 mm × 30 mm) and numbers of mat. Add the required amount of catalyst and mix it thoroughly and gently. Place the reinforcement mat on the surface. Impregnate the reinforcement with the resin. Using a hand roller, distribute the resin uniformly into the fiber reinforcement. Place the subsequent reinforcement layers and impregnate with resin until a suitable thickness is built up. The laminate is allowed to cure at room temperature. After curing chip off the laminate from the surface, wash thoroughly to remove the release agent, and then cut the edges.

A.8 Bulk Molding Compound Preparation

Typical composition of bulk molding compound (BMC) is given in Table 4.3. A typical mixing schedule for BMC would be as follows:

1. Blending resin, catalyst, release agent, and pigment for 10 min
2. Mixing for 5 min sigma mixer after adding 50% filler
3. Mixing for 10 min after adding remaining filler
4. Reinforcement addition and mixing for 3 min

A.9 Compression Molding of BMC

Clean the die halves and apply silicone oil on the inside surfaces. Place/fix them on the platens of press. Allow them to heat to the required temperature. Place the required quantity of BMC on the bottom die. Close the die and apply pressure by moving the platens. Keep it for a few minutes. Open the die and remove the product.

A.10 Determination of Glass Fiber Content in FRP

Cut a small piece ($\sim$50 × 50 mm^2) of glass fiber reinforced plastics (GFRP) and weigh it accurately. Put it in a furnace and increase the temperature to 575 ± 25°C. Keep it for 30 min. Carefully take out the burnt material and put it in a desiccator. (To avoid the fiber loss during handling, the sample can be taken in a ceramic crucible.) Take the weight again after it is cooled to room temperature. This weight corresponds to the weight of fiber. From the initial weight of the composite, the weight percentage of fiber can be calculated (ASTM-D3171). The weight percentage can be converted to volume percentage by using the relationship given in Chapter 1.

Index

A

Advanced curing processes, PMCs
  advantages and limitations, electron beam, 229
  conventional thermal, 229
  electromagnetic radiation, 229
  energy transfer, 229–230
  exothermic chemical reactions, 230
  glass fiber-reinforced polyester composites, 230, 231
  ionizing radiation, 228
  Maxwell–Wagner polarization, 230
  microwaves, 229, 230
  production rate, 228
Advanced winding machines
  basic motions, five-axis filament, 201
  commercial applications, 201
  end closures, 202
Aliphatic polyamides, 126–127
Alumina-based fibers
  advantages, 62–63
  continuous, 63
  description, 62
  α-Fe_2O_3 particles, 63
  metallo-organic and inorganic compounds, 63
  mullite, 63
  polymorphic forms, 63
  properties, commercial oxide, 64
  silica addition, 64
  sol-gel process, 62
  transition alumina, 64
Alumina matrix composites, 401
Aramid fiber-reinforced aluminum laminates (ARALL), 232
Aramid fibers
  advantages, air gap, 55
  aerospace applications, 56
  aromatic polyamides, 53
  creep rate, 58
  dry-jet wet-spinning process, 54, 56

high-modulus yarns, 57
Kevlar and Nomex molecules, 54
liquid crystal solution, 54
metal and ceramic matrices, 93
molecular arrangement, 55, 57
molecular rods, liquid crystalline state, 54, 55
para-aramids, 54
preparation, Technora fibers, 54
properties, 56–58
sensitivity, ultraviolet radiation, 57
solvent, 100% sulfuric acid, 54
spinning, 54
structure, Kevlar fiber, 55
tensile strength, 56
vibration damping property, 56–57
Asbestos cement, 453
Autoclave process
  automatic tape-laying, 190–192
  composites, 187
  description, 187
  heating provision, 186
  low-viscosity resins, 187
  molding, 186, 192–194
  prepregs, *see* Prepregs
Automatic tape-laying
  computer-aided design, 190
  description, 190, 191
  fabric-reinforced prepregs, 190
  indexed mylar templates, 191
  thermoplastic prepregs, 192
  unidirectional fiber prepregs, 191
  woven fabric prepregs, 192

B

Bast fibers, 38–39
Bulk molding compound (BMC)
  compression molding, 598
  formulation, 213
  preparation, 597
Butyl rubber, 135